Carefree Through 1001 French Locks

Michael Walsh

Other recent books by Michael Walsh

Sequitur - To Cape Horn in Comfort and Style
Published 2013 - ISBN: 978-09919556-0-2

Carefree on the European Canals
Published 2014 - ISBN: 978-09919556-4-0

Carefree Through 1001 French Locks

Copyright © 2015 by Michael Walsh

Cover by Edith Gelin & Michael Walsh

All rights reserved. No part of this book may be reproduced in any form by
any electronic or mechanical means, including photocopying,
scanning, recording, or information storage and retrieval
without permission in writing from the author.

ISBN: 978-09919556-7-1

Published by Zonder Zorg Press

www.zonderzorg.ca
michael@zonderzorg.ca

Printed in the United States of America
on Sustainable Forestry Initiative® (SFI®) Certified Sourcing papers

Table of Contents

	Acknowledgements	vi
	Introduction	vii
1.	Setting the Scene	1
2.	A Look at the Locks	8
3.	Historic French Barges	18
4.	Historic Dutch Barges	24
5.	Pleasure Cruising on French Canals	27
6.	Buying a Canal Boat	31
7.	Researching History	48
8.	Taking Possession	54
9.	To Friesland for a Refit	58
10.	Planning a Metamorphosis	63
11.	The Mid-Life Refit	68
12.	Monitoring the Refit from Afar	74
13.	Moving Aboard and Settling In	89
14.	Which Way to Go?	97
15.	Southward Toward France	101
16.	Into France	110
17.	Into the Champagne	118
18.	Enjoying Champagne	129
19.	From the Champagne to the Burgundy	150
20.	Down the Saône	168
21.	Down the Rhône to Avignon	177
22.	Exploring Avignon	193
23.	Onward to the Midi	204
24.	The Canal du Midi	216
25.	Wrapping-Up for the Year	235

26.	Return to France	241
27.	Up Canal From Carcassonne	246
28.	To Castelnaudary	253
29.	Over the Hump to Toulouse	262
30.	Onward From Toulouse	278
31.	Moissac to Buzet	291
32.	Heading Up the Lot	310
33.	The River Baïse	335
34.	Down la Garonne to Bordeaux	349
35.	Up Canal de Garonne to Toulouse	359
36.	Back Across the Midi	368
37.	Up the Rhône	394
38.	Up the Saône	417
39.	Canal du Centre	428
40.	Roanne and the Loire	436
41.	Canal du Nivernais	443
42.	Canal de Bourgogne	462
43.	Dijon - Capital of the Burgundy	492
44.	Preparing for Winter	502

Acknowledgements

Most of the photos in this book were shot by the author and his mate Edi Gelin. Many of the maps in the book were prepared by the author using NASA satellite images. Older maps and images were found in the public domain. The author is grateful for the following: The maps on pages 1, 2, 98, 99, 100, 169 and 502 are from the Voies Navigables de France (VNF) website. The map on page 6 is adapted from one on Wikimedia Commons by Thomas Steiner. The illustration on Page 8 is adapted from one by Cmglee on Wikimedia Commons. Photos of models and drawings on pages 25, 59, 60 and 61 were shot of displays in the Fries Scheepvaart Museum, Sneek. The photo of the luxemotor on page 26 is from a model in the Scheepvaartmuseum, Amsterdam. The photos of the Linssens on page 28 and 31 are from the Linssen website. The photo at the top of page 32 is from the Aqualine website. The photo of Owlpen on page 33 is by Ad Meskins on Wikimedia Commons. The photo of La Cité on page 97 is by Jean-Pierre Lavoie on Wikimedia Commons. The map and diagram on page 216 are by Pinpin on Wikimedia Commons. The map on page 235 is from the City of Carcassonne website. The photo at the bottom of page 260 is by PHGCOM on Wikimedia Commons. The map on page 439 was adapted by MatthiasKabel on Wikimedia Commons. The top map on pages 492 is by G CHP on Wikimedia Commons. The map at the bottom of page 492, on the left on page 493 and the one on page 494 are by Marco Zanoli on Wikimedia Commons. The map at the top of page 493 is by Sémhur on Wikimedia Commons. The map on the lower right on page 493 is adapted from one by FlyingPC on Wikimedia Commons. The second image on page 503 is adapted from one on the H2O website.

Throughout the book, screenshots of the TomTom© and Navionics© apps on the author's iPad are used to illustrate points in the narrative. They are meant as illustrations only and are not intended for navigation. Similarly, photos of selected portions of Éditions de l'Écluse Fluviacarte and Éditions du Breil Guide Fluvial are used to illustrate the text and are not meant for navigation, but rather to illustrate the usefulness of these guides.

Introduction

This book begins with an outline of the history and geography of inland navigation, starting with a broad view and then narrowing to focus on the rivers and canals of France. An overview of canal types is followed by an examination of the various styles of locks found in France and their different modes of operation. A section touches on some of the many types of boat that have worked these waterways through the centuries and how they were used. This is followed by a look at the variety of current vessels suitable for cruising in France, such as modern motor cruisers, replicas of classic barge designs and converted ex-commercial barges.

A large section gives hands-on detail of and insights into the search for, and the selection of a classic Dutch barge in the Netherlands. There are many details on making an offer, commissioning a survey, contracting the purchase, doing a major refit and then taking the barge southward into France.

The meat of the book is in the sections dealing with actually cruising through the inland waterways of France. Here will be found both overviews and details on the broad variety of cruising waters. Much of the detail in this four hundred page section is taken from our cruising logs, condensed, edited and illustrated with photos we have shot along the way. The intent of this narrative is to give the reader a flavour of what it is like cruising these waters and it is by no means meant as a definitive guide, but rather as sketches of some of the things that are possible.

Michael Walsh
Vancouver, Canada
January 2015

Dedication

This book is for you, Edi. Your steady hand on the tiller, both of Zonder Zorg and of our carefree course through life, makes adventures like those described here not only possible, but so much more enjoyable.

Chapter One

Setting the Scene

Across Europe there is a complex network of waterways. The ocean and seas surrounding the continent are interconnected by a web of canals linking watersheds to each other, enabling navigation far inland. There are hundreds of canals making it possible to cross Europe by boat from the North Sea to the Mediterranean, from the Irish Sea to the Black Sea, from the Mediterranean to the Atlantic.

In 1992 continental Europe adopted a classification system for its inland waterways. This was created by the European Council of Ministers of Transport, which rated the canals ECMT Class I to Class VII. The classification was based on the dimensions of the waterway structures, such as lengths and widths of locks, clearance under bridges and the available depths. At the smallest end of the scale, Class I corresponds to the historical Freycinet gauge of France dating to 1879. At the upper end, the larger river classification sizes focus on a waterway's ability to handle mulit-barge convoys propelled by pusher tugs.

Carefree Through 1001 French Locks

In France today there are ninety-four navigable rivers and canals with a total length of 8800 kilometres and 2165 locks. Most of these waterways are interconnected in the northeastern and the southwestern parts of the country and the remaining 500 kilometres are isolated waterways in western France. About twenty percent of the French waterways are Class IV and above, mainly the major rivers. Of the remainder, the vast majority are Class I and lower with minimal commercial traffic and ideally suited to pleasure cruising.

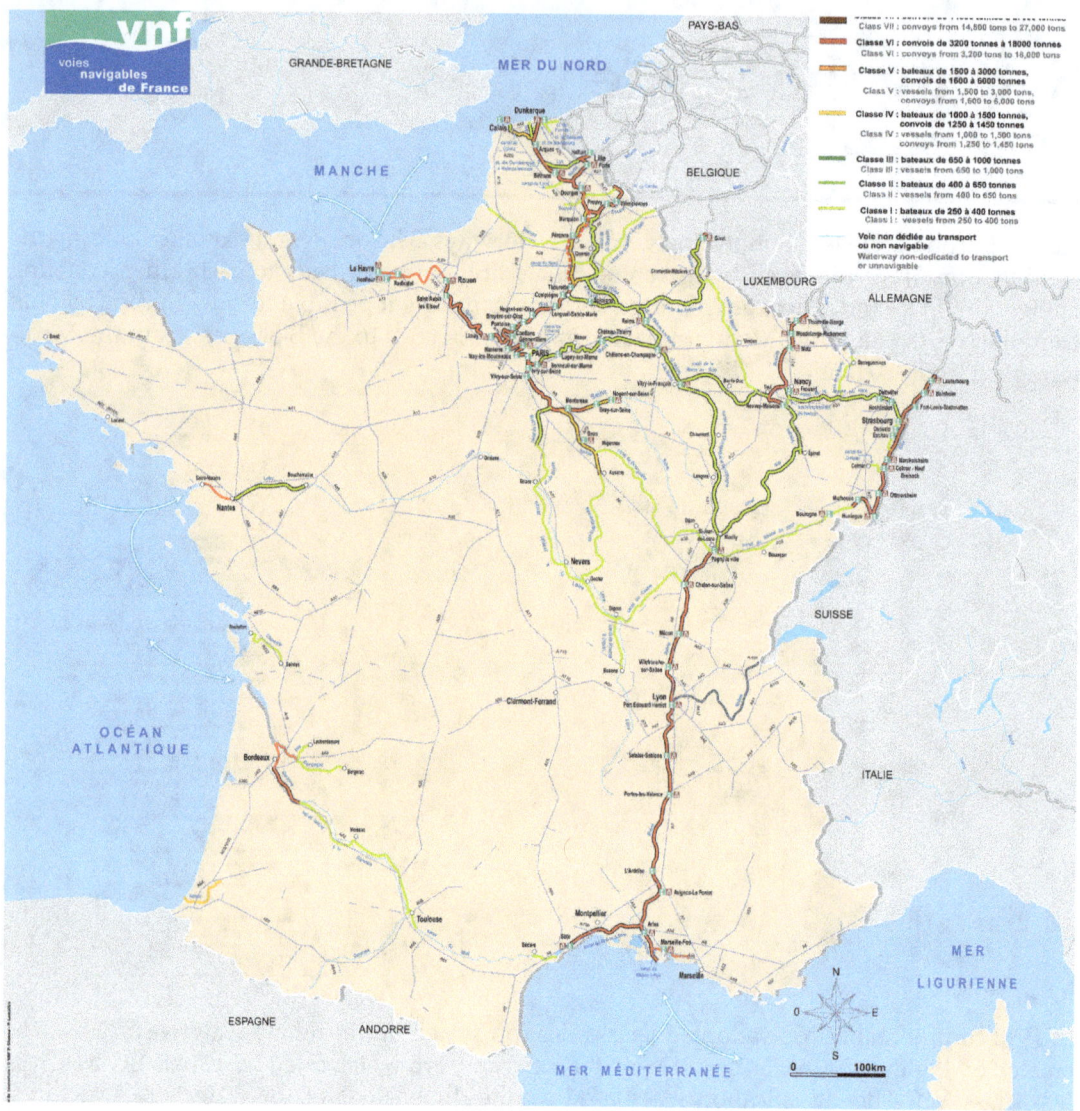

More than three-quarters of the French inland waterways network is under the jurisdiction of Voies Navigables de France, commonly referred to as the VNF, which was established in 1991 to replace l'Office national de la navigation. Their mandate includes maintaining and operating their portion of the waterways and the environment bordering them.

Setting the Scene

As commercial freight traffic declined along the waterways after the Second World War, there was a slow increase in the number of pleasure boat users. Accelerating this increase through the 1970s and 80s was the rapid rise in the popularity of rental boats. Their growth reached a plateau in the late 1990s and their numbers have declined around thirty percent over the past decade and a half. Meanwhile, the number of private pleasure craft continues to slowly increase, making the current pace on the French canals more peaceful than at any time since the mid 1980s.

To further set the scene, it is necessary to look at the various types of inland waterway. The most fundamental is the tidal river, such as the Garonne, which from its mouth below Bordeaux, is navigable upstream for 90 kilometres, or the tidal Seine, which is a major transportation route allowing seagoing vessels to navigate 124 kilometres upstream toward Paris from the English Channel at Le Havre. On both of these rivers the tidal currents are very strong, so close attention must be paid to the tides, ensuring that the passage is coordinated with favourable currents.

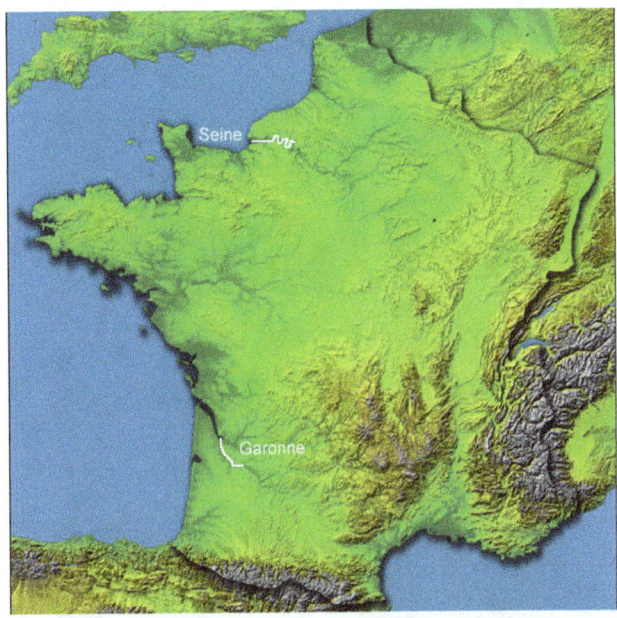

A second type of waterway follows along a non-tidal river that has been tamed and controlled by locks. These locks step the traffic up or down past cataracts or rapids in the river and offer relatively still water in the pounds between successive locks. The locks are often located on a short canal by-passing the dam or weir that is built across the river to maintain the water level upstream. The Rhône and the Saône are good examples of larger lock-controlled rivers. The Rhône takes traffic from the Mediterranean 310 kilometres northward through twelve locks to Lyon. From Lyon, navigation on the Saône continues for another 407 kilometres through twenty-four locks to Corre. On a much smaller scale are the rivers Lot and Baïse in the southwest.

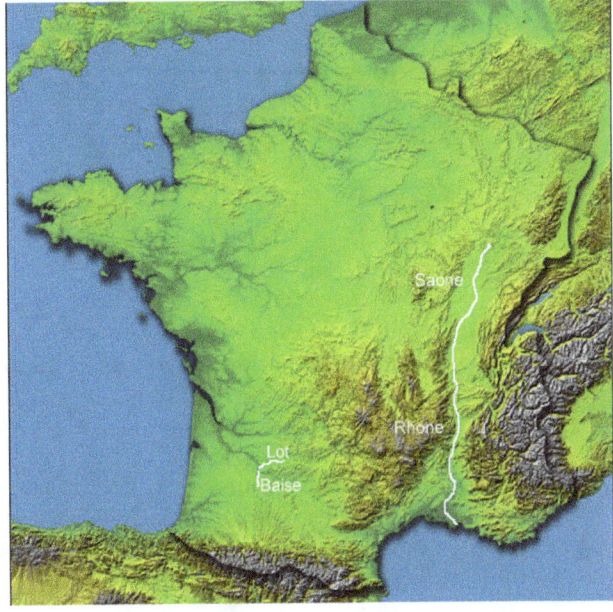

A third type of waterway is a lateral canal, which is an artificial trench that is dug beside the course of a river and fitted with locks. These are built primarily because the course of the river is too sinuous or steep to be easily controlled by weirs and locks within the river bed. The river provides the water to fill the canal and operate its locks. On some lateral canals, in sections where the gentleness of the river allows, the bed of the river is used for navigation and the canal locks into and out of the river. An example of a lateral canal is Canal Latéral de la Loire, which with the branch to Roanne, follows the Loire River for 252 kilometres through 47 locks. Another example is the Canal de Garonne which connects Toulouse to the tidal Garonne at Castets-en-Dorthe, a distance of 193 kilometres with 53 locks.

A fourth type of waterway is the summit canal, which is constructed to cross a pass and connect two watersheds. Because of the necessary climb out of one valley and down into another, many more locks are normally required. The major challenge with a summit canal is to provide a sufficient source of water above the summit to feed the upper pounds so that the canal remains navigable even in the driest seasons. The first summit level canal in France was the Canal de Briare, completed in 1642 using 32 locks in 54 kilometres to connect the basins of the Loire and the Seine. In 1681 le canal royal du Languedoc, which since the French Revolution has been called the Canal du Midi, was completed to connect Toulouse to the Mediterranean using 91 locks over its 240 kilometre length. It rises 189 metres from the Mediterranean to the summit pound then descends 57 metres to the junction with the River Garonne at Toulouse. On both these canals, the final rise to the summit pass is sufficiently gentle that it was practicable to dig the canal directly across the saddle of the pass.

More commonly with a summit canal, the final rise is so steep that it is easier to dig a tunnel through to the other side beneath the top of the pass. On the Canal entre Champagne et Bourgogne there are 114 locks in 224 kilometres with a rise of 239 metres to the summit pound at 341 metres above sea level. The canal then passes through a tunnel

Setting the Scene

4820 metres long before descending steeply 156 metres through 43 locks in 69 kilometres. The Canal de Bourgogne is another example of a summit level canal. It is the highest canal in France, reaching an elevation of 378 metres above sea level, where it passes through a 3350 metre tunnel. It is also the most lock-intense canal in France, requiring 189 locks in its 240 kilometres.

The oldest known navigational canal is the one that was built approximately 2300 BC to bypass the First Cataract of the Nile at Aswan. The first recorded use of canal locks was in the Canal of the Pharaohs, the first Suez Canal. The canal had been started in the late sixth century BC, linking the Red Sea to the Nile, but it is thought to have remained unfinished until the third century BC when Greek engineers under Ptolemy II devised flash locks to deal with the height differences between the sea and the river. With a flash lock, as the gates are opened, a barge flows through the gap in the weir with the current. Moving upstream against the current requires much more time and effort and spills much more water downstream.

Later, the Romans built many canals, mainly for drainage, irrigation or water supply; however, several were for transportation. In the late first century BC, Fossa Augusta was built to link Ravenna to the Po estuary. Also in the first century BC, in Gaul they linked Narbonne to the Mediterranean and dug canals around Arles for military transport. In 12 BC, also for military purposes, they built Fossa Drusiana to link the Rijn to the Flevomeer, today's IJsselmeer in Friesland. In 47 AD Fossa Corbulonis, a 35-kilometre link was dug to join the Rijn and the Maas without having to go out into the Nordzee. The Foss Dyke in England is thought to have been built around 120 AD by the Romans to connect River Trent to Lincoln. It is likely the oldest canal in Europe that is still in use..

The first recorded use of a pound lock, or chamber lock was in 984 in China. To negotiate a steep decline, two flash locks were built within a short distance of each other. The engineer, Qiao Wei-yo realised he had just devised a system to allow boats to move upstream as easily as down. The pound lock is standard on all modern canals and uses gates at each end of the lock chamber to raise or lower the water taking boats to the level upstream or downstream. Leonardo da Vinci is credited with inventing miter gates in the late fifteenth century. This refinement uses the upstream water pressure to hold the gates tightly closed.

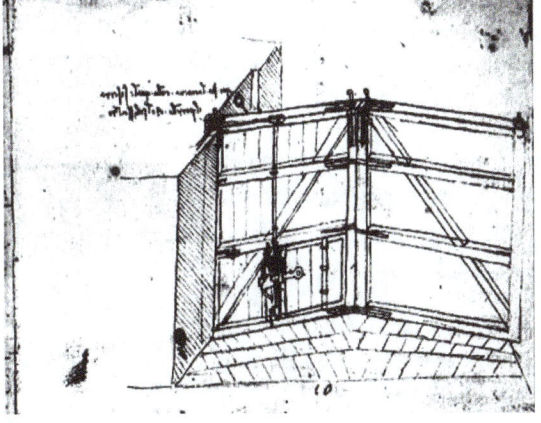

The first pound lock in Europe is believed to be the one built in 1373 in Vreeswijk, Netherlands where the Merwede Canal from Utrecht joins the River Lek. It was used to adjust for the different levels between the river and the canal. In the Netherlands, canal building was primarily done to drain the polders and create dry land, but in so doing, an extensive and efficient inland waterways network evolved and the ease of transportation spurred economic development.

The Rivers of France

In more hilly topography than found in the Netherlands, navigation had been limited to the river valleys, either in the rivers themselves or in works constructed alongside to bypass steep sections. The first canal to break out of a river valley and head over a pass, in what is called a summit level canal, was the Grand Canal of China in the late sixth century AD. It used flash locks.

We have seen that the first summit canal to use pound locks was the Canal de Briare in France, completed in 1642 to connect the basins of the Loire and the Seine. With the completion in 1681 of le canal royal du Languedoc (the Canal du Midi) connecting Toulouse to the Mediterranean, canal construction had reached its maturity. Along its course the canal had passed through the first canal tunnel in Europe and over three major aqueducts spanning river valleys.

During the Middle Ages the transport of goods inland by water cost a tiny fraction of transport overland. A pack horse or mule could carry an eighth of a ton over long distances and on a soft road it could pull a cart of little more than half a ton. The same animal could pull a load of thirty tons on a barge.

The Industrial Revolution spurred a rapid expansion of canal networks across Europe. In Britain, the navigable waterways network grew to keep up with the increasing demand for industrial transportation. A system of large pack horse trains had developed, but few roads at the time could stand up to the heavy wheeled vehicles required to move large amounts of heavy materials. Canal boats were very much quicker, could carry large volumes, and were much safer for fragile items. As the canal network expanded, transportation of raw materials improved, prices fell and this continued to feed the rapid pace of the Industrial Revolution.

From the late eighteenth century, new designs and technology allowed canals to be improved. Where earlier canals had contoured around hills and valleys, later ones went straighter. Locks took canals up and down hills and longer and higher aqueducts spanned broader and deeper valleys. Longer and deeper tunnels pierced the ridges and passes.

In France, canal building continued, with the Canal d'Orleans completed in 1692 to add another link between the Loire and Seine valleys. During the second half of the eighteenth century many canals were begun. In 1789 the Canal du Loing completed the link between the Loire

Setting the Scene

and the Seine and in 1794 the Saône was linked to the Loire by completion of the Canal du Centre. The other works in progress were interrupted by the French Revolution and very little happened with them through the Napoleonic era.

With the Restoration after Napoleon's defeat, came a huge public works program. In 1820 François Becquey, Director of Bridges and Highways proposed an immense program of 126 canal and river improvement projects totalling over 25,000 kilometres. Many of these projects were discarded or postponed, but by 1822 Becquey had pushed bills through the French legislature authorizing the finance, construction and operation of ten new waterways. Another 2000 kilometres were built. By the time of the abdication of King Louis Philippe in 1848, France had more than 10,000 kilometres of navigable inland waterways. During the last third of the nineteenth century, another 3000 kilometres of canals were upgraded or completed and 4000 kilometres of rivers had improvements to navigation.

Among the major works: the Rhine and Saône were linked by a canal over the southern Vosges in 1834, the Seine and Saône were linked in 1842 by the Canal de Bourgogne, the Marne and Rhine were linked in 1853 over the northern Vosges, and in 1856 the Canal Lateral à la Garonne eased navigation from Toulouse to Bordeaux. Also constructed during the first half of the nineteenth century were canals leading northward into Germany, Belgium and the North Sea. A canal linking the Marne and the Saône was begun in 1870 and completed in 1907. At this point, the French network reached its greatest extent with 12,778 kilometres of navigable inland waterways.

From the mid-nineteenth century, the expanding railway network began replacing canals.. As rail transportation became more advanced, land transportation became cheaper and faster. To deal with the decline in canal usage, the engineer, Charles-Louis Freycinet as Minister of Public Works in 1879 began a standardization of the waterway network that had been built over the previous three centuries. He saw that the large variety of lock dimensions was impeding efficient commerce, so he began a rebuilding project that by 1913 had refitted over 1500 kilometres of canals to standard dimensions. These have been called Gabarit Freycinet, the Freycinet Gauge and call for lock dimensions to handle barges with maximum dimensions of 38.50 meters length, 5.05 metres beam and 1.8 metres draft. Low bridges were rebuilt to allow for a minimum clearance of 3.5 metres.

After World War I the commercial transportation in much of Western Europe increasingly went to rail and road and the economic viability of many of the canals steadily declined. Their decreasing use meant reduced revenues for maintenance and they fell into disrepair. Many became non-navigable and were closed. Some were filled-in. In France between 1926 and 1957 nearly 5000 kilometres of navigable waterways were closed or abandoned.

There remain today in France ninety-four navigable rivers and canals with a total length of 8800 kilometres and 2165 locks. Of these, 8300 kilometres are interconnected in the northeastern and the southwestern parts of the country. More than three-quarters of these are Class I and lower with minimal commercial traffic and are ideally suited to pleasure cruising.

Chapter Two

A Look at the Locks

The type of lock that is used almost exclusively these days on rivers and canals is the pound lock. These have gates at each end of the chamber to control the level of water and they operate as follows:

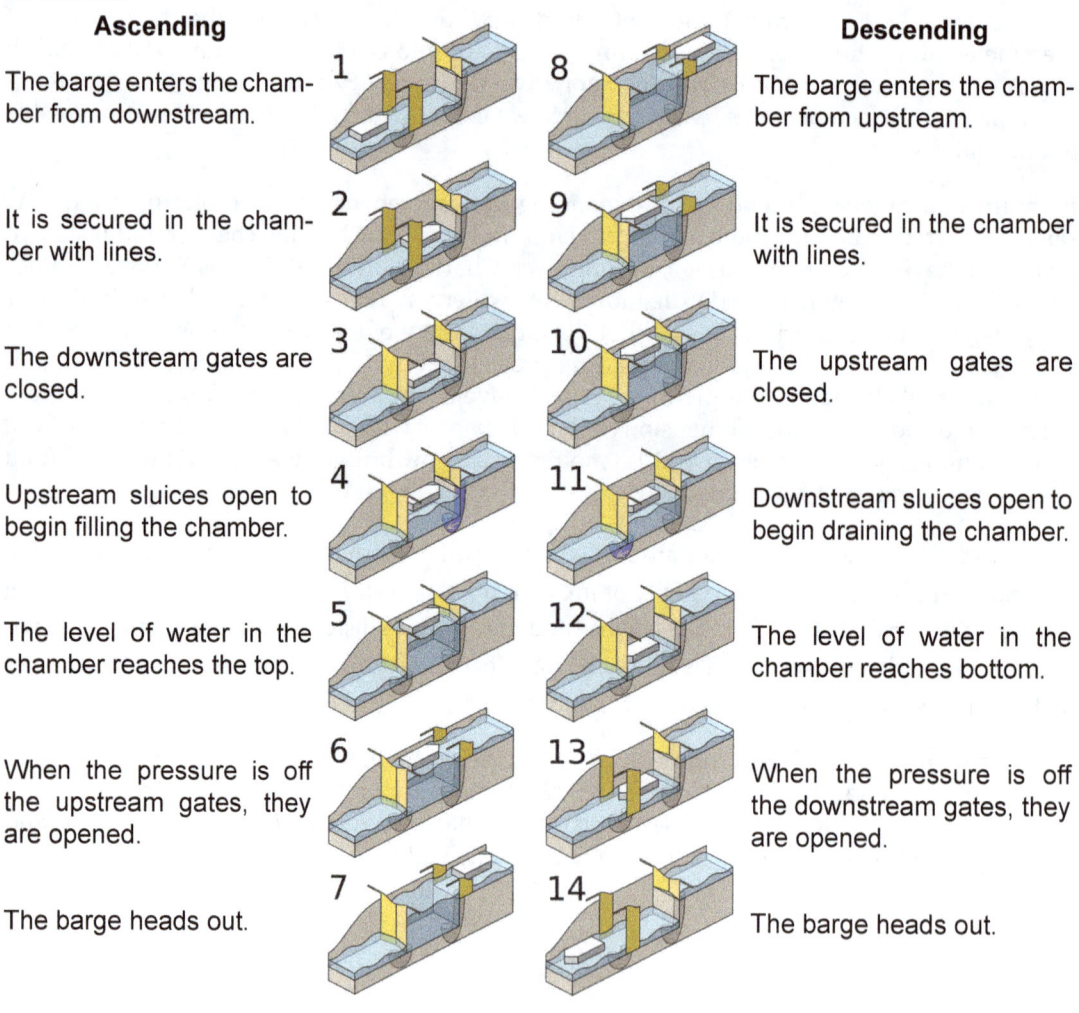

Ascending

The barge enters the chamber from downstream.

It is secured in the chamber with lines.

The downstream gates are closed.

Upstream sluices open to begin filling the chamber.

The level of water in the chamber reaches the top.

When the pressure is off the upstream gates, they are opened.

The barge heads out.

Descending

The barge enters the chamber from upstream.

It is secured in the chamber with lines.

The upstream gates are closed.

Downstream sluices open to begin draining the chamber.

The level of water in the chamber reaches bottom.

When the pressure is off the downstream gates, they are opened.

The barge heads out.

A Look at the Locks

As well as there being different types of canals, there are different types of pound locks and different methods of operation. Lock size is one of the differences; at the small end of the scale are the locks of Canal de l'Ourcq running eastward from Paris, which are only 3.2 metres wide and offer only 0.8 metres depth. At the upper end of the size scale is Écluse François 1er, which takes deep sea vessels into the port of Le Havre from the tidal waters of the Channel. It is 400 metres long and 67 metres wide. In another dimension, Écluse Bollène on the Rhône has the highest rise and fall of any lock in France at 22.5 metres. Its chamber is 190 metres long and 11.4 metres wide. However, the majority of the locks in France are the standard Freycinet gauge, built to take barges 38.5 metres long, 5.05 metres wide, 1.8 metres draft and providing a clearance of 3.5 metres under bridges.

The standard French lock has parallel, vertical-sided chamber walls. However, the locks of the Canal du Midi were designed with curved sides with a very slight slope to the walls. This design was adopted to add strength, and it seems to have worked. The original stonework from the 1670s and 80s is still mostly intact and remarkably preserved.

The entrances to the Midi locks are 5.8 metres wide, but inside the chamber the width increases to a maximum of 11 metres at the centre, allowing two 20-metre barges to lie side-by-side.

Another type of lock chamber that is occasionally seen is the sloped wall. To make it easier and safer to use these, many have been fitted with floating pontoons that ride up and down on rails attached to the sloping wall. On the pontoons are mooring bollards, and the method of use is to simply moor the barge as one would on a float in a marina. The one pictured here is along the Yonne River between Auxerre and Migennes.

L'écluse Ronde d'Agde gives another type of lock. This one is near the Mediterranean end of the Canal du Midi, where it is necessary to cross the Hérault River. Because at this point, both the river and the canal are near sea level and at similar elevations, there was no way to provide the canal with an aqueduct over the river. To solve the problem, in 1680 a round lock was built with three sets of gates. Coming in through the set of gates to the east, is up-bound canal traffic and river traffic from upstream of the weir. Once the gates close, the traffic can choose either to be lowered into the downstream portion of the river and head out the gates to the south toward the Mediterranean, or to be lifted to the continuation of the canal and exit through the gates to the west. This was a very elegant seventeenth century solution to the problem and it continues to serve well to this day.

Regardless the shapes of the locks, it is necessary to stabilize the barge in the chamber while the water level is being changed, and this is particularly important in the up-bound direction. The turbulence of the incoming water wants to move the barge around, sometimes rather violently and this movement needs to controlled.

A Look at the Locks

To assist in the control of the vessel during the ascent or descent, locks are fitted with a wide variety of devices. Mostly, these are fixed bollards along the lip of the chamber. In some locks these are plentiful, well-spaced and in good condition. In other locks, there are missing bollards, almost always exactly where they are most needed, thus requiring the juggling of the barge into position to use what exists and even then having to resort to creative line-handling.

When heading upstream through higher locks, it is often nearly impossible to toss a line over the bollards from the deck of the barge, let alone see the bollards from below. This requires the crew to disembark before the lock and walk up to receive the tossed lines as the barge enters.

When there is no easy place to disembark, the chamber ladder is used to access the rim of the lock. The ladder is dirty and slimy, since it is regularly in and out of the water as the lock fills and empties. In this photo we can see from the line of slime on the wall that the fill level of the chamber is nearly two metres below the rim. This added to the three metre rise of the lock makes the likelihood of ringing an unseen bollard with a line toss from the deck of the barge extremely remote.

To complicate the situation even further, often with locks near populated centres or in areas of high tourism, there have been safety barriers placed around the chambers. Most often these

interfere with the easy and safe use of the locks by boaters. This photo shows how a safety barrier on the Canal du Centre makes it nearly impossible to use the only upstream bollards in the lock. The placement of the barrier precludes throwing a bight of line up from the deck to ring the bollard. To use it requires the crew to climb the ladder to the lip of the chamber and then to move into a very awkward and dangerous position to place a line over the bollard. We continue to tell the lock keepers that these safety barriers are dangerous for boaters. It seems that theoretical engineers with no canal boating experience are responsible.

In some of the higher locks, in-chamber arrangements are provided for mooring. Among these is the vertical series of mooring points in recesses up the wall. These are generally spaced about a metre and a half apart, so two are always within reach. It is easier and safer to use two lines, placing the next one before removing the existing one. With this type of mooring point, it is possible for the lines to become jammed both when ascending and descending.

As with all line handling in a lock chamber, it is dangerous to tie or cleat a line, even temporarily. It can be easily forgotten and quickly become a potentially deadly weapon as the barge tightens against it to an explosive breaking. Even if the line doesn't break, a jammed, cleated or tied line can rip out or damage the boat's mooring bitts or cleats. The boat can hang-up in the lock and require the filling or draining cycle to be stopped and reversed.

A Look at the Locks

Another type of in-chamber mooring is the vertical pipe. With these, the mooring line, or a bight of it is passed around the pipe and held. As the barge moves up or down in the chamber, the line slides along the pipe. Care must be taken to closely monitor the lines, ensuring they continue to move up or down with the barge. We have found many of these arrangements with heavily corroded pipes that offer great resistance to the free sliding of the line. Even the coating of slime on the pipes does little to overcome this friction. The lines need to be jiggled, flipped and jostled frequently through the process to prevent binding.

Also, it is essential that the integrity of the system be tested before committing the load of the barge to it. It may not be securely attached
to the lock. We have seen several with loose anchors on the lock rims. One on Canal de Garonne nearly collapsed on us. There are also many with their pipes missing, making them useless.

Along the series of locks leading down from the aqueduct across the Garonne River at Agen, the lock infrastructure is in deplorable condition. Most of the pipes are missing from the vertical slots, safety fencing makes it very difficult to access the few bollards on the chamber rim, and deep fissures and cracks in the masonry of the chamber lip make jamming a mooring line a distinct possibility. Creative mooring and very close attention to the lines is necessary throughout the entire locking process.

In some locks the lack of suitable bollards or the unsuitable placement of them, makes motoring against a spring the only viable mooring arrangement. This entails leading a line forward from the bow or stern, or from both and then engaging the engine astern to pull against the lines while using the helm to keep the barge against the chamber wall. This can also be done by motoring forward against lines led aft.

In some locks with difficult access to the chamber rim, fixed lines are suspended down the sides of the chamber to be used for securing the barge. In this photo taken in the lock below the town of Clairac on the Lot River, we used the suspended lines, but their placement put our rudder in the way of the downstream gates when they closed. There was no convenient alternative but to motor forward against the line while keeping a very close watch on the line dangling down the opposite wall, lest its submerged end be pulled into our turning propeller.

Not all lock mooring arrangements are as cantankerous as these, many of the very large locks have floating bol-

A Look at the Locks

lards set in vertical bays in the chamber walls. Most often these are spaced at intervals of 30 metres or more, making securing to one at the bow and one at the stern very difficult and impracticable for most pleasure boats, which are generally much shorter than this. The easiest way for a vessel shorter than about 25 metres is to stop the barge centred on the bollard and take the bow and stern lines to the bollard. Always ensure that the bollard is free to move in its slot before committing to it. This can be done by carefully adding some body weight to it with a foot.
Closely monitor the movement of the float throughout the locking process, being prepared to quickly adjust the lines or remove them if the float jams in its ways.

Besides the mooring arrangements, variation among locks is also found with the gate designs and their opening mechanisms. Gates are the watertight doors which seal the lock chamber from the upper and lower pounds. On most smaller locks, the gates are a pair of mitred doors hinged to the sides at each end of the chamber. Most commonly the doors are mitred so that when closed, the pair meets at a broad angle like a chevron pointing upstream. The pressure from only a slight difference in water-level will squeeze the closed gates tightly together. This reduces the leaks between the doors
and prevents their being opened until water levels have equalised. Here on a Canal du Nivernais lock, the gates have just been swung closed with the balance beams, the weight of which balance the non-floating portion of the oak gates to take pressure off their hinges.

In this photo of slightly more modern lock gate system on the Canal du Nivernais, the water in the pound has just arrived at the level of the upstream pound, releasing the pressure against the gates and allowing them to be cranked open. Here the opening mechanism is a pair of columns each with a hand crank that drives bevel gears that turn a shaft with a pinion on its lower end that moves a rack that is attached to its gate.

The downstream gates of a lock need to be taller than the upstream gates since they have to seal off the full chamber, whereas the upper gates need only hold back the upper pound. The lower gate's height is the sum of the lift of the lock and the depth of water downstream of it. In some locks the lift is so high that a full height gate is impracticable. Here is an example from the Canal de Roanne à Digoin in which a solid end wall closes the portion above the downstream gates. The chamber fills to just below the top of the wall.

In this lock the doors are opened and closed using a pair of hand cranks on columns set on the end wall. Because of the weight of the doors the shaft and bevel gear system is run through reducing gears to give greater mechanical advantage, making the cranking process very long.

Another type of gate found on the deeper locks is one that lifts to allow the traffic to pass under it. Pictured below is Écluse Bollène on the Rhône. At its upstream end, the gate sinks beneath the water, allowing the traffic to pass over the top of it.

A Look at the Locks

The gates of the larger locks are electrically or hydraulically operated. It would be nearly impossible to move such massive gates by hand. During the past two decades, many of the locks on the Freycinet gauge canals have also been automated. On many of these canals, it is the boater that operates the locks. This is done with a variety of systems, ranging from remote control units to lock-side control panels.

The photo on the left, above shows the hand-held remote control box that is issued on the Canal entre Champagne et Bourgogne. The photo beside it shows one of the new user-operated automatic control pylons installed at lock side near the summit of the Canal du Midi. While the locks along the Canal du Midi have all been automated, all except a very few are still operated by lock keepers.

This photo shows one of the earlier forms of remote control that still exist on many of the French canals. A dangling rod is suspended above the canal a few hundred metres before the lock. Giving it a proper twist sets the lock cycle in motion and usually gets a response from the traffic lights at the lock. Depending on the canal, the response can be a flashing orange light and/ or the light changing from red to red and green, indicating that the lock is being prepared, or to green indicating that the lock is ready to be entered. Sometimes there is no response from the lights and the boater is left guessing whether the signal was received, whether the lights are malfunctioning or whether the entire system is malfunctioning. The solution is to wait a while, then moor and walk up to the lock for a look. There are call boxes in the lock that connect directly to VNF.

Chapter Three

Historic French Barges

The English word barge traces its origin to 1300 from the Old French, which had adopted it from colloquial Latin. The word originally referred to any small boat and the modern meaning didn't evolve until around 1480. The word bark, meaning a small ship, entered the English language around 1420 from the Old French word barque, which had come from the colloquial Latin barca at the beginning of the fifth century. By the seventeenth century, the word bark was being used to refer to a three-masted ship, and it often used the French spelling, barque to distinguish this meaning. Both colloquial Latin words had likely evolved from the classic Latin barica and from the Greek baris, meaning Egyptian boat. The Coptic word bari meant small boat. The Ancient Egyptian word ba-y-r, hieroglyphic ⛴ or 𓃀𓄿𓇋𓂋 ⛴ referred to a basket-shaped boat.

In the context used in this book, a barge is a flat-bottomed vessel of shallow draft that was designed for navigation along a river or canal. They evolved differently in each region, their designs dictated by local navigational conditions and the nature of the cargoes they carried.

Throughout the Middle Ages in France, most of the heavy inland transportation was along the courses of its many rivers. Shallow boats of simple design and construction drifted and sailed downstream with the current and were sailed, pulled or poled back up. Often, where the current was too strong for the upstream voyage, crude barges were cheaply built to last only the downstream trip. On arrival with their cargo, they were broken-up and sold as lumber or firewood.

Different regional uses spawned dozens of designs, such as: the marnois of the Marne, upper Seine and Yonne valleys; the chaland from the Loire; the sisselande from the Saône and Rhône; the courpet in the Dordogne; the chalibardon on the Adour; and the sapine of the Allier. Very few examples of these have survived, but in recent years enthusiasts have built working replicas of some of the old designs.

Two Chalands unloading at Orléans on the Loire

With the coming of the canals, barge designs quickly evolved to take advantage of the still waters made possible by the locks. The barges grew in length and width to match the dimensions of the locks and bridge cuts they needed to transit. Their draft increased to the maximum depth available on the canal. In many regions of France, cargo barges became rather boxy in appearance, but there were exceptions.

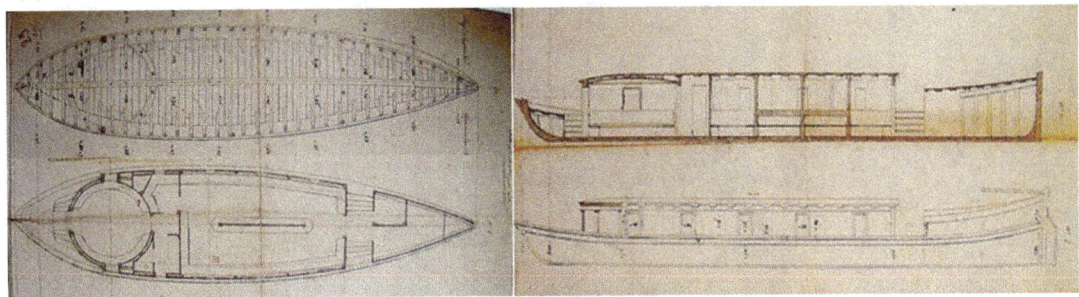

With the 1681 opening of le canal royal en Languedoc, (now called le Canal du Midi), the demand for fast passenger service was quickly met. Post boats were designed and built for the opening and they began providing a fast, efficient transportation system between Toulouse and Agde.

To save time spent in locks, a total of forty boats shuttled in the pounds between the locks and passengers changed boats at the ends of the pounds. Fresh horses at the trot hauled the post boats along at eight kilometres per hour in vessels that could accommodate fifty passengers, divided between the first class lounge in the front and the communal room in the centre. Passengers took their beds and meals ashore in the hostels along the canal, while the helmsman had accommodations in the vessel's stern.

The 240 kilometre journey from Toulouse to Agde took four days. Boat designs and management of the system evolved and by 1834 the boats were travelling at ten kilometres per hour and the trip was cut to three days.

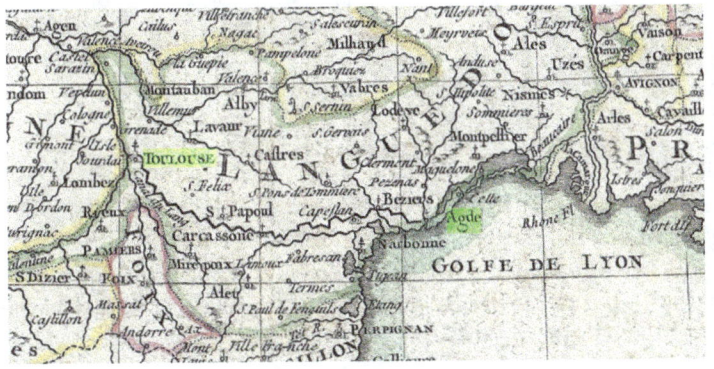

With later canal construction, the service expanded to Beaucaire on the Rhône and Agen on the Garonne upstream from Bordeaux. At its peak, the passenger volume reached 100,000 per year. In 1858 the opening of the railway almost immediately killed the post boat service; it didn't quite make two centuries.

Sailing was one of the methods of propelling a barge along a river or canal, but in vessels with flat bottoms and little or no keel, this was limited to downwind sailing. In contrary winds, the barge had to either be poled or hauled. The current in the lower 400 kilometres of the Loire flows to the west and the prevailing wind blows upstream from the west. This enabled barges to run downstream with the current and sail back up in a following wind.

In canals and rivers with no current or gentle currents and few favourable winds, hauling was done by men. With family operated barges, often the women and children would do the hauling, while the man tended to more exacting tasks like relaxing, smoking and steering. It takes some effort to get the barge moving, but in calm conditions, little effort is required to keep it moving at two kilometres per hour.

More often in France, horses were used, and sometimes oxen. Moving along at six kilometres per hour through the placid waters of a canal pound in this fashion was peaceful. Towpaths had been built alongside the canals and through the bridge cuts and the only complications were when meeting a barge coming in the opposite direction. Tow lines had to be detached from harnesses and reattached once the barges had drifted past each other.

On rivers with stronger currents, such as the Saône and the Rhône, barges like the sisselande were steered using long sweep oars. For better control when heading downstream in swift current, they often used steering sweeps both fore and aft.

To do the re-ascent against the current required very large teams of horses. Depending on the current, this could mean from twenty to as many as forty horses, and these required one driver per two or three horses.

Animal haulage on the faster sections of the Saône and the Rhône quickly diminished around 1840 with the expanding use of steam tugs. In a few hours these powerful machines could haul a string of barges upstream over a distance that would have required two hundred horses and seventy to a hundred men a day or more to accomplish. The Industrial Revolution was in full swing.

On the more gentle rivers and on many of the canals, less powerful steam tugs began being used to pull long trains of barges. In much of Western Europe there was a ready adoption of these new methods, though an exception to this was in France, where the old ways, the gentle and quiet ways were maintained by many.

Motorized tractors began appearing on the towpaths to replace animals. Various versions were also devised of electric motors running suspended from overhead cables.

In some of areas, such as Cambrai, rail-mounted traction machines were introduced. Then in 1926 la Compagnie Générale de Traction sur les Voies Navigables (CGTVN) was established in the north and east of France. The company operated electric tractors running on rails on the towpaths. The transition to the mechanical haulage was resisted by bargemen, many of whom saw it as a disruption to their gentle ways.

However, the traditional barge operators were being challenged on another front. Since the turn of the century, engines had begun being installed in barges. Initially this was usually a retrofit to an existing wooden barge, but gradually new iron or steel barges, loosely called automoteurs, were designed specifically for power. However, unlike in other areas of Europe, the transition to motor barges in France was slow.

In 1935 there were still 1500 working bateaux-écurie, horse barges along the French canals. Most of these were built of wood and they were equipped with horse stables and a loading ramp to move the horses between the barge and the bank. By the late 1950s horse haulage was still common along the canals, but the horses were being gradually replaced by electric traction along the towpaths. The northern and eastern canals were the first to retire horses, but in the centre of France, horses stayed on despite the powered tractors and the convenience of powered barges. The last bateau-écurie docked in 1969. The self-powered barges gradually took over, and because of a lack of business, in 1972 la Compagnie Générale de Traction sur les Voies Navigables filed for bankruptcy.

Historic French Barges

The style of barge most common in France in the late nineteenth century and through the first half of the twentieth was the péniche, which was based on the Flemish Spits. This 1908 photo, taken during a chomage, when the canal was drained for maintenance, shows its boxy shape.

The standard péniche had evolved to take full advantage of the maximum space available in the canal locks. The new gauge for canals that was implemented in 1879 by the French Minister of Transport, Charles-Louis Freycinet, had been based on the péniche dimensions.

At the end of the nineteenth century there were about 12,000 commercial canal boats in France. Of these, 8000 were the péniches in northern France that were employed with hauling coal to Paris.

The péniche was based on the Flemish spits, which the French called bélandre, It had evolved from the Dutch bijlander, whose name meaning *near land*, referred to its coastal use. Its gracefully curved ends, sweeping sheer and tumblehome bulwarks were the results of how wood could be bent in building the hull.

That such a graceful ship as the bijlander evolved in Belgium and France into the mundane, utilitarian péniche, is a major reason why modern canal boating enthusiasts gravitate to Dutch barges.

Chapter Four

Historic Dutch Barges

The geography and conditions in the Netherlands caused inland boats to evolve very differently from those in France. The bijlander, from which the French bélandre and then the péniche descended, was a flat-bottomed vessel used for coastal trade. Variations of it are known back to about 1500.

From the bijlander the tjalk evolved in the seventeenth century. Tjalk is a rather generic term used to designate a long, narrow and shallow barge shaped somewhat like a shoebox with rounded corners. Most tjalks had a mast rigged with a gaff, they carried a bowsprit, and because of the shallow waters of the Netherlands, they had flat bottoms and no keel. To counter the press of the wind when sailing, they were fitted with leeboards. Their gracefully rounded ends and pronounced sheer lead to their being likened to Dutch wooden shoes.

There were many different styles of tjalk, their variations being based on the particular conditions and intended uses where they were built and sailed. They were broadly classified by style, such as paviljoentjalk, which had a raised after deck with accommodations below or a roefschip, which had a small cabin aft of the hold and a lower aft deck, or a dektjalk, which had neither cabin nor raised deck. They also were classified by the region where they were built, such as Groninger tjalk, IJsseltjalk, Friese tjalk or Hollandse tjalk. Tjalken were generally 20 to 30 metres in length with beams of 4 to 5 metres.

The skûtsje was a smaller tjalk, shorter and narrower to trade in the shallow lakes and narrow, winding canals in southwest Friesland. They were generally between 15 and 20 metres in length and had beams 3.4 to 4 metres. The skûtsje was developed between 1855 and 1860 and they were initially built of wood. Then in 1887 riveted iron was introduced, and during the first decade of the twentieth century, the yards gradually converted from iron to steel. A total of 870 iron and steel skûtsjes were built until the final one in 1933.

With the closing of the Zuiderzee in 1932, the commercial usefulness of the skûtsje dropped. Some had engines fitted, but even then, they could not compete with the purpose-built motor barges. During the ensuing years, many skûtsjes were scrapped, some were converted to homes and some to pleasure

or racing vessels. There are currently about 90 of these historic vessels restored to full sailing trim and competing in regular races in Friesland.

The larger tjalken also declined in usefulness as the new breed of motor barges took over the commerce. Like the skûtsje, many were scrapped, but hundreds of them have survived. Many live on as woonboten, houseboats permanently moored along the canals, while others have been converted for pleasure cruising. In most towns around the IJsselmeer there are restored sailing tjalken offering a variety of trips, tours and charters.

Other unpowered sailing barges included various versions of the aak, such as the Hasselteraak, Lemsteraak, boeieraak, and klipperaak. Like the tjalk, these had leeboards for sailing. With their hulls not designed for motor propulsion, they too became commercially obsolete and other uses were found for them. Unfortunately, for many, this meant the scrap heaps.

In the early 1920s the old non-motorized barges became uncompetitive after the introduction of the luxemotor, the modern classic of Dutch ship design. Its sharp, plumb entry and its stern designed for proper water flow to the propeller made it an efficient and fast barge. Adding to its attraction and giving the barge its name were the luxurious accommodations in the stern.

With their larger hold capacity, much better efficiency and greater comfort, luxemotors quickly became the standard on the canals of the Netherlands, rapidly replacing the older aak and tjalk hull styles. They were built in lengths from 18 to 30 metres, though many were lengthened in the 1950s in an attempt to compete with the falling price of truck freight. Even with lengthening, the design became increasingly commercially unviable in face of competition from more modern barges.

Other types of motor barge enjoyed some limited success during the reign of the luxemotor. The steilesteven, with a fatter stern and its wheelhouse all the way aft, provided an increased hold capacity. Another was the kitwijker, which like the steilsteven was designed originally without an engine. Neither ever achieved the popularity of the luxemotor, and they too were eclipsed by improved designs after the middle of the twentieth century.

Retired examples of the luxemotor, the stilesteven and the kitwijker are also popular choices for makeover to liveaboard and cruising barges. With the tjalks and aaks, they form the most popular group of barges for conversion to pleasure cruising on the canals of Europe.

Chapter Five

Pleasure Cruising on French Canals

I first became aware of the European canals in 1966 when I was serving in France with the Royal Canadian Air Force. In the Lorraine, Champagne, Franche Compté and the Bourgogne I saw old brown péniches moving slowly along the canals, I watched some of the last horses on the towpaths and I frequently paused to watch barges passing through locks. I saw no pleasure boating.

After I resigned my commission from the Royal Canadian Navy in 1981, I set-up in the wine business. For two decades we made very frequent visits to Europe while searching for wines to import or later while conducting wine and food tours. During these trips we always spent a week or two or longer in France and never failed to search-out the canals and watch the activity along them.

In 1984 we rented our first canal boat in France from Blue Line in St-Jean-de-Losne in the Burgundy. It was mid-Spring and we had the Canal de Bourgogne, the Canal du Centre and the Saône almost completely to ourselves. There was minimal commercial traffic, it was off-season for the few hotel barges that existed at the time and we saw very few other rental boats and even fewer private cruisers during our three weeks. The lock houses on the canals all had resident lock keepers, with whom we paused to chat while assisting them with the gates and sluices. We bought fresh eggs and garden produce from them and learned of interesting places to visit and favourite dining spots. We were hooked.

I had not been able to find a travel agent in Canada with any knowledge of French canal boat rentals, so I had to write to Blake's in England to book our boat. Since then the number of companies offering canal boat rentals rapidly increased and then slowly declined. There are currently a dozen large companies and about three dozen smaller companies with canal boat fleets in over a hundred locations throughout France. It is still an important, though shrinking and consolidating tourism sector.

The latest Voies Navigables de France (VNF) report shows that there were forty-seven rental companies offering a total of 1604 boats from 121 rental bases throughout France. These numbers are down fifteen percent over the decade. The report states that almost fifty percent of the rental activity is in the south and southwest and that seventy-two percent of the rentals are for a week or less. My recommendation to avoid the biggest crowds of the most inexperienced boaters is to avoid the high season in the Midi.

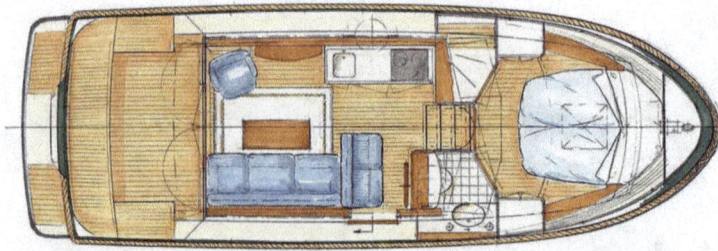

In 2014, a typical weekly rental fee for a boat for two people was €1300 in low season, around €1700 in shoulder season and close to €2000 in high season. This Dutch-built Linssen 28 Sedan is a fine example of a two-person rental boat.

Larger boats for four, six, eight, ten and more people cost more, but the small incremental increase per additional bed makes multi-couple rentals very popular. A 14.9-metre boat like the Linssen Euroclassic 149 shown here, with four cabins all with ensuite showers and toilets rents for €2700, €3300 and €3750 in low, shoulder and high seasons, about half the rate per person than for the smaller boat.

One of the great attractions of the rental boats is that they require no boating experience or knowledge, no boating skills or training, no license or qualifications to operate. A brief introduction to the operation of the boat is given at the beginning of the rental period, but effectively, with very little boating awareness, most renters are set free to learn boat handling for themselves.

Another popular way to cruise the canals is aboard a hotel barge. Though a few were purpose-built, most are converted freight or general cargo barges, gutted and re-fitted with high-quality interiors. There are currently seventy-six hotel barges in France with a total of more than a thousand guest beds. They range greatly in size, accommodating from four to twenty-four guests and they also vary in the quality of the accommodations, amenities and dining provided. Most are fully catered and have organized tours and activities ashore as the barges slowly cruise their itineraries along the canals.

Pleasure Cruising on the French Canals

Many of the largest ones are converted péniches or spits and among the smaller hotel barges are converted tjalken, aaken and luxemotors. A common feature is a high staff-to-guest ratio, with the smaller barges often providing higher comfort and individual attention. Weekly rates generally run in the €3000 to €4000 per person range. Some barges with more basic accommodation and self-catering cost under €2000 per person, but for a grande luxe hotel barge, like the Alouette pictured here, the weekly rates are around €5000 per person.

According to the latest published VNF study, the hotel barges handled a total of 19,200 guests, 75% of whom were foreigners, mainly Anglophone with a large portion from North America. Assuming an average charge of €4,000 per person per week, this would have generated an annual sales revenue of nearly €75 million.

Most of the hotel barges follow a set itinerary throughout the season, doing a cruise in one direction along the canal one week and retracing the route in the opposite direction the next. A few are more pelagic, their itinerary being along a canal or series of canals with guests joining in one location and disembarking at the end of the week in another as the barge works its way through one or more regions. Some of the smaller barges are known to write their itineraries based where their early-booking clients want to explore and then connect the dots and fill-in the gaps with later bookings. The larger hotel barges fill the majority of their cabins through specialty travel agents.

River cruises form another category of inland pleasure cruising. In recent years the popularity of cruises on the Rhône and Saône has rapidly increased. In the latest report sixteen operators were running thirty-eight river cruise ships with a total 4976 passenger beds and they carried a total of 222,850 passengers for 910,900 nights, up more than ten percent in a year. Many of river cruise itineraries are week long; however, there are also three, four and five day itineraries. Most of the prices are in the range of €90 to €150 per person per night, though some are much higher, so it is conservative to say that the operators took in revenue in excess of €100 million on the year.

The final category of commercial pleasure cruising is in the Bateaux Promenade, the short trip boat commonly called Bateaux Mouches and most associated with the Seine in Paris. A simple tour through central Paris lasts seventy minutes and costs €12.50 per adult, with reduced fares for children and groups. Lunch cruises are offered on the weekends at €55 and dinner cruises every evening offer menus at €99 and €140. The latest VNF report shows 9,989 990 Bateaux Promenade passengers in France, seventy-seven percent of them in Paris. Annual revenues are in excess of €150 million, and since Paris is the most visited city in the world, it is safe to say that the vast majority of this is from outside France.

Carefree Through 1001 French Locks

I have gone into some boring facts and figures in looking at these various commercial operators. This is because I think it important to point-out the significance of their contribution to the French economy. The report shows that more than €400 million was spent by guests aboard the broad variety of commercially operated pleasure boats on the inland waters of France. In each of the categories except self-drive rental boats, the number of passengers or passenger nights was up and continues to rise strongly. In addition, there are the jobs generated in the supply, maintenance, operation and administration of the fleet of some two thousand vessels and their bases. The French authorities are not likely to allow this revenue to be endangered.

Not mentioned in the VNF report were the privately owned and operated pleasure boats, though according to other VNF figures, there are currently over 20,000 active private boats on the inland waterways, fifty percent of which are foreign owned. This private fleet is nearly ten times the size of the entire commercial pleasure fleet, but its impact on the economy is considerably less. In annual canal usage fees it generates less than €1 million. An average of €5000 per boat for annual moorage, maintenance, food, drink, supplies and services would provide another €100 million to the economy, just one quarter of the commercial pleasure fleet revenue. Because of this comparatively low revenue generated per boat, €5000 compared to €190,000, the private boats receive very little support with infrastructure and facilities.

However; owning one's own vessel on the waterways offers the most freedom and flexibility. In 2000 I bought my first canal boat in France. Lady Jane was a 14-metre Dutch motorkruiser built of steel in Groningen in 1970. After a refurbishing refit, we cruised throughout northeastern France on the Saône, the Canal du Rhône a Rhin, the Canal du Centre, the Canal de Bourgogne. We wandered up through the Franche-Comté, crossed into the Champagne, spent time in Paris, explored the upper Seine, the Yonne, Loing, the Briare and the Loire.

After six years on the French canals, I grew restless and wanted some wild and remote sailing. While I was still young enough, I wanted to sail through Patagonia and around Cape Horn. I sold Lady Jane and had a new 15-metre cutter-rigged sloop built and christened Sequitur. In 2009 Edi and I sailed south from Vancouver and after three years exploring South America, we rounded Cape Horn and headed up the Atlantic. Along the way, in dealing with stultifying Third World bureaucracies, sparse resources and a hurricane and three Force 11 storms at sea, we realized that we were approaching our best-before-dates for wild cruising. We decided to look at a more sedate and gentle style of boating.

Chapter Six

Buying a Canal Boat

Because I had previously owned a Dutch motorkruiser, our initial thoughts were on buying another one. From what I had seen during my previous years of canal boating, most of the motor cruisers on the French canals had been built in the Netherlands. The Netherlands, with only one quarter of the population of France, has several times as many used boats available than does France; the wisdom almost universally expressed among canal cruisers is that the Netherlands is the place to buy.

As we worked our way up the Atlantic from the Falklands to Uruguay and then around the coast of Brazil, whenever we had an Internet connection we combed the online listings on Apollo Duck, the major international listing site used by both brokers and private sellers. Searching for used powerboats in the Netherlands yielded over three-hundred listings; narrowing these to boats between 14 and 20 metres took the list to fifty-one.
Refining the list further to boats less than twenty years old pared it down to twenty-nine.

As we looked at the listings, we were linked to many brokers' sites, and on these we found other listings. Because most of the online listings have multiple photos, we were easily able to form opinions on boats that weren't suitable and narrow the list further. We looked at many possibilities and dreamed.

Somewhere along the way our thinking evolved to buying a new boat and we began investigating that side of the market. Most intriguing to us was the line of boats being built by Linssen in Maasbracht. The company was established in 1949 and has been building a range of very high quality motor yachts that are well-suited to the European waterways. I recalled having admired those I saw on the French canals a decade previously.

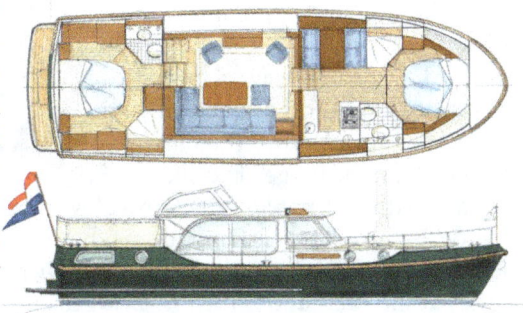

As we sailed across the Caribbean, pausing to relax in Barbados, Martinique, Guadeloupe, Saint Croix and Puerto Rico, our Internet research went into full speed. Since we were thinking of a new boat, we investigated the possibilities of having a new barge built to a classic Dutch barge design. We found many companies building modern versions of the luxemotor or styles adapted from it. Among those that interested us in England were Branson, Piper and Will Trickett and in the Netherlands we found Euroship and SRF interesting. Most intriguing to us was Aqualine, a British company that set-up a factory in Gdansk, Poland to use the skills from their long history of shipbuilding and to take advantage of the lower labour costs.

During the exploration process, our thinking slowly morphed toward buying a converted ex-commercial barge. We pored over the online listings of the many brokers in France, the Netherlands and Britain. We looked at the listings for converted aaks, klippers, luxemotors, steilstevens and tjalken.

As we grew more familiar with the various barge types, we were increasingly drawn to the tjalk, the most typical of all the Dutch small ships. The name is from the old Dutch language and was first mentioned by Witsen in Architectura Navalis in 1690. This painting of a tjalk by Ludolf Bakhuizen is dated 1697. Tjalken are flat-bottomed sailing cargo ships with shallow draft ideally suited to use in the shallow inland and coastal waters of Friesland and Holland.

On the Internet we saw a photo of a beautifully restored tjalk named Owlpen, which convinced us that we wanted a similar barge. The shape of the tjalk, with its tumblehome gunwales and pleasing sheer, is reminiscent of the Dutch wooden shoe. Because of their flat bottoms they were fitted with leeboards instead of deep keels, and they could easily dry-out on the beaches with the tides. Most tjalken had easily lowering masts so they could pass under the low bridges and then re-erect the rigs to continue sailing.

Buying a Canal Boat

As we sat in St Augustine Florida preparing Sequitur for sale, I began compiling a spreadsheet of suitable tjalken to consider. I began contacting brokers in the Netherlands and refining the list and by the time we had arrived in Friesland in July 2012, I had a complex spreadsheet of data from seventeen tjalken we had chosen to look at. Most of the ones we had selected had been built in the 1890s or during the first decade of the twentieth century, and they ranged in length from sixteen to twenty metres. There are much more rigorous licensing requirements for vessels over twenty metres in Europe, so we wanted to stay under that length.

We had arranged for a rental a car at the airport and had booked accommodation in a holiday apartment with a well-equipped kitchen. There was a large supermarket a five minute walk away, so shortly after we had arrived, we went shopping for groceries. We had appointments through their various brokers to view several barges on our list, and we were awaiting responses from four other brokers.

Our first appointment was in Leeuwarden on Saturday morning. After breakfast, our car's GPS took us to the canal alongside our first tjalk. Lady Gazina is 19.98 by 4.60 metres and was built in 1907. She still had her complete sailing rig in working condition.

Her interior was well finished, and fully set-up for comfortable living. However, the systems and galley equipment were very dated and the interior layout was not to our liking. We would have to gut the entire interior and start over. Her high asking price reflected her condition and equipment, much of which we would have to throw away. Also, her generous headroom gave her a very high air draft at 3.5 metres. This meant she would scrape on many of the bridges in France and would be too high for the Bourgogne, the Nivernais and the Midi.

On Saturday afternoon we drove to near Koudam on the shores of the Alde Karre, near the IJsselmeer for an appointment to see the next tjalk on our list, a 19.90 by 4.15 metre vessel built in 1910. From the outside she appeared to have long since lost the love of her owner. This was strongly confirmed by a visit to her interior. The owner explained the wet bilges to be from a leak she had not yet tracked down, and the stains and rotting floorboards forward pointed to other leaks from above and below. She was very quickly scratched off our list.

Making our trip worthwhile, though was the sight of a skûtsje sailing past in a stiff breeze on the Alde Karre. We were reminded that skûtsje racing is a very active and keenly followed sport in Friesland, where dozens of extremely competitive antique skûtsjes regularly compete in organized races.

Our next appointment was southward in Lelystad to see a 19.80 by 3.80 metre tjalk that had been built in 1903. She had a full working rig in apparently good condition. Her interior; however, was absolutely

basic and this was diminished further by carpentry of a very poor standard and signs of poor maintenance. Her interior would be easier to gut than the first barge we looked at, but she simply did not feel right.

With our day's appointments completed, we drove back by way of a stop in Harlingen. We walked along the canal in the centre of town, in awe of the beautiful old buildings and admiring the obvious pride of the home owners. The canals were filled with boats moored along both sides. Among them were several classic boats: tjalken, Lemsteraken, klippers and skûtsjes.

In the outer harbours were many commercial tjalken, and we watched as one of them arrived and unloaded its day tour passengers and another appeared to be stocking-up for a more extended trip with another group. We counted fifteen restored klippers and tjalken, most over a century old, that appeared to be in the charter trade.

We had an early dinner back at the apartment and quickly went to bed to continue working-off the jet-lag from our eight hour time change. On Sunday we had no appointments, so we went exploring, hoping to find a suitable tjalk with a Te Koop sign, a *For Sale* sign.

We also went looking for one of the listings on our spreadsheet, about which we had not received any response from the broker. This 19.98 by 4.49 metre tjalk was built in 1897. After some questioning, we finally tracked it down alongside a farmhouse outside of Kollum. We barged-in and met the owners on their patio and they gladly showed us their ship. We could find nothing positive to say about it.

We then drove to Warten to look at boats in the marina at Boten en Meer. After sharing coffee with us, Auke van der Meer, the owner of the marina and its associated brokerage showed us around. He had nothing suitable onsite, but we were very taken by a beautiful little 5.4 by 2.2 metre boeier built of oak in 1941 for a Nazi general during the occupation.

Auke was the listing broker for one of the tjalken on our spreadsheet, but its owner had moved the barge to a boatyard in Heeg to have it stripped-out to a bare hull in preparation for a new fit-out. It is a 19.66 by 4.22 metre Pavjonentjalk built in 1898 and still with its full rig, but this had been removed for the work. The yard is closed on Sundays, so we arranged to meet Auke on Monday morning for a visit.

On Monday we liked what we saw. It was stripped-out, the engine and machinery were removed, the deckhouse had been cut away, the poured concrete ballast had been jackhammered out and work had begun with some bottom re-plating.

We talked in broad terms of various refit possibilities with Auke and the yard manager. The barge was still sufficiently original that it could be easily restored and accepted for listing as an historic vessel. This would allow free moorage at museum harbours throughout the Netherlands. We looked at the leeboards and other bits and pieces of rigging that had been removed, and then we went back to Warten with Auke to talk details.

One clog in the works was that, since he had begun work, the owner had rather steeply raised the asking price from what we had seen in the listing. Another detraction was the estimate of nearly a year to complete the fit-out. At least we had found a tjalk that we hadn't rejected.

On Tuesday morning we drove across the Afsluitdijk to North Holland for an appointment with Peter Rood of Scheepsmalelaardij Enkhuizen. Peter had four of the listings on our spreadsheet, two of which we had already viewed through him on Saturday and Sunday. He had made appointments for the other two for us for later in the afternoon and our meeting with him was on our way south to see them. We also had an appointment to see a tjalk from another broker. We sat in Peter's canal front office, sharing coffee and giving him a better understanding of what we were looking for.

One of the better ways to search for a boat is to engage the services of a professional broker. The boat brokers work for, and their commissions are paid by the seller. Their job is to find clients for their listings, since they make their commissions only on completion of a sale. A good broker will eagerly take on serious buyers and search for suitable boats for them. They have their own listings, but they also know what is available in the market from the other brokers and they stand to split commissions if they make a sale of another broker's listing.

First on our list was a 19.80 by 4.20 metre live-aboard in Amsterdam. We found it comfortably finished, but completely wrong for our tastes. We would have to gut it and start over, and because of its level of finish, the asking price was high, making it unreasonable to consider.

The second tjalk on our day's list was in Aalsmeer. It was a 1908 skûtsje of 16.38 by 3.44 metres, the second smallest on our spreadsheet. The pretty little barge had been bought by her current owner in 1975. He had converted and lovingly maintained her for 37 years. Her sailing rig was ashore in a shed, where the owner had been restoring it until he became too ill to continue.

The interior, while rather smaller than we had set out looking for, oozed charm and showed a distinct pride of ownership. The owner talked lovingly of many years of boating with his children and now his grandchildren.

We now had a second barge on our list to consider; how very different they were from each other.

The final barge of the day was a 19.14 by 4.10 vessel built in 1902. It was in Loosdrecht and was owned by a marine engineer and his wife. He works in the boatyard where they are moored. The tjalk was fully fitted-out, rigged for sailing and was very well maintained, as can be expected from being owned by a professional boat maintenance person.

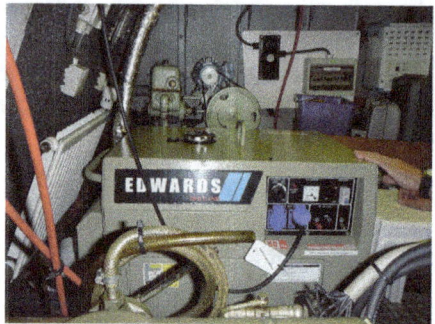

I was impressed with the engineering fit-out, by the completeness of the systems and equipment and by the orderliness and cleanliness of the barge. Its price reflected the quality. While the interior is not completely as we would like, its modernization refit would not be such a huge task.

All of a sudden, we found ourselves with three choices, each dramatically different from the others. During the 160 kilometre drive back to Leeuwarden we had much to discuss. Essentially, we had already made up our minds which one we wanted. However, as we drove I played devil's advocate and represented the other two, offering ever decreasing arguments in their support.

Buying a Canal Boat

We arrived at our apartment at twilight, and before beginning to cook dinner, I sent an email to the broker with an offer on Nieuwe Zorg, telling him we would like to meet in his office on Wednesday to formalize the offer. On Wednesday morning we received an email from the broker informing us that the seller had accepted the amount of our offer, and that he would await our formal written offer.

At 1100 we met with the broker. Among other things, we went over lists of recommended surveyors and insurance agents and looked at possible yards to haul-out for the survey. I was delighted to see on the short lists the names of the surveyors and insurers to whom the previous evening I had emailed requests for quotes. These names I had harvested from the list of links on The Barge Association site. We had rejoined the Association, having let the membership lapse after sale of Lady Jane in 2006.

With paperwork well underway, we headed back up to Leeuwarden, packed our belongings and moved out of the apartment. The week's rental was up, so we moved south to an apartment in Volendam. We had found a rather modern place in a location much more central to our anticipated activities over the following week or so. It has a well-equipped kitchen and a second-floor balcony overlooking a large, active marina.

On Thursday morning we went on a walking exploration of old Volendam, through the labyrinth of narrow back streets and then along the old waterfront. There we admired the traditional Volendamer kwaks, the wooden fishing boats similar to Zuiderzeebotters.

These kwaks fished the Zuiderzee for shrimp, eel, anchovies and herring in the nineteenth and early twentieth centuries, and at its peak in 1910, the Volendam fleet numbered over three-hundred kwaks. With the closing Afsluitdijk in 1932, the former Zuiderzee waters around Volendam became the IJsselmeer, and as the waters slowly lost their salt content, the fishery and the local liveliehood changed.

Carefree Through 1001 French Locks

On Friday we drove up to Enkhuizen, where we proofread, amended and eventually initialled and signed the formal offer to purchase Nieuwe Zorg. The HISWA (Netherlands yacht brokers' association) standard contract was clear and concise, as were the nearly four pages of appended special conditions, which had been compiled in face-to-face, email and phone consultations among the owners, the brokers and us.

Peter then gave us an insurance application to complete, and he faxed it to the broker, phoned to see if all was in order, had us redo one page and re-faxed it. Our quote for insurance on Nieuwe Zorg came in at less than eight percent of our just cancelled insurance on Sequitur. It looks like the Dutch canals are a much lower risk than the Chilean ones.

Back in our apartment on Friday afternoon, using Skype and the reception office's fax, I ordered our bank to send a wire transfer of the fifteen percent deposit in Euros to the broker's Stichting Derden Geld (third-party or trust account). Meanwhile Peter had coordinated our chosen surveyor, a haul-out yard and Nieuwe Zorg's owners into a very smooth three-day schedule of events beginning on Monday.

We were up early on Monday to drive to Hoopddorf to the EuropeCar agent to arrange for an extension of four days on the rental car. Then we continued on to Aalsmeer, arriving at Nieuwe Zorg at 1002, two minutes after the appointed time. We surprised the owners Henk and Madij, who been told that we would be arriving at noon. This was to be the only glitch in the complex schedule involving eleven people, four locations, three days and two languages.

We relaxed onboard having coffee with Henk and Madij and waiting for their daughter, Jacqueline to arrive. Edi and I were feeling comfortably at home onboard as we increasingly realized how suitable Nieuwe Zorg is for us.

We reflected on the meaning of her name: Nieuwe Zorg translates to *New Concern* or *New Worry*, and we thought that she should instead be called Zonder Zorg: *Without Concern, Without Worry, Carefree* as we are.

Jacqueline eventually arrived and provided excellent translation for Henk and Madij. We headed out of the little port and down the canal toward the Westeinderplassen, scooting easily under a low bridge, while other boats lined-up waiting for it to open. Nieuwe Zorg's air draft is only 1.95 metres, which gives her a great exploratory range. Henk gave me the tiller and I quickly gained a feel for how easily she handles.

We arrived at Kempers, and Henk being unfamiliar with the marina layout, headed in through a narrow entrance and weaved his way between lines of moored boats toward a couple of people on the bow of a cruiser to ask directions. The way was back from where we had come, and I was pleased with how easily Hank handled the barge in the confined spaces. She appears very responsive, and I noted she has a left-turning screw, which in astern gear gives the stern a kick to starboard. This has always  been my preferred turn, since it makes very easy mooring with the favoured starboard-side-to. We found the proper entrance and came to bollards on the wharf short of the travel lift.

Marine engineer Rutger Versluijs arrived onboard and began a systematic inspection of the engine and the machinery spaces. He paused from time to time to explain to me what he was doing and what he was finding. First with the engine and generator off, and then with them running, Rutger continued inspecting the machinery spaces for over half an hour, including the engine cooling and exhaust systems, the through-hulls, the electrical systems, the generator, the batteries, the heating system, the propane tank  installations, the engine mounts, the transmission, the thrust bearing, the stern gland and its associated greaser, and so on. During this process, he used an electronic temperature sensor to take readings from a variety of places on the engine and its ancillaries.

He then had Henk motor out onto the lake and we did a half hour sea trial, including ten minutes at full throttle with Rutger down in the engine room with his flashlight and instruments. Back alongside, Rutger performed battery load tests and did inspections of the galley and heads systems. It was early evening by the time we were finished with first part of the survey; the second and third parts, the hull and the rigging were scheduled for Wednesday.

We were delighted with the thoroughness of the surveyor, and with his findings thus far. We drove back to our apartment in Volendam, where we dined well and slept well. On Tuesday morning we arrived back in Leimuiden for the scheduled 0800 haul-out. The weather the past few days had been clear and warm and forecasts had this fine spell continuing for several more days. It was glassy calm as Henk motored Nieuwe Zorg the short distance into jaws of the travel lift.

We admired the efficiency of the haul-out process. It consisted of one man with a push-button remote, calmly and methodically doing what we have seen teams of three and more chaotic and scrambling men perform in Vancouver, Callao, Puerto Montt and St Augustine.

Henk told us that Nieuwe Zorg had last been out of the water in 2005, when he had re-plated her bottom. We had expected to see a heavily fouled bottom, encrusted with mussels and other growth. Instead, there was only light vegetation and a few small crops of mussels. As Nieuwe Zorg hung in the slings, the lone operator power-washed her bottom.

The scales on the travel lift showed that she weighs 18.52 tonnes, which, for the benefit of the metrically-challenged, is about 20.4 tons.

With a touch of thumb pressure, the operator single-handedly nestled Nieuwe Zorg onto the stands as he adjusted the wooden blocks and wedges to evenly distribute the old girl's weight. The one-man operation was much calmer and took less time than any of Sequitur's haul-outs.

After taking another series of measurements of her interior, we left Nieuwe Zorg resting on her stands and headed north again. We drove to Alkmaar in central North Holland and explored the range of furniture and kitchen appliances in the crowd of large home decor stores across the canal from the historic centre of town. Shopping craving satiated, we headed across the bridge and strolled through the old town until a canal-side restaurant tempted us to pause for lunch.

Back in our apartment in Volendam on Tuesday evening I prepared a huge pot of mussels. Not the ones from Nieuwe Zorg's bottom; we had found some great-looking fresh mussels in the market in Alkmar. I began with a butter sauté of julienned garlic and shallots and diced red, green and yellow peppers, then added the mussels and a generous splash of beer. The steamed mussels were garnished with chopped parsley and served with melted butter and fresh baguette slices accompanied by a bottle of Champagne Veuve Clicquot. It was a delicious, if maybe a bit premature way to begin celebrating our pending new adventures.

On Wednesday morning we arrived back at the Kempers yard at 0800 to find that Rutger had already begun the bottom survey. He had completed the port side and was halfway back on the starboard with his electronic sounder. Unlike previous versions of this instrument, which I had seen with Lady Jane's surveys in France in 2000 and 2006, this one does not require removal of the bottom protection layers.

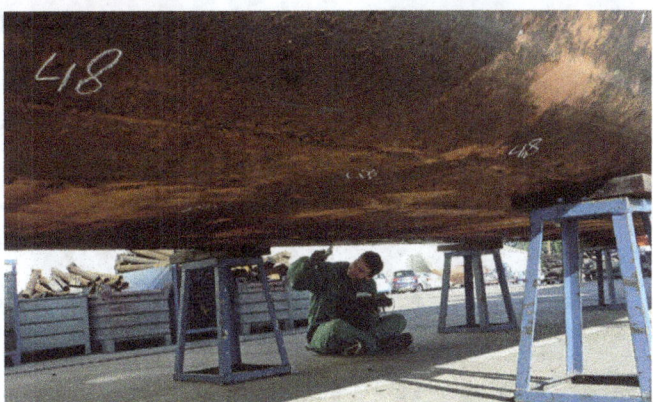

Application of a conducting gel to the hull's surface and then applying the sensor to the gel gives an instantaneous reading of the metal thickness. This is a wonderfully non-destructive process, much better than the former necessity of scraping or grinding away the coatings, and dramatically better than the earlier need to drill holes through the bottom, take measurements and then weld closed the holes.

Something that I had quickly noticed when the barge was hauled on Tuesday also caught Rutger's attention. Henk had added a second raw water intake for engine cooling to handle the expected dirtier water before he went to France a few years ago. I saw that the unconventional installation was vulnerable to being compromised or knocked off by large flotsam as well as by canal banks and lock sides. I was pleased to see Rutger think the same way and condemn it.

Rutger found the rudder bearings to be well within tolerance, but found the propellor shaft with a rather large amount of play. The tolerance is two percent of the shaft diameter, so he applied his dial indicator to the shaft as Henk pried the propeller up and down against the skeg. The readings calculated to exactly two percent, and Rutger explained that in a case like this, the owners and buyers should share the replacement expenses. We agreed.

The aluminum anodes showed erosion, not excessive, but sufficient to indicate that they had been doing their duty protecting the more noble parts of the boat. Rutger declared that these and the remainder of the underwater were gear good.

We next went aboard and inspected the inside of the hull under the floorboards. The flash on my camera made this process much easier than contorting behind a flashlight and trying to get a viewing angle. The photos showed a dry and dusty environment with virtually no rust on the century-old riveted iron bottom.

The only place where there was any appreciable amount of rust was in the well beneath the mast tabernacle. The foot of the mast, with its 1100 kilogram counterweight swings through a slot in the foredeck as the mast is raised or lowered, and the cover over the deck slot is very prone to leaking. Rutger called this the weakest part of a skûtsje. He hammered and sounded and declared it still good. Though not critical, Rutger recommended we scale and treat the area. We both remarked that there were no bilge pumps in Nieuwe Zorg, neither here nor in the engine room; there had apparently been no need for over a century. This had been noted in her sale listing.

Buying a Canal Boat

As we were finishing the hull inspection, Peter, the broker arrived. We all met: seller, broker, buyer and surveyor, and we received the verbal survey report. There were a few things for Henk to correct, repair or replace before the completion date. There was nothing major:

- un-seize, clean and grease the starboard leeboard winch;
- repair a broken cog on the port leeboard winch;
- relocate the secondary raw water inlet and blank-off the old hole;
- replace the propeller thrust bearing rubbers;
- repair a leak in the cooling water pump; and
- repair the interior heating system.

There were two items that we agreed will be shared 50/50:

- replace bow-thruster battery bank; and
- replace propellor shaft stern bearing.

Henk and Edi and I agreed to have Peter investigate the costs involved in the shared items, and we would then come-up with an amount to deduct from the balance owing on the purchase. We arrived at an easy agreement and then we got into our four separate cars and Henk led us around the Westeinderplassen to Hoofddorp, to a warehouse where he has been storing Nieuwe Zorg's rigging. The leeboards had been newly refinished.

We were aware from the listing that some fittings and pieces of rigging were missing and that there is no jib, only a mainsail. We saw that the sail was in good condition, still crisp and as far as we looked, with intact stitching. The mast has a deep split running about half its length, and this had been exacerbated by an earlier repair attempt. Henk had invited a carpenter to attend the inspection, and after a quick examination, we all agreed that it was uneconomical to repair. The boom is in far better condition, though missing a few fittings. In our contract, the seller

is to install the leeboards and deliver the remainder of the rigging onboard Nieuwe Zorg before the closing date. We told Henk that he could exclude the mast from this, but to include the mast counterweight. It was mid-afternoon by the time we had all shaken hands and had driven off in our separate directions. Everything had gone smoothly; totally zonder zorg.

Edi and I drove northward, past Volendam and out across the Afsluitdijk to Friesland and then further northward to Harlingen. We had been researching yards to do our proposed refit on Nieuwe Zorg. Among other yards, we had looked at the marvellously modern, but not-yet-completed new facilities at Kempers, and had met with Willem Dokkum, the works manager. We were now on our way to meet with Lex Tichelaar, a partner at Scheepsbouw & Reparatie Friesland, known better as SRF.

As we left our car and walked across the yard toward a couple of men talking, one of them stepped out to meet us. We asked where we could find Lex, and he said right here, pointing to his chest. Lex is a robust man, very alive and immediately likable.

He showed us around, beginning with a huge loft full of shelves laden with a vast assortment of old boat parts and fittings. Next he showed us a similar sized hall full of new parts, fittings and hardware. We saw a mast and spar building room, in which they build solid and laminated wooden spars and masts. We followed him through huge hangars with a broad range of ship construction, repair and refit projects in progress in wood, steel and composite. He showed us the cabinetmaking shops, the steel fabrication shops and we completed the loop up in his office sharing coffee. We liked the place; they appear passionate in their restoration and refit of traditional vessels, employing as appropriate both the old and the fully modern shipbuilding techniques and equipment.

Lex confirmed that they had space inside for our skûtsje in October to begin a restoration and modernizing refit over the winter. He asked for the current location of Nieuwe Zorg so he could look at her to gain a better understanding as we discuss work over the coming weeks. We left SRF convinced that it fully met our requirements to do a major refit on Nieuwe Zorg.

Buying a Canal Boat

We were now committed to the purchase of an antique skûtsje that had been built in 1908 in Gaastmeer, just inland from the Zuiderzee. With a spare day until our flight back to Vancouver, on Thursday we played tourist. Among other things, while we were browsing along the Volendam waterfront, we had our portrait taken dressed in traditional nineteenth century Zuiderzee boating clothing.

On Friday we drove to the airport, dropped the rental car, checked-in and boarded a the first leg of our flight to Vancouver. We had driven just over 4250 kilometres in a very relaxing two weeks, during which time we had found a wonderful old boat. We were delighted with the ease of the entire process, and particularly with the clear, concise and efficient system used by the brokers, surveyors and insurers in the Netherlands.

There were a few items arising from the surveys that the owner must address before the closing date, which we had set in our offer to purchase contract as 01 September to allow six weeks for the comfortable completion of any arisings from the survey. Also, before the closing date we need to transfer the balance of the purchase price to the broker's trust account.

Chapter Seven

Researching History

With the details taken care of and having six weeks before our possession date, I began researching skûtsje history in general and that of Nieuwe Zorg in particular.

On the brokerage offering she was described as having been built in 1908 by the Wildschut yard in Gaastmeer. From our contract to purchase, I found that the sellers, Henk and Madij had bought her in 1975 and Henk later told us that her name had always been Nieuwe Zorg.

I searched online and found a website: www.skutsjehistorie.nl, where there is a listing of registrations for all the iron and steel skûtsjes that had been built. I sorted the list alphabetically by ship names and scrolled through the Nieuwe Zorgs. There are 40 of them plus one Nieuwe Zorgen listed as having been registered between 1895 to 1923, none of them built by Wildschut, nor even in Gaastmeer.

I next sorted the list by registration date and scrolled to 1908. Of the sixty-eight skûtsjes built in Friesland in 1908, four of them were named Nieuwe Zorg, but none of these are listed as having built in Gaastmeer, nor by Wildschut. The two 1908 registrations that had been built by Wildschut in Gaastmeer, Hoop op Zegen and Zeldenrust, had measurements that were different from those of our Nieuwe Zorg.

I then sorted the list by building yard and scrolled to the Gaastmeer section. There I found a listing for De Nieuwe Zorg built in 1901. Things weren't making sense until I noticed that the date sequence was awry; the 1901 listing was sorted with the 1907s, which is where her Registry Number, S831N placed her.

48

Researching History

I resorted the list by Registry Number, and De Nieuwe Zorg's S831N fell in among the 1908 listings.

I opened the details of the S831N line and found her second measurement from 1941. These are the exact measurements listed for our Nieuwe Zorg in the brokerage listing: length 16.38 metres and beam 3.44 metres. She was calculated in 1941 at 31.971 tons, and the owner was listed as Douwe Albert Visser of Stavoren.

I emailed the webmaster of the site and pointed out the apparent error to him, and asked for confirmation that it actually is an error. A few days later I received a reply from Frits Jansen, the webmaster confirming that the date had been mistyped when compiling the spreadsheet from the original ledger. He told me that it would soon be corrected on the website. Frits attached a photo of a spread of pages of the Sneek registration ledger showing the listing for De Nieuwe Zorg.

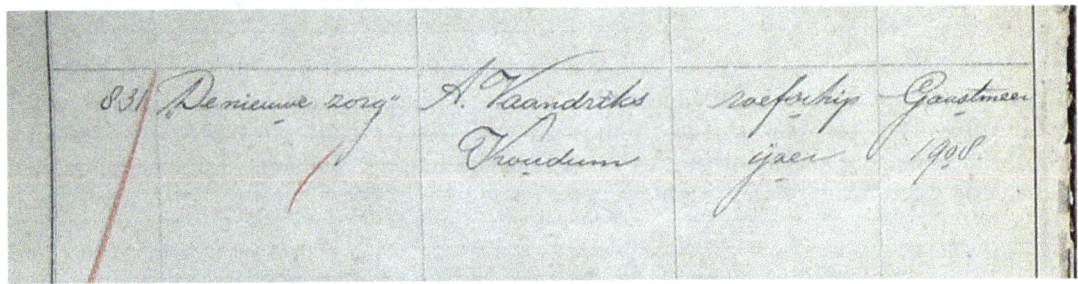

I blew-up the high-resolution photo to show the details clearly: 831 | De Nieuwe Zorg | A. Vaandriks / Koudum | Roefschip / yaer | Gaastmeer / 1908.

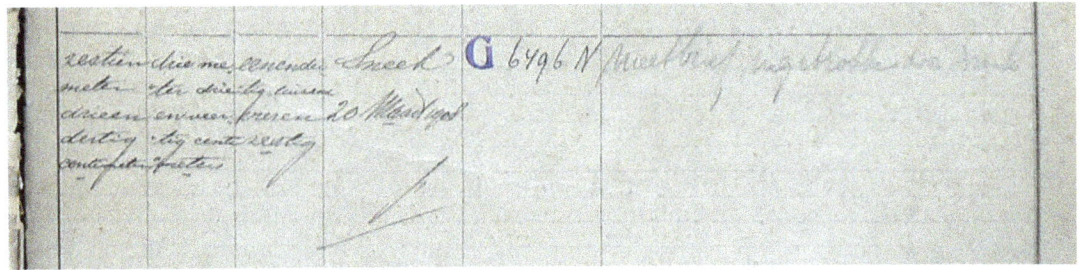

On the facing page, the registration continues with her measurements all written-out in words. Her first measurement on 20 March 1909 gave her length as 16.33 metres and her beam as 3.43 metres. A later measurement is referenced with new registration number, G6496N. Under Notations, the final column, is an undated entry indicating the registration was cancelled.

The listings showed that the first owner was Age Vaandriks of Koudum. I searched the registry and found that Age had commissioned a second ship from a different yard in 1925, a 19.15 X 3.9 metre skûtsje named Dankbaarheid. To its details page had been added a photo of Age, and with further digging I found that he was born in 1876, so would have been 32 when he had the Wildschut Brothers build De Nieuwe Zorg.

Between 1857 and 1953, four generations of the family Wildschut had a shipyard in Gaastmeer, a short distance from Sneek in southwestern Friesland. Beginning with ship repair and maintenance, they soon grew to building wooden sailing tjalken, and as the demand for more nimble barges grew, they developed the predecessor of the skûtsje.

At the end of the nineteenth century Wildschut began building in iron, and then from 1904 they appear to have specialized in iron and steel skûtsjes, building eighteen of them in the ten years leading up to The Great War. This photo shows the Wildschut facilities in 1907, around the time that De Nieuwe Zorg was being built. The yard also built tjotters, tjalken, klippers, palingaken and a variety of smaller inland water craft.

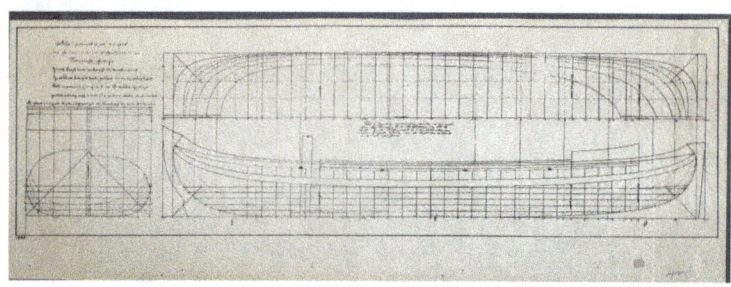

The Friesian skûtsje had been developed between 1855 and 1860 and they were initially built of wood. Then in 1887 riveted iron was introduced, and during the first decade of the twentieth century, the yards gradually converted from iron to steel. A total of 870 iron and steel skûtsjes were built until 1933. There are many line drawings of skûtsjes in the Fries Scheepvaart Museum in Sneek. This one is of built in 1908 by Auke van der Zee in Joure.

They were initially used in the eel fishery and in hauling peat and composted manure from inland Friesland to the coast and along the Zuiderzee. The worldwide depression starting in 1929 decreased the commercial usefulness of the skûtsje and the closing of the Zuiderzee in 1932 sounded its death knell. There were only four skûtsjes built in all of Friesland after 1928, and with the last one in 1933, their production ceased.

Researching History

During the ensuing years, many skûtsjes were fitted with engines, but even with power, they were too small to compete with new designs, particularly the luxemotor. By the end of World War II, like most tjalk type barges, skûtsjes were commercially obsolete, and were either abandoned, were sold for scrap or were refitted as houseboats.

The houseboat shown in this photo was Nieuwe Zorg in 1975 when Henk and Madij bought her. The snapshot was among the folder of papers that came with the boat. Also in the folder was the bill of sale showing they had paid 15,500 Guilders for her.

Some Skûtskes were converted to sailing yachts, retaining the mast and rigging, removing the small roef aft to make a cockpit and turning the cargo hold into living accommodation with a jachtenroef. Our skûtsje is one of these.

Henk was a professional welder and he spent three years refitting Nieuwe Zorg, starting with scrapping the wooden superstructure and gutting the interior. He added a graceful jachtenroef, which is seen in this 1978 photo. He told us that he had copied the lines from De Groene Draeck, the Lemsteraakjacht that the Dutch people had given to Princess Beatrix on her eighteenth birthday in 1956.

A few skûtsjes were maintained in their original form and used in inter-town races. As the popularity of these races grew, more and more houseboat and yacht conversions

were reverted to their original configurations. There are currently about ninety of these historic vessels restored to full sailing trim and competing in regular SKS and IFKS races in Friesland.

Wildschut's twenty skûtsjes are a small portion of the 870 iron and steel skûtsjes that were built by some thirty Friesland yards between 1887 and 1933. Of the twenty Wildschut skûtsjes, we have accounted for fourteen survivors, with ten of these now in full sailing trim. Of these, six have the distinction of having been registered as racing Suûtsjes in one or the other of the two leagues: Sintrale Kommisje Skûtsjesilen (SKS) or Iepen Fryske Kampioenskippen Skûtsjesilen (IFKS). Another is the representative skûtsje in the Zuiderzee Museum in Enkhuizen and another is a registered historical sailing ship. The skûtsjes that Wildschut built are held in very high esteem, and we are proud to own one.

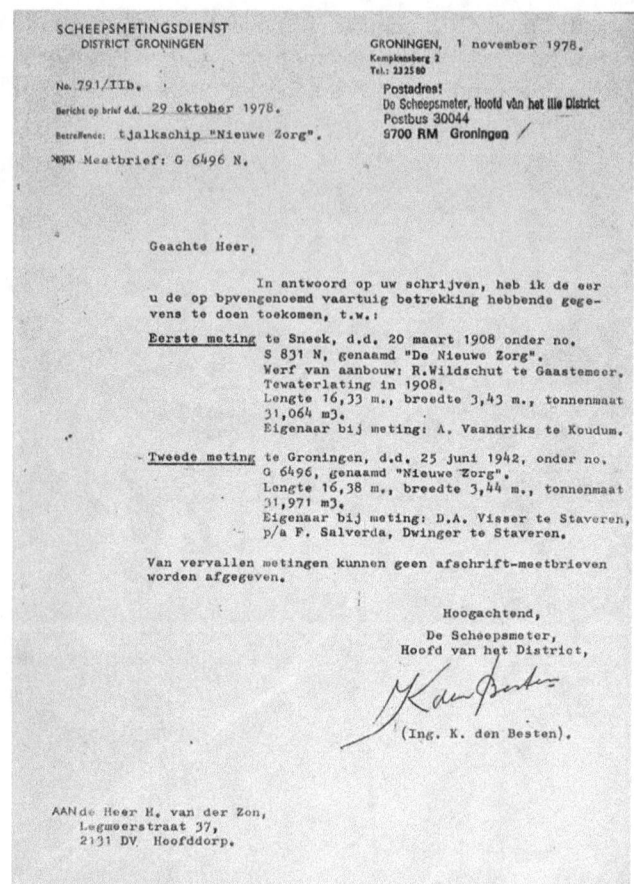

We still had some gaps in our skûtsje's history. The new registration number, G6496 noted on the second page of the Sneek ledger matches a new owner, Douwe Albert Visser of Stavoren.

Among the papers in Henk's file aboard Nieuwe Zorg was a 1978 letter he had received in response to a query on the registration history. It confirmed the measurement data we had already found and confirmed the new owner as D.A. Visser and listed his address as: p/a F. Salverda te Stavoren (in care of F. Salvera at Stavoren).

We tried to trace D.A. Visser or Douwe Albert Visser as he is listed on the De Nieuwe Zorg page at www.skutsjehistorie.nl but we kept coming up with modern versions of the name. Our online searches are seriously affected by there currently being an Albert Visser and two Douwe Vissers as very competitive skûtsje racers as the skippers of three of the fourteen competing skûtsjes in the annual SKS championship series.

1n August 2013 Douwe Visser, Jzn. skippered his crew in the De Sneker Pan to his third consecutive championship in the fourteen-day, eleven-race series. This was his eighth championship since taking-over as skipper in 1989. Complicating our search further is that a Douwe Visser, Azn. is also an SKS skipper, and he won the series championship twice, 2005 and 2009 in the skûtsje, Doarp Grou. In 2011, 2012 and 2013 he was runner-up to his cousin Douwe. Further confusing search engines is that Albert Visser, Jzn, brother of the Sneker Pan's skipper, had for years been the skipper of Twee Gebroeders, the SKS skûtsje from Drachten, and in 2013 moved to take over as skipper of the Lemster skûtsje. With all the notable skûtsje involvement of these modern Douwe and Albert Vissers, Google couldn't get us back beyond very recent history.

Again I emailed the webmaster of www.skutsjehistorie.nl and again Fritz quickly replied with the family tree: Douwe Albert Visser, 1902-1997 married Tjitske Taekesd Salverda 1906-1998 in 1928. They had five children: Albert, Taeke, Jappie, Immertje and Trijntje. Taeke was a skipper in the SKS on d Halve Maen from 1982-1984. Albert was the father of Douwe Azn Visser who is now skipper on Doarp Grou in the SKS. Jappie is the father of Douwe Jzn Visser, who is now skipper on Sneker Pan in the SKS, and Albert Jzn Visser who is now skipper on Twee Gebroeders in the SKS.

Frits continued with his genealogy: Tjitske the grandmother also comes from a skipper family; her father Taeke Salverda was a fisherman from Stavoren. Her sister Willemke married Klaas Keimpesz van der Meulen, an SKS skipper and four-times champion. Their children Keimpe, Taeke and Ype van der Meulen are or where skippers in the SKS.

So now we know the relationship between Nieuwe Zorg's second owner and the current successful SKS skippers. He was grandfather or great uncle to more than a third of the skippers of the current fleet. The family members have won the SKS championship eighteen times since 1945, when the organization was established.

We are in communication with the Vissers and van der Muelens to fill in the blanks we have in Nieuwe Zorg's history between 1941 and 1975 and to see if we can find out what the skûtsje Nieuwe Zorg was used for. In 2014 Douwe Azn Visser won the SKS championship, his cousin Douwe Jsn Visser came second and brother Albert Visser came fourth.

Chapter Eight

Taking Possession

Toward the end of August, we wired the balance of our skûtsje's purchase price to the yacht broker's trust account in Enkhuizen, and at the same time, we wired the insurance premium to the agent. With all the purchase details taken care of, we began organizing our return to the boat in Aalsmeer, Netherlands.

Although Aalsmeer is at the end of the runways of Amsterdam's Schiphol Airport, Air Canada does not fly there. They do; however, fly to Brussels, about 200 kilometres from Aalsmeer. All the flights to Brussels were overbooked at the end of August; however, on the Tuesday after Labour Day there were flights with many open seats, so short notice booking on our choice of them was easy. Through the broker, we easily arranged to postpone the possession date to 5 September.

There is something about Tuesday and Belgium, so Edi booked flights to Brussels via Montreal for Tuesday the 5th of September and I reserved a car to get us from the airport to Aalsmeer and to use during the first few days for shopping after we arrive.

When I was reserving the car, I clicked the GPS option and was shocked to see that even with my Hertz #1 Club Gold, which still triggers Air Canada Super Elite discounts from a former life, the GPS charge from Brussels for the four days came to just over €70, including taxes. I decided to look at options. When we were in the Netherlands in July, everyone referred to TomTom when talking of driving or walking directions. Online in the Apple App Store I found TomTom with maps for all of Western Europe for my iPad for $89.99, about €72.50.

While I was in the online App Store, I decided to look at Navionics charts for Europe to see if there is coverage of the inland waterways. We had been delighted with the quality of the Navionics charts for iPad and

Taking Possession

they became a very important adjunct to our chart-plotter aboard Sequitur as we navigated from La Punta Peru southward along the Chilean coast, through the archipelagos of Patagonia and Tierra del Fuego, around Cape Horn and up the eastern coasts of the Americas. Many times the iPad was our primary source of navigational information. I found the Navionics Europe HD application for $64.99, and some quick investigation showed me that the lakes, rivers and canals are all charted. I downloaded a copy to my iPad.

We each packed a piece of checked luggage at 22.5 kilos and balanced our four carryon pieces at exactly 10 kilos each. The total of 85 kilos is much less than the 224 kilos we had lugged back to Vancouver from St Augustine when we had left Sequitur two months previously. However, our current trip was planned for only five weeks, sufficient time to take possession, move aboard and become familiar with the skûtsje as we slowly took her from Aalsmeer, North Holland to Harlingen, Friesland for her facelift.

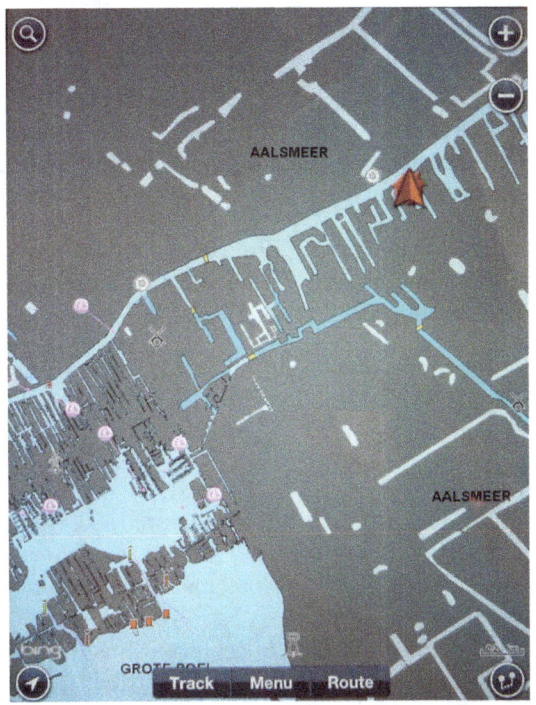

On Tuesday morning we wheeled our luggage the two blocks to the SkyTrain station and rode to the airport. Because of a late arrival of the inbound plane, our flight was delayed leaving Vancouver, and we had run to make the connection in Montreal. Of course, our Vancouver flight arrived at one end of the terminal and the Brussels flight departed from the other end. Fortunately, Trudeau Airport is rather compact, so the run was not so bad.

Customs and Immigration in Brussels were a snap, our luggage had made the tight connection and our car was waiting for us at the Hertz gold counter. The TomTom worked well on my iPad, and shortly before noon, we arrived at the marina in Aalsmeer. Henk was aboard Nieuwe Zorg ready for our noon appointment.

During the past few weeks he had completed the work arising from the survey, which included mounting the leeboards. They look wonderful. We did a walk-through and a systems familiarization, and then we signed the transfer papers, formally taking possession.

Carefree Through 1001 French Locks

Henk and Madij's attention to detail, their care and their pride were all so obvious from the moment we first saw Nieuwe Zorg. She took possession of us, and now we had reciprocated.

In July, after we had contracted the purchase of Nieuwe Zorg, we had encouraged Henk and Madij to take her out cruising during the five weeks until the completion date. They had spent three weeks cruising with their extended family, during which time Henk must have spent many hours maintaining and fine-tuning the skûtsje; she was in even better condition than when we had left her after the survey.

We then spent a delightful couple of hours with Henk learning the idiosyncrasies of the vessel's systems and listening to stories from their thirty-seven years of restoring, upgrading, maintaining and enjoying this wonderful historic skûtsje.

After Henk left, Edi and I lay down for a nap to try to catch-up from having been up and travelling for over twenty-four hours and to start adjusting to the eight hour timezone change. We were expecting Lex and Klaas from SRF to arrive onboard sometime in the early evening to discuss preliminary ideas for Nieuwe Zorg's refit and for her restoration to sailing.

Lex and Klaas arrived shortly before 1800, waking me from my slumber on the salon couch as they arrived. We showed them through the skûtsje and threw ideas around with them for the better part of two hours. They took measurements, made notes, did calculations and we all scratched our heads and pondered the possibilities. They left with a good idea of what we wanted and a better understanding of how to achieve it than they had gleaned from the emailed list of items we had sent them a couple of weeks previously.

A little before 2000 we rushed off following directions on my iPad's TomTom to the closest Albert Hein for some groceries and cooking utensils. It was a smaller version of the huge Dutch supermarket chain, and they had no cooking utensils. From the store's information counter we were given the address of a much larger Albert Hein store in Amstelveen, which we punched into TomTom and allowed the program's mellow voice to lead us there. Their cooking equipment supply was limited, so I punched in Ikea's address in South Amsterdam and TomTom took us there with 45 minutes of shopping time left before the 2100 closing.

From Vancouver, we had brought sheets, pillow cases, duvet cover and towels, plus cutlery, stemware and our favourite coffee cups and saucers, so we needed everything else. As we hastened around the circuit, we loaded into our cart an assortment of pots and pans, cooking utensils, plates, bowls, pillows, a duvet and whatever else we could think of before the closing announcements began. We were back in the car with our load of goodies shortly before 2115 and following TomTom back to the larger Albert Hein. With a bit more than half an hour until closing, we set a more relaxing pace selecting sufficient groceries and supplies to settle us into living aboard until, after some much-needed rest, we could conduct a more organized laying-in of basic goods.

During the following days we alternately measured and remeasured spaces aboard, rested, went on shopping trips and drew preliminary layout plans for Nieuwe Zorg's interior makeover. As much as we love the craftsmanship and design of Nieuwe Zorg's interior, it is dated. It is a third of a century old, as are her systems and her construction technologies. Our intention was to totally strip-out her interior down to bare riveted iron and start over. The makeover includes making the bedroom the full width of the hull, rather than two-thirds the width. Also, we wanted to move the shower to the centreline to take advantage of the higher headroom there. Because of the low curved roof, using sliding doors instead of hinged ones will make access to the bedroom and heads much easier.

The current galley had extremely limited countertop space, the counter being mostly taken-up by the gas cooktop and the sink. We thought of adding an island to increase the countertop workspace, but there was insufficient room, so we reworked the design to accommodate under-counter convection-microwave oven, dishwasher and fridge-freezer. Because of the easy and frequent availability of fresh produce along the inland waterways of Europe, we won't need two fridges and two freezers like we had in Sequitur; a single under-counter combo unit will be sufficient.

For breakfast on our third morning aboard, Edi prepared some pain perdu au jambon. We had settled-in quickly, and we began organizing for a slow trip northward to Harlingen, Friesland for Nieuwe Zorg's facelift. At 104 years of age, the old girl is past due for her hundred-year makeover.

We figured that along the way, as we became more accustomed to the old lady, our ideas of what we wanted and how we wanted them would slowly evolve.

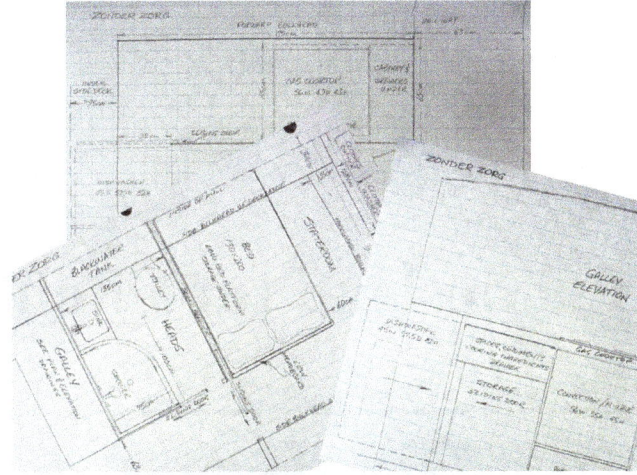

Chapter Nine

To Friesland for a Refit

Six days after we had taken possession, we were ready to head north to the land of the skûtsje. I hauled out the portfolio of ANWB charts that Henk had left aboard and began looking at possible routes. The first leg was easy; turn right and follow Ringvaart van de Haarlemmermeerpolder to Rieker plas, through it to Nieuwe Meer and along it to Nieuwemeersluis, through the lock and into Schinkel of Kostverloren vaart, which is followed northward to the Noordzee Kanaal.

I also looked at the route on the Navionics app on my iPad. It confirmed that the lowest bridge along the route was 2.44 metres, so with our air draft of 1.95 metres, we would not have to wait for any bridges to open for us.

It was drizzling on Tuesday morning when I called Henk; he had asked to be there when we cast off and headed out. Henk arrived and suggested to me that I put on some rain gear, otherwise I'd be drenched by the time we got to Amsterdam; the erected canopy is too high to pass under some of bridges. He dug down into a locker and pulled out a suit that he had left onboard for me. A little past noon we slipped and headed out as Henk watched his cherished skûtsje motor slowly away. For just over thirty-seven years, more than half his life, he had poured his heart and soul into Nieuwe Zorg, and now she was leaving. We all felt lumps in our throats and had wet cheeks.

To Friesland for a Refit

Over the following three weeks we wound our way northward through Amsterdam and Edam to Hoorn and then south across the Markermmer to the Randmeeren and around them to towns and cities of the old Zuiderzee coast. From there we pressed onward to Friesland with stops in Lemmer, Sneek, Grouw and Leewarden before arriving in Harlingen for our refit rendezvous with SRF.

Along the way we visited many museums using our museum pass. Among these was the Zuiderzee Museum in Enkhuizen to see another of the fourteen surviving Wildschut skûtsjes. We also visited the Fries Scheepvaart Museum in Sneek where we spent hours in the skûtsje rooms examining displays.

There are many superbly detailed models on display, which show the skûtsje's design features. I spent a long time closely examining these and took dozens of photos for later study. In one display case is a skûtsje alongside a tjalk, and this close juxtaposition makes it very easy to compare and contrast the designs. The skûtsje is obviously more sleek; it is narrower and has a shallower hold. The leeboards are broader and shorter to allow very shallow water navigation. To make-up for the narrowness, the roef is slightly longer and a bit higher.

Among the exhibited models is a very finely rendered skûtsje half-section, which clearly shows the internal structure of the ship. At the after end of the cargo hold is a watertight bulkhead. This forms the forward part of the roef, the ship's accommodations. Often it was the home of the barge owners and their family.

It is flat-sided with a curved top, two shuttered windows on each side, a pair of portlights aft and a companionway topped with a sliding hatch. Growing out of the forepart of the roef is a crutch to cradle the boom.

In the museum is a restored skûtsje roef, which shows the typical layout and style. The interior design makes maximum use of the small space and it is very well appointed and finished. In the centre of the forward bulkhead is a fireplace, which serves for both cooking and heating. On each side of the fireplace is a storage cabinet, each with twin doors. The wood panelling and trim appear to be ash or beech, which is lightly varnished.

A table sits in the centre of the room and around it are chairs for the family. There are cabinets along the sides, which make full use the spaces under the side decks. Overhead, in the centre of the roef is a pigeon box skylight, which affords lighting as well as cooling and ventilation when required. An oil lamp hangs from the ceiling and there are decorative lace curtains on the windows and portlights. Leading aft from the room, tucked under the aft deck is a bed platform, extending back into the round of the stern. In the stern are two circular portlights adding some light, and sometimes there was a pigeon box skylight.

A framed drawing on a nearby wall in the museum shows the scale of the interior of the accommodation space. It also shows how important and multifunctional the central table was to the room. In this space lived the family that owned and operated the barge. The skûtsje was both their home and their business.

To Friesland for a Refit

Their business consisted of finding loads to carry. Often times this was cut peat to be carried from inland across the shallow lakes and through the narrow sluices to seaside ports. Sand, gravel and compost were also regularly carried, as were eels, both fresh and smoked.

The skûtsje was designed as a sailing vessel, and it was fitted with very broad, but relatively short leeboards to better navigate the shallow waters. Other tjalk-types had narrower, longer leeboards that extended deeper into the water. The mast on a skûtsje is fitted with a counterweight to make easy the process of lowering the rig to pass under low bridges and of re-erecting it once clear.

When there was no wind, or when the wind was adverse, the family donned harnesses and pulled the barge along the canal. As small engines became economical, some skûtsjes were fitted with auxiliaries. With the coming of purpose-built powered barges, such as the luxemotor, the usefulness of the skûtsje declined.

A few kept on working, some were scrapped, some were converted to houseboats, some were converted to pleasure yachts with extended roefs over the hold and some were preserved and maintained for racing. Skûtsje racing is huge sport in Friesland, with two leagues: the prestigious SKS and the newer IFKS.

Near the entrance to the skûtsje rooms in the museum is a trophy case, and among the trophies is a retired SKS Championship trophy. It shows the names of the series champions from 1988 to 2007. On the trophy's plaque, Douwe Visser of Sneek is shown as the SKS champion for 1995, 1996, 2002, 2003 and 2007. Douwe won the championship again in 2011, 2012 and 2013.

In the 1930s and 40s his grandfather, Douwe Albert Visser had raised his family aboard our skûtsje, Nieuwe Zorg. Another of his grandchildren, Douwe Visser of Grou won the SKS championship in 2005, 2009 and 2014..

Carefree Through 1001 French Locks

From Sneek we continued along through Grouw and Leeuwarden to Harlingen, arriving on 2 October, four weeks after we had taken possession of Nieuwe Zorg. We had spent three wonderful weeks bringing her 298 kilometres from Aalsmeer, Noord Holland. During the following few days we gradually refined our plans for the proposed refit, which would restore her exterior close to her 1908 appearance and her interior to twenty-first century systems and conveniences. As much as we liked the historic appearance of the interior in the museum in Sneek, we had decided on going modern.

Chapter Ten

Planning a Metamorphosis

On the following afternoon Wychard Raadsveld, one of the four partners in Scheepsreparatie Friesland came aboard to continue gathering information from us to gain a better understanding of what we wanted done and how. We were shortly joined by Lex Tichelaar, another of the partners we had earlier met, and the four of us continued the long process of planning Nieuwe Zorg's second century facelift.

We discussed ways to gain interior space. My head brushed the ceiling in the heads and the bedroom. Among the ideas were to use thinner insulation in the ceiling and to try to lower the floor, but the possibility and effectiveness of these needed to await the gutting of the interior to see what the old girl's underpinnings were. If we could gain some headroom, we could move the shower all the way over to the port side and reduce the length of the room by 15 centimetres. By also rearranging the forward bulkhead in the master cabin and putting the bed athwartships, we could fit a bed with a 200cm X 140cm mattress, possibly even a 160cm one.

With all of the shuffling to fit-in what we wanted, we still needed to find an additional metre or so of interior length. The easiest way to gain this was to shorten the engineering spaces. Wychard remeasured our engine room and machinery spaces to see how much room we could liberate by installing a smaller engine. Nieuwe Zorg had a six-cylinder Perkins, which is rated at 86 kW. Besides being nearly half a century old, the engine is much more powerful than the 50 or 60 kW that the skûtsje requires. A smaller engine will allow us to reduce the length of the machinery spaces and to lengthen the living spaces.

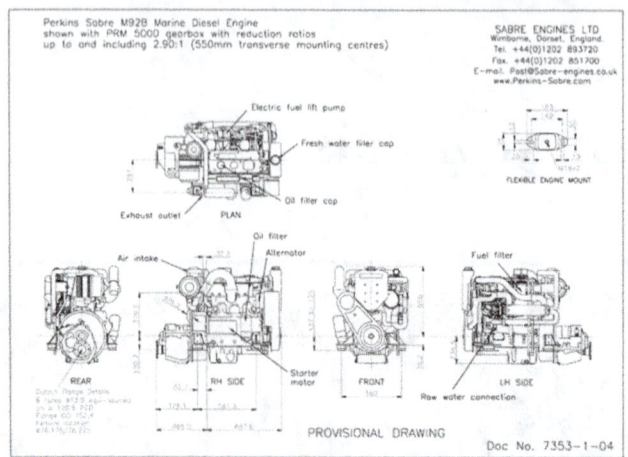

Wychard proposed the Perkins-Sabre M92B with a PRM 5000 gearbox. At the propeller shaft, it develops 50 kW at 1500 rpm and 64 kW at 2400 rpm. The engine is a normally aspirated, low revolution unit with its maximum torque of 320 Nm produced at 1400 rpm. At that speed, the fuel consumption is 4.5 litres per hour and at canal cruising speed, 3 litres per hour. With these specifications, it appears to be the ideal power unit for our skûtsje.

The first step in determining the amount of living space we will have available is to plan the space required by the mechanical installations. Wychard took more measurements in the after spaces as he gathered information to begin designing the layout of the equipment. The engine and transmission, the fuel tanks, the generator, the battery banks, the isolation transformer, the inverter-charger, the water heating and the central-heating systems all need to be laid-out in the most compact arrangement that is consistent with easy inspection and maintenance access.

Edi and I walked the 2.5 kilometres into the centre of Harlingen each day to absorb the wonderful town with its many sixteenth and seventeenth century buildings and its historic harbours, canals and bridges. Harlingen received its city rights in 1234 and rapidly grew in importance as a trading port. It was first mentioned in port registers of England in 1311.

Among the fascinating structures is the Raadhuistoren, the old city hall tower built in 1587. Running through the central part of the old city are pedestrian-only shopping streets, which add immensely to the charm of the area.

We did our daily food shopping in the markets along the way and in the Jumbo and Albert Hein supermarkets. We continue to be amazed at the wonderful selection, the great quality and the low prices of everything from bakery items, meats, fish, dairy and fresh farm produce and of course, wine.

Planning a Metamorphosis

After a week of planning, measuring, head scratching, re-drafting, proposing ideas, listening, discussing, thinking and dreaming, we received an itemized quote from the office. It covered the removal of equipment and gutting the skûtsje down to an empty hull, moving the engine room bulkhead aft, lifting the roef and moving it aft, re-engineering and rebuilding the interior, installing plumbing, lighting and appliances, re-rigging for sail, finishing and painting. We were ready to leave things in the hands of SRF and head home.

On one of our walks through the centre of Harlingen, we saw in the window of a toy hospital a teddy bear of the same vintage as Nieuwe Zorg. It was in for a facelift and make-over, just as was our skûtsje. We wished both of them success.

On Monday and Tuesday we had packed into three large and one huge storage bin everything in Nieuwe Zorg that we wanted to keep. Because all our cooking and dining goods were packed, we went out to dinner for the first time in more than a year. Food aboard Sequitur and Nieuwe Zorg had been so good that we were never tempted to dine out.

We got up at 0400 on Wednesday, packed the bedding, towels and the last bits of stuff into our luggage and drove a rental car the 325 kilometres to the airport in Brussels for our 1040 flight. We had a long wait in Montreal for the connection onward to Vancouver, where we arrived near midnight. Fortunately, our 69 kilograms of checked luggage hadn't make the flight, and after an easy registration with the baggage agent, we walked our carry-on bags to the SkyTrain and to our loft without the huge encumbrance of heavy luggage. The wayward luggage was delivered to our door the following day.

During the following weeks we continued to refine our ideas for the interior. I downloaded a free version of Google SketchUp and taught myself how to use the software to create some three-dimensional drawings and then to look at the implications of various layouts.

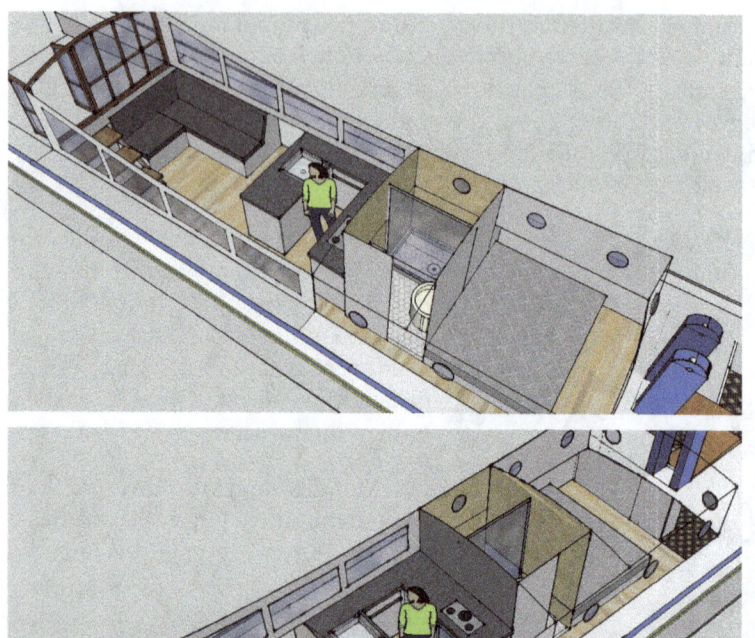

I played with ideas to see how much additional room we will gain by incorporating the hallway with the master cabin and by getting rid of the clothes closet. With this layout, there is room for a full-size European bed with room to walk around forward of it and beside it; whereas, in her current configuration there was room in the bedroom for a 130cm x 200cm bed and for only one person at a time to stand on the tiny patch of floor.

Instead of a closet taking valuable floorspace, we will use the space under the starboard side deck for our clothes storage. The side-by-side washer and dryer will sit beside the mast step and extend forward under the foredeck.

Though the space planned for the heads is smaller in area than in the existing configuration, by getting rid of awkward jogs and using a narrow-profile sink, will be able have a larger shower stall, better ergonomics and more useable floorspace. In the galley we wanted to install granite countertops and I played with ideas to expand the work space by adding two wings. The wing along the port side, partially under the side deck would house the dishwasher, while the new transverse wing would have the fridge-freezer and the sink. The fridge-freezer door would be in the end of the wing, with its hinges to the left for easy access and to reduce space conflict.

We looked at a shorter settee in the salon, 200cm instead of the current 260cm, which will easily con-

vert to a full-size guest bed. This reduced length, combined with the space liberated from the engine room and the heads, allows the larger galley and master cabin. The graceful compound curved jachtenroef provides the main accommodation space. It extends a total of 8 metres of the 16.38 metre length of the skûtsje.

Planning a Metamorphosis

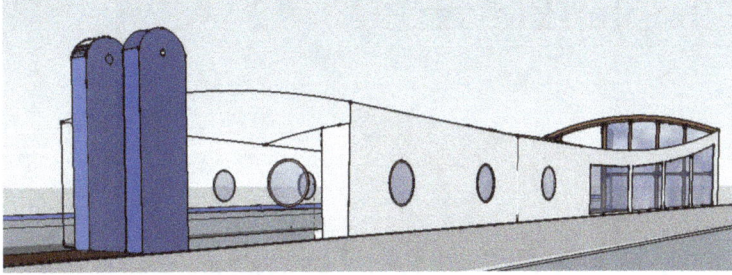

Forward of the master cabin, under the foredeck, there is a space nearly 4 metres in length. The headroom here varies from 100cm to 120cm, and we will utilize the space mainly as storage and occasionally as overflow guest accommodation. Access is through a door in the forward part of the master cabin and through a hatch at the forepart of the foredeck. Our bicycles can be easily lowered into the space by removing the cover over the mast slot.

The windows, of course will be double-glazed. The ones that I had drawn here may have been optimistically large; however, we wanted them to be as large as possible consistent with proper engineering, offering a very bright and airy feeling to the new interior.

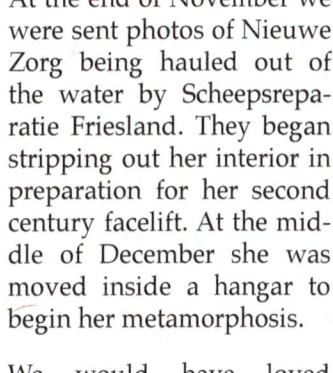

At the end of November we were sent photos of Nieuwe Zorg being hauled out of the water by Scheepsreparatie Friesland. They began stripping out her interior in preparation for her second century facelift. At the middle of December she was moved inside a hangar to begin her metamorphosis.

We would have loved to have been aboard our skûtsje to decorate her for Christmas, but with her in Harlingen under the plastic surgeon's knife, and with us 7700 kilometres away in Vancouver, I instead resorted to PhotoShop to do the decorations.

Chapter Eleven

The Mid-Life Refit

In Vancouver we kept busy doing presentations on various aspects of our three-year voyage down the west coasts of the Americas, around Cape Horn and up the east coasts. Among the diverse venues were some of the Canadian Power and Sail Squadrons, yacht clubs and the Bluewater Cruising Association. We were experiencing our first Winter in several years; our cruising had us in the Southern Hemisphere for their Springs, Summers and Autumns and we commuted north to Vancouver for three months of Summers as each Winter had happened south.

In mid-December, Edi and I were presented the Cape Horn Award by the Bluewater Cruising Association, only the third time that this award had been presented since the Association was founded in 1978.

We were sent some progress photos from Friesland. Our skûtsje had been gutted. The interior that had been so lovingly installed, maintained and decorated for the past thirty-seven years by Henk and Madij was completely removed. She was taken down to bare 1908 riveted iron so that we could install completely new electrical, mechanical and plumbing systems. With the many changes to systems, equipment and technology over the decades, it makes much more sense to begin with a clean slate, than it does to modify the existing installations.

The Mid-Life Refit

Corner braces were installed on the inside of the roef to stabilise it and tangs were welded in place along its lower sides. The forward portion of the cockpit deck was cut out About a metre aft, a new watertight bulkhead was fabricated to separate the expanded accommodation area and the reduced machinery space. The top portion of this bulkhead forms the new position of the roef, just forward of the hatch over the engine room.

The roef was cut between the first two portlights and the cut was extended to the rear along a line just above the side decks. With the roef detached, it was slung aft about a metre and stabilised in its new position with the tangs.

The new positioning of the roef gives us not only a longer living space, but also raises the interior clearance by a few centimetres, allowing me full standing headroom throughout the salon, heads and bedroom.

The cockpit is a bit shorter now, but in its former incarnation at over 3.5 metres, it had been rather longer than we saw necessary. A new smaller four-cylinder engine will replace the former larger six-cylinder one and the machinery spaces will be completely reorganised to make efficient use of the still very generous area beneath the cockpit.

To see the progress to date and to give us a better idea of the newly expanded interior space, at the beginning of February Edi and I flew back to Europe, rented a car and drove northward to Harlingen. Work appeared to be progressing well on raising and lengthening her roef; most of the steelwork and welding had been completed.

We were very pleased with what we saw. With the additional 11 centimetres in height and the extra 1.15 metres of length, the interior space seemed so much larger. Of course, with the interior stripped-out, the space appeared even larger.

We were delighted with the condition of the 105-year-old riveted iron. It looks as good as new, and if an observer didn't know it was over a century old, it could be easily mistaken for rather recent construction. After spending Monday evening getting over our jet-lag, we spent much of Tuesday and Wednesday measuring, sketching and planning interior layout details, and in meetings with Klaas and Wychard discussing mechanical, electrical and plumbing systems, cabinetry, appliances and interior materials and finishes.

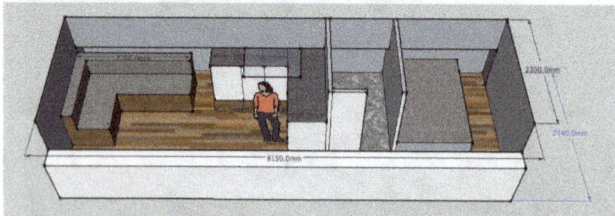

With exact measurements of the interior, we could now do more detailed planning of the interior layout. Among the many things we decided was to forego the third wing of the galley counter and reshuffle the appliances. We determined the countertop style and colour, chose the wood for the interior flooring, selected the exterior paint scheme, redesigned the salon windows, picked the portlight styles, numbers and placements. We discussed plans for an anchor winch, a spud leg, the engineering spaces layout and many other details.

On Thursday morning, leaving Klaas and Wichard to work-out the engineering details of the engine room, galley and heads, we brushed the snow off the car, scraped the ice off the windshield and headed south in search of a warmer climate. We paused for the night in Meaux and Cognac, France, Pamplona and Morella, Spain before stopping for a few days in Calpe on the Costa Blanca, where temperatures were in the upper teens and low twenties. The view from our balcony on Valentine's Day had the sun rising directly out of the Mediterranean over Peñón de Ifach.

The Mid-Life Refit

Suitably warmed, after four wonderful days we drove the back roads to Andora, where Edi crossed-off her 85th country and I did my 79th. We then continued on into France to Carcassonne and Narbonne to look for winter moorings for our skûtsje for the end of the coming cruising season.

We headed slowly back to Friesland with overnight stops in Gigondas, Beaune and Reims for tour and tasting appointments with Louis Bernard, Jaffelin and Roederer. Suitably refreshed, we continued northward through Belgium and back to Harlingen to see what work had been accomplished in the two weeks since we had headed south.

The first thing we saw was that the salon windows had been removed and the holes had been enlarged and reshaped. Also, the three various sized and styled portlights on each side of the forward part of the roef had been removed, their holes filled and two larger new portlights let in on each side of the house. The steelwork on the house had been completed and ground. We were very pleased.

In the bottom of the engine room, a part of the hull had been cut away and a well constructed to accommodate the new engine and transmission. Even though the new engine is smaller than the old, because of the lengthening of the house by 1.15 metres, we are moving the engine mounts aft by that amount. This necessitated the construction of a bubble beneath the existing hull so that a proper shaft angle could be maintained.

On the exterior, there will be a slight widening of the keel leading to the stern bearing and the rudder skeg. To compensate for the small disruption in water flow around the fatter keel, a larger 22 x 20 propeller has been ordered. I specified a left-hand screw, preferring its characteristics for coming to and leaving moorings starboard-side-to.

The scheepstimmerman had begun work on the interior by laying the subfloor and he told us he would shortly begin stringing battens for the interior sheathing. On the foredeck, the two vented lockers for the gaz bottles had been constructed, one on either side of the mast tabernacle.

Also on the foredeck, pads had been welded in place to take the new anchor winch. The winch, which is a new-built replica of an antique design, meant to take a wire rope around its spool. It is hand operated and has a free-fall feature, which allows quick deployment in an emergency.

We next went to the carpentry shop to look at the new frames for the enlarged salon side windows. The timmerman was just glueing the last pieces on the sixth and final mahogany frame when we arrived. The other frames were standing on the floor on either side of his workbench, waiting for their double-glazed inserts. We were very pleased with the progress, and with the quality of the work that had been done.

The Mid-Life Refit

We met with Klaas and decided on a few more details. With most of the design decisions made, we felt easy leaving Nieuwe Zorg in the hands of SRF and heading back to Vancouver.

On Saturday we drove southward to look for a place to stay within an easy drive of the airport in Brussels. It was threatening to snow, and we wanted an easy drive on Sunday morning for our 1030 flight back to Canada. Just past Antwerp we left the highway to head into Mechelen to find some accommodation. We drove for many kilometres through the very bland city and its urban sprawl and saw no hotels, inns or any other form of lodging. As we continued through mundane linear communities, we became increasingly convinced that people mustn't stop in Belgium; we could find no public accommodation. Finally, after more than an hour of urban and suburban traffic, we found an Ibis Hotel in the centre of Leuven, about 20 kilometres from the airport.

Sunday morning as we swept the snow off the car, we were pleased that we had stayed so close to the airport. As it was, it took us nearly an hour to drive the short distance. The flight was nearly an hour late departing because of weather, so we missed our connection in Montreal and had to wait over four hours for the next flight to Vancouver, which was also weather-delayed. Finally, at just past 2300 we arrived home after more than 24 hours in transit. We were travel weary, but very upbeat from our wonderful three weeks of adventures in Europe.

Chapter Twelve

Monitoring the Refit from Afar

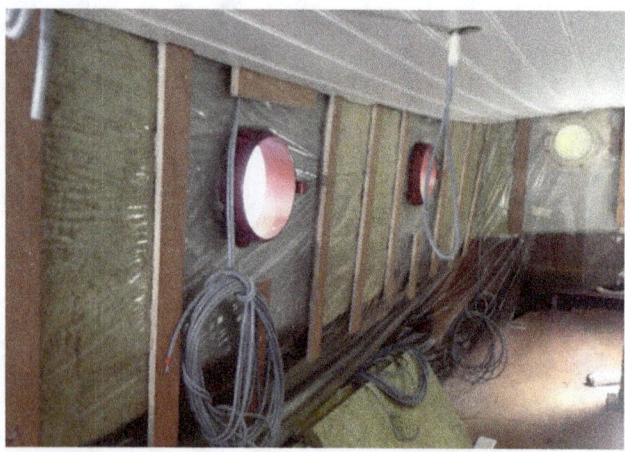

We were sent frequent photographic updates on the progress of the work. The first week of March the subfloor had been completed, the interior electrical wiring had been run, the ceiling had been installed, the insulation and vapour-barrier was in place and the interior sheathing battens were fitted.

Work was also obviously progressing in the engineering spaces, with the fabrication of plumbing parts for the hot water system, which will incorporate waste heat from the engine and generator as well as input from the diesel-fired central heating furnace.

The following week photos showed that the new mahogany window frames had been dropped into the roof and installation of the interior sheathing had begun. We had chosen off white panelling with minimal wooden trim to add to the lightness of the interior and to the sense of spaciousness. I have always equated dark wood panelling to the interior of a coffin.

The new Fischer-Panda generator had arrived and was unpacked and readied for installation. It is one of the new Hybrid Power models and it is rated at 6000 Watts DC. It is designed to be interlinked with battery bank, inverter and shore power through intelligent controls so that the most efficient power source, or combination of sources, is used to meet changing power needs.

Most of the time the smaller amounts of 240V power will come from the batteries through the inverter. The battery levels are automatically monitored and recharged from shore power, or

when not connected, then automatically by the generator. By intelligently monitoring the battery bank, the generator's operating time is significantly reduced, reportedly by more than 70%. In reading through the manual, I see there is an available easy override of the automatic system, which I will likely use most of the time.

Among the email exchange back and forth with SRF was a discussion on the cook top layout. I wanted a large wok burner and at least two, preferably three other burners. Wychard searched the catalogues and came up with this one from Candy. It fit our requirements perfectly, but not being familiar with the brand, I researched it on the Internet. It is a privately-owned Italian company that owns Hoover, Kelvinator and so on. I decided to dig further.

Rather than installing overpriced and under-performing 12 or 24 volt yacht or camper appliances, we had decided to install domestic 240 volt ones. We needed to find an under-counter fridge-freezer, a dishwasher, a microwave-convection oven and a washing machine and clothes dryer, so I looked at online reviews. Candy placed extremely high in customer satisfaction, with many of their appliances receiving 100% ratings and most of the remainder above 95%. We decided to install Candy.

The photos we received the last week of March showed the progress of work. The first one gave a general view of the interior looking aft through the framing for the bulkhead between the galley and the heads. We could see that the sliding doors had been fabricated and installed on the cupboards under the side deck. Also, the new window frames had been removed to be fitted with double glazing.

In a view looking forward we could see that the bulkhead between the heads and the master's cabin has also been framed, and beside the worker's head, we caught a glimpse of the toilet. Again for ease of maintenance, convenience of use, and attractiveness, we had chosen to install a domestic version, rather than a cantankerous, ugly and over-priced boat one.

We chose a bulkhead-mounted model and a Saniflow, which we see beside it in this photo. Because of the difficulty of taking plumbing through the ancient floors of many European buildings, the wall-mount and the Saniflow pump are commonplace. The pump automatically moves the waste to the blackwater tank. This is dramatically better than the marine toilets with clunky fixtures, hand pumps and the attendant complexity of levers, valves and clogs.

Moving aft into the salon, we saw that the hot water radiators for the heating system have been installed. Beside the radiator in the port aft corner of the salon we could see a spaghetti plate of wiring. This is lead through from the engine room and will eventually be connected to the circuit breaker panel, to the generator control panel and to the switch panel.

Monitoring the Refit From Afar

The framing of the salon settee had begun. Under the seat cushions and behind the seat backs will be large areas for storage. The interior work appeared to be progressing well and we began pressing for a completion date at the end of April. SRF talked of early June; there was still much engineering to do and all the painting.

The batch of photos that arrived in the second week of April showed more details of the work in progress. One showed a detail of the port forward corner of the galley and the bulkhead separating it with the heads. The line in the upper centre of the picture is the drain from the blackwater tank going through a very robust gate valve and thru-hull. Also shown are some of the electrical runs and copper plumbing from the hot water tank.

The above shot showed us that more of the freshwater plumbing had been installed, with runs for the dishwasher and the galley sink and continuing through the bulkhead for the shower, washbasin and toilet. Drains, traps, vents and valves were also in place.

This view shows two SaniVac systems, the one on the left under the galley sink will handle the graywater from the sinks, dishwasher and shower, and the other on the right will serve the toilet. Also in the photo we saw the propane lines had been led for the cooktop.

Here we see that the installation of the galley cabinetry had started. Because there are few if any square corners in a boat and also because of generally smaller spaces than found in a house, the modern pre-fabricated cabinets that are found in kitchen shops will not fit. Nearly all boat cabinetry has to be custom built in place.

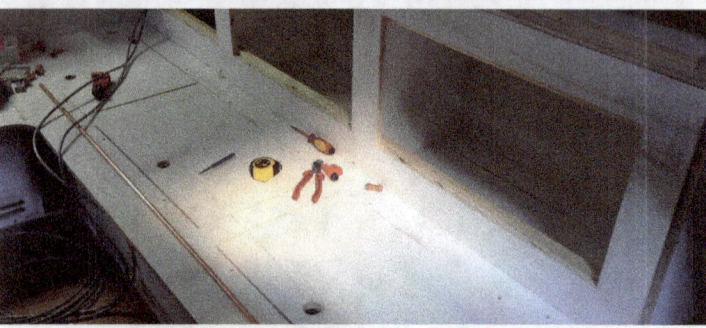

In the view looking aft from the galley, we could see that the construction of the salon settee was nearly complete. The photo also shows the huge storage spaces that will be easily accessible beneath the seat cushions and behind the backs.

The old rear window had been removed to be replaced with a new double-glazed version. Through the window is a crane hook and sling, lowering the engine into place for a fitting session and to more accurately map-out the remainder of the engine room. Klaas told me that the engine was then removed so that they could begin the installation of the other engineering components: generator, inverter, battery banks, furnace and hot water heater.

Monitoring the Refit From Afar

She was scheduled to be moved to the painting hall on 30 April so completion of work, both inside and outside was moving ahead at a pace to have her ready for then. On 9 April the timmerman began installing the appliances in the galley and we were sent another batch of photos, which showed the oven and dishwasher had been placed.

On 10 April more photos arrived to show that the cabinet doors had been hung. It looked like the timmerman had been busy and the galley cabinetry is awaiting only drawers and countertops. We liked the clean appearance of the cabinet fronts.

Other photos showed us that meanwhile workers were scraping the paint on the roef back to bare steel, getting the old girl ready for her appointment with the painters. The existing paint layers there were not compatible with the two part finish we had chosen.

Another photo showed the completed bed platform in the master's cabin. On its top we saw mattress vent holes and some inspection hatches for access to the domestic water tank.

While we were in Friesland in February, we had visited the recommended upholstery shop to select fabrics for the salon settee. With the interior now ready, we emailed to tell them they could now visit and measure the bed and the salon settee for mattress and cushions.

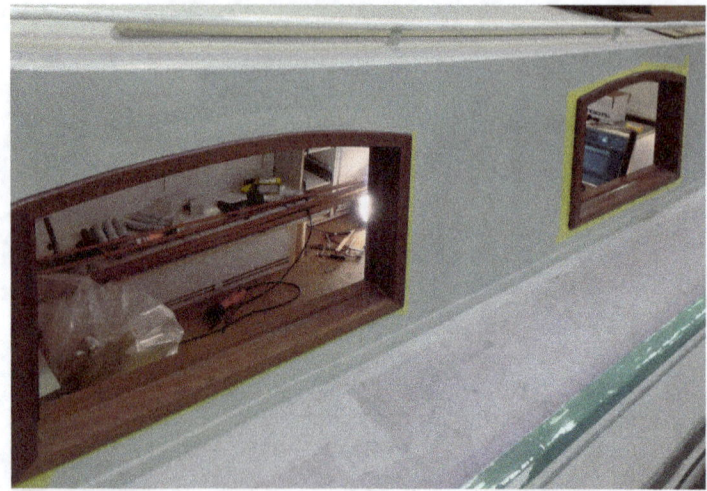

On 23 April we received some more photos. We could see that the scraping of old paint had been completed and that the roef has been primed. Through the salon windows we could see that work was continuing on the interior.

The iroco planks had arrived and were stacked on the top, ready for installation as our new floor. When we were at SRF in February, Klaas had shown us a large selection of flooring samples. We were attracted to the look and feel of the iroco and Klaas told us that with a good treatment of oil, it was very simple to maintain. We like simplicity.

Looking inside we saw signs of much activity. Trim had been tastefully added, leaving a very bright and clean-looking interior, without the ponderousness of too much wood. While we like wood, too much gives a boat interior a sombre coffin-like feeling. We're still much too young for coffins.

Some nice detailing had been added to the area around the settee. Here we see the vent on top of the central heating radiator along the aft bulkhead of the salon.

Monitoring the Refit From Afar

Work was progressing at a very pleasing pace. In the middle of the final week of April we were sent the following schedule:

- This week we are finished with the woodwork.
- Next week we start the interior painting.
- Following week we start the outside painting.
- Techniek we try to finish by the end of May.
- So I think you can come over the first week of June.

This is one month later than we had originally anticipated, but we must remember that it is a boat; things always seem to take longer with them. Without today's easy communications it would be rather difficult to monitor a boat's refit from 7700 kilometres away. Fortunately we have been able to keep an eye on her while we remained in Vancouver.

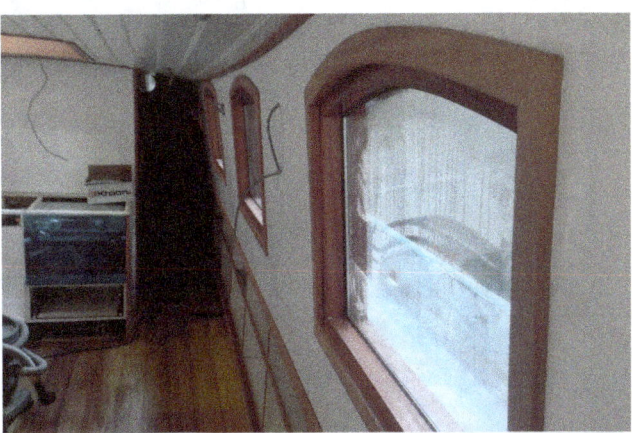

With this schedule we received another batch of photos. Among them we saw that the new iroco floor had been laid. We could also see that the wood trim had been finished and that the double glazed windows had been installed in the frames. We were delighted with how attractively her interior had evolved.

On 3 May we received another group of photos. We saw that the new aft window had been installed and that some of the wiring had been pulled from the engine room to the place where the new monitoring and control panel will be fitted.

From the cockpit, we could see how well the newly built window frame matches the elegant curves of the roef. Its inward opening feature will allow a nice cross breeze with the opening skylight above the galley and the six opening portlights forward. We could see that it was still awaiting the completion of the lattice work of double-glazed panes.

Another interior shot shows the sliding pocket door leading to the heads. We designed it as a sliding door because a hinged door would take way too much room. The previous interior had used hinged doors and they conflicted with the curved ceiling and would not open fully. It is the only door in the interior of the barge, we having decided on an open plan to give us better use of the space. All along the starboard side are storage cupboards set in beneath the side decks. The photo shows them with their sliding doors removed for the installation of the trim.

On the technical end of things, we see that the stern tube has been fitted and is awaiting the propeller shaft, which is awaiting the installation of the new Perkins M92B engine and PRM transmission. In the engine room machinery and equipment was being placed around the periphery of the space, leaving room in the centre for the engine.

We received photos showing the new generator being lowered into the after end of the engine room. It was then moved aft and onto the bed that had been prepared for it. Other views showed that the Marine Booster diesel furnace had been installed as well as the Victron inverter charger and an 876 amp-hour, 26.4 volt battery bank made up of 2.1 volt cells.

On 7 May some new photos arrived showing her still in the work shed with the masking and priming of her roef looking complete and the hull draped in plastic sheeting. She appeared to be ready to have the first of the top coats applied to her roef.

Monitoring the Refit From Afar

There was a photo looking aft over the mast tabernacle, showing us the graceful curves of the newly-extended roef. Another view showed us details at the mast tabernacle. Between its uprights, a caisson for a spud leg is unobtrusively fitted in place. When deployed, the retractable spud leg will extend a little bit more than two metres below the bottom of our skûtsje.

With her bottom three-quarters of a metre below waterline, this will give a quick and convenient mooring to the canal bottom in water depths of two-and-a-half metres or slightly more. Since most small canals are less than two metres in depth, many of them much less, the spud leg be useful for remaining in position while waiting for a lock or a bridge and will also be ideal for securing alongside the usually rough and irregular canal banks that are found in most places except the organised moorings.

The spud leg is just a little more than five metres aft of the bow, so mooring lines ashore from the bow and stern cleats will have more than ample leverage to hold her aligned to the bank. Our three-metre aluminium gangplank, which attaches to the gunwale at one end and has large casters on the other, will easily bridge the gap to the canal bank.

The placement of the spud leg between the pillars of the tabernacle precludes our fitting a mast. While our initial intention had been to re-rig the skûtsje for sail, we decided that we were getting to an age where sailing her would be too much of a chore. With traditional rig, a crew of four is considered minimal, so we would need to take on crew when we wanted to sail. Additionally, our thinking was that for the few times we would have sailed her, the clutter of folded mast, boom and gaff and all the associated shrouds, stays, sails and other rigging would be in the way on the canals. Our voyage through Patagonia and around Cape Horn had satisfied our thirst for sailing. We decided to forego the mast and go for a more relaxed boating adventure.

During the second and third week of May a few more batches of photos arrived. These showed the continuing work in the engine room. The installation of the generator appeared to have been complete. It is at the after end of the engine room, centred above the propeller shaft, and thus will be easily accessible through the aft engine room hatch.

Clockwise from the generator we see the start battery for the generator.. Next along from this is the Maritime Booster diesel-fired central heating and hot water system. This system is linked to thermostats that control both the domestic hot water and the central heating. The unit also has a burner to provide hot water on demand.

Running along the port side up to the forward bulkhead is one of the two interconnected diesel oil tanks. To monitor the fuel level there is a sight tube. Around on the forward bulkhead is the raw water strainer for the main engine and next to it we see that a switchable pair of Racor diesel fuel filters had been installed. These and the strainer are readily accessible through the hatch.

Continuing clockwise across the bulkhead is the raw water strainer for the generator. Around the corner in front of the starboard diesel oil tank we see the Victron MultiPlus 24/5000/120. This 5 kilowatt pure sine wave inverter-charger provides many automatic functions, including up to 120 amps of charging to the 24 volt bank of house batteries as well as for the bow-thruster battery and the start batteries for the engine and generator. Beneath the unit is the engine start battery.

Monitoring the Refit From Afar

Aft of the Victron is a very beefy switch and breaker panel linking the inverter-charger to the 876 amp-hour, 24 volt house battery. We chose to go with gel cell because of its greater availability of power, greater forgiveness and lower maintenance.

In front of the battery bank is the propeller shaft's stern gland greaser. Its location appears to make it very convenient to give the crank a turn or so after each day's run. Immediately aft of it is the heat exchanger to heat domestic hot water with waste heat from the engine and generator. It appeared that the only things missing were the engine and its transmission.

The Perkins-Sabre M92B is a 4.4 litre, 4 cylinder in-line, vertical 4 stroke water cooled and naturally aspirated diesel engine with a fresh water heat exchanger. Among the things I like about the engine is that its maximum power and torque are developed at low revolutions. The torque curve is rather flat, with its maximum developed at only 1400 rpm, where the fuel consumption is about 4.5 litres per hour. With a properly matched propeller, this will allow us to cruise at the 8kph speed limits of most small canals at around 1200 rpm or so, while burning about three litre per hour.

On 21 May Klaas sent us some photos showing the engine being lowered into the space that had been left for it. It was then fitted onto the mounts that had been prepared for it, and in short order, the raw water cooling, exhaust, fuel and electrical systems were connected. our skûtsje again had mechanical power.

The next day we again received photos from Harlingen. These showed us that the custom-made Marlan® countertop had been installed. Marlan® is a Dutch product that is non-porous, dirt repellent, stain resistant, hygienic and food safe. Unlike the Corian®, which we had in our sailboat, Sequitur, Marlan® is also flame resistant and has a high resistance to scratching.

We had initially wanted granite countertops in the galley; however, once Klaas showed us samples of and specifications for Marlan®, and explained the seamless fabrication and shaping, our decision to go with Marlan®, even though it is more expensive than granite, was easy. To coordinate with our iroco flooring, mahogany trim and the pale cream cabinetry and bulkheads, we chose the Himalaya Bronze colour. The cooktop was ready to nestle into its position above the convection-microwave oven. In his email, Klaas told us that the interior would be completed at the end of the week and the iroco flooring would be oiled the following week.

On Friday 24 May we were sent some photos showing the progress with the painting. We had chosen an ivory colour for the roef, for us a more pleasing and slightly warmer colour than the previous white. We were told the first finish coat had been applied a few days before. The rubbing strake, gunwale tops, winches, bollards and other deck gear are in black. Our thinking here was that with these being the most commonly chipped, chafed and worn areas as the barge is operated, black would be the easiest colour to match and the simplest to apply touch-up without looking too patchy.

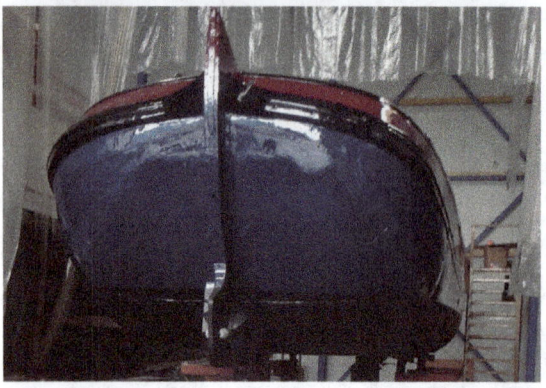

The photos we received 25 May showed how well our other colour choices work. For her hull beneath the rubbing strake, we chose a navy blue, and for the topsides we decided on a rich burgundy. We opted for these darker colours on the hull because the previous white too easily showed rust weepings. We thought that we could find better things to do than frequent cosmetic touch-ups.

The photos showed that the first finish coats were all on as were some of the top coats. The decks were waiting for the surrounding paint to dry, so that they could be masked in preparation for painting with a non-skid layer. A few days later, on Thursday 30 May we received another batch of photos showing the completed paint job. We thought that our skûtsje had a rather fine complexion for a 105-year-old lady.

Proudly wearing her new colours, she was transferred onto a low crawler and rolled out of the hangar and into the yard to be prepared for re-launching. She was moved to the edge of the water and was slung in the travel lift.

For the first time in six months her bottom was wet again. We had decided that at this moment we would rename her **Zonder Zorg**. Zonder Zorg is Dutch for *Without Worry, Without Care, Carefree*. We feel that this name more aptly describes us and our style of boating than did Nieuwe Zorg, which means *New Worry* or *New Concern*.

Before she was even out of the slings, work resumed aboard. There were still many details to be attended to make her liveable. It was already Thursday afternoon and were scheduled to arrive on Monday to move aboard. We were assured that we would be able to move in upon arrival. Even then, we knew there would be small things needing completion, tweaking, adjusting and refining. We were looking forward to settling-in and making Zonder Zorg our home.

Chapter Thirteen

Moving Aboard and Settling In

On Sunday morning, 02 June we rolled our ten-piece luggage train the block and a half from our Vancouver loft to the SkyTrain for the ride to the airport. Our flights took us without hitch via Montreal to Brussels and we breezed through Immigration, with Edi's new Dutch passport allowing us to do the very short *EU Passports* line; the *Other Passports* line looked thirty or forty minutes long. Our luggage arrived promptly and our car was waiting for us at the Hertz Gold Counter.

Within three hours we were looking across the canal at Zonder Zorg, looking wonderful in her new colours. Her leeboards had not yet been installed, and we could see work underway with hatches open fore and aft, an entry door standing on the side deck and the carved artwork leaning against a shoreside hangar wall.

After a quick visit aboard, we went up to the office to announce our arrival to Klaas and to get an update from him on the progress of the work and to find out where our stored bins were. We were shown to a third-floor loft in one of the hangars, where from our pile of belongings we selected

two bins in which we could see tableware, stemware, cutlery and cooking utensils and vessels. We lugged them down to the boat. We then hauled our ten pieces of luggage onboard from the car and drove into town shopping for ingredients for dinner and breakfast.

Back onboard we unpacked and began organising. The rough sorting was easy with all the sliding doors removed from the fronts of the compartments. There are twenty-two compartments running all the way along the starboard side. With six each in the salon, galley and bedroom and four across from the heads, sorting by place of use was simple. The five drawers and two cupboards beneath the galley countertops easily swallowed everything we threw at them and still seemed almost empty. The bins beneath and behind the settee are huge and they remained nearly empty.

While Edi continued to sort, I prepared dinner. There were no cooking pots in the bins we had brought down from storage, and the hangar was locked for the night. I decided against driving into town to look for a small pot, so with just a wok and a sauté pan to work with, we skipped the green beans amandine and had a simple meat and potatoes dinner. At 2200 we dragged ourselves to bed, thirty-one hours and 7700 kilometres after we had sprung out of bed in Vancouver. The new mattress was very comfortable and we were very soon asleep.

The next morning we saw Sfinx sitting in the slip around the corner from us. She was one of the three tjalks we had winnowed onto our short list from the seventeen we looked at. Looking at her and at Zonder Zorg now, we are delighted with our decision. Looking at the tangle of rigging on Sfinx, we are thankful that we decided not to restore Zonder Zorg to sail; the rig would have been in the way most of the time for the style of boating we want to do now.

Moving Aboard and Settling In

Through the following four days, as the yard workers tended to some of the remaining small details aboard, we hauled more things down from the storage loft, We also shopped around in Harlingen for an espresso machine, a bean grinder, a toaster and other comforts of home. On Wednesday and Thursday we made trips to Ikea in Groningen and each time came back with a stuffed car. Thursday afternoon we bought two lightly used bicycles in a shop in downtown Harlingen, and in the evening we tried them out with a grocery shopping trip into town. On Friday we returned the rental car to the Hertz office in Leeuwarden and took the train back to Harlingen.

On Saturday morning we headed out in Zonder Zorg for a shakedown cruise to try out her various systems. We found a very simple and practical way to secure the bicycles aboard, with a fender on the mast tabernacle and a bungee cord. The sluit cover makes a perfect wheel chock. The bicycles came fitted with wheel locks, which also incorporate a robust cable lock.

We headed along the Van Harinxma-kanaal toward Leeuwarden under scattered clouds in gusty crosswinds. Among the many boats we saw were a few sailboats on beam or broad reaches along the canal. Zonder Zorg's new larger diameter propeller gives her helm a much more stable feel. Where Nieuwe Zorg's tiller needed to be tended full time, and with considerable effort, now the tiller can be let go for short periods and is much easier to handle.

We motored through Faneker and shortly beyond it we turned off the main canal into a narrow one heading southward. The sign at the canal entrance confirmed the notations on the chart that gave the maximum bridge clearance of 2.4 metres, but with Zonder Zorg's low air draft, we easily passed under.

We wound our way through pastures and tiny villages for a few kilometres, and then turned off the small canal into a tiny one and secured alongside in a tiny wilderness park. Earlier in the week we had paid €12 for an annual membership in De Marrekrite, a recreation group that provides 3500 free mooring spots at 285 locations throughout Friesland. For the fee we received a map to all the locations and a burgee which allows us a year of free moorage. We settled-in and relaxed. It may take us a while to become accustomed to this pace, but we are confident that with practice, we eventually will.

We arrived back in Harlingen midday on Monday after our weekend shake-down cruise. With us we had a short list of arisings that needed attention. While Zonder Zorg is a hundred-and-five years old, that is on the outside only; inside, she is completely new. As with many new installations, there are tweaks and adjustments necessary that are discoverable only with use. We found several. One of these was the hand sink in the heads; in the standard Dutch fashion, it was plumbed for cold water only. The easy fix was to add a mixer to its feed line under the galley sink.

Another item that needed attention was that a right-hand propeller had been installed instead of the left-hand one that I had specified. I much prefer a left-hand screw because of the ease of handling when going alongside starboard-side-to. When the engine is put astern, the stern kicks to starboard, nicely pulling it alongside. When leaving, the forward turning screw pulls the stern to port, moving it nicely away from the wall and allowing a clean departure.

Moving Aboard and Settling In

We motored Zonder Zorg around to the travel lift to be hauled and have the propeller exchanged. The lift was quick and very efficient. With Zonder Zorg out of the water, it gave us an opportunity to again familiarize ourselves with her underwater architecture. The bottom of the rudder extends a metre and a half aft of the pintles, but is protected by being ten centimetres above the skeg. The propeller is also protected from shallow bottoms by the skeg.

The retaining nut was heated and undone. The propeller was then heated and thumped off the tapered shaft end. While the shaft was cooling down, the crew went for their koffie break. With the shaft cool, the new 22 X 20 LH screw was introduced to the tapered shaft, heated and snugged up with the retaining nut, which was then locked in place with a setscrew into a shaft dimple.

While the barge was out of the water, we had a good look at the original registration number stamped into her side in 1908. The S831N signifies that she was the 831st ship registered in Sneek in the Netherlands. Above the number is a line of original rivets joining iron plates in her century-old hull.

Also while on the hard, we had a new through-hull installed just below deck level on the port bow. This is to reroute the water tank breather line, which had been badly routed and had allowed water to overflow into the bilges when we had filled-up the fresh water tanks.

After the very quick and efficient work, we went back into the water and motored around to work wharves on the south side of the site. Shortly after we secured, a Lemsteraak rafted outboard. We counted seventeen boats being worked on in the water and on the hard. Plus inside the hangars there were another eleven vessels being built or refitted. Every space was full. SRF is an extremely busy yard with over fifty workers and seemingly more work than they can handle in what was widely reported as a continent-wide slowdown. Many of the other yards were struggling to find work; some were completely idle.

On Zonder Zorg work immediately resumed, with such things as adding re-enforcement webs to the main engine room hatch and installing sound-deadening insulation to the undersides of both hatches. We had a locking latch installed, so that we can secure the engineering spaces when we leave the boat.

While work progressed, we kept ourselves busy fine tuning and adding interior details. Being now the full width of the interior, the bedroom is spacious and comfortable, with lots of convenient hanging points. The washing machine and the clothes dryer are nicely tucked-in under the foredeck; convenient to get at, but out of the way. We gave them several loads to work with.

On our first trip to Ikea we had bought a MemoryFoam pad to put on top of our mattress, and this made the very comfortable bed even more luxurious.

Moving Aboard and Settling In

We love the clean simplicity of the interior, and the wonderfully convenient cabinets behind sliding doors all along the starboard side. There is more than enough storage.

The arrangements of fixtures in the heads has allowed for a very spacious shower, with more than ample elbow room. The water heating system provides hot water for as long as the two large water tanks and the supply of diesel fuel hold out. The toilet is well placed and being wall-mounted, it takes very little space. The electric flush through the Saniflow is quiet and efficient.

The galley is very well laid-out, with everything ergonomically arranged and conveniently placed. We have thoroughly tested the dishwasher. At 45 centimetres width, it is just large enough to swallow in a single load everything that we use to prepare and enjoy breakfast, lunch, snacks and dinner.

The fridge sits conveniently under the port side-deck and is sufficiently large to store several days of food. Although with shops and supermarkets being very plentiful often within a short walk or cycle from the canals, we don't need to stock-up like we did during our years of wild and remote sailing.

The seven-month refit had been completed and the last few adjustments and details were being addressed as we settled-in and made Zonder Zorg a home. Of course we tried-out the new galley repeatedly and it has consistently produced wonderful meals. It is such a joy to work in and the appliances far exceed our expectations.

By late afternoon on 18 June the most of the items on the list had been tended to and completed. There were just a couple of things still unfinished, and the workers were confident that they would be completed the following morning. The last of the testing would be done and everything would be ready for us to head out to begin our cruising.

In the evening we again tested the galley to ensure everything was working properly. It produced another splendid meal. We were looking forward to many years of zonder zorg cruising in Zonder Zorg.

Chapter Fourteen

Which Way to Go?

On our way back from Spain toward Friesland in February, we had stopped in the Midi region of France to see if we could arrange winter moorage. To hedge our bets, we applied to the capitaineries in both Carcassonne and Narbonne. In mid-May, after three months of communicating and negotiation, a place was confirmed for Zonder Zorg in the Port de Carcassonne.

The Port is located at the edge of Ville Basse, which is referred to as the New Town. New because it dates only to 1247, when Louis IX founded it across the River Aude from La Cité, the fortified citadel on the hill that was built by the Romans dating the first century BC.

Carcassonne is home to about 50,000 people. It is about mid-way along the Canal du Midi from the Mediterranean to Toulouse, and is within a few kilometre from being the most southerly part of the French canal system. A major portion of its businesses is tourism, centred around La Cité and the Canal du Midi, both of which are UNESCO World Heritage Sites. Wine growing is also a major focus here, it being in the heart of the Languedoc-Roussillon wine region, with many famous areas, such as Minervois, Corbières and Limoux nearby. We looked forward to a pause there.

However, Carcassonne is about 2000 kilometres south of Harlingen, across the Netherlands, Belgium and France. We had a long way to go and we had a very large variety of routes from which to chose to get us there.

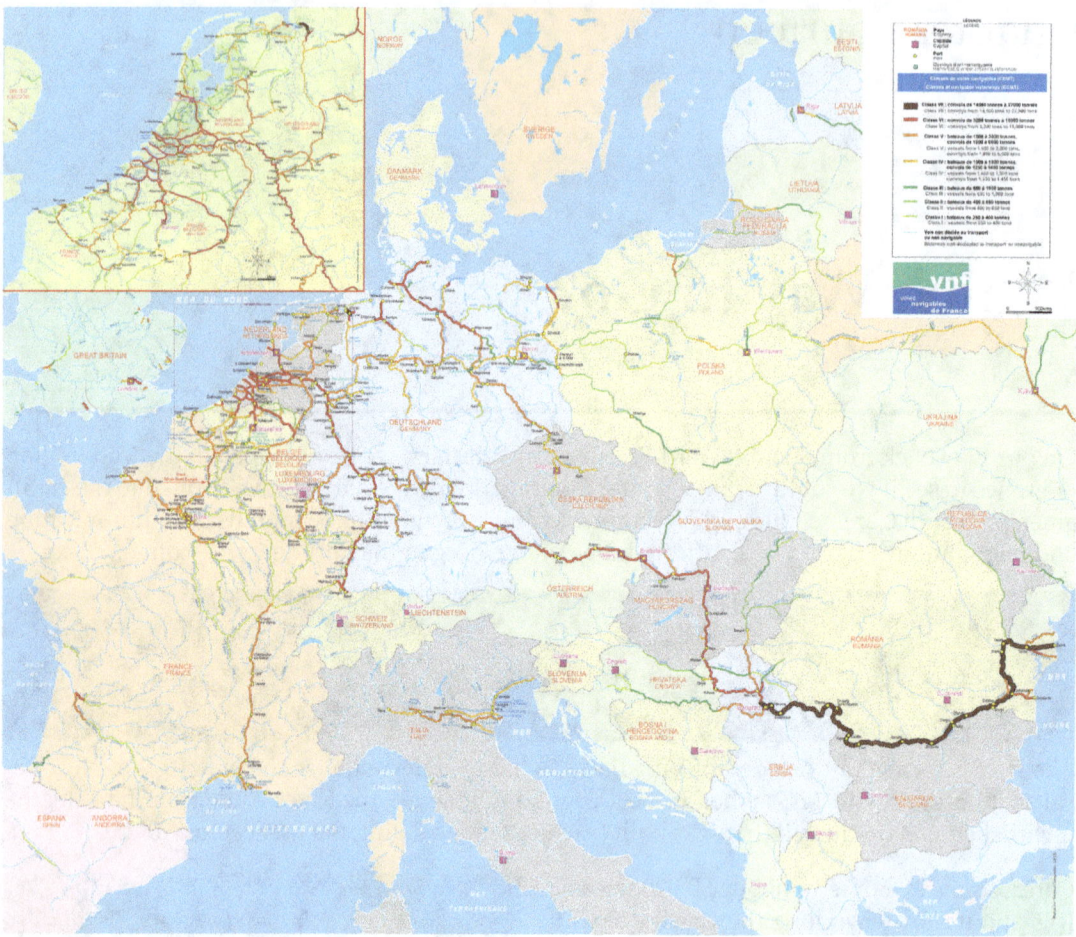

There are over 45,000 kilometres of navigable rivers and canals in Europe. That is more than a global circumnavigation's worth of inland waterways to explore. In France alone, there are 8,800 kilometres.

Our intention was to do a thorough shake-down of Zonder Zorg by wandering around the canals of Friesland. Once we were satisfied that all systems were working properly, we thought to follow a meandering route through the remainder of the Netherlands to Maastricht on the Maas at the border with Belgium. The river changes names at the border of Walloonia, the French portion of Belgium; it becomes la Meuse. This is the shortest route through Belgium and it is reputed to be very pleasant in its southern portions. There wasn't time on this trip to explore in Flanders, the Dutch-speaking portion of Belgium, where the pleasure boating excellent.

Once we entered France at Givet, we would be faced with many choices of routes to take us through northeastern France, but it was a rather safe bet that whatever way we chose, our first destination would be the Champagne. Onward from the Champagne, we could opt to continue via Paris, with the decision of whether to go by way of the Marne or the Aisne, Oise and Seine. From the Seine above Paris we could choose to continue on the Loing, Briare, Loire and Centre to the Saône, which entails 554

Which Way to Go?

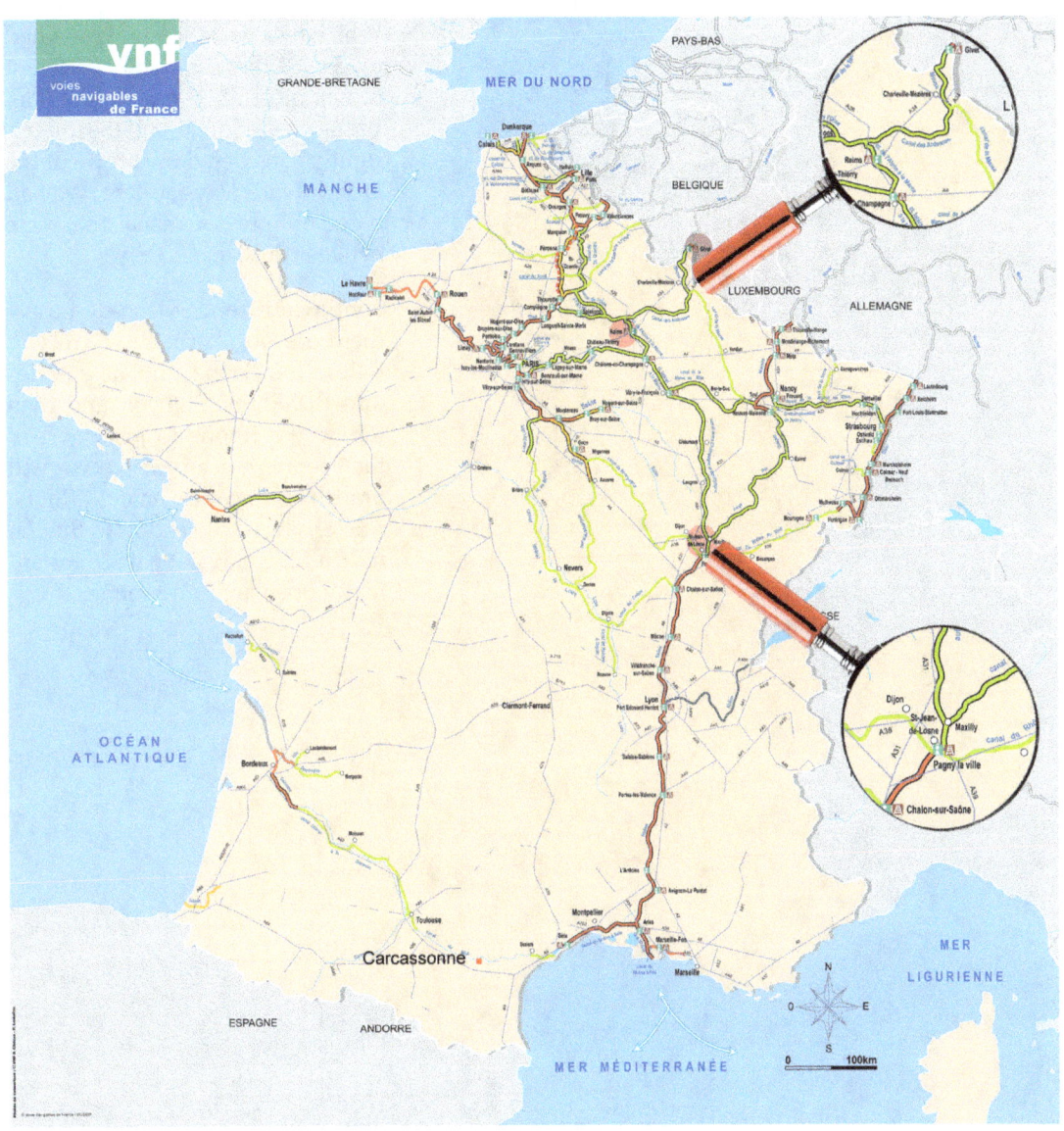

kilometres and 157 locks; or possibly take the more scenic Yonne, Nivernais, Loire and Centre, which is a bit longer and has a few more locks; or we could take the Yonne and Bourgogne to the Saône, which is shorter at 401 kilometres, but involves more work with 214 locks.

Alternatively, if we decided to give Paris a miss on this trip, we could follow the Aisne a la Marne, the Lateral a la Marne and the Canal entre Champagne et Bourgogne to the Saône, or we could do the Marne a la Rhin and Canal des Vosges to the Saône. An additional eastern alternative is to continue along the Marne a la Rhin into the Alsace, then southward and take the Rhône au Rhin to the Saône. The route we took depended on many factors: how much time we had, how many locks we felt like doing, what history and culture we wanted to see, and what we wanted to eat and drink.

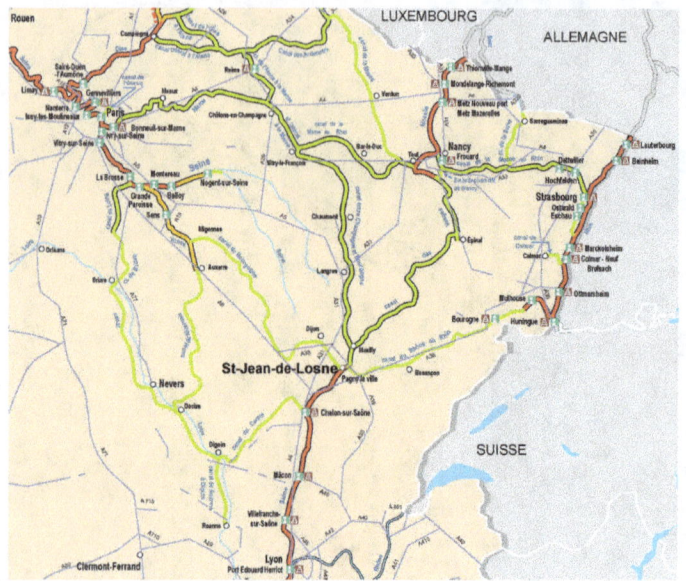

Whichever route we took, we had to get to the Saône. Six different routes meet at or near St-Jean-de-Losne on the Saône, and the town is rightly considered the centre of the French waterways system. We had based our previous canal boat there between 2000 and 2006.

Southward from St-Jean-de-Losne, there are no alternatives. We will follow the Saône until it empties into the Rhône at Lyon and then follow the Rhône to just short of the Mediterranean, where we will branch off into the Canal du Rhône a Sète, which will lead us to the Canal du Midi and along it is our winter moorage in Carcassonne.

Chapter Fifteen

Southward Toward France

On Wednesday morning, 19 June we went up to the office of Scheepsbouw & Reparatie Friesland to bid our farewells. We had the rare good luck of finding three of the four partners, Lex Tichelaar, Wychard Raadsveld and Klaas Koolhof in the office at the same time. After we had thanked them for the wonderful work they had orchestrated, I posed for a photo with them. I am not a short person, but with the Dutch having the tallest average height in the world, these three men at 195 to 200 centimetres tall towered over me. Their work seems to tower over that of other yards in our experience.

We flashed-up Zonder Zorg's new engine, and at 1015 slipped our lines and headed out. Over the following six weeks we followed a sinuous track through the Netherlands. As we wound our way through Friesland, we visited Workum, Hindenloopen, Heeg, Gaastmeer and Lemmer and along the way we shook-down our new skûtsje. Among the trials we had to perform was the operation of the galley, and our meals deliciously proved it to be working wonderfully.

We paused in Gaastmeer to look at the town where our skûtsje had been built in 1908. We found the modern version of Wildschut, the company that had built it and met a descendant of the original family. In the single slipway of his yard was a skûtske casco from the same era as ours, although it had been built by a different yard. It had been completely stripped-out and primed and was waiting for someone to buy it and have it fitted-out to their own specifications. We thought of our having gutted our Nieuwe Zorg and rebuilding her as Zonder Zorg from a similar state. Except for the jachtenroef on our skûtske, which we had to cut off, extend aft and heighten, there was little difference. The biggest difference is that ours was from the Wildschut yard, one of the most revered builders of skûtskes.

We found a wonderful mooring in the centre of Lemmer and paused for two days to allow the engineers from SRF to come and repair minor diesel and coolant leaks on the Perkins engine.

From Lemmer we left Friesland and crossed the Flevoland to pause in Kampen and then down the IJssel to the Randmeeren and along to Elburg. There we were visited again by an engineer from SRF, this time to replace a blown connection on the supply side of the 5 kW inverter. While we were in Kampen we were running the dishwasher and washing machine when Edi

began using her hair dryer. We smelt hot, then burning insulation before I ran to switch off the breaker. In Elburg the engineer found that, while the circuit was protected with a 5000 Amp breaker, the lead to it was only rated for 3000 Amps. He replaced it.

We continued around the former Zuiderzee coast to Spakenburg, where not long after we had secured we watched a long line of wooden botters parade past. About four dozen of these marvellously restored and maintained old fishing boats, many of them dating from the late nineteenth century call Spakenburg home. Most of these were now heading out for an evening sail on the Eemmeer. On some, crews were busily raising their sails as they went by us.

We watched the parade go past and then walked along the dike to its end at the edge of the meer. We stood watching for a long while as two and a half dozen antique wooden sailing vessels, ex-fishing boats, played with each other, tacking and gybing in the light evening breezes. As we retraced our steps along the dike to Zonder Zorg, we thought of watching evening sails on Vancouver's English Bay, then immediately dropped the thought.

In the following days we crossed Markermeer to the Noordzee Kanaal and along it past Amsterdam to Zijkanaal, which led us southwestward to Noorder Buiten Spaarne and onward into Haarlem. We spent a wonderful eight days in Haarlem and found a mooring next to De Waag in the very heart of the historic city. From there we visited museums and art galleries and took many walking explorations through the streets that wind past marvellous historic buildings.

We continued along to Leiden and then to Gouda, where we paused to enjoy the city for six days. Among other things, we enjoyed the superb sixteenth century stained glass windows in Sint Janskerk, which is more than half of the stained glass surviving from that era in the Netherlands. We also stocked-up on a large variety of Boerenkaas, the genuine Gouda cheese. We bought enough to last us well into southern France and to augment the marvellous French cheeses.

From Gouda we continued eastward, pausing for the night in a peaceful setting in Montfoort. We continued to test the galley to ensure it provided us with wonderful dining. It did not disappoint.

Continuing our way eastward through the provinces of Utrecht, Gelderland and Noord Brabant, we stopped to absorb the history and culture for two to four days in each of Gorenchem, Heusden and Veghel. We then headed across Zuid-Willemsvaart toward the province of Limburg. A night's pause in Helmond and a few hours the following day brought us to Maasbracht on the Maas River, where we paused for a couple of days to explore nearby Thorn, which from the twelfth to the late eighteenth centuries was the smallest Principality in the Holy Roman Empire.

From Maasbracht we headed south up the Maas to Mastricht and slightly beyond to Pietersplas, where we spent a week. This is very fine and peaceful marina, an easy three-kilometre pedal south of Maastricht and just two kilometres from the Belgian border. We had done somewhat of a zigzag route through the Netherlands, covering 597 kilometres from Harlingen. A fuller description of our 2012 and 2013 wanderings and explorations through the Netherlands is in my 2014 book: *Carefree on the European Canals*.

We were sad to leave the wonderful cruising of the Netherlands, and promised ourselves that we would be back after we had completed the last few waterways in France that we had not yet cruised.

Southward Toward France

First we had to get through Belgium and our chosen route was the most direct. The Maas becomes la Meuse at the border and flows through Wallonia, the southeastern portion of Belgium. The language and culture here are French, as opposed to the Dutch language and culture in Flanders, the northeastern portion of the country.

On 11 August we slipped from our mooring in Pietersplas and headed up the Maas the short distance to the border and within another kilometre we were in the chamber of the first lock. The lock is huge and fortunately it is fitted with floating bollards. We were lifted 13.96 metres to Canal Albert, which we followed through a dirty, drab and crumbling industrial canyon for seventeen kilometres to the next lock, which took us up to more blandness along the five kilometres to the river, la Meuse. This is a huge waterway with huge commercial barges, and the infrastructure before the locks and in them is not set-up for pleasure boats. We saw no border check point.

We continued along to the city of Liege, which from the river looked very uninviting and we were not in the least tempted to stop.

Beyond Liege the grim industrial canyon continues nearly unbroken. Although there are a few isolated breaks of more pastoral setting, nowhere along the river thus far had we lost sight of the huge factories and industrial complexes that seemed to be rendering the surrounding limestone hills into various elements and compounds. In front of many of them we saw huge barges loading bulk materials.

In the late afternoon we secured to bollards alongside the wall in Port de Plaisance de Corphalie, across the river from three nuclear power plants at Huy. We had come 53.7 kilometres into Belgium through mostly drab scenery and had not yet seen any border checkpoint. Le Capitain in the marina office gladly took our money and was not at all concerned with boat's papers. The view upriver looked less industrial and this coincided with what we had read: La Meuse gradually becomes less industrial and more picturesque upstream from Huy".

The following day we motored through Huy and after a two more locks and 36 kilometres, we arrived in Namur and took moorage in Port de Plaisance de Jambes for a week. We were 89.8 kilometres into Belgium without seeing a checkpoint, nor being asked for our papers. From here it is only 43.3 kilometres up la Meuse and into France.

Since we were two-thirds of the way through and had not yet been formally welcomed into the country, we decided to do it ourselves. For dinner we had panfried fresh Icelandic filet de fletan with sautéed Friesland potatoes, Belgian haricots fins almandine and sliced Roma tomatoes with shredded basil. To mark the welcoming ceremony, we accompanied dinner with a bottle of Cava Ferriol Brut.

We had paused here to be within an easy commute to Brussels International Airport. Dad had passed away in the Spring at ninety-nine and Mom a couple of years earlier at ninety-four. Their ashes still needed to be scattered and my extended family had agreed to come together on 18 August at the old family home at Shediac Cape, New Brunswick, Canada.

Southward Toward France

We booked flights departing Brussels at 1020 Thursday morning, 15 August to Montreal with a connection to Moncton. After six different transportation modes, we arrived in the wee hours of Friday morning. On Monday morning, family visits and duties completed, we headed back, arriving aboard Zonder Zorg twenty-five hours after we had popped out of bed. It was Tuesday and we were back in Belgium, relieved, though somewhat travel weary.

Early on Wednesday morning we slipped our lines and continued up la Meuse toward France. The scenery continued very pleasant with many fine buildings, some grouped in small villages and many standing as solitary riverside estates. Along the way we passed several impressive chateaux and some fortified hilltops.

There are nine locks between Namur and the French border, their chambers taking barges up to 98 metres long and 11.8 metres wide. The lifts vary from 1.7 to 2.89 metres, all but two of them being over 2 metres. The bollards along the tops of the chambers are broadly spaced for the large commercial barges, with no intermediate securing points for smaller craft.

Some overly zealous safety bureaucrat had caused protective fencing to be recently installed, making it impossible to toss lines over the bollards from our barge. Edi had to climb the slimy ladder up the chamber wall dragging the bow line with her, while I stabilised the barge in the chamber. After she had looped the bollard and passed me the end of the line to secure the bow, we moved aft and I tossed her the stern line to repeat the process.

This system worked satisfactorily; we had plenty of time to stabilise in the chamber before the operator began closing the downstream doors and then gradually opening the sluices to flood the lock. However; it was beginning to wear on Edi, who was nursing cracked ribs from being overeagerly hugged by my nephew as we were saying our farewells in New Brunswick.

As we were midway through the gates of écluse du Dinant, they began closing. We rushed to secure. There was no reachable bollard for our stern, so we did a long bow line and used it as a spring, motoring forward onto it as the lock keeper opened the upstream sluices all at once and sat back in lockside chairs with a buddy drinking beer and watching the antics of *les maudits plaisanciers hollandais*, the damned Dutch pleasure boaters. Fortunately, there were only three more Belgian locks before France.

We survived the abuse and motored into Dinant, which looks like a wonderful place to stop and spend some time exploring. However, we were by that time so turned-off by Belgium, that we continued without a pause.

We were still looking for the Belgique declaration station, when we passed a Douane sign. Just beyond it we saw a temporary building near the canal bank that could have been the reporting post, but its shutters were closed and it was very obviously not in use. Looking back from across the border, we could see no signs indicating boaters should stop. We suspect that the Walloons had decided to shutdown the reporting post, their attitude and infrastructure having scared away most of the pleasure boaters.

Southward Toward France

It no longer mattered; we had left Belgium and we were in France. Our experiences in Belgium were all in Wallonia, the southeastern portion of the country. From all that we had heard and read, pleasure boating in the northwestern part of the country, Flanders is much more pleasant. The topography there is similar to that of the Netherlands and they encourage pleasure boating and provide facilities. In Flanders an annual permit for Zonder Zorg would be €100, which by all reports is a very good value for the services received. On the other hand, there are no fees charged to pleasure boaters in Wallonia and the services provided seem worth that amount.

We shortly arrived at the next lock, the first one in France. L'eclusier took our lines with a hook on a long pole and placed them over the bollards for us. When we were secured he activated the lock and when we had reached the top, he motioned to me to come into the office. I took our thick portfolio of licenses, certificates, documents and other boat related papers. He didn't want to see any of it, he wanted only my name and the name of the barge to write in his log and he asked me whether we had a vignette. I told him that I had earlier bought the annual one for €534 and downloaded and printed it and that it was in the Zonder Zorg's window. He didn't want to see it

He gave me a remote-control clicker to activate the automatic locks in the Departement d'Ardennes, saying we would give it back at the final Ardennes lock. We continued on to Givet, where shortly before 1800 we stopped for the day.

To celebrate our arrival in France I sautéed some filet de fletan and we enjoyed it with mushrooms, fine green beans, pan-fried potatoes and sliced tomatoes with shredded basil.

We anticipated that the accompanying bottle of Champagne and the full moon were the first of many that we would be enjoying aboard Zonder Zorg in France.

Chapter Sixteen

Into France

We had secured along the wall just above bridge in Givet. On Thursday morning we walked across the bridge to the town of Notre Dame and headed through it and out beyond, following directions to the BricoMarché, the French answer to HomeDepot. We needed a floodlight for the two tunnels that were just ahead of us at Ham and at Revin. The FluviaCarte states: "these tunnels have no illumination, so a floodlight is needed". We had found several great rechargeable lights at Canadian Tire the previous week in New Brunswick, but their transformers were all 120 volt only, not 120/240. After a three kilometre walk we arrived at the shopping plaza at noon to find that the BricoMarché was just closing for lunch. Back onboard, I dug out two LED flashlights and decided they were sufficient. The longer of the tunnels was only 500 metres.

At 1350 we slipped our lines and continued upriver under the ramparts of le fort de Charlemont. The fort was built by Charles V in 1555 during the invasion of France into the Spanish Netherlands as a means of securing the Meuse. Three thousand workers very quickly built defences that held up to attack when still unfinished. The works were subsequently added to by William of Orange, then the Spanish. In the seventeenth century the fort was remodelled by Vauban.

The size and complexity of the works tell of the strategic importance of the site. The Meuse has been a major trading route back into prehistory.

We arrived at the next lock, Les 3 Fontaines, which was manned, and again l'eclusier took our lines from the lip of the chamber with a hook. We were now in the Freycinet gauge locks, designed to take barges 38.5 metres long by 5.05 metres wide and 1.8 metres draft. Outside her fenders, Zonder Zorg has about half a metre of room on each side in the chambers of these locks.

Immediately out of the lock we came to Tunnel de Ham, and as we approached it, we could easily see the light at the far end of the tunnel. Having no floodlight seemed now to be of even less concern. I simply stood on the centreline with the tiller in the small of my back and aimed the mast at the light at the end of the tunnel. As long as I kept her relatively well centred, Zonder Zorg adjusted automatically to the water pressures from the sides and almost self-steered. After an easy 500 metres we were through.

Within another half kilometre we came to the next lock, which thankfully at 3.2 metres in height, was manned. Madame l'eclusier came out and took our mooring lines on a hook and placed them over bollards. We thanked her for her assistance as we were leaving the chamber. She confirmed that we had a remote-control, telling us that the next fifty or so locks are all automatic and unmanned.

We headed out of the lock and along the 800 metre derivation toward the river. The chart showed a *pont levis*

méchanisé, a mechanised lift bridge just before regaining the river, and I wondered whether the remote control worked for it. As we approached, I saw there was no concern; the bridge had been jerry-rigged in a partly open position to give the required 3.5 metre clearance. This clearance was one of the standards for the canals of France that was laid down by the 1879 Freycinet gauge.

Probably the vehicular use of the bridge is now so light that it makes more sense to keep it raised and rigged as a pedestrian crossing, and lower it only when required.

Once we were back out into la Meuse, Edi brought up a tray with a selection of salami, paté, pears, olives, cheeses and bread for us to nibble on as we continued upstream.

The waterway is mostly in the river, with short canals, called derivations leading to and from locks that step the water level past rapids or cataracts. Between the locks there is very little current. After twenty-four kilometres and six locks, at 1802 we stopped for the day at the Halte Nautique de Fumay.

On Friday morning, after a leisurely breakfast, we walked to the tourist information office and made use of their free wifi connection to catch-up. Then shortly before 1300 we slipped our lines and continued upstream. La Meuse describes a rather winding course through this area, and during the first four kilometres after leaving Fumay, we steered every point of the compass. At some bends, like at Ham and Revin. tunnels cut through a narrow neck to shorten the route.

After two locks we came to the Revin Tunnel, the second unlit tunnel for which we were warned to have a floodlight. Since the warning was written in the chartbook, lights have been installed, though likely these are to illuminate the cycle path that appears also to have new guardrails.

After we had cleared the tunnel and the lock following it, Edi went below and prepared a tray of nibblies for lunch and we enjoyed them as we made our way toward the next lock.

As we waited for ecluse Dames de Meuse to cycle, we were entertained by swans posing with an arched railway bridge as a backdrop.

At 1728 we secured alongside at the halte nautique de Laifour. It is a very small village, too small to have a bakery. Since everyone in France seems fixated on fresh bread every day, there was a placard next to the mooring informing visiting boaters that bakery goods be reserved the day before at the tabac. I walked up to small tabac across from the Mairie, which also had a *depot de pain* sign. I ordered four croissants and a baguette for the next morning and madame said they would be available from 0800.

We relaxed onboard in the peaceful little moorage, enjoying glasses of a beer we had picked-up in Namur: Blanche de Namur, biere sur lie, the 2012 winner of the title "World's Best White Beer".

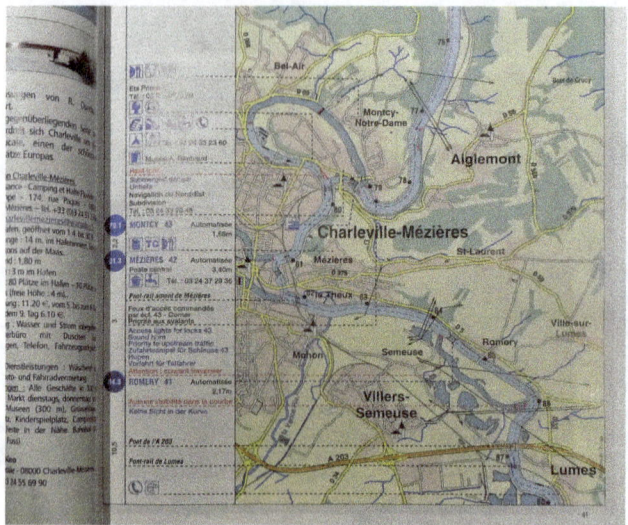

On Saturday morning it began raining just as I returned onboard with the bread and croissants. After slowly enjoying espressos and croissants in the dry of the cabin, we donned our rain gear and continued up river.

Our destination for the day was Charleville-Mézières, our first large community since Namur. Mézières was founded in the ninth century and half a century ago it amalgamated with the new town across the river, Charleville, which dates only to the seventeenth century. With la Meuse taking a very winding course through the city and offering over six kilometres of riverfront, we were looking forward to a pleasant scene. Among the places we wanted to see is Place Ducal, reputed to be one of the most beautiful squares in Europe.

Twenty-seven kilometres and four locks along from Laifour we began winding our way through industrial slums and crumbling canal-side infrastructure, which continued unrelieved by any pleasant scenery for over four kilometres. Adding to the glumness was the persistent rain. We passed through ecluse Montcy and followed signs to le Port de Plaisance, the pleasure boat port and found the mooring places filled with residential barges. We continued along to the marina, where we saw only a dozen or so boats scattered around the eighty mooring pontoons. We also saw notice that the marina was restricted to boats under 14 metres in length.

We turned around just above the weir and found a spot on a concrete abutment on Ile du Vieux Moulin below a pedestrian bridge. There were already two boats on it and the remaining space was sufficiently long to take almost two-thirds of Zonder Zorg's length, leaving the stern hanging out. As we were securing, the heavy rain turned into an intense thunderstorm. During a break in the downpour, I took a short exploratory walk on the island, turning quickly back as the skies opened again. I saw little more than dense, wet vegetation. Thus far we had seen nothing of the city except crumbling industrial areas or a solid screen of unkempt heavily vegetated riverbank, beyond which we could see no attractive buildings nor any pleasantness. While we sat comfortably below in the torrential downpour, I searched the Barge Association forum and waterways guide for information on Charleville-Mézières. I found warnings of bicycle thefts from boat decks, of loud late night drunken partying by local youth in the park next to the moorage and of a barge having its moorings undone in the middle of the night and left to drift downstream toward the weir. Because of the heavy rain we felt safe from vandals, so we spent the night. Nonetheless, the uneasiness persisted and we decided to move on first thing the next morning.

Into France

At 0841 on Sunday, 25 August we slipped and continued up la Meuse, arriving after two more kilometres of bland scenery at écluse Mézières, just as the lock system was turned on for the day at 0900. The lock has a lift of 3.4 metres, and the flood wall at the top of the chamber adds a similar height. There are no bollards in the walls of the chamber, so while I stabilised the boat, Edi took the bow line up the slimy ladder, which had been soaking in the filled chamber overnight. The rungs were all bent and twisted out of shape, likely from barges using them to secure on the way up.

She secured the line to a bollard and then tried to find a way to take the stern line. The new lock house had been built directly on the edge of the chamber and safety fencing had been placed to make it safe for tourists, but dangerous for lock users and impossible for boaters to toss a line the six or more metres up to a bollard. Lying next to the lock house was a long pipe with a hook on its end, but it was too heavy for Edi to safely use. I resorted to using the ladder rungs to secure the stern, just as their distortions attest to many having done in the past.

When Edi had activated the chamber, all the sluices opened at once, sending great torrents of water flooding in. The lock seems to have been designed by non-boaters.

We continued along the river for another fifteen kilometres and two more locks, where at 1227 we passed through ecluse Meuse and entered le Canal des Ardenes. After one more lock, at 1330 we secured on the canal bank just beyond Pont-à-Bar. There are bollards along both sides of the canal for 250 to 300 metres, but they were filled with the fender-festooned boats of a rental base on the one side and half the other side was taken-up by a boat repair company's works-in-progress. The remainder of the bollards were occupied by long-term and semi-derelict boats. We dropped the spud pole to secure the bow and pounded a steel pin into the bank to take our stern line.

It appeared that many of the boats occupying the mooring bollards beyond the rental base and the service yard were lived-aboard by employees of the two companies.

In a light drizzle on Monday, we raised the spud pole, retrieved our mooring pin and at 0810 we headed up the canal, our departure timed to have us arrive at the first lock as it opened for navigation at 0900. As we motored, Edi went below to prepare some breakfast panini, which we enjoyed with cups of espresso.

While we ate we watched the passing scene as we glided through tranquil rural villages and past pastures of grazing cattle. The drizzle had abated to light mist and there was a growing warmth from the sun as it worked at burning through the dissipating clouds.

The first lock was followed immediately by a right-angled turn into a second lock, which led into a short tunnel leading through a narrow ridge. We had left la Meuse and entered Canal des Ardennes. On its way up it follows a small river, la Bar, which wanders off eastward in a long loop of a dozen or more kilometres around the end of the ridge. While it is doing that, the locks, tunnel and canal take about half a kilometre to make the same progress as the river.

Fourteen kilometres and two more locks brought us to ecluse Sauville, the final up lock on the canal. From this point, a 9.6 kilometre *bief de partage*, connecting reach or summit pound would take us to the beginning of a steep descent down into the vallée de l'Aisne.

At 1315 we secured to the wall at the halte nautique in Le Chesne. This consists of a 40 metre indentation in the stone walls on each side of the canal next to the narrow bridge opening. There is no fee and the moorage is supplied with free electrical outlets, water points, garbage bins and recycling containers. We decided to stay a couple of days, relaxing and getting ready for the twenty-eight locks in the next leg.

Chapter Seventeen

Into the Champagne

For two days we enjoyed the free moorage in La Chesne taking advantage of the free shore power and water. Except for a series of thunderstorms rolling through on Tuesday, 27 August, the weather was generally fine. We had an artisanal bakery just across the street and a good 8 à Huite supermarket across the first bridge and along half a block.

The town is important enough to have two bridges across the canal, and on a couple of occasions we stood on the one above Zonder Zorg watching the passing commercial barges. There was a regular shuttle of barges past us, with the same half dozen coming and going every few hours, so we knew there was something going on not far along the canal.

We casually explored the small town and relaxed onboard. We often sat in the cockpit to snack or dine on the wonderful items we had found in the market.

We were gathering energy for the next leg of our journey. A kilometre along from us was the first of a series of 26 locks that would drop us in rapid succession more than 80 metres down to the Aisne. From there another two locks and 8 kilometres would take us to the village of Attigny, the next stopping place.

Into the Champagne

Wednesday, 28 August dawned clear and already warm. It looked like a splendid day to attack the series locks. At 0851 we slipped our lines and continued along the bief toward the first lock, arriving just after the system had been switched on for the day. This was our first down lock since we entered the Maas at Maasbracht in late July, and it was a joy to be able to simply drop the mooring lines over the bollards as we entered the chamber.

As we exited the lock, we saw the next one little more than a hundred metres further on. When we entered the second lock, we saw a commercial barge rising in the third. While we were being lowered, the commercial began leaving, and by the time our doors had opened, it was in the centre of the very short bief.

We slowly danced around each other and then headed into our respective locks.

These locks were built to the dimensions devised by Charles de Freycinet in 1879 when he was the French Minister of Public Works; he later served four times as Prime Minister of France. The chambers are 39 metres long, 5.2 metres wide and 2.2 metres deep to accommodate barges with a maximum size of 38.5 metres by 5.05 metres by 1.8 metres. Because the width and depth of a laden barge are within a few centimetres of the size of the lock, it takes much longer for it to pump its way into an up-bound lock or out of a down-bound one. It is like a piston in a cylinder. We had already settled into our lock and set it in motion while the commercial was still churning its way into its lock.

The fourth lock is only 600 metres from the first. We continued in rapid succession through locks 4, 5, 6, 7 and 8, admiring along the way the old lock houses. Now that the locks have been automated, there is no need for a resident lock keeper. Many of these houses have been rented out, but in remote areas, many remain vacant. The old lock house at écluse 8 appears to be in good condition, but it is still waiting for a tennant.

We arrived at écluse 9 with the normal red and green lights showing, indicating that it was being prepared for us. We backed and filled in the approaches for ten minutes with no action from the lock. There are no mooring bollards in the waiting area, so I approached the grassy bank, dropped the spud pole and took a stern line to an old telephone pole across the towpath.

I walked up to the control building, pushed the intercom button and was immediately answered. I said: *"Nous descendons à l'écluse 9, mais il est en panne"*, We are descending at Lock 9, but it is not working. I received an immediate *"J'arrive"*, I am coming.

We had been at the lock for three-quarters of an hour by the time the roving lock keeper arrived. An up-bound commercial was by this time in écluse 10, so we needed to wait for it to finish there and then be lifted in écluse 9. Meanwhile a down-bound commercial had caught-up to us and was waiting just upstream. Since we move through the locks much more quickly than the Freycinet barges, we were asked to go ahead as soon as the up-bound cleared. We entered and watched the two commercials dance past each other astern of us in the narrow bend.

Into the Champagne

We continued along uneventfully, quickly passing through the locks. Increasingly the lock houses were lived-in and well cared for. At écluse 14 we had passed the midway point down the flight.

In the chamber I had selected a set of bollards that placed our stern rather close to the upstream gates. This would normally have been no problem, but the level of the upstream water was several centimetres higher than the gates and as we descended in the chamber, an increasingly high waterfall emerged, its spray wetting our cockpit and nicely cooling me in the heat of midday.

As we arrived in Éclise 15 we saw a sign on the downstream gates advising of *dragage dans la bief*, dredging in the pound. Below us we saw a Freycinet commercial alongside a scow with a mechanical shovel on it. The shovel appeared to be just finishing loading the barge with dredges from the canal bottom.

As we left the lock, the Freycinet slipped from the scow and headed upstream and we snaked a course leaving the Freycinet to port and the scow to starboard as we aimed for the next lock chamber.

I figured that we were now beyond the source of all the commercial activity we had seen upstream and that it was unlikely that we would be seeing any more commercial traffic for a while.

Shortly after 1400 we arrived at écluse 26, the last of the chain of automated locks.

At 1421 we had locked through and entered l'Aisne. It had taken us five and a half hours to transit the twenty-six locks and nine kilometres distance from Le Chesne, including nearly an hour and a half waiting at écluse 9. The FluviaCarte says: "It takes 7 hours to go down the automatic set and the next shops are at Attigny a day's navigation away". The phrasing of this was a bit confusing to me, but it was clarified when I saw Attigny was only ten kilometres along from the bottom of the set.

After less than a kilometre we passed through a low lock taking us from the river and back into la Canal des Ardennes. We hadn't eaten since breakfast, so after we had passed through, Edi prepared a platter of open face sandwiches; fresh lox and capers on chaource slathered multi-grain baguette slices. We devoured these and dishes of olives as we followed the canal's meander through a lush forest.

At 1531 we secured alongside in the halte nautique d'Attigny. A quick visit to the supermarket two blocks away gave us the fresh ingredients for a delicious dinner. We relaxed on-board, pleased that the day had gone so easily.

On Thursday morning we slipped our lines at 0842 and continued downstream, arriving écluse Givry at 0900 as it was activated for the day. It was a slightly hazy day, which caused the sun to play wonderful light on the passing scene.

Into the Champagne

After we had passed through the lock, Edi went below and prepared a large pan of pain perdu au lardons for our breakfast. We enjoyed this with cups of espresso as we followed the winding canal along the 7.9 kilometres to the next lock.

The calmness of the day and the gentle morning sun offered wonderful scenes as we slowly made our way alone along the canal. We were surprised by, but delighted with the lack of traffic. We had the canal to ourselves, and except for the commercial activity around the dredging at Montgon, this has been the case since we left the Netherlands. We have shared only five locks and at our evenings' stops, there have rarely been more than two or three other boats, if any at all. Most of these have been Dutch flagged and heading back to the Netherlands.

Among the bushes along the banks we had spotted a cotoneaster plant filled with red berries, but we were well past it before Edi thought some branches would make a nice addition to the potted hydrangea we had in the bucket at the bow. I kept a sharp lookout for the next bush, and after a couple of kilometres, we found one to nose into.

A little further along the canal we spotted an apple tree heavily hung with reddening apples. I quickly reversed and manoeuvred the bow under the branches, and as we slowly drifted, we quickly picked all we could reach. Edi brought a boathook and we used it to bend a few more branches into reach. The apples were at a point of ripeness where they almost fell into our hands. We left with about ten kilos and continued along the canal.

The canal winds through mostly rural countryside, with a mix of pasture and forest along its banks. It is a very pleasant and peaceful route with an occasional tiny village, a distant steeple and a few impressive private estates. After four locks and 18.8 kilometres, at 1220 we secured to the wall in the halte nautique de Rethel. The moorage is clean and well organised, and it is only two hundred metres from a supermarket and the beginning of a rather large shopping district.

We walked through the town centre up the hill to église St Nicolas to admire the intricate stonework on its windows and doors, and on the way back stopped in the markets for fresh

supplies. After a leisurely breakfast the next morning, we slipped our lines at 0935 with our intended destination: Asfeld, a small village twenty-two kilometres and four locks further along.

We motored through pleasant forests with the calm water covered by dust and pollen. It gently moved aside as we glided through, leaving a beautiful pattern in our stern. The guide showed Asfeld as having a simple moorage, but when we arrived shortly after 1300, we found it rather grubby and barren, so we decided continue on. The next village that showed a mooring possibility was Guignicourt, sixteen kilometres and two locks further along. Within three kilometres we had passed through the last lock of the canal, écluse Vieux-les-Asfeld. We had come to the end of le Canal des Ardennes and entered le Canal Latéral à l'Aisne.

Edi brought up a tray of cheeses, breads, olives and sliced meats for us to nibble on as we made our way along the 6.8 kilometres to the next lock.

As we nibbled we were mesmerised by the reflections in the water as they made wonderful designs with the pilings along the canal banks. Shortly before 1600 we passed straight through Guignicourt, seeing no reason to stop there. The moorings are semi-derelict and are situated next to a sugar refinery and some huge grain silos. They offer no facilities and no charm. The town is a kilometre away and across two bridges. We continued on.

Into the Champagne

At 1602 we secured alongside the bank at the entrance to la Canal de l'Aisne à la Marne. We slipped in between a Frecinet péniche and a British cruiser with a metre or so to spare each end. There were other mooring places available, but they were all occupied by the complex sprawling of gear of a dozen or more anglers. I had long ago learned that you do not disturb the anglers; overnight vandalism to boats seems directly related.

About an hour after we arrived, Hensie, a deeply laden péniche arrived and asked the British boat to move, explaining it was in the only deep moorage available. The British boat skipper refused to move, indicating the commercial site further along, so Hensie's skipper gently rafted onto him, dropping his spud pole to secure the bow a quarter metre off.

His bow overlapped Zonder Zorg by about five metres, so he could use our uncluttered foredeck for easy access to and from the shore. We later learned that Hensie's load was for the industrial site about two hundred metres further along the canal with ample deep moorage, but the bargee said it was too dusty over there and mooring there would mean having to clean the barge.

At 0840 on Saturday morning we slipped our lines and backed out from under the bows of Hensie and around those of Medea, the péniche that was moored astern of us. A hundred metres along we came to the entrance to Canal de l'Aisne à la Marne. We could see no control device to activate the lock, which the FluviaCarte listed as automatic, nor could we see any indication there was a lock keeper.

After we had nosed up to the lock and had made it very obvious that we wanted to enter, the red light changed to red and green, indicating the lock was being prepared for us. We assumed that the lock keeper at écluse Berry, the next lock along Canal Latéral à l'Aisne only a hundred or so metres away, has controls for both locks.

The first lock house appeared to be abandoned, though it still seemed to be in reasonable condition. We were again in up-bound locks, so we changed our routine. As I slowed the barge in the chamber, I threw a stern line up the 2.5 metres to loop a bollard at the top of the wall, ran out three or four metres and secured the line to stop us. Leaving the engine in gear and pulling on the spring, I then climbed the ladder and went forward to take the bow line from Edi, who was still nursing broken ribs from a nephew's overzealous hug, and couldn't toss a line up to the bollard.

It was drizzling lightly as we worked our way up the canal and we seemed to be the only people around. Then around a bend came an unladen péniche. We waved acknowledgements as we passed. A couple of locks further on we had to wait while another down-bound commercial worked its way through the chamber.

126

We passed through a few small villages, but mostly the canal sides were agricultural land, predominantly wheat and sugar beet. From the fifth lock, we saw the lock houses had been rented out and they were very nicely maintained and decorated. The drizzle became intermittent and there was some detail coming to the bottoms of the clouds.

After nine locks and eighteen kilometres we came to the industrial outskirts of Reims. Five kilometres further along, we were still in unattractive surroundings as we approached the Relais Nautique de Vieux Port, the pleasure boat marina. Its maximum size is listed as 15 metres, too small for Zonder Zorg, so we continued past. Besides, it is very noisy and exhaust fumed with the Autoroute directly across the canal and busy stop-and-go urban traffic alongside the marina.

At 1352 we secured a kilometre further along on a stone wall with good bollards. Directly beside us was a very busy traffic lighted intersection, and as we sat there we could not relax. We decided to have lunch and then continue on up the canal.

Finally, after four more locks and another ten kilometres, we arrived in the town of Sillery, and at 1633 we secured to the wall directly in front of the Capitainerie in Relais Nautique de Sillery. The relais is a clean and rather new facility, with aluminium finger floats taking sixteen boats up to fifteen metres in length. Along the wall there is room for four or five barges of sixteen to thirty metres. We chose the place that was the closest to the wifi antenna.

Sillery is located on the western slopes of Montaigne de Reims and is one of the seventeen Grand Cru villages of the Champagne viticultural area. Our winding route from Harlingen had taken us in ten weeks through 128 locks and along 996.2 kilometres of canals, rivers and lakes to cover a straight line distance of 452 kilometres.

After we had settled in, I visited the Capitainerie to check in and pay for a week's moorage and to get maps and information on the town, its surroundings and its facilities. There is a large, modern Intermarché supermarket just along the canal from the marina, and among the things I had learned during my decades as a wine importer and wine writer, is that where there is great wine, there is great food.

We easily found wonderful ingredients for a dinner to celebrate our arrival in Champagne. We were also celebrating that it was the last day of August; in the following days, France would go back to work after their *mois de congé*, their month of holiday, when much in France shuts down, or at least seems to.

Chapter Eighteen

Enjoying Champagne

We relaxed onboard and tended to chores. We also went on walking explorations of the area. Just along from our moorage is one of the nineteen *nécroploes nationales*, French national military cemeteries, which occupy a total of thirty-five hectares in the area east of Montagne de Reims. They are the final resting places for 114,120 members of the French military during World War One. This was one of the major battlegrounds of the war from its beginning to its end. More than eight percent of the total French military dead from the Great War are buried here.

On Tuesday morning we caught the local bus into Reims. For €2.50 each we bought return tickets, which were good for an hour in each direction on all lines, including the new streetcar. Among other things, we wanted to visit Notre Dame. The current structure was begun in 1211 and it replaces a series churches on the site dating back to the late Roman era. In 496 Clovis, the first King of France was baptised in one of the former churches on the site. This led to twenty-five French kings, from Louis the Pious in 816 to Charles X in 1825 being crowned at Notre Dame de Reims.

The Cathedral is one of the finest examples of Gothic architecture, and its design uses flying buttresses to allow the huge windowed walls to soar to great heights. It was severely damaged during World War One and the restoration is ongoing. The more important pieces among the statuary in the entrance portals have been restored, including the Smiling Angel.

Most of the stained glass windows from the thirteenth century were destroyed by German artillery shelling; however some remain. The Rose Windows over the main portal are the most prominent of these. They are considered as superb examples of the stained glass art.

Also, high up in the nave and the choir some windows from the 1240s have survived.

Modern windows have gradually replaced temporary fillers. This trio by Marc Chagall, one of the great pioneers of Modernism, was completed in 1974.

The interior of the Cathedral is huge, and eight hundred years after it was built, it still ranks among the highest church naves in the world. The light strikes the interior in a seemingly endless variety of designs and the feeling inside is of a monument, rather than of a place of worship.

Outside, facing the Cathedral rides a bronze Jeanne d'Arc, the teenager who turned the tide for the French in the Hundred Year's War. She was condemned as a heretic and burned at the stake at the age of nineteen. The Church said "Oops" twenty-five years later and admitted a mistake. She was later canonised and is one of the Patron Saints of France.

A few blocks away is Eglise St Jacques, which was begun about 1190, two decades before Notre Dame. Its interior is wonderfully serene, and unlike Notre Dame, there is an overwhelming feeling of being in a spiritual place.

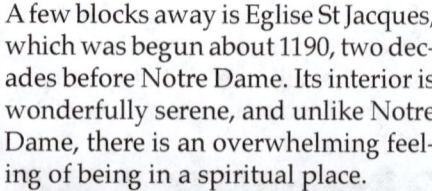

We were captivated by the peacefulness of the place.

On Wednesday morning I walked a few blocks to Case à Pain, broadly reputed as the best bakery in the region. We enjoyed a splendid breakfast of Brie de Meaux and Forme d'Ambert with fresh croissants and cups of espresso. The croissants proved to us that the high reputation of the baker is valid.

We relaxed and explored, but mostly relaxed and waited. We were waiting for a technician to arrive from the French distributor of our Perkins engine. Since our first outing from Harlingen in early June, we had experienced a problem with leaking of diesel oil and coolant from the engine as well as some oil from the transmission. We had Dutch technicians come to us from Harlingen when we were in Winsum, Workum, Lemmer, Ens, Elburg and Haarlem. Then we had them come from the distributor in Utrecht while we were in Gouda and Gorenchem. Now in Sillery we were waiting for the first French technician to arrive.

We continued to celebrate our arrival in Champagne with more Champagne. For dinner I prepared pan-seared coquilles St Jacques, basmati rice and green beans amandine and we enjoyed these with a bottle of Champagne Canard Duchesne.

As we were savouring our way through the bottle I thought of my friend and former business partner, Philip Holzberg, who in the 1980s and early 90s had done a splendid job of representing and promoting Canard Duchesne in Canada. He had gone on to own a vineyard in the Burgundy and a wine chateau in Bordeaux, before his tragic death a couple of years ago. Cheers, Philip!

Late on Friday morning Julien arrived after a two hour drive from Beauvais, and from the series of photos I had emailed, he quickly found a faulty connection on one of the fuel injector lines. He had brought spares and he had soon fixed the fuel leak for the ninth time. From stains, he tracked down the coolant leak to a faulty connection in the return line to the coolant reservoir. He had no spares to for this, but was able to cobble together a temporary remedy with a piece of hose and some clamps. After engine run-ups and testing, he cleaned a couple of litres of fluids from the bilges and took them for disposal.

He left mid-afternoon, telling us he would have the proper parts ordered and would return to install them the next week, wherever we happened to be along the canals. We needed to keep heading south; we still had a long way to go to reach our winter moorage in Carcassonne.

We still had over a thousand kilometres to go to get to Carcassonne and a little over two months to do it before scheduled work on the Canal du Midi would hamper and possibly prevent our progress. Because of timing, we decided to give Paris a miss on this trip and to head southeastward from the Champagne, taking the Canal Lateral à la Marne and the Canal entre Champagne et Bourgogne to the Saône. Not only was this shorter, but it entailed fewer locks.

However; we still had some more of the Champagne to explore before we left, so we decided that the following morning we would continue around the Montaigne de Reims and head a bit west to Épernay, the Capital of the Champagne.

On Saturday morning, 7 September I walked over to Case à Pain for some pain chocolat and croissants, which we then enjoyed in the cockpit with cups of espresso. As we relaxed over breakfast, the sky slowly filled with cloud and the forecast rain showers began. We had the patio umbrella up, so we continued our late breakfast.

At 1015 we slipped from the wall in Relais Nautique de Sillery and continued up the canal. We glided along through wheat fields with a backdrop of Champagne villages nestled among their vineyards on Montagne de Reims.

It was drizzling lightly, but this did not deter the serious fishermen, nor joggers nor even a young woman on her horse, all of whom used the old tow path beside the canal. Along our planned route for the day were three locks leading up to the *bief de partage*, the summit pound, in the middle of which is Souterrain de Mont-de-Billy, a 2302-metre-long tunnel under a rib of Montagne de Reims. From the tunnel, the waterway leads down eight locks to the end of the canal. The lock houses we saw on the way up were occupied and well-maintained.

The intermittent drizzle had turned to persistent light rain by the time we arrived at the entrance to the tunnel. Its signal light was red and we could see the interior lighting was on, indicating a boat was inside. We didn't know whether the boat was coming or going. If it was going we had a long wait; if it was coming the wait was much less.

There are no bollards along the new concrete wall in the tunnel approach, so we eased over and dropped the spud pole. Within a few minutes faint details emerged in the shadowy darkness of the tunnel mouth. The details filled and soon became the bows of a commercial péniche completing its transit.

Immediately the commercial barge had passed the sensors at the mouth of the tunnel, the traffic light switched from red to green. Then as its stern cleared the mouth, the tunnel's interior lights went out. We waited for the suck and turbulence of the passing barge to subside, and then raising the spud pole, we headed toward the entrance.

As our bows crossed through the beam of the sensors, the interior lights came on. Faintly in the distance we could see the point of light at the far end of the tunnel, 2.3 kilometres away. The towpath beside the canal continued into the tunnel.

The entire interior of the tunnel is lined with stones laid in a graceful barrel vault and the construction was completed in time for the opening of the canal in 1866. During World War One, after the Germans artillery had disabled the canal, the French used the tunnel to hide barges loaded with artillery pieces. Looking back we could see the entrance slowly getting smaller. The interior was dry, with none of the commonly found dripping from the tunnel roof.

We took advantage of the respite from the rain. Edi went below and quickly put together a tray of bread, cheeses and olives for us to enjoy during the twenty-odd minute passage through the tunnel. It was not her usual fancy spread; the table top was wet from the rain and would have soaked a tablecloth and our dry time was not anticipated to be sufficiently long to allow her to do a fancy arrangement. Lunch was wonderful, made even more so by the setting.

Zonder Zorg handled superbly through the tunnel, balanced between the effects off each wall and easily seeking the centre. A little bit of guidance with the tiller from time to time was all that was required. After twenty minutes we were at the end of the tunnel and ready to head back out into the rain.

We started down the six-kilometre-long series of eight down-bound locks and about halfway along we came to a malfunctioning one. As we approached, we had as usual turned the control rod dangling over the canal. It lit the orange flashing light at the gate and turned the signal from red to red and green. It did everything that was automated to happen except filling the lock and opening the gates for us. We secured to a bollard on the bank and I called Ch22 on the VHF. Mme. l'éclusier said: "J'arrive".

The next lock malfunctioned also. It had allowed us in, but after I had lifted the activation bar the doors would not close. Again I signalled the lock keeper and she quickly came to set things in motion again. She seemed to know exactly what to do, as if these were regular occurrences, which I suspect they were so late in the season. They are likely on the VNF list of winter maintenance.

At 1607 we finally made it down to the end of Canal de l'Aisne à la Marne, a T-junction with Canal Latéral à la Marne. We turned to starboard and headed downstream. The first lock along the canal, the one in Tours-sur-Marne malfunctioned, so I needed to call VNF again. It took nearly half an hour for a different lock keeper to arrive and lock us through, and by this time it was getting very tight for us to make it through l'écluse de Mareiul before it closed for the day at 1700. Slowing us along the way was the second movable bridge we had seen since the Netherlands, but it responded quickly to our twist of the

dangling wand and we needed to pause for only short while.

We arrived at the signalling wand for l'écluse de Mareiul at 1650 and watched as our twist was acknowledged by a flashing orange light. However, the red light went out and stayed out; we got no red and green to show the lock was being prepared for us. We later learned that, even though the system is listed as operating until 1700, any new locking-through is switched off ten minutes before that; we likely missed it by seconds.

We secured to the bollards on the quai at the approaches to the lock and shut down for the day. We were in the countryside a kilometre short of the village of Mareiul-sur-Ay and four kilometres short of the town of Ay. When I checked the engine after our day's run, I again saw diesel oil and coolant puddles in the bilge. I was not happy.

Sunday morning we arose to a world covered in heavy dew with the sun rising mists off the canal. Beside us, next to the lock house the geese were warming themselves in the first rays of the sun, and above them on the slopes the grapes began another day of ripening. With only about two weeks until harvest, they need all the sun they can get.

Edi had the diced bacon smelling wonderfully and was just about to add the eggs to the pan when Mme.l'éclusiere arrived to ask us if we wanted to lock through. I told her we were getting ready for breakfast and were thinking of going through in half an hour. She said the lock wasn't working properly and it needed help, but that she would be away for over an hour. I told Edi to turn the bacon off and hold the eggs. Within two minutes we were entering the lock.

We passed through Mareiul-sur-Ay, and not tempted by the rather Dog-patch-looking moorings, we decided to continue on to Ay. From my visit here in 2004 in Lady Jane, I remembered a long quai in the centre of town, though I did recall very shallow water alongside. The new FluviaCarte showed the quai, but did not mention of the shallowness alongside. They left it to us to discover with our 0.95 metre draft. Fortunately, Zonder Zorg's skeg is about 10 centimetres deeper than the tips of her propeller blades.

I easily poled off with a boathook, and we continued along to écluse de Ay, which we passed through without any need of assistance. This automatic lock was working, though with its very slow emptying, we suspect there is a problem with the upstream paddles. We secured alongside the wall on the bollards just below the lock. Edi reheated the bacon and put on the eggs while I pulled some espressos. We had arrived in Ay.

For the first time since Namur, Belgium we dug the bicycles out of the forepeak. We needed fresh produce and meat or fish for dinner, so we crossed the canal bridge and continued along about three hundred metres to the Leclerc supermarket, which I recalled from our previous visit here a decade ago. The store was still there, but it is closed on Sundays. We turned around and headed the three kilometres into Magenta, across the Marne from Epernay, where we found the Leader Price open. It is located in a cheap neighbourhood and has very low quality produce and meats to suit its main clientele. Nonetheless, we did manage to put together a small basket to see us through until the better markets open on Monday.

After lunch we walked into Ay to explore. At the entrance to the town is the wine-country-mandatory decorated wine barrel. Ay is one of the best of the Grand Cru villages of the Champagne and its specialty is growing the two red grapes of the region: Pinot Noir and Pinot Meunier.

Ay is home to over forty Champagne producers. Signposts throughout the town point directions to those nearby. It is a small town, so most of the signposts show half to three-quarters of the producers.

The town is long and narrow, ranging along a strip between the canal and the base of the slopes of Montagne de Reims. We walked the three blocks through to the beginnings of the vineyards. The grapes were in good tight clusters and most had taken on the deep blue-purple colour of maturity.

Some bunches in less sunny exposures were still in the process of changing colour. Still others had not yet begun ripening. It has been a rather patchy growing year, with frost, rain and hail disrupting the flowering, fruit set and the ripening.

This is not unusual for the Champagne, a region that lies at the northern extreme of the viticulture zone. The vines are tough and resilient and over my four and a half decades of visiting the region, I have seen that the growers and producers are even tougher and more resilient.

We walked along, passing the bottoms of the corduroy slopes of Montaigne de Reims and back in among the buildings that abut the vineyards.

We passed a wonderful series of open gates giving glimpses into the domains and lives of *les viticulteurs*, the grape growers. During my decades in the wine trade, I consistently saw that the art and science of grape growing are the most important aspects in the production of fine wine. Wine of fine quality cannot be made without fine quality grapes. The winemaker can do nothing to improve the quality that comes from the grower; rather, the winemaker's task is to prevent as much as possible any lessening of the quality, shepherding the natural processes as the grapes become wine.

We walked along streets that seemed completely lined with Champagne producers, some large, some small, some famous, others near unknown outside their own small circles. It was Sunday and everything was closed. Edi and I seemed to be the only activity in town.

We headed back down into the centre of town, past the church as darkening clouds threatened more rain. As we scurried to beat the rain back to Zonder Zorg, we hurried through an older section with many half-timbered buildings from a few centuries back. We promised ourselves to take more time in the following few days to further explore Ay and its surroundings.

Back onboard, the rain passed and we relaxed. Within a week we had seen the temperatures plunge from a daytime high of 38° when we were in Sillery to 19° this week. Autumn was closing in on us and the tree above us sent a continuous rain of reminders onto our decks.

We had a private visit scheduled with Champagne de Venoge in Épernay on Wednesday, and our thoughts had been to head down the canal four kilometres, pass through the lock at Dizy and then head a little over five kilometres up the branch of the Marne to Port de Plaisance d'Épernay. This would entail about two and a half hours of motoring. Moorage in Épernay was listed as €2 per metre per day, so €34 a day on top of the €14 for the ten litres of diesel consumed getting there and back to our free moorage in Ay. Edi suggested that for the €48 we stay in Ay, take taxis to and from Épernay and buy a bottle of Champagne with the change.

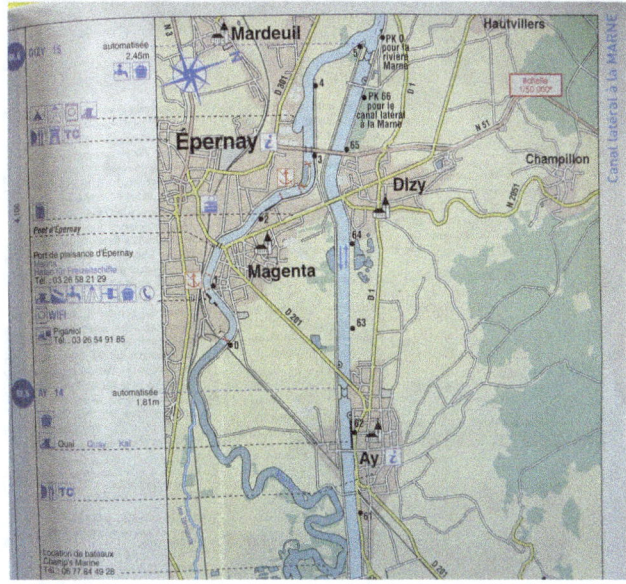

With the trees above us accelerating their defoliation, we decided to move Zonder Zorg back up through the lock to the quai above it. It has no overhanging trees. We watched as a Freycinet gauge peniche filled the lock. These barges are within a few centimetres of the same dimensions as those of the chamber, and we watched as the skipper put the twin rudders hard over so that the lock doors astern could swing past them and close.

When the lock was clear we slipped our lines and I headed down canal about 200 metres to a point where the canal is about 20 metres wide, sufficient for me to wind Zonder Zorg around and in the process, twist the lock signalling wand. Meanwhile, Edi had walked up to the lock to receive the lines when I entered. In short order we had made it through the lock and were secured in a much less leafy place. We swept away the thick layer of leaves and washed the boat, then settled back into relaxing and enjoying Aÿ.

On Wednesday morning I phoned to reserve a taxi for 1440 to take us from Zonder Zorg to Epernay. This timing was calculated to get us there about five minutes before our 1500 appointment with Champagne de Venoge. The taxi arrived at the barge twenty minutes early, so we arrived at the doors of de Venoge nearly half an hour ahead of schedule. This left us with too little time to take a meaningful walkabout. Besides, it was threatening to rain again, so I rang the bell.

Our hostess and guide, Emma Dawe-Coz found no problem with our earliness. She is a lively young woman, born in French Guyana of a British mother and French father, and is nearing completion of her wine degree in Reims. As a part of these studies is serving an internship in the export marketing offices of a Champagne house.

As we walked through the tank rooms, she explained the background of the house of de Venoge. It was founded by Henri-Marc de Venoge, a Swiss who had earlier travelled to Italy and discovered wine. In 1825 he settled in the Champagne in Mareuil-sur-Ay, where he set-up in the wine trade. In 1837 his business became Champagne de Venoge. His wines very quickly rose to prominence and he played a dominant role in the rapid expansion of the Champagne business. He experimented with disgorgement and the refinement of other steps in the Champagne process. In 1864 de Venoge launched Cordon Blue, a symbol of nobility since the sixteenth century, and which has since become the synonym of refinement and of the art of living. De Venoge is credited as the originator of the illustrated label, a departure from standard at the time; black lettering on white paper.

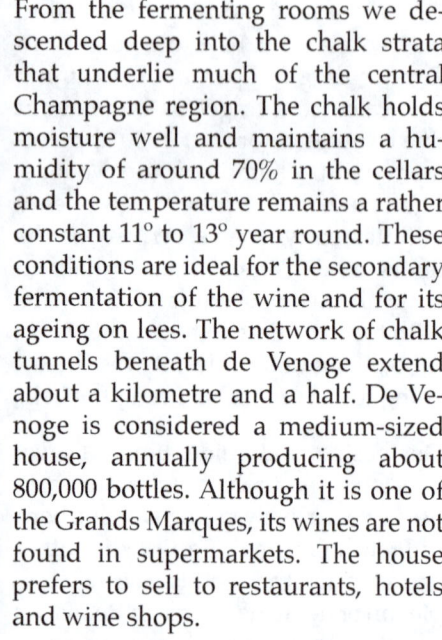

From the fermenting rooms we descended deep into the chalk strata that underlie much of the central Champagne region. The chalk holds moisture well and maintains a humidity of around 70% in the cellars and the temperature remains a rather constant 11° to 13° year round. These conditions are ideal for the secondary fermentation of the wine and for its ageing on lees. The network of chalk tunnels beneath de Venoge extend about a kilometre and a half. De Venoge is considered a medium-sized house, annually producing about 800,000 bottles. Although it is one of the Grands Marques, its wines are not found in supermarkets. The house prefers to sell to restaurants, hotels and wine shops.

We walked past lines of *pupitres*, the riddling racks that are used to move the sediment from the side of a bottles to its neck. The sediment is mainly dead yeast cells from the second fermentation, the fermentation in the bottle that gives Champagne its bubbles. At de Venoge, hand riddling is still done on its cuvées prestiges and its vintage wines.

We then came to the reserves of older vintages. There are variously sized stacks of the what appeared to me to be the whole gamut of the house's vintage-dated brands: Louis XV, Grand Vin des Princes, Blanc de Blanc, Blanc de Noirs, Brut Rose and Cordon Blue, and these were in varying size formats.

I found a bin of 1961, among my favourite vintages. The scrawled label on the stack showed that only forty-five bottles remained.

A bit further along was a barred gate, much like the image of a prison door. Emma told us the room beyond held the older vintages, but it was too dark inside to see anything but forms. I thrust my camera through at full arm's reach and shot some random flash photos. One of the photos showed some stacks around the corner to the right and across on the opposite wall. The labels that are readable show bottles from 1921, 1911 and 1893, two of the vintages younger than our skûtsje Zonder Zorg and one older.

We continued our circuit of the caves, slowly making our way back to the staircase and then back up them to the bottling, disgorging and labelling lines. We continued to the tasting room.

While we looked at the displays of current and past packaging, Emma prepared a tasting for us. She offered us a range of their Champagnes, beginning with the Cordon Blue Extra Brut and progressing through to the Rosé Brut Réserve. Our favourites were the Blanc de Noirs Marquis and the Rosé Brut Réserve. We particularly liked the Rosé, with its hints of strawberries on the nose, delicate cranberry-strawberry palate and long, elegant fruity finish. This is finest non-vintage rosé that I can recall tasting in my many dozens of cellar visits in the Champagne since my first visit here in 1967. At about €35, it is definitely on our buy list.

We emerged from De Venoge after a splendid two hours and walked into the centre of Épernay to continue our search for proper Champagne glasses. Aboard our sailboat, Sequitur we had used some of our hand-blown hollow-stem crystal from home, and thinking we would quickly find similar ones in Europe, we hadn't brought any with us to Zonder Zorg. Now in the capitol of Champagne, we thought that we would surely find some. Before we left De Venoge, we had asked where we would best look and we visited the all shops recommended, and finding no stems even close, we asked for further recommendations. They yielded nothing better than the supermarket stems we had bought in the Netherlands. Empty handed, we continued on to the huge Carrefour supermarket, where we filled our huge bags to overflowing and took a taxi back to Zonder Zorg.

On Friday the thirteenth of September, my sixty-ninth birthday, we were interrupted mid-breakfast by the arrival of Julien, the technician from the Perkins distributor. The parts he had ordered had arrived and he did the two hour drive from Beauvais in a little over ninety minutes to come and install them and to thoroughly clean the mess in the bilges, which had come from the various leaks.

After he had replaced the jerry rigged coolant return line with the proper part, Julien tracked-down the new coolant leak that I had reported after our trip from Sillery to Ay. The coolant reservoir overflow had simply spilled over the engine and into the bilges. He added a line to the overflow spigot and led it to a plastic bottle he strapped to the raw water inlet at the bottom of the bilge. This is not as elegant as an expansion tank, but at least now any overflow will have a better place to go.

Enjoying Champagne

After Julien had finished the installations and cleaning of the bilges, he did a long engine run, both free turning and under load at low, medium and high revolutions to see if there were any more problems.

Finally, with the repairs, cleanup, testing and warranty servicing paperwork done, Julien headed back toward Beauvais and we began retracing our route back up the Canal Latéral à la Marne. As we again passed the halte nautique of Mareiul-sur-Ay, it looked even more unattractive than it had the previous week on our way down.

Of more interest to us were the vineyards planted to the brink of the chalk cliffs above the town of Mareiul. In years past I had marvelled at vine roots poking out of the cliff face six and more metres below the vineyards. I had several times used this exact location to bring my small groups to discuss Champagne viticulture as I conducted wine tours through France.

After two locks, we came again to Condé-sur-Marne, where instead of heading back up the Canal Aisne à la Marne, we continued straight along Canal Latéral à la Marne. We had come to the end of the Champagne viticulture area; the land around us was now flat and the vineyards had been replaced with fields of wheat and sugar beets. Though we were still in the Champagne-Ardennes Region, we had left what oenophiles think of as The Champagne.

The lock controls along this section of the canal continue with the dangling wand. By this time Edi was an expert at grasping and twisting the wand as we passed by.

From four kilometres back we had seen a large white object at the entrance to the lock, and as we approached, it had become clear that it was a very poorly designed overflow sluice. It was directing a large volume of foaming water across the canal just before the entrance to the lock chamber. As we neared, I tried to assess the strength of the crosscurrent so that I could compensate. Previous experience told me that as our bow reached the narrow current, it would swing to port, but to not correct; the stern would be almost equally swung as it met the current.

The trick is to correctly estimate the amount the barge would be offset and to then compensate by steering that amount off centre on the approach. At this lock, with the sluice so close to the entrance and with its stream so strong, there was little margin for error. To approach too slowly would place us too long in the current; to approach too quickly would risk serious collision damage to the barge if the estimate of the current's effects were wrong.

My estimates were close, and with some heavy engine, tiller and bow-thruster work, we ran through the current and its resultant eddies and countercurrents and made it into the chamber with only a few fender bounces. Once inside we saw that the lock had another trick to play on us. The mooring bollards are set in small concrete bays, with only about fifteen centimetres space to allow a tossed line to fall into place. Complicating this, the safety geeks had placed railings to protect the public and to further hamper the boater in looping the bollards. To make the situation even less comfortable, it began to rain again.

After two more locks and a total of 35.8 kilometres, at 1630 we backed onto a five metre finger pontoon in the halte publique in Châlons-en-Champagne and dropped the spud pole to secure the bow. When I did my post-run engine room check, I was disappointed to see a small puddle of fluid in the bottom of the bilge. Hoping that it was merely residue from the previous leaks that had finally made its way down from puddles in the engine's nooks and crevices, I decided to let it be for the moment, and monitor it in the coming days.

Enjoying Champagne

To cheer things up, in the evening I prepared seared St-Jacques and served them with mushrooms sautéed in butter and Armagnac over basmati rice with haricots verts fins amandine and roma tomatoes. We celebrated my sixty-ninth birthday with a bottle of Champagne Canard Duchêne and bid a formal farewell to The Champagne.

Chapter Nineteen

From the Champagne to the Burgundy

It was raining on Saturday morning as we raised the spud pole, slipped our lines and continued up canal. It continued to rain, varying from heavy drizzle to heavy downpour as we worked our way up eight locks and along 34.1 kilometres of canal to Vitry-le-François.

For the first while we passed rather bland canalside scenery, much of it grain elevators; this area is the breadbasket of France, with huge fields of grains, mostly wheat. Along the way there were remnants and ruins of industrial efforts from the early nineteenth century as France emerged from the turmoil of the Revolution and the Napoleonic era.

The rain had eased to a fine drizzle at 1600 when we reached the end of the Canal Latéral à la Marne. At the T-junction we turned right and started along Canal entre Champagne et Bourgogne. As we motored, we saw the banks on both sides of the canal lined with commercial péniches, with not a single space available. Within a kilometre we arrived at the Port de Plaisance de Vitry le François, where fortunately, one of only two spots in the port capable of taking Zonder Zorg was empty, so we quickly manoeuvred to fill it.

From the Champagne to the Burgundy

We wore around, backed into the space and secured alongside at 1620. According to the sign on the gate, the Capitainerie closes at 1600 and is closed on Sundays, so we had no way to pay for our overnight moorage unless we stayed until Monday. Memories of the town from a previous visit in 2003 in Lady Jane, plus the fact that everything would be closed on Sunday, quickly ruled out a two day stop.

In a steady light rain on Sunday morning, 15 September we slipped our lines from the bollards and turned to starboard to continue up Canal entre Champagne et Bourgogne. Ahead of us was a climb of 239 metres with 71 locks in 155 kilometres. This would take us to the summit pound and a 4.8 kilometre tunnel through the top of the pass, before starting our descent of 156 metres over 43 locks and 64 kilometres to the Saône at Heuilley. We were hoping that, once we reached the summit and started down into the Burgundy, the near-steady rain of the past week would have stopped.

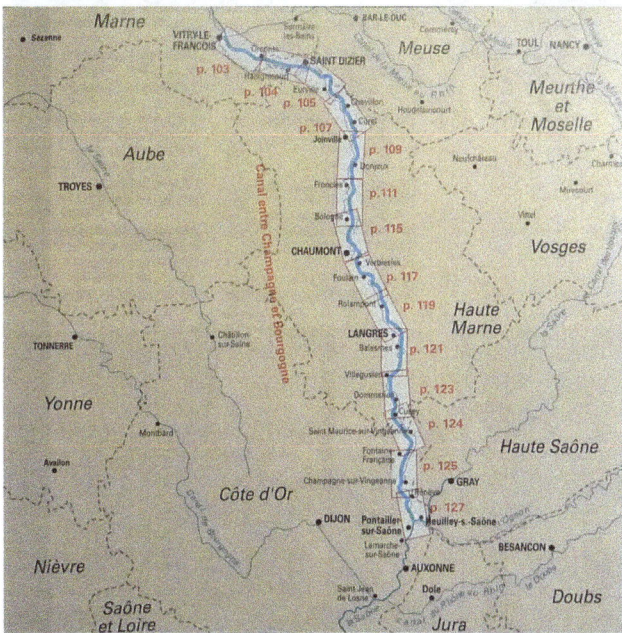

Within a few hundred metres we were out of the rain under the broad railway bridge, waiting for the first of 71 up locks to be prepared for us. After we had entered, Mme l'éclusiere took our lines with a hooked pole and placed them over bollards at the top of the wall 3.5 metres above us. When we had risen to the top, she gave us a *telecommand*, a remote control device to activate 51 of the locks and half a dozen of the lift bridges along our way to the summit. She also gave us brochures explaining the automated locks and the operation of the nineteen manual locks and seven manual lift and swing bridges on the way up, as well as details on the 43 locks down the other side.

We headed out of the lock as the skies began to clear for the first time in many days. Our remote worked as advertised at the next lock and we managed to loop the bollards at the top of the lock wall. The first dozen and a half locks have a lift of a few centimetres either side of 3.15 metres, so with Zonder Zorg's deck half a metre above water, the tops of the bollards are not much more than a metre above our heads, making it rather easy to toss a loop of line around them.

The locks in this section of the canal fill right to the brim, lowering the height of the line toss, but making fenders useless in protecting the hull from the masonry of the lock's rim. It is not so bad going up; the loss of fender protection is only for the last few minutes while waiting for the gates to open and then motoring out of the lock. I remember these locks from our trip down them in 2003. Entering them required precise manoeuvring, and with a lightly gusting wind, it became nearly impossible to enter without some scrapes. Lady Jane received a few paint blemishes at waterline during that descent.

Our remote control functioned well until we came to a lock without the "Click Here" sign three hundred or so metres before it. I had noticed that directional signs along the towpath had been turned around, apparently as a juvenile prank and the missing transponder post was likely another. I let Edi off on the bank and she went walking back along the towpath, clicking as she went. The pranksters must have pulled out the signpost and its transceiver and threw it into the bushes; however, Edi's persistent clicking of the remote finally elicited a response from the signal lights at the lock. They turned from Red to Red and White. It appeared that the pranksters had also smashed the green lens.

A few locks further along, our twelfth of the day, the remote refused to activate the lock as we repeatedly clicked at the sign on our way by. I dropped Edi off at a mooring dolphin and she

From the Champagne to the Burgundy

went back along the towpath again, but even standing beside the sign, she could get no response from the lock's signal lights. I called the VNF on the cell phone and reported the situation, adding that I suspected the batteries were dead in our remote control. We secured to the dolphin and waited. In about twenty minutes a VNF van arrived with a fresh remote control, and it easily triggered the lock to cycle for us.

It was 1730 by the time we had made it through the lock and into the town of St-Dizier and we moved quickly along the 1.8 kilometre bief to make it through écluse St-Dizier before it closed for the day. The lock functioned flawlessly and were quickly locked through. Our guide shows a mooring quai immediately beyond the lock, and we had been planning on using it for the night; however, the neighbourhood looked seedy to us and there were scattered groups of young men lolling about and appearing to be looking for mischief. Knowing that this is an economically depressed area, our long experience and our guts told us rather forcefully to continue along the canal.

After three kilometres we arrived at Pont Levis de Marnaval and glided through the narrow gut. This is actually two bridges in series, the first a swing bridge, which is normally open and it is immediately followed by a bascule lift bridge. Beyond these bridges is écluse Marnaval, on which the signal lights were out as we arrived, indicating it was closed for the day.

We secured to the spud pole and a pin pounded into the bank in a serene rural setting just downstream of the lock. On Monday morning Edi walked our gangway to the bank and started back down the canal toward the signal sign. She cheered as she was immediately successful at triggering the lock.

Carefree Through 1001 French Locks

We were running low on both water and diesel oil and needed to replenish. On our FluviaCarte I had highlighted the places that showed these were available. The next water shown was in écluse Chamouilley, three locks and five kilometres further along. An hour and a half later when we arrived at Chamouilley, there was a VNF éclusier there to take our lines and explain to us that the automatic mechanism was not working correctly and that he would manually bypass the system to lock us through. I asked about water, and he said he had never seen any available at this lock. We both then searched and proved the FluviaCarte wrong.

He told us there was water at the lift bridge house at Bayard, three locks further along, and that he would be accompanying us along the way to assist with two lift bridges and a lock on which the automatic systems were not fully functioning.

While we motored in the steady rain, l'éclusier rode the towpath on his Moped, passing us to open the bridge or work the recalcitrant lock mechanisms.

At 1315 we secured to spud pole and a pin pounded into the bank, just downstream from the manual lift bridge in Bayard and its bridge keeper's house. Once we had secured, l'éclusier led me to the water he had promised. It was a simple faucet in the utility sink in the room he used as his waiting place while tending his series of locks and bridges. It had no threads and I thought of jury-rigging a cuff adaptor, but we weren't that needy of water. He suggested we simply fill-up some bottles; I told him we needed close to a thousand litres.

From the Champagne to the Burgundy

The good news was that there is a gas station in the supermarket lot just 300 metres from Zonder Zorg's mooring. The gauge level had been down below the sight glass, and I estimated about 30 litres remaining. It was below the level of the intake for the hot water and central heating system. Hot water was not a problem, since the engine heated the tank when we motored and gave more than a day's supply; however with the persistent rain and cool weather, interior heating would be welcomed. I made four trips with our two jerry cans on the trolley and added 160 litres to the tanks.

At 0955 on Tuesday we raised our spud pole and retrieved the mooring pin from the bank. We had told l'éclusier that we intended continuing up canal at 1000, and he had arrived a few minutes early to push the buttons to halt traffic and lift the bridge for us. He then scooted up the towpath and hand cranked the next lift bridge for our arrival.

Then he continued along to the next lock to take our lines as we entered and to override the controls on the malfunctioning automated system.

He scooted on past us after we had left the lock and opened the manual lift bridge half a kilometre further along. He bade us farewell as we passed through the bridge, telling us that the all the locks and bridges from écluse 51 through 23 were automatic and, as far as he knew, properly functioning.

A few hundred metres further along as we entered Joinville, we passed a quai overflowing with boats, including two British barges. One was a replica luxemotor, the other a replica narrowboat. We continued on past about a quarter kilometre to an empty place on a public quai, where we had been told there was free water. At 1355 we stopped to fill our tanks, which took over an hour. By the time we were watered, it was past mid-afternoon, and since we had already done 15.9 kilometres, 7 locks and 6 lift bridges, we decided to pause for the night.

Not long after we had stopped for water, the two couples from the British boats walked along and saw the free quai and quickly moved their boats along to moor astern of us. They had been on a quai with charges for mooring and for water. With our tanks full of water and more readily available, I flashed-up the generator and we ran three loads of laundry through the washer and dryer.

At 1020 on Wednesday we slipped our lines and continued up canal. The two British boats had departed before us, so unless there were down-bound boats, all the locks would be filled when we arrived and we would need to wait for them to cycle down. At the next few locks, we again saw the marvels of incompetent lock design, with overflow sluices dumping a gushing crosscurrent just short of the lock entrances. Also here we saw locks overflowing their rims when filled. We are amazed that nearly a century and a half after these locks were designed and built that these flaws still remain.

The chambers of the locks are deeper along this section, and instead of the 2.85 to 3.15 heights we had been seeing, the walls were now 3.4 to 3.85 metres high. This made it rather difficult to throw a line up from our low decks to ring an unseen bollard whose position is marked with a faded paint stripe at the top of the wall. At the next of these higher locks, while we waited for it to prepare for us, I nosed into the bank and Edi hopped ashore with a boathook and walked to the lock to take the lines.

By the time we reached écluse 41 it had begun raining again, making it our twelfth consecutive day with rain. The guide showed a water faucet at the lock, but we could see no sign of any. Immediately upon heading upstream from this lock, the canal crosses a bridge over the Marne and then immediately passes a lift bridge. Over much of its length to the summit of the pass near the source of the Marne, the canal follows the general course of the river using its waters to fill the locks and pounds.

We arrived in Donjeux relieved to see that the two British boats had decided to stop. This meant that our progress would be quicker with the locks empty from the passage of a downbound commercial barge that we had just met. We continued up canal.

About half of the lock houses along this stretch of canal appeared to be occupied. Some were completely plain, with no apparent effort to dress-up the setting. Others were well decorated and the one at écluse 38 was a screaming example of kitsch run wild. There was a small replica of Michelangelo's David mixed in with gaudy gnomes, storks, pink flamingos, some Bambi characters, Grecian urns, squirrels, frogs, snails and so on.

At least two of the gnomes were a tad risqué. We wondered at the taste of the tenant and pondered her sanity as we watched her watching our reactions from the upper window.

In the early afternoon we paused before a lock to wait for a barge to cycle down, which meant the following locks would be in our favour as we approached them.

Then at 1540 we secured alongside in the halte nautique of Froncles, which is part motor-camper parking and

part boat moorage. We had come another 23.6 kilometres and had worked through nine more locks. A short while after our arrival, the captain came by and collected our €4 fee for moorage. Since we had paid for it, we plugged into the shore power and topped off our water tanks.

During a lull in the rain in the late afternoon I walked through the small town in search of the reported supermarket. The town had certainly seen better days, but it appeared that the metalworks, Les Forges had been on hard times for a long while. There had been an attempt to bring some commerce to the old area by the chateau, but most of the spaces look to have been vacant for a long time. I found the supermarket just as the skies opened for another downpour, so I took my time shopping, giving the rain plenty of time to abate.

At 0910 on Thursday we slipped and continued up canal, me in Zonder Zorg and Edi on bicycle. We had decided that it would be easier to handle the lines from the top, and instead of dropping her off before each lock, it seemed easier for her to pedal between the locks, which are spaced here at an average of about two kilometres apart. The speed limit in the canal is eight kilometres per hour, so she had no problem arriving at the lock in advance of Zonder Zorg.

Along the way we passed one of the few fishermen we had thus far seen in France that was not enraged that we had disturbed his fishing by having the audacity to use a boat on the canal. To us it seems that many of the canalside fishermen think that the canals had been built for fishing, not boating, and that boaters have no right to disturb their lines. This wooden man was an exception.

The locks were in our favour, and we moved along smoothly, with Edi using a large carabiner on the end of a line to haul up the mooring lines and drop them over the bollards.

From the Champagne to the Burgundy

Our eleventh lock of the day, écluse 26 at Condes was a complex one. Almost immediately out of the lock is a tunnel. Then directly from the tunnel the canal crosses on a viaduct over the Marne and at the end of the viaduct there is a lift bridge.

At 1406, having done 25.8 kilometres, twelve locks, three lift bridges and one tunnel, we secured alongside at Port de la Maladière in Chaumont. Shortly after having arrived, we pedalled across the bridge and back down the canal a kilometre to the huge new Leclerc supermarket to restock our supplies.

In the late afternoon we were visited by the VNF area manager, who asked whether we were leaving the next day, and if so, when and how far we wanted to go. The two locks after Chaumont are automatic, but from there the following nineteen are manual and we needed to be accompanied by an éclusier. We told him we would be at the first manual lock for its opening at 0900.

Shortly after 0800 on Friday we slipped and followed a tiny Belgique boat with four huge people aboard. The boat was no more than eight metres long and barely more than two metres of beam and it listed almost dangerously when any of the crew moved to one side or the other.

We followed the little boat into the first lock, just a few hundred metres along from our mooring place. There we were amused with their coordinated balancing act as they kept the boat more-or-less upright while getting their lines to the lock bollards. We learned that the two couples had been boating together as friends for a dozen years and we tried to imagine the boat's cramped interior arrangements to accommodate their four very ample bodies.

The second automatic lock followed within a kilometre and a half, after which there was a three kilometre pound to the first manual lock. During the motor between the first two locks, Edi had laid-out everything for breakfast, and then as we left the second lock, she began cooking. Very shortly she emerged into the cockpit with my plate of pain perdu à jambon garnished with sliced tomatoes and avocadoes. A minute later she brought up a hot cup of espresso.

We finished our breakfasts shortly before we arrived at the third lock, where we were met by a tiny elderly woman. She hand cranked the gates closed behind us, closed their paddles, went to the upstream gates and cranked open the paddles there to flood the lock.

As Zonder Zorg's foredeck reached the lock rim, we had offloaded Edi's bicycle. Once we had reached the top of the lock, I cranked open one of the upstream gates as madame l'éclusiere did the other, then after I had motored out, I watched as she closed the gates and dropped their paddles. She soon drove past on the towpath toward the next lock.

From the Champagne to the Burgundy

Edi was in hot pursuit on her bicycle along the towpath to take our lines as we arrived in the locks. This process worked well for us, for the two couples in the Belgian boat, and because Edi helped crank the gates, it also helped madame l'éclusiere. Besides the handwork on the locks, there was a vertical lift bridge that needed cranking up. Here Edi could do nothing but watch as the bridge opened and we passed under.

At our first lock after noon, madame l'éclusiere was replaced by a young man, who continued the process she had commenced. Thus we worked our way up the canal for 29.6 kilometres and through fifteen locks and one lift bridge, until at 1440 we came to our spud pole and a pin pounded into the bank in Rolapmont. The rain had stopped a little before we arrived and the skies had begun clearing, promising better weather than we had seen in two weeks. This was our thirteenth consecutive day with rain.

Moored astern of us was the 13.5 metre Lemsteraak, Nettie. She had been built in 1911 not far from where Zonder Zorg had been built three years earlier and she still had her traditional sailing rig. Aboard was a young Dutch couple, at least younger than us. They were on their way back to Amsterdam, which they had left in April to navigate up the Rhine and the Main, then across the canal to the Danube and down it to the Black Sea. They then sailed to Istanbul, through the Bosporus and the Greek Islands, around the Agean coast and around the Mediterranean coasts of Italy and France. From there they motored up the Rhône and the Saône and over the Plateau de Langres to here.

On Saturday morning the sky was the clearest we had seen it in two weeks. We bade farewell and success with the remainder of their voyage as Nettie continued down canal and we continued up.

We had organised with l'éclusier to have the next lock ready for us at 0900. There were five more manual locks and a manual swing bridge that needed to be worked for us. Following us into the locks was a French couple from St-Jean-de-Losne. He is an electrician for one of the boatyards there, and because his expertise, he is in too much demand during August when nearly everyone in France takes their holidays, he and his wife take their major annual boating trip in September.

By 1100 we had smoothly made our way up the first four manual locks and then through the swing bridge at Jourquenay and were in the final manual lock. As we were locking up, I asked l'éclusier what the traffic was like at the Balesmes Tunnel, and he phoned the control point to find out. There was a péniche chargée on its way up from the other side that we would need to wait for, likely for up to two hours.

We thanked l'éclusier for his work and we headed off to do the final three locks to the summit pound. These are all automatic, they all worked properly and shortly past noon we were at écluse 1, the seventy-first and final up lock of the canal. In seven days we had come 152 kilometres and up 239 metres through 71 locks 15 movable bridges and a tunnel. From here there is a 10.1 kilometre summit pound with two tunnels, the first being only about 150 metres long. The second tunnel is 4820 metres long, nearly five kilometres, and it passes nearly beneath the source of the River Marne.

As we motored along the bief the two kilometres to the first tunnel, we passed below the ramparts of the citadel of Langres 130 metres above us. The ramparts are nearly three kilometres long as they encircle the ancient city that dates at least to the Roman era. As we motored, the clouds thickened and it looked like we would have our fourteenth consecutive day with rain.

At 1241 we arrived at the traffic control displays at the entrance to the Balesmes Tunnel. They were very confusing; the gate was up, the panel was alternating between *Tunnel Open 243 Go, Tunnel Ouvert 243 Passez* and *Tunnel Offen 243 Durchfahrt*. However; the light was red. I radioed the control centre for clarification, and was told to ignore the open gate and the multilingual signs telling us to go. They are wrong and we must wait for the green light.

With a port stern line, I looped a bollard on a duc d'albe as we slowly passed, then swung in toward bank on it out of the channel, stopped and dropped the spud pole and tightened in the stern line. Our wait extended beyond the two hour estimate, then grew to two and a half hours. At two and three-quarter hours we were still waiting, then the light went green.

We had a green light, an open gate and a panel telling us in three languages that the tunnel was open and to proceed. We waited and wondered about communications and safety.

As our wait neared three hours, the faint blur we had seen deep in the tunnel had grown into a barge. We continued to wait as it painfully slowly pumped its way out of the tunnel and along the narrow cut. It was a deeply laden Freycinet gauge barge that was making about 1.5 kilometres per hour. I kept the spud pole down and the line attached to the dolphin as the barge approached and passed us. Even at its extremely slow speed, its propeller was churning a huge suction to overcome the thin slice of water beneath its deeply immersed flat bottom.

Finally, at 1546 we raised the spud pole, slipped our line and started into the narrow channel leading to the tunnels. The first one was quick and easy, almost like training wheels for the real thing.

As we entered Balesmes at 1600, we saw that the towpath had been blocked-off with a locked door and hoardings to keep out anyone but the authorised. We were relieved to see a series of green lights running down the length of the tunnel as far as we could see. Also, we could see no navigation lights of boats coming toward us down the tunnel.

Faintly in the distance we could see the light at the end of the tunnel, nearly five kilometres ahead. The speed limit in the tunnel is 4 kilometres per hour, but Zonder Zorg does not naturally move that slowly. At idle she moves along at somewhere around 5.5 kilometres per hour and I decided to avoid the hassle of shifting in and out of gear to stay at 4, and I simply ran at idle.

From the Champagne to the Burgundy

The oval of light at the end of the tunnel gradually grew as I watched the distance markers along the tunnel wall. Every hundred metres there is a plaque affixed to the wall giving the distance from the southern entrance in decimetres, so watching them was a welcome distraction from tedium of steering a course centred between the walls. It also helped prevent auto hypnosis.

The tunnel is 8 metres wide and running alongside the channel is a towpath of 1.5 metres. This gives a navigable width of 6.5 metres, so we had a tad above a metre each side for wiggle room. After 45 minutes the oval of light had morphed into an ellipse, the bottom half being the reflection in the water of the upper half.

At 1651 we emerged into daylight again. We had done the transit in 51 minutes, averaging 5.67 kilometres per hour, about 40% over the speed limit. We were relieved to see no flashing blue lights of traffic police as we motored out.

At the end of the summit pound, within two kilometres of the tunnel, the route begins a steep descent, with eight locks of 5.13 to 5.28 metres height in about three kilometres. It is an automatic series, with each lock triggering the next. I calculated that we had enough time remaining before lock closure to run the series, so we set off. At 1840 we were in écluse Percey, the eighth and last of the series, and from there we motored the two kilometre bief to Villegusien-le-Lac. At 1855 we secured to bollards on the bank just short of the next lock. It had been a long day, with 17 locks, 2 tunnels and 1 lift bridge.

Sunday dawned totally clear, and after a leisurely breakfast, at 0956 we slipped from the bollards and continued down the canal and through écluse Villegusien, lock 9 of the 43 that would take us down to the Saône.

The day remained clear and warm as we enjoyed the play of light through the trees and the reflections off the water. Shortly after noon we came to écluse Choilley, number 16 in the series to find it refusing to open for us. I called the VNF office on the radio and received the familiar response: *"Quelquen arrive"*, someone will arrive. Someone did eventually arrive, and after a 37 minute wait, we finally were allowed into the lock.

We continued without problem from there, working our way down through another fourteen locks to écluse Pouilly, lock 28 of the 43 to the Saône. At lock 24 we had passed signs indicating we had entered the Burgundy. As we left the Pouilly lock, its traffic lights went out, indicating the system was closed for the day.

We continued along the 2.7 kilometre bief to just above écluse St-Seine, lock 29 where at 1829 we secured the stern to a bollard on a tiny concrete jetty and dropped our spud pole to hold the bow off the rubble bank. We had come another 30.6 kilometres and 21 locks.

At 0958 on Monday morning I raised the spud pole, slipped our line and headed toward the St-Seine lock. It looked to be another fine day, already with a dome of blue having burnt through the top of the morning mists. The automatic eye caught our approach and the lock functioned flawlessly. We continued down the flight of locks, ten of them in eleven kilometres without problem, then at the eleventh lock of the day, the automatic system refused to function.

After a pause to wait for the arrival of the itinerant éclusier from VNF, we continued along. Two more locks malfunctioned and we got to see more of the roving éclusier. In the early afternoon we passed under the impressive stone viaduct at Oisilly. It was built in 1866-67 to carry the railway 4820 metres across the valley of le Vingeannot, the river that the canal follows down from the Plateau de Langres.

Two more locks malfunctioned on our way down to the Saône. On one, the hydraulic ram to one of the downstream doors appeared seized, and it needed to be detached and the door hand cranked. It was getting late in the season, and the locks were obviously begging for some winter maintenance.

After another lift bridge and three more locks, we came to the final lock of the canal. We deposited our telecommande in the slot and once it was digested by the automated system, the lock cycled down and let us out into the kilometre-long bief to the end of the canal.

At 1531 we arrived at the Saône and joined the line behind a rental boat, waiting for our first lock on the river. In nine days we had worked our way through 114 locks, 71 up and another 43 down to cover the 224 kilometre

length of the Canal entre Champagne et Bourgogne. Along the entire route, we had seen no rental boats; I knew that was about to dramatically change.

Chapter Twenty

Down the Saône

As we waited for the lock, astern we could see a steady parade of bumper boats zigzagging downstream toward us. We had entered an area of easier navigation and the realm of the rental boats. These get their well-deserved labels of *bangy boat*s or *bumper boats* by their manoeuvring antics in and out of locks and moorings. They require no license or qualifications to drive and are amusing to watch, but because of their unpredictability, they are dangerous to be near. Unfortunately, they constitute the vast majority of the pleasure boats in the Burgundy.

Once we had locked through, we headed down the Saône, which varies between 60 and 100 metres wide along this section. Its current was rather gentle; even after all the recent rains, it was only about two kilometres per hour, but it helped us along.

All the mooring space was occupied as we passed Pontailler-sur-Saône and Lamarche-sur-Saône, so we continued down river. Actually, there was plenty of room for ten or more additional boats to moor, but the rentals had spaced themselves out eight to a dozen metres apart, with no gap long enough to squeeze-in Zonder Zorg. Experience told me that getting one or two of them to move and make space is nearly impossible. It was early evening by the time we reached Auxonne and we found the same gap-toothed boat arrangement on the public moorings there, so we went into the commercial marina of Port Royal.

Down the Saône

We were a tad tired from the 49.1 kilometres and 17 locks, so we slept-in a while on Tuesday. Then at 1040 we slipped from the marina float and continued downstream through the next lock and arrived at St-Jean-de-Losne at 1300. The public moorings on the Quai National were full, so we motored past.

St-Jean-de-Losne is considered as the hub of the French canal system; from here boaters can head out in six directions. It is on the Saône, which can be followed south to the Rhône and Mediterranean and across the Midi to the Atlantic, or followed north through to the Lorraine and the Moselle into Germany or Belgium, or across to the Champagne and Paris or onward to Belgium. The town is at the junction with the Canal de Bourgogne, which can be followed across to the Seine and down to Paris. Four kilometres upstream of the town is the junction with Canal du Rhône au Rhin, the canal up over the Vosges and into the Alsace and the Rhine. Downstream about 55 kilometres is

the junction with Canal du Centre, which can be followed westward across to the Loire and then northward to Paris. We had based Lady Jane here for four of our six years a decade ago.

We turned off the river toward the entrance to Canal du Bourgogne and continued to starboard beneath the bridge and into the basin.

At 1315 we secured to a visitor's float at Blanquart. We chose it because both the guide and the advertising in the guide note that it has wifi, and we had not found an Internet connection in nearly two weeks. The signal is very weak and the few times we managed to connect, there was no bandwidth to even load emails.

We stayed for two days, replenishing our larders from the two nearby supermarkets. I also spent some time pumping-out the oil from the bottom of the bilge and cleaning the area. I pumped out one-third of a litre of lubricating oil using our Pela oil change extractor and then set

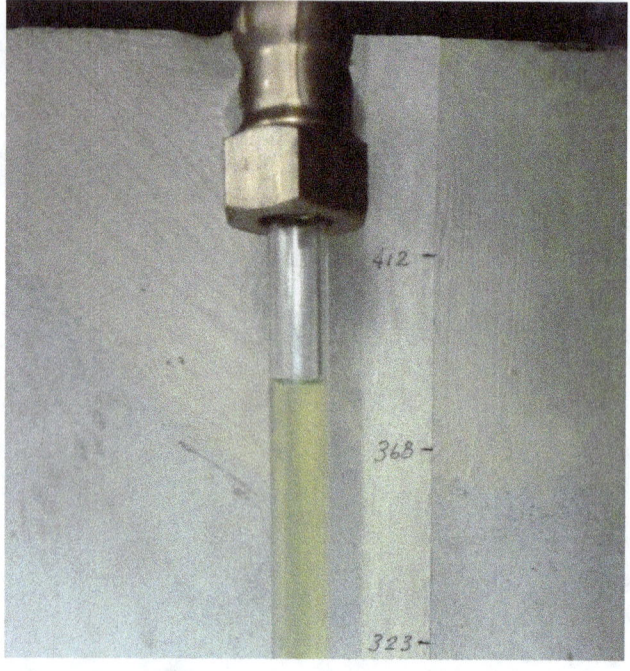

about trying to find its source. I had suspected the PRM gearbox, since the initial drops of oil were beneath it and there were stains on its bottom. The level on the gearbox dipstick was a bit below the low mark, so I topped it up. After I searched in vain for any sign of where the leak might have been, I thoroughly wiped-down the entire engine and gearbox and laid absorbent pads beneath them to quickly show any new leakage.

Midmorning on Thursday we motored out to the fuel float and took on 300 litres of diesel. Before fuelling, the gauge showed that we had about 80 litres of diesel remaining in the tanks and after fuelling the accuracy of my calculated marks was proven. A winter project now is to add more permanent and more easily readable markings.

After fuelling we continued down the Saône aided along by the current. An hour and a half later we came to our first big lock since Belgium. It is 185 metres long and 12 metres wide and is one of

five similar sized locks that will take us down to the Rhône at Lyon. Four kilometres before the lock I contacted l'éclusier on VHF and announced our arrival in twenty minutes. He said he would have the lock ready for us when we arrived. With drops of 2.5 to 4 metres, they are not high, but they are equipped with intermediate bollards in the walls to make locking easy. We were quickly through.

As we headed downstream we passed many birds along the river banks. There were grey herons watching the water for their next meal and white ibis doing the same. Swans were rooting around in the shallows and sharing the water with thirsty cattle.

Mid-afternoon we left the Saône and motored up the Doubs to look for a mooring for the night in Verdun-sur-les-Doubs, a kilometre upstream. When we arrived at the port we saw it was arranged as a sort of modified Mediterranean moorage, stern ties without the bow anchor or buoy. With our rudder extending a metre beyond the stern, I did not fancy backing in. I thought of nosing in, then thinking of  having nothing but 8 to 12 metre bangy boats to hold our 17 metres and 18 tonnes vertical to the quai and broadside to the river current, I decided to move on.

We headed back out into the Saône and continued downstream. Shortly after passing through a second lock, we saw space on a float in front of a restaurant in Gergy. As we approached, we saw a body hanging on a scaffold above the float. On closer examination it was a sculpture of a curvaceous female pirate climbing the rigging of a non-existent ship, apparently as part of the decoration theme of the restaurant. The restaurant and float are part of a camper  facility, which was closed for the season. At 1549 we secured, having come 43.4 kilometres. We walked the short kilometre up the hill to downtown Gergy for some fresh provisions for dinner.

Later, just as I was turning the cutlets over in the pan for their final three minutes, the flames on the stove went out. Our propane tank, which we had started when we arrived aboard on 3 June, had finally run out. It had lasted us seventeen and a half weeks. I went forward to change bottles, having had the forethought to buy a spare tank while we were in Sillery. I had chosen one of the new transparent fiberglass Butagaz Viseo tanks. These hold only 10 kilograms of fuel, rather than the 13 of the aluminium tanks. Their full weight of 16.5 kilos compared to the 26 of the aluminium, makes them much easier to strap on a bike for an exchange. Being able to see the level of the remaining fuel is a huge plus.

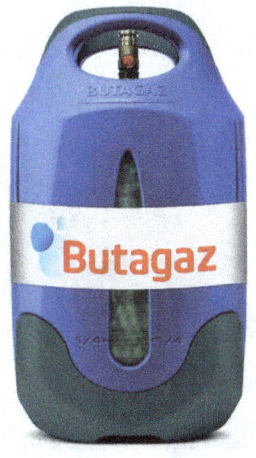

Our regulator didn't fit the new bottle. The sautéed potatoes were done, as were the Brussels sprouts and mushrooms, but the cutlets were raw on one side. I put them on a covered plate and zapped them for a minute in the microwave and plated a very nice dinner.

Late morning on Friday we slipped and continued down the river. Along the way we were overtaken by large commercial barges and we met others heading upriver. Also, there were many bangy boats. These are most readily identified from afar by the ridiculous number of fenders attempting to protect their sides from the results of erratic boat handling. This one, if there is the same arrangement on the other side, has thirty-two fenders, including three across the stern.

At noon we passed the entrance to Canal du Centre, which heads westward to the Loire across the bottom of Burgundy's Côte d'Or. We continued down to the city of Chalon-sur-Saône and under its fifteenth century Pont St-Laurent.

At 1222 we secured to the visitor's float in the port de plaisance. We were met by a staff member, whom I told we needed to buy some parts for our gaz system. He told us that a two hour stop is without fee, but for more than that it would be €9 for up to six hours and a ridiculous amount for overnight. I said we'd be less than two hours.

I quickly went ashore with our boat's regulator and searched among the gaz fittings in the hardware section of the huge Carrefour supermarket. I found nothing compatible. I found an Internet connection, surfed to the Butagaz website, found their new Viseo fibreglass tank, looked at the photos and saw that it comes with a special regulator. Apparently the

Down the Saône

woman in the Intermarché in Sillery had omitted to include ours when I had bought the tank. I walked over to the fuel pump check-out booth in the supermarket parking lot, which is the most common place to exchange gaz bottles, and told the clerk that the Viseo regulator had been forgotten when I had bought a new tank.

With no questions asked, she gave me a regulator. Back onboard, I connected it to our line, clipped it to the bottle and Edi put the kettle on for tea. At 1426 we slipped and continued past the swans and back out into the Saône.

As we approached the commercial port, a 190 metre barge was winding around and was perpendicular to the canal bank. I could see its bow was moving left, but as a courtesy I called it on VHF Ch 10 and asked permission to cross its bows. I received a friendly affirmative.

Within minutes the deeply laden barge had completed its turn and was beginning to overtake us. In a few more minutes it was past us and moving at double our speed down the river.

We met other large barges heading upriver, both deeply laden and empty. In the late afternoon we locked through écluse Ormes, which because of our radio call was ready for us when we arrived.

At 1800 we secured to the wall on mooring rings, just upstream of the Pont de Tournus. Because of the current, we had passed under the bridge, then turned around and stemmed the current back upstream to the quai.

Saturday is market morning in Tournus, so soon after rising we took advantage of it and shopped the stalls for breakfast goods and some fresh items for dinner.

We found some wonderful multi-grain bread at a small stall. Avocados, green beans and tomatoes were bought at another. A fishmonger sold us a thick dos de cabillaud and so on.

At 1022 we slipped, turned in the stream and continued downriver. Moored toward the southern end of the city centre was a huge cruise barge of rather graceless design, built more to cram into the 195 by 12 metre locks than it was for looks.

An hour later we were passing the place where in 1984 we had rescued an unfortunate couple from their sailboat. They were headed to the Mediterranean in their new twelve-or-so metre boat on their maiden voyage and they had found a shoal in a bend. They had apparently misinterpreted the channel markers in a narrowing.

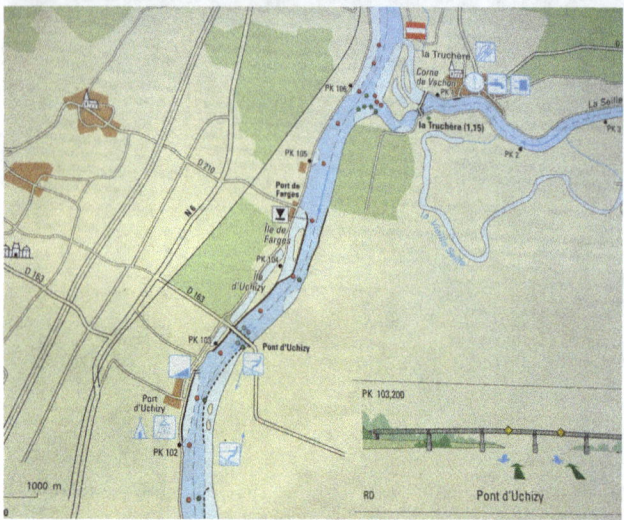

It had been April and the river current was strong. When we had arrived on the scene, the boat was heeled over beyond its gunwales by the current and it was hard against the bank of the shoal. We passed a towline and tried to pull the boat off, but the current was far too strong for that. We took their anchor and set it at full rode length out into the channel, but their windlass was not strong enough. Finally, as the boat began flooding, we took them and a few of their valuables off. At their request, we had landed them on a ramp just upstream at Port d'Uchizy, where they said they would phone for assistance. We told them we would report the incident when we reached Macon. As we had passed back downstream, we had seen the boat awash and rolled onto its side well onto the shoal from the channel.

Back to the present, we continued through Macon and at 1614 we secured to a float in Montmerle-sur-Saône. We had come another 59.9 kilometres and had passed through another lock. About an hour after we had arrived, the Capitain de Port came by to collect €10 for moorage, which included water and electricity, so we filled our tanks and plugged in.

Around midnight there was the noise of heavy machinery outside, and powerful searchlights illuminating us and the river bank. I looked out through our windows to see a solid wall of windows slowly gliding past. A lock-filling cruise barge was gradually coming to a set of dolphins and a landing stage just upstream of us.

In the morning I headed ashore to the market just after busloads of cruisers had been whisked away to their daily tours. It was raining as I walked the short distance to the covered market to see what there was to entice me. I found nothing that I hadn't seen better the previous evening in the supermarket across the street, so I crossed the street.

At 1110 we slipped from the float, turned and continued downstream. We passed under some wonderful old suspension bridges, dating to the first half of the nineteenth century, like this one at Beauregard.

Passerelle de Trévoux is particularly attractive. It crosses the river at what was once the capitol of the independent principality of Dombes, which had broken away from the kingdom of Arles in the eleventh century and had its own parliament until 1762.

In a steady rain we continued down the Saône and into the suburbs of Lyon, the third largest city in France. We wound our way around the bends and under the bridges in the strengthening current as we neared the confluence with the Rhône. As advised by the guidebook and riverside signs, we closely monitored VHF Channel 18 to listen to other boats reporting their progress up and down the river.

By 1600 the heavy rain had eased to a light sprinkle as we passed under Basilique Nôtre Dame de Fourvière and began passing the downtown core of Lyon. At 1636 we secured to a stub of a finger float in the newly renovated Place Nautique du Confluence at Kilometre 1 of the Saône. The construction was so new that the capitainerie was still not finished, so it was housed temporarily in a utility trailer. I walked up to pay our fees, which were €22 per night, including all services. I asked for the wifi code and was told that, while the wifi worked well throughout the marina, the connection to the Internet did not. The connection varied from very poor none at all. We have continued to see this in France; they appear to have a major problem with connectivity.

Nonetheless, I paid for two days, primarily since we needed a break. In four days we had come 214 kilometres from St-Jean-de-Losne and we were just one kilometre from end of the Saône. Besides, just across the passerelle above Zonder Zorg's stern is a vast new shopping mall anchored by a huge Carrefour. In addition to the break, we needed a shopping fix.

Chapter Twenty-One

Down the Rhône to Avignon

After a two day break in Lyon, at 1049 on Tuesday the first of October we slipped from the float in Place Nautique du Confluence and re-entered the Saône to run its final kilometre to the Rhône. The architecture along the final kilometre is a cross between creative, experimental and wild. An orange building has as its exterior curtain wall a coarse filigree of metal plates, and in one corner there is a five storey high crater that seems to be the shockwave impact from an asteroid that appears to be just about to collide with it. The architect likely has a more artful explanation of the design

Looking back upstream we saw the shopping centre next to the port, with a Novotel occupying the upper floors of the complex. It and the building next to it would be considered radical designs were it not for the orange creation next along and the others just beyond it.

To add to the wild, just along from the orange building is a green one with two huge craters in its riverside façade. Next to it is a four storey black shoebox-shaped structure rather daringly cantilevered way out over the river from a base of ten or twelve metres height.

Next to the shoebox is an adolescent's rendition of a space station. It is difficult to get a scale, but I estimate the building to be 40 metres high. It is still under construction.

From the signs on the riverside hoardings, the building is the new site of the Musée des Confluences. It is sited appropriately, on the narrow point of land between the Saône and the Rhône.

At 1109 we passed the bifurcation buoys and entered the Rhône. The current seemed to be about 5 kilometres per hour as we moved along at 14 with turns for 9. With this much current, when heading downstream there is little margin for error in the bends and through the bridge holes. Fortunately the channel is generally wide and the bends are mostly gentle.

In less than four kilometres we came to our first Rhône lock, écluse Benite. It was ready for us as we approached, I having used the VHF to advise l'éclusier of our arrival from twenty minutes back. L'éclusier took our details over the radio: name of boat, flag, length, beam and draft, number on board, departure port and destination. There are fourteen locks along the river to the Mediterranean and we need to pass through twelve of them.

At 190 metres, they are 5 metres longer than those on the Saône, but at 11.4 metres width, they are 60 centimetres narrower. The big difference; however, comes with their height. Where the Saône locks range from 2.5 to 4 metres, the twelve locks on the Rhône average 11.9 metres in height and they range up to 22.5 metres.

Fortunately, all these locks have floating bollards in their walls, so it is easy to secure the barge for the ride. In fact, these huge locks are easier to negotiate than are the smaller, Freycinet gauge ones of which we had done some 250 during our first six weeks in France. One thing I always do with a floating bollard, once I have a line around it, is put some of my weight on its top to see if it moves easily. If it is stuck or jammed with detritus in its guide rails, it is difficult to scramble along to a working bollard after the water has begun moving in the lock chamber.

We were lowered nine metres very smoothly and effortlessly; it was our easiest lock since the Netherlands. We paused a while after the giant sluice gate had opened before we headed out, giving it the opportunity to drip off much of its water before we passed under it.

Outside and back into the river we were again being swept along with the current. It was a bit slower in this stretch and we moved along at a rate of 12 to 13 kilometres per hour. The river is managed by La Compagnie Nationale du Rhône, La CNR, which was set-up in 1933 with three main objectives: improvement of the navigation conditions; creation of irrigation canals and the installation of a series of hydroelectric power stations to utilise the strong flow of the river. On the riverbanks are vineyards in terraces climbing the steep hillsides.

We were entering the wine growing region of the Northern Rhône. The hillsides above the riverside roads and behind the dotting of little villages are covered in terraced vines wherever the slopes face an adequate portion of the day's sun and are not too steep. On many of the hilltops are ruins.

Our next lock, écluse Vaugris was ready for us as we arrived, and again it was a very simple procedure locking down. As we left the lock and moved back out into the current beneath the dam, we were alongside the vineyards of la Côte Rôtie, which translates to the Roasted Hillside. These vineyards are reputed to be among the oldest in France, with wine known to have been grown here at least since Roman times and likely earlier.

Among the more famous vineyards here are Côte Blonde and Côte Brune on the steep hillsides above the village of Ampuis. The story goes that a wealthy vineyard owner gave his two finest plots to his two daughters, one a blond and the other a brunette.

Four kilometres downstream from Ampuis begins a pair of sharp bends in the river at the town of Condrieu. Before the serious works done by the La CNR to reduce the current, this was a major navigational hazard. In the middle of the bends were les Roches de Condrieu, the rocks which brought to grief many boats.

On the steep slopes above the town are the terraced vineyards of Condrieu. The Viognier vine is at home here, making some of the world's finest examples of that variety. The vines are believed to have been brought to the area by the Greeks in the sixth century before Christ.

We passed under le Pont de Condrieu, which bridges the narrows, and we headed around the next sharp bend. Navigation has been made so much easier through this once fearsome place.

Down the Rhône to Avignon

In the middle of the bend, above the suburb of la Maladière, are the vineyards of Château Grillet, often erroneously called the smallest Appelation d'Origine Contrôlée in France.

At the next lock, écluse Sablons we had to wait while a lock-filling cruise barge was lifted and headed upstream past us. We have been seeing three or four of these each day on the rivers since we passed downstream from Chalon-sur-Saône. If we see that many in the distance we cover in a day's cruise, there must be several times that many on the Saône-Rhône system. It was into October and the 180-metre floating resorts all seemed to be filled with passengers.

We entered our third lock of the day, and again had a very easy descent. This one has a 12.2 metre drop, which shows how much the engineers had raised the water level through the bends and past les Roches de Condrieu.

Downstream about four kilometres from écluse Sablons, on the riverbank in front of the village of Champagne, is an old and now retired cable ferry pylon. For centuries before the bridges were built, lines were stretched across the river between anchors such as this. A ferry was attached to a pulley on the line and, by using a rudder to angle the side of the ferry to the river, the current pushed the ferry across the river.

In the old pylon at Champagne we saw a staircase of projecting stones winding its way up the tower. These stone steps were used to reach the top of the tower to string and adjust the cable. They reminded us of the steps we had used on our climb of Wayna Picchu from Machu Picchu in 2010 during our three-year sail around the South American coasts.

Three kilometres downstream from the pylon at Champagne is le Pont d'Andance, the oldest suspension bridge still in use in France. It was built in 1827, engineered by Marc Séguin, renowned as a pioneer in wire rope design and fabrication. To test his cable strength theories, he built a small suspension bridge over La Cance, a narrow tributary downstream a few kilometres.

In 1825 he built his first major suspension bridge across the Rhône, joining the towns of Tain l'Hermitage and Tournon. That bridge was partly destroyed in 1944. Le Pont d'Andance was his second bridge across the Rhône and it still carries regular traffic.

Six kilometres further along, at 1800 we secured to float just upstream of Pont St-Vallier. It has no facilities, but it is a solid new pontoon of about 30 metres length in a very quiet area. It is so new that it does not appear in the latest edition of Du Breil Guide Fluvial from 2011. We had come 76.1 kilometres and had passed through three locks.

On Wednesday morning we slipped at 1030, turned in the current and continued down river. Half a dozen kilometres down we passed beneath the ruins of Tour d'Arras and its surrounding vineyards. We were four kilometres from the next lock, écluse Gervans and I radioed l'éclusiere to inform her of our arrival in twenty minutes.

She told us there was a commercial following us down the river and that it would arrive a few minutes after us. We were to wait for it to enter and then follow it in. We arrived at the lock and slowed to drift along as we waited for the barge to overtake us.

As we approached the commercial mooring dolphins, we saw people with luggage waiting on the catwalk. They motioned to us and we quickly understood that the barge was going to pause on the dolphins to pick them up. We accelerated and moved further along and out of the way of the barge.

We had dawdled because the waiting float for pleasure craft had a bangy boat plunked exactly in its middle, leaving insufficient room for anyone else at either end. We moved along to it and asked the skipper to move either forward or astern to allow us to moor. He seemed afraid to move, likely having bounced around and struggled to get there. He folded his arms, thrust out his chin and refused to move his 10 metre boat from the centre of the 25 metre float.

We motored forward and came to rest against the crash barriers, hoping the huge barge didn't need their assistance with his manoeuvring into the lock. There were no horizontal beams to take our fenders, nor was there anything we could use as a temporary bollard. We hung onto the pilings to stabilise Zonder Zorg against the suction and wake of the passing barge. The skipper appeared to have properly appraised the situation and he motored dead slow past us and into the lock chamber.

After the barge had entered, we had the amusement of the bangy boat bouncing around in the chamber trying to moor as two more pleasure boats, which had been following the barge, joined us. One was a Belgique cruiser, the other another bangy, which offered us further entertainment.

After the lock had taken us easily down 11.5 metres, we all paraded out and continued down river. Within five kilometres we swept around a bend and headed east past the town of Tournon on the Right Bank.

Linking Tournon with its neighbour Tain-l'Hermitage across the river, is la Passerelle de Tournon, built in 1825 by Marc Séguin as his first suspension bridge across the Rhône. After its near destruction during World War Two, it was rebuilt and is now used for foot traffic only.

Across the bridge and above the town of Tain-l'Hermitage, on steeply terraced slopes are the famous vineyards of Hermitage. Because of the bend in the river, the vineyards face south for the maximum daily sun. This, added to the fact that slope of the hill is almost perpendicular to the summer sun, the vineyards receive as much direct sunlight during the growing season as is possible here at 45° North.

The Syrah is the grape planted here, and the conditions allow it to show at its finest, making some of the greatest Syrah wines in the world. So renowned were the wines from Hermitage, that the Australians and other New World wine producers called their Syrah grapes Hermitage. After nearly two centuries of name abuse, trade laws have now finally stopped that practice.

As we swept around the next bend, which is just over two kilometres along, and again headed south, we looked back at hill of Hermitage. It is such a small area for such a famous wine.

Another dozen kilometres brought us to écluse Bourg-lès-Valence. The same four pleasure boats paraded in astern the commercial, with us last. We quickly secured and watched the remainder of the bouncing acts. After another easy 11.7 metre drop we all paraded out and downriver for five kilometres through Valence.

In the centre of Valence, under Pont Frédérique Mistral, a huge barge was sunk pointed downstream against the wall on the Left Bank. The bridge is at the end of a sweeping 45° bend to starboard, and it appears the skipper misjudged the current and was swept to port and ended up on the wrong side of the channel beacon and the bridge pier it marked.

There was a work scow with a pollution boom moored alongside on spud poles. It appeared to have been a very recent accident. As we motored in the erratic wakes of the bangy boats, we hoped that the skipper of the sunken barge hadn't been trying to avoid a meandering rental boat.

On the southern side of the city the two bangy boats dropped-out of the parade and entered le Port de Plaisance de Valence-Épervière. The remainder of the parade continued down river, along the way meeting another huge river cruise ship.

Another hour and a dozen kilometres brought us to écluse Beauchastel, where we and the Belgian boat followed Camaël into the lock. After an 11.82 metre drop we headed out and continued down the river following Camaël, while the Belgian boat stopped on the waiting float at the downstream side of the lock.

A short distance along we met a large unladen barge putting up a very large wake as it headed upstream against the current. A few minutes later we passed under Pont de la Voulte and looked at the mooring that is indicated in the guide. I did not like the openness of the wall and the way the wake of the barges ahead and astern of us rolled along it. It looked like a rough place to stop, so we continued along.

We met another floating resort as we made our way along the eighteen kilometres to the next lock. It was a long bief, and ahead we could see Camaël slowly pulling away from us, until even in the long strait sections we could no longer see her. We knew that we had fallen too far behind her to have her wait for us at the next lock.

At écluse Logis Neuf we had the chamber all to ourselves, which really gave the space an added feeling of size. The drop of 11.7 metres went quickly and without hitch, and we motored out to run the three kilometres to our prime mooring spot for the day at Cruas. The little port is designed for small boats and has a tricky crosscurrent entrance past shoals, but it has space for three boats of fifteen metres length, which we might squeeze onto. As we arrived, I could see all the larger spaces were occupied, so we continued along.

Just downstream is la Centrale Nucléaire de Cruas, which looked benign in the early dusk. We continued along in the fading light looking for a place to stop for the night. With the current running at five kilometres per hour, and with the uncharted waters beyond the channel markers, we could not simply approach a bank and moor.

Down the Rhône to Avignon

Five kilometres downstream, as we approached the eleven kilometre derivation leading to the next lock, I looked at the possibility of mooring at the canoe haul-out in the short arm above the Barrage de Rochemaure. However, prudence and poor chart symbols in the guide made me continue on past.

The sun set, but I was not concerned about navigating in the dark. We had our Navionics charts on my iPad and with its built-in GPS, we had an accurate plot of our position and track. The track was following perfectly our path along the derivation, which leads eleven kilometres to écluse Chateauneuf. I switched on our navigation lights and we continued along confidently.

In the fading light we met a lock-filling barge convoy on its way up river. We crept along the final half kilometre to the lock, using primarily the iPad to help pick out shapes around us. Most of the locks we had been in thus far on the Rhône have had a similar layout in their upstream approaches, so we knew what to look for. We found the pleasure boat float, but plunked exactly in its centre was a boat of about nine metres length. There was nobody aboard, and not wanting the liability of moving the boat along so we could share the float with it, at 1958, after adjusting our fenders, we rafted onto it and shut down for the day. We had come 88.5 kilometres and had passed through another four locks.

We were awakened shortly after 0800 on Thursday morning by an irate man who was demanding to know why we had rafted onto his boat. He could not understand that he was hogging the entire float and leaving no space for others. I told him we would be ready to move in about fifteen minutes and he fumed.

Carefree Through 1001 French Locks

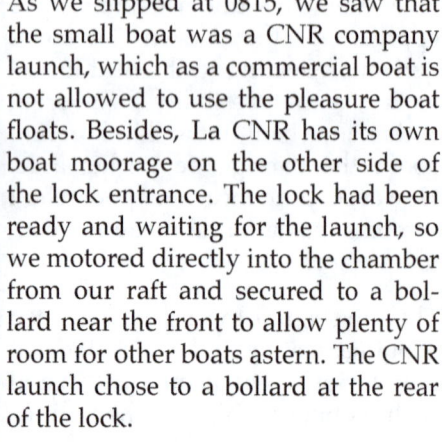

As we slipped at 0815, we saw that the small boat was a CNR company launch, which as a commercial boat is not allowed to use the pleasure boat floats. Besides, La CNR has its own boat moorage on the other side of the lock entrance. The lock had been ready and waiting for the launch, so we motored directly into the chamber from our raft and secured to a bollard near the front to allow plenty of room for other boats astern. The CNR launch chose to a bollard at the rear of the lock.

After a 16.5 metre drop, we waited for the gate to lift and the barriers to open. When the light turned green, we slowly motored out, as is required by regulations. The CNR boat was impatiently nipping at our stern all the way out. Once we were clear of the lock entrance, it sped past us, cut across our bows and went across to the pleasure boat float. We continued downstream.

We were a little over a kilometre further downstream when the launch came speeding down toward us and cut across our stern. As the skipper saw me aiming my camera, he put his hand up in an apparent attempt to block the shot. It appears we had discovered a CNR employee with a company boat and an attitude problem.

A short while later we watched as two huge cruise barges danced around each other changing mooring places. These cruise ships operate on fixed schedules, and with the rapid increase in their numbers in recent years, it appears that available moorage facilities have not followed apace. This juggling looked like the rearrangement of a raft to prepare for departure.

Down the Rhône to Avignon

Soon we entered le Défilé de Donzère, the Donzere Gorge, a deep and narrow corridor through limestone hills. Because of the force of the water through the defile, the early boaters called it Robinet de Donzère, the Donzere Faucet and it was one of the most respected stretches of water along the old Rhone. It is crossed by Pont de Robinet, a wonderful old suspension bridge that was built to a Marc Seguin design in 1847.

Within a kilometre downstream we came to a more modern structure, la barrage du garde at the beginning of the 28 kilometre derivation canal that takes traffic around a series of the former navigational hazards, including Le passage de Pradelle, Le Banc Rouge, La Désirade and Pont Saint-Esprit. These hazards were finally bypassed in 1952 with the opening of écluse Bollène, which with its 22.5 metre drop is the highest lock in the European network. Bollène even tamed the waters in Le Robinet.

A short distance into the derivation there are crossover signs on the canal banks, indicating that boats must cross to the other side of the channel. This is to assist large barges in manoeuvring through a pair of 80° bends. As I crossed over to the port side of the channel, I was hoping there wasn't a bangy boat coming upstream unaware of the meaning of the signs. Fortunately, the only traffic was a large pushed convoy, and we met green to green as required.

When I contacted l'éclusier at Bollène, I was told there was an up-bound commercial in the lock and a down-bound waiting for it. We could join the commercial in the lock. As we approached the lock we met the up-bound Cartoon, and saw ahead of us Bonova, the barge that we would follow into the chamber. The timing was such that we had to slow only at the last minute to allow Bonova to settle into position. We quickly came to a bollard and were ready for a 22.5 metre descent.

As the walls grew around us we marvelled at the engineering that had gone into taming the Rhône and making it much more safely navigable. I thought of boats that, before the taming of the river, had to run through a series of rapids, rocks and bends as the river dropped thirty metres in thirty kilometres. We reached bottom, the gate opened and Bonova headed out. We were left in the cavernous space by ourselves, which added to its sensation of size.

As we motored out, the enormity of the lock continued to impress us. The hydroelectric generating complex at Bollène is huge, particularly when viewed from the downstream side. La CNR manages thirty-three power stations, which together generate 16 billion Kilowatt hours of electricity, about four percent of the French consumption.

We passed under the ramparts of medieval fortified castles at Montfaucon and Roquemaure. Along the way we met a steady stream of commercial barges heading upstream.

Some graciously posed for us beneath the ruins of Château de l'Hers. Château de l'Hers dates to the twelfth century and sits on a small rocky outcrop on the Left Bank of the river. It once regulated navigation on the Rhône, but today the commercial barges and pleasure boats pass by giving it only admiring looks, rather than taxes and tolls.

Beyond Château de l'Hers on a narrow ridge a few kilometres east of the river sit the ruins of Châteauneuf-du-Pape, the Pope's New Castle. The château was built as a residence in the fourteenth century by Pope Jean XXII, the second pope of Avignon. The vineyards that he had planted around the château are today the centre of the famous Châteauneuf-du-Pape wine growing area. The intensity of the wines come from the fist-sized stones throughout the coarse gravel and clay of the ridge, which hold the sun's heat well into the evening and continue to warm the vines into the night.

Edi created a wonderful platter of open-face sandwiches for lunch and we ate along the way.

Eight kilometres downstream we came to écluse d'Avignon. At 9.5 height metres it is a very large lock, but rather diminished in our eyes after having so recently come through the one at Bollène.

We were quickly down and passing through the suburbs of Avignon. At Villeneuve-lès-Avignon we passed beneath the ruins of Tour Philippe-le-Bel, which was once part of the western access control to the twelfth century Pont Saint-Bénezet, the famous Pont d'Avignon.

We continued down to the junction of the two branches of the Rhône, turned in the current and headed up the eastern branch into the centre of Avignon. On the quai, just short of the Palais des Papes was a line of moored cruise barges.

Ahead of us was Pont Saint-Bénezet and on our starboard bow were Nôtre-Dame des Doms and Palais des Papes. The Palais, which was built in the fourteenth century as the papal residence, is the largest Gothic palace in Europe.

We continued around the end of the remains of le Pont d'Avignon and half a kilometre upstream of it, at 1605 we secured to mooring rings on Quai de la Ligne. We had come another 75.5 kilometres and had descended through four more locks.

In three days we had come just over 240 kilometres from Lyon through swiftly flowing current and busy commercial traffic. We needed a break.

Chapter Twenty-Two

Exploring Avignon

Late on Friday morning, after a sleep-in and a leisurely breakfast, we walked through the wall at Port de la Ligne and into the old city. We wound our way through a maze of streets, many just barely wide enough for small cars and others too narrow even for that. We were headed for Les Halles, the covered market.

Inside we found a huge selection of stalls, offering the full range of fresh items that would be found in a huge supermarket. There are fruit and vegetable stalls, one of these with a great selection of specialty mushrooms: chanterelles, cèpes, trompettes, morilles, pied de mouton, girolles and so on. There are several butchers and charcuteries and there is a line of fishmongers along the side of the hall. There is a stall with a vast selection of fresh and dried herbs, nuts, legumes, dried fruit, candied fruit, salts and peppers. Of course there are the cheese stands and the bakers and the wine vendors. We chose some fish and some produce, including a selection of wild mushrooms for the evening's dinner.

After we had dropped-off our purchases onboard Zonder Zorg, we headed back through the old city to le Palais des Papes, the fourteenth century seat of the Catholic Church.

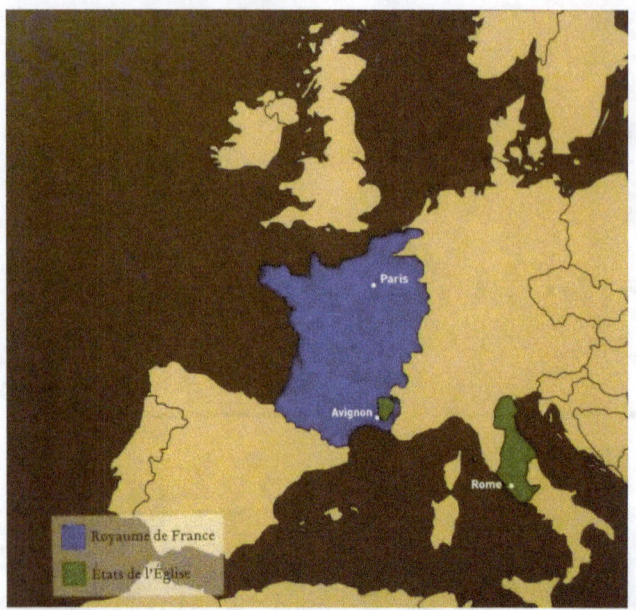

In the thirteenth century, the papal court was itinerant, rarely staying in Rome and residing in several cities in its states. At the beginning of the fourteenth century, bitter disagreements between King Philip the Fair of France and Pope Boniface VIII culminated in 1303 with an attack against the Pope in his palace at Anagni. In 1305 the archbishop of Bordeaux was elected pope as Clément V. He remained in France, primarily to adjudicate the fate of the Knights Templar, whom the King of France, being in deep financial debt to them, had capriciously accused of heresy. In 1309, still with no fixed address, the Pope stayed in Avignon and in Comtat Venaissin. He systematically promoted French bishops to cardinal, and by the time of his death in 1314, he had created a small French majority in the Sacred College, ending the Italian domination. The next six popes elected were consequently French and they chose Avignon as their permanent seat and built the Palais des Papes.

Avignon was the property of Charles II of Anjou, Count of Provence, King of Naples and Sicily and therefore vassal of the Pope, to whom he offered support. Avignon was also next to Comtat Venaissin, a Church possession since the late thirteenth century. This was a peaceful and stable territory at the confluence of the Rhône and Durance rivers and close to the Kingdom of France, to which it was linked by Pont Saint-Bénezet, the famous Pont d'Avignon. Lying at the crossroads of major land and water routes, Avignon was ideally located in the heart of Christian Europe.

For two years after the death of Pope Clément V, the College deliberated on choosing the next pope. Finally, after intense negotiations, in 1316 they chose the seventy-two year old Jean XXII, intending him as a transitional pope in view of his advanced age. He lived for another eighteen years and transformed Avignon into the capitol of Christendom. During his reign, the Curia increased from 200 to 500 and an additional 1000 lay officials were working in its support.

Exploring Avignon

We paid our admission and entered the huge complex of buildings. The construction of the palace had been the largest building project in the Christian world in the fourteenth century. Work was begun in 1335 under Pope Benoît XII, and by the time of his death in 1342, the Old Palace had been built.

Huge additions, called the New Palace, were begun under his successor, Clément VI. Construction paused for a four year break because of the Great Plague of 1348. It was then mostly completed in 1352. It was then and is still now the largest Gothic palace in Europe.

The construction accounted for approximately one quarter of the annual expenditure of the Pontifical treasury and involved monthly salaries for a workforce of 850. Locally quarried limestone was used and its purchase and transport accounted for sixty-five percent of the cost. Reinforcing the vast complex are ten tons of iron embedded in the masonry in an innovative technique. In the complex are 15,000 square metres of floor space.

Inside the first courtyard, le Cours d'Honneur, we saw continuing excavation of the grounds, with seemingly the same temporary scaffolding I had seen on a previous visit in the mid-80s. Either there is some serious archeological work being done, or this thirty year dig is simply an example of the speed of French public works.

We strolled around the huge courtyard getting a feeling of the scale of the place and marvelling at how well

preserved much of it is after two-thirds of a millennium. It has survived not only time, but also the ravages of political and religious wars, revolution and the confiscation and destruction of Church properties.

The architecture is rather bland and slab-sided, typical of the Gothic style, but there are some very pleasing features. We headed inside, following the clearly laid-out tour route and on our handheld audio guides listening to commentary about what we were seeing.

In the lower kitchen we saw across the entire end of the large room a fireplace, in which food was cooked. Set in one end of it was a much smaller domed oven, reminiscent of the ones we had seen in the sixteenth century Monestario Santa Catalina in Arequipa, Peru on our wanderings there.

On the second floor of la Tour du Pape, the Pope's Tower, is le Trésor Bas, the Lower Treasury. There we saw stones in the floor lifted to reveal hidden vaults. These vaults, one in each corner of the room, were discovered as recently as the mid-1980s. According to documents, the vaults had held Church archives transferred from Rome to Avignon by Pope Benoît XII. Also concealed below the stones in ironclad chests with multiple locks, would have been gold and silverware and sacks of gold and silver coins. Only the Pope, the Chamberlain and the Treasurer could enter the heavily guarded treasury.

Exploring Avignon

On display nearby were examples of the silver coins that had been struck by eight of the Avignon popes: Clément V, Jean XXII, Clément VI, Innocent VI, Urbain V, Grégoire XI, Clément VII and Benoît XIII. Only one of the Avignon popes did not strike any coins; Benoît XII considered that his predecessor had issued a more than sufficient quantity.

The coins on display were either gros d'argent or demi-gros d'argent, which translate to bigs of silver and half bigs of silver. The gros d'argent pictured here had been issued by Clément VI.

In le Trésor Haut, the Upper Treasury, recent restoration work has uncovered original painted ceilings and upper walls from the fourteenth century. The high room apparently had dropped ceilings installed over the centuries, likely to make the room easier to heat.

The tour led out into the gardens behind la Tour du Pape. In these gardens the popes had a menagerie of exotic birds and animals and plantations of imported trees. The animals and trees are gone, but from the space we could look up at the Pope's Tower in the Old Palace, completed by 1342.

Along to the right is the New Palace, which was completed in 1352. We walked along the base of its walls and went back inside to continue the tour.

We climbed a long circular staircase. This was part of the new kitchen tower built in 1342 by Clément VI, which featured several stories for storing and preserving provisions.

At the top of the tower is the Upper Kitchen. This has an eighteen metre high pyramidal chimney above its centre. Beneath the chimney had been the hearth surrounded by a large rectangular structure with several levels of spits, on which large quantities of food could be simultaneously cooked. Dishes were then taken directly along to the Grand Tinel for final preparation in the Dressoir. The Grand Tinel was the papal banqueting hall. Its interior had been decorated with frescoes in 1345 by Matteo Giovannetti. The Pope held sumptuous receptions there, sitting separately along the south wall on a cathedra that was bedecked with multicoloured fabrics and surmounted by a canopy. He dined off priceless crockery, gold and silverware and looked out over the guests seated at tables around the walls.

In 1413 the Grand Tinel was gutted by fire and for a long time it was known as the burnt room. Its roofs and terraces were rebuilt between 1414 to 1419. From the Grand Tinel we made our way up onto roof edge walk, from where we looked across Cour Benoît XII to the bell tower on Palais Vieux.

Exploring Avignon

From another portion of the roof edge walks we had a fine view across the courtyard of the tower of the Old Palace.

Back inside we looked at displays of some of the original hand-painted floor tiles. These had been found during excavations and restorations both here in the Palais des Papes and also at Châteauneuf-du-Pape, the Pope's new castle about fifteen kilometres north of Avignon.

We came to a room that had been richly decorated with frescoes depicting outdoor scenes. Most impressive to me is the wall with a hunting scene featuring a falconer with his two dogs.

Another impressive frescoed wall depicts a scene of people using various nets and lines to catch fish in a pool. The colours are still vibrant after more than six hundred and fifty years.

The colours on the painted ceiling are also bright, so I suspect that it and the walls must have been protected behind covers of some sort as the palace went through a number of uses after Benoît XIII, the last Avignon pope left the palace in 1403.

Photography is forbidden in the room, likely to prevent accidental flash use damaging the frescoes. The room is guarded, so I had gone to the next room and set-up my camera to shoot inconspicuously in front of the guards. Back in the room, I managed to get a few acceptable shots.

We entered the fifty-two metre long Grand Chapel, which at the time of our visit was spoiled by an exhibition of horribly garish modern attempts at art. Even with the distraction, we were able to appreciate the scale and grandeur of the vast room.

From there we went up and out onto another set of roof edge walks. This one, above la Cour d'Honneur offered a great view across the courtyard to la Tour du Pape. We went back inside and down to the huge Tribunal de la Rota, which is at ground level, beneath the Grand Chapel. It was here that Tribunal of the Apostolic Causes met to deliberate. Their judgments were without appeal. On the walls and vaults above where the Tribunal sat, Matteo Giovanetti in the 1350s had painted a series of frescoes depicting the Last Judgment.

There are still vestiges of frescoes on the walls and in one vault there remains a bright fresco depicting a series of saints.

From the 1403, when the past pope left, until the French Revolution, legates and then vice-legates governed Avignon and the Comtat Venaissin on behalf of the Pope. They altered and redecorated apartments in the palace, destroying certain elements. In 1791 the people of Avignon expelled the Pope's last representative in Avignon and Comtat Venaissin became a part of France. Palais des Papes remained vacant for almost two decades, and then from 1810 to 1906 it served as a military barracks, a horse stable and a prison.

As we walked back to Zonder Zorg for a very belated lunch, we were amazed that the palace has survived as well as it has.

After our lunch, in the late afternoon we walked along the riverfront to Pont Saint Bénezet, le Pont d'Avignon of history, fable and song. The fable goes that bridge was the inspiration of a young shepherd named Bénezet who came from Burzet, a hamlet in Ardeche. In 1177 he arrived in Avignon as an adolescent, claiming he had been told by God to build a bridge across the Rhône. The young teen was a leader of men; his words stirred crowds and he encouraged the people of the town to build the bridge. He organised fund raising and travelled broadly collecting alms and donations.

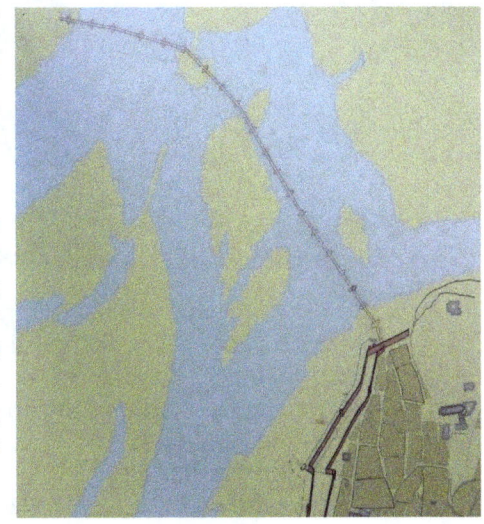

The bridge, which was completed in 1185, was an extraordinary undertaking for the time. It spanned 915 metres with twenty-two arches and in doing so, it crossed the fastest and most powerful river in France. The course of the Rhône at Avignon has been constantly changing, with bars and islands appearing and disappearing. The spring thaws bring catastrophic flooding. Huge bridge-destroying trees are washed downriver in spates and after storms.

All that remains of the bridge today are four arches with a chapel in the pillar between the second and third arches. It has a long history of political and environmental destructions and subsequent rebuilding. In 1226 the King of France in a bitter dispute destroyed all but two arches of the bridge, from the French end all the way to the Bénezet Chapel, all that he claimed belonged to him. It was likely quickly rebuilt, since the records show that people of Avignon were officially rebuked by the Pope in 1237 for having done the work despite Royal prohibitions. In the fourteenth century, the Avignon Popes strongly supported the maintenance of the bridge. Finally, at the end of the seventeenth century, the repeated reconstruction of the bridge was abandoned.

We put our passes from Palais des Papes into the turnstile slot and were allowed in. A bridge led across from the Avignon ramparts to the toll tower.

The construction of the ramparts of Avignon had been begun by Pope Innocent VI in 1355 to strengthen defences and they remain well preserved today. Over four kilometres around, they enclose an area that made Avignon at the time the second largest town after Paris. At the toll gate, fees were collected for use of the bridge, the amount depending on whether it was foot traffic, horse and rider, hand cart, wagon or herd of livestock. Our fees had already been collected, so we continued through the toll tower and across the drawbridge onto the bridge. At the third pillar we descended a flight of stone steps on its side and arrived at the chapel.

Exploring Avignon

Across from the simple stone chapel is a balcony on top of the downstream buttress of the pillar. These triangular buttresses had been built on both the upstream and downstream sides of the pillar cribbings to break the river's current. Research has determined that the method of bridge construction was to prepare a stone casing on a large wooden raft, float it into position and begin overloading the casing with stones until it sank. Under the weight of the masonry, the wood remained on the bottom.

Looking back along the bridge to Avignon, we see Palais des Papes just inside the ramparts beyond the end of the bridge. The popes there ran through the process of canonising the teenaged Bénezet for his miraculous bridge.

In 1965, during restoration work, Compagnie Nationale du Rhône discovered wooden planks and beams beneath the four remaining cribbings. Their thickness varied from 20 centimetres to 1 metre 10 and the wood was carbon-dated between 290 and 530 AD. Across the river at the Villeneuve end of the bridge, the cribbings include fir beams dating between the ninth and eleventh centuries. There were bridges here before the young Bénezet came to town and the fable of Saint Bénezet has a few more holes in it.

As we strolled back in along the bridge, looking upstream at Zonder Zorg, we pondered what we had seen and learned during this very full day in Avignon.

Chapter Twenty-Three

Onward to the Midi

At 1008 on Saturday, 5 October we slipped our lines from the rings on the wall of Quai de la Ligne, turned in the current and headed back down river. The skies were clearing from an overnight shower as we glided past les Roches du Dom and headed around the stub of Pont St-Bénezet.

Le Pont d'Avignon posed gracefully for us in the midmorning sun as we approached. It then took on an ominous appearance as we passed round its end and fell under the eye of le Palais des Papes and ghosts of popes past. They probably sensed our thoughts on their impoverishing the guilt-ridden people so that they and their vast court could live in sumptuous splendour, gaiety and decadence.

Just downstream we began passing the long line of cruise barges that lined the quais. As we passed we wondered whether the German couple had found their barge before it had left. They had stopped us to ask where the boats were. We told them to head out through the gate and follow the river downstream. They asked what is downstream. When we told them it was the direction the water is flowing, they wanted to know how to tell. We gave up. As we passed the barges, the current swept us along, adding about six kilometres per hour to our speed.

We rejoined the main branch of the Rhône and continued downstream, having no problem determining which way it was. We passed under the impressive Pont TGV Méditerranée, the twin bridges that carry the high speed train to and from the Mediterranean coast.

Below the mouth of the Durance we met an upbound laden barge pushing a huge wall of water as it bucked the current, which was sweeping us along at above 16 kilometres per hour. I had the engine turning at 1400 rpm to give us nine kilometres per hour through the water so we could maintain proper steerage.

Because of our radio call, l'éclusier had prepared the lock for us by the time we arrived at écluse Vallabrègues. This was our last Rhône lock, and as with all the previous ones, the ride down was easy.

After a descent of 11.7 metres we emerged and waved a farewell to l'éclusier. Except for the rude and aggressive CNR launch operator at écluse Châteauneuf, all the staff at the locks had been very professional, polite and friendly.

Almost immediately out of the lock, we began seeing Château de Beaucaire ahead on the Right Bank, and across the river from it in Tarascon, Château du Roi René.

Château du Roi René is a superb fifteenth century structure commenced in 1400 and completed in 1435. It was built by Duke Louis II d'Anjou to replace the fort built earlier by Duke Charles I d'Anjou on the site of an ancient Roman castrum.

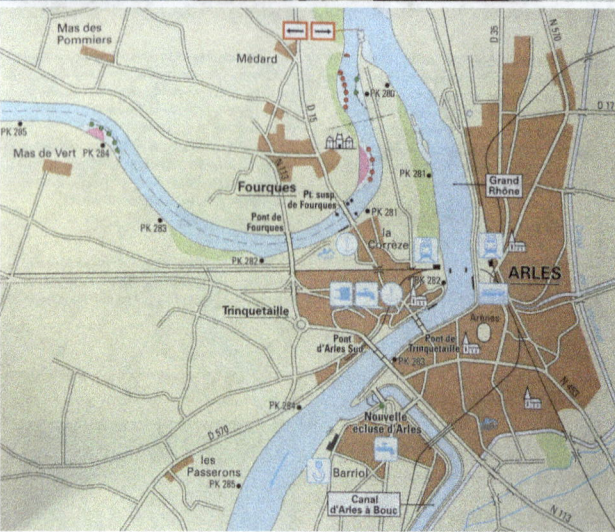

In the early afternoon a southerly wind picked-up, and blowing upstream against the swift current, it kicked-up a short, steep chop. We butted into the chop, making tall spouts of spray. Further downstream we passed a dredge and barge. It was working at a necessarily ongoing operation along the channels of the river. The changing currents constantly move banks and shoals and even add new ones.

We met what appeared to be a low-budget cruise barge. It was one of the earlier ones from a couple or more decades ago, from a less popular time before huge and garish had become the fashion.

As we approached the junction with le Petit Rhône, we were still making about nine kilometres per hour through the water and the current was moving us along over the bottom at above sixteen. In a current like this it is important to be aware of drift in the river bends. As we neared the junction I made sure we were properly lined-up well in advance.

There was no problem; we easily made it into the small branch of the river. We had left the domaine of La CNR and had again entered waterways run by VNF. As we left the main stream of the Rhône and entered le Petit Rhône, our speed began dropping. Within a few hundred metres we were down to about ten kilometres per hour and into considerably easier steering. Our intention was to stop on a quai that is marked on the charts in both the FluviaCarte and the Guide Fluvial. It is shown just downstream of le Pont suspendu de Fourques in the Arles suburb of la Corrèze.

Onward to the Midi

For us it was an ideal place to stop overnight and visit Arles, a town which dates to Greek settlements in the sixth century BC. I slowed as we approached, in preparation for manoeuvring alongside the quai and as we passed under the bridge, the only thing that looked like a moorage was a two-metre-long stage on two pilings with another piling three metres further along. The current from the bridge abutments was swirling in eddies in front of the tiny landing. We decided to visit Arles another time.

We continued down le Petit Rhône, paying close attention to the channel beacons. Some marked clearly visible rocks or patches of crud, others marked shoals made obvious only by the dead trees lodged on them, but most marked unseen shallows.

Instead of our thoughts of pausing for lunch on the quai, Edi brought up a platter of open-face sandwiches for me to enjoy at the tiller, while she ate hers in her lounge chair on the foredeck.

Mid-afternoon we came to the junction of le Petit Rhône and le Canal du Rhône à Sète. We took the righthand fork and left the river for the canal.

Barely a hundred metres down the canal we came to écluse St-Gilles, which was ready for us because of my earlier radio contact. The lock is large, 80 metres long by 12 metres wide, but the guides give no height. When I asked l'éclusier, he told me it was a drop of 44 centimetres. This is a long way from the 22.5 metre drop we had at Bolène.

Immediately out of the lock we began seeing derelict boats hauled-out onto the canal banks. It appears that the VNF have been doing some clean-up of sunken, derelict and abandoned boats.

As we made our way along the edge of the Camargue, the vast delta formed over the ages by the various river courses at mouth of the Rhône, we began seeing chevaux camarguais, the famous Camargue horses. Today, most of the wild horses in the delta have been captured, and in the area bordering the 85,000 hectare parc naturel régional de Camargue, breeding Camargue horses has become an important industry.

As we continued along the canal, we saw so many derelict boats and barges that we thought we were going through a boat graveyard. We could not believe that so many had been allowed to deteriorate to such a degree without intervention of the authorities. The canals seemed lined with derelicts, both afloat, sunk and appearing ready to sink. The sheen of oil on the water marked the recent sinkings; around the long sunk, the slicks had mostly dissipated. These derelicts are a hazard to both the environment and navigation.

Onward to the Midi

We continued along the canal, taking in the late afternoon the branch leading into Aigues-Mortes. At 1800 we came to the spud pole and a pounded pin on a grassy bank about 700 metres short of the centre of town. We had passed a supermarket next to the canal and we had stopped about a hundred metres beyond it, in the first available place. From Avignon, we had come 83.7 kilometres and had passed through two locks. While Edi straightened-up Zonder Zorg, I walked back to the supermarket for fresh provisions for dinner.

At 1000 on Sunday morning we raised the spud pole, recovered our pin and continued along the canal into Aigues-Mortes. The town had been built by King Louis IX in the thirteenth century to give France a Mediterranean port. In the main square is a statue of the King, later canonised as Saint Louis, setting out on the Crusades.

Dominating the town is tour de Constance. Its six metre thick walls were once used to imprison religious and political dissidents. It now houses the tourist office. Around the corner to starboard is a low railway bridge, with a clearance of only a metre beneath it. It is normally open, unless there is a train scheduled, when it closes ten minutes before the scheduled time. We hoped none was due.

We were in luck, and we passed through the open bridge and headed up the branch leading back to le Canal du Rhône à Sète. Since it was Sunday, a common change-over day for the bangy boats, we met a mix of wildly zigzagging novices having just picked-up their boats and more stable skippers with a week of practice.

After about nine kilometres we began a section of the canal that runs along a narrow sandbar with the Mediterranean on one side and the étangs on the other. The étangs here are slightly salty ponds isolated from the sea by bars and low dunes generally rimmed with salt marshes. Pink flamingos are a common sight here, as are, unfortunately, derelict campers, huts and hovels.

We continued to pass derelict boats moored along the canal banks. Amazingly, the boat pictured here had not been abandoned; we saw people living aboard. To us they appeared to have fallen a little behind on their maintenance and they have an rather strange sense of tidiness.

As we passed the rental base in Carnon, we were relieved to see so many bangy boats moored. The marina appeared full and they were rafted two deep along tha canal, meaning that only a small portion of their fleet was out jeopardising other boat traffic.

Five kilometres further along, at 1250 we secured with spud pole and a stern line to a bollard in le halte nautique les Quatre Canaux. After we had secured, we pedalled the kilometre into Palavas-les-Flots, a Mediterranean beach resort, where we found it a dead post-season community that likely would not have enthused us even in high season, so we pedalled back to relax aboard Zonder Zorg.

Shortly after 1000 on Monday we raised the spud pole, slipped and continued along the canal. A short distance along, we passed a long line of outboard skiffs, mostly 5 to 6 metres in length, which were moored in front of small houses on the thin bank between canal and étang. This was the étang fishing fleet.

Onward to the Midi

Not far beyond the tiny community, the bank separating the canal from the étang narrowed to a metre or less, and in some places had disappeared altogether.

Over its top we could see nets hung on pilings and frames, and here and there, a fisherman in a skiff working trapped fish out of the nets. Herons and gulls dot the banks using a more basic method of fishing.

Across the canal, on the Mediterranean side is a towpath. This is well used by hikers and cyclists, including some that pause to try their luck at catching any fish that had escaped the nets and birds.

We passed another tiny fishing community, a few houses on a fattening of the bank. These communities are on narrow islands of canal bank and their only method of access is by boat. All the way along the canal we had seen the banks have been eroding and gradually disappearing.

The speed limit is 8 km/h and this is prominently posted frequently along the canal. The local fishermen, whose homes and livelihoods depend on the banks, speed by in their runabouts at 20, 25 and more km/h, putting up huge wakes. They all seem to be in a great rush to get to and from their nets before their banks erode.

At 1200 we secured to bollards on the quai just downstream of the lift bridge in Frontignan. The mechanism of the lift bridge is in deteriorated condition, so it is opened only twice a day, ten minutes each at 0830 and at 1600. We had decided to arrive early, do some provisioning and top off our water tanks. Under a hinged cover in the quai, I found the water tap shown in the guides. It was dusty and covered in cobwebs, and when I put a wrench to the seized faucet and turned it, nothing came out.

We walked across the bridge and along to the moorings on its upstream side. There we found modern electrical and water pylons along the quay. A minute before 1600, nearly all the boats on both sides of the bridge had engines running in anticipation of the opening. Not far behind schedule, the bridge opened and the boats swapped sides. We secured to the quai and found the modern pylons not working. The water and electrical outlets worked on tokens, and the token machine was out of order. A phone call to the Capitainerie in the commercial port brought a man who switched on the pylons, bypassing the need for tokens, and we all enjoyed free water and shore power.

Frontignan is famous for its sweet white wine: Muscat de Frontignan, a vin doux naturel made from the Muscat Blanc à Petits Grains in a technique that dates to the thirteenth century.

At 1000 on Tuesday, 8 October we slipped our lines and continued up canal. The sky was completely clear, there was not a breath of wind and the temperature was already approaching 25°. On our way out of Frontignan, we motored slowly past several more boats threatening to sink. One wooden one was so nail sick that I was expecting planks to fall off as we slowly passed.

Onward to the Midi

Very shortly we came to the end of le Canal du Rhône à Sète and entered le Bassin des Eaux Blanches, a lobe of l'Étang de Thau. Thau is a shallow lake of brackish water about seventeen kilometres long. Filling nearly half its area are oyster farms. These are on the northern side of the lake and a buoyed navigational channel runs across their southern edge. As well, there are navigational channels through the oyster farms to the three towns on the north shore: Bouzigues, Mèze and Marseillan.

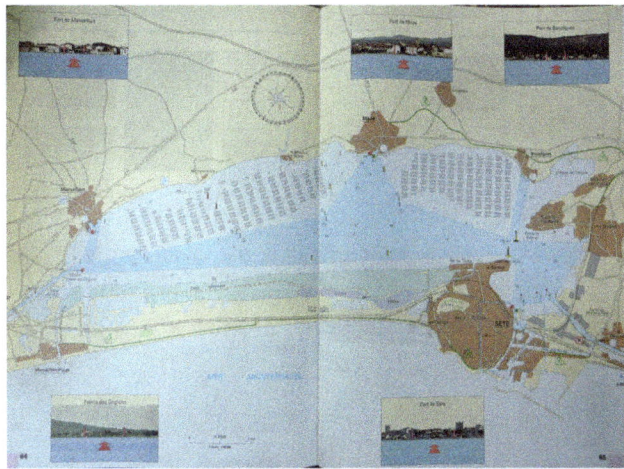

I headed on a course to take us between the cardinal buoys at the narrows. Once we had put them in our stern, I steered on a lump of land that was just rising over the horizon. From the chartlet in the Guide Fluvial and the chart on the Navionics app on my iPad, I determined this was the hill behind Pointe des Onglous, which is at the entrance to le Canal du Midi. We cut across the lake, ignoring the dogleg of the buoyed channel and saving a bit of time exposed to the building wind.

About an hour and three-quarters into the crossing our course brought us to the channel along the edge of the oyster farms. The annual harvest from these is around 20,000 tonnes, about ten percent of total annual consumption of oysters in France.

Shortly after reaching the buoyed channel, we passed the entrance to the fishing port of Marsiellan across from a channel out to the Mediterranean. With its access from the sea, Marsiellan is a popular yachting port and its mooring fees reflect this. The fees listed in the guide made our decision not to stop an easy one.

As we approached the lighthouse on the breakwater at Point des Onglous, the steep wind chop gradually abated and then we found lee.

At 1225 we left the waters of Étang de Thau and immediately began motoring past a long line of derelict boats. Many of these appeared to have been abandoned and many were sunk at their moorings.

The majority of the sunken vessels were sailboats, mostly small sloops under eight metres in length. With the proximity to the sailing waters of Étang de Thau and the Mediterranean, these boats were likely seeking the free and apparently unregulated moorage of the canal. This is the beginning of the Canal du Midi.

None of the sinkings appeared to have been recent occurrences. We were amazed that they remained untouched by the VNF, with seemingly no concern for the pollution of the water nor of the view. Most had been stripped of all their rig and gear, but one still had a cradled mast.

Onward to the Midi

Not all the derelict, abandoned and sunk craft were sailboats. As we moved beyond the easy reach of the sea and the étang, the sailboats had left space for power boats to be abandoned. Among the derelicts were also cheap and graceless liveaboard conversions that are prime candidates for future dereliction.

Four and a half kilometres along the canal from Étang de Thau, we came to the first lock. It was closed for lunch until 1330, so we paused for a bite to eat ourselves, hoping that as we continued beyond the lock we would see a more pristine Canal du Midi.

Chapter Twenty-Four

The Canal du Midi

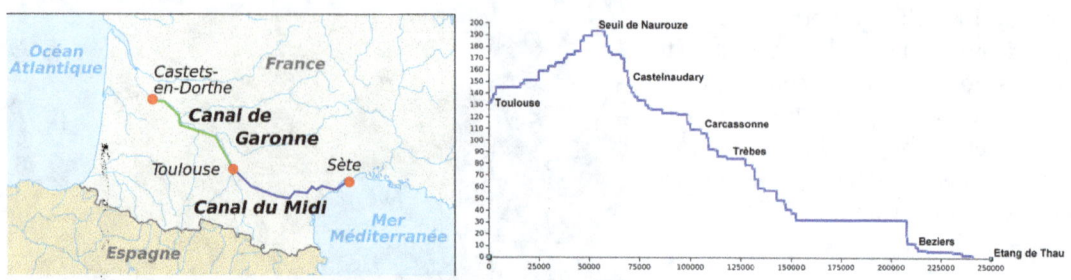

The Canal du Midi connects the Étang de Thau on the Mediterranean to Toulouse. In doing this, it rises 189.4 metres through 74 locks over a distance of 183.5 kilometres to a 5 kilometre summit pound. From there it descends 57.18 metres through 17 locks in 51.6 kilometres to Toulouse, where it connects to the Canal de Garonne, which leads past Bordeaux to the Atlantic.

Construction of the canal was begun in 1666 and it was completed in 1681. The inspiration and driving force behind its construction was Pierre-Paul Riquet. At the time it was widely praised as one of the great achievements of the seventeenth century. Initially it was named Canal royal en Languedoc, but with the Revolution in 1789 it was renamed Canal du Midi. In 1996 it was named as a UNESCO World Heritage Site, described as: "one of the most remarkable feats of civil engineering in modern times".

While Edi prepared a quick lunch for us, I walked up to the lock to look at the bollard arrangement and to plan our passage through it. Most of the locks on the canal have curved sided chambers, though a few have been rebuilt with parallel sides. The original dimensions of the chambers are 29.2 metres long with a maximum width of 11 metres at the centres. The entrances past the doors are 5.8 metres wide, the design principal being to allow two 20-metre barges of 5 metres width to easily fit into the lock simultaneously. The average height change of the locks is 2.1 metres.

Shortly after 1330, madame l'éclusiere finished her lunch break and pushed the appropriate buttons on her portable box to open the gates and allow us in. The rise of this lock is only 1.51 metres, so looping the bollards was easy, though with a broken mid-point bollard not replaced, we were rather strung-out with our lines. We made it through our first curved lock unscathed and motored out with a better idea of what to expect.

Very soon out of the lock we passed through a section of canal where all the plane trees had been cut down. In 2006 the first outbreaks of coloured canker were detected. This disease is caused by a microscopic fungus, now believed to have come to France during World War Two in the wood of US Army ammunition boxes. The number of infected trees reach 83 in 2008 and 153 in 2009.

Selective tree-felling campaigns have been conducted to try to stop the spread but with no effect. In addition there is no effective treatment against the disease. By 2011, a total of 1,338 diseased trees had been identified in 211 separate areas.

There is now a fifteen-year program to remove all 42,000 plane trees along the Canal du Midi and replace them with other species. Costs will run to a quarter billion Euros. They had been planted in the 1830s to help stabilize the canal banks and the plane had been selected because its roots grow in dense tangles and don't mind wet feet.

As we approached écluse de Garde de Prades we finally came to a section of canal with seemingly healthy trees; at least they were all standing and all still had leaves. The écluse de garde serves as a flood gate and is normally open unless l'Hérault is in flood. We passed through and out onto the river, which for a short stretch serves as as the navigational channel. After about a kilometre we turned off into an arm leading to l'écluse ronde d'Agde.

This lock was built in 1680 with a perfectly round chamber to allow barges to enter and choose to be lifted and go straight ahead into the continuation of the Canal du Midi, or to be lowered and turn left into the river Hérault below the weir. It was a wonderful piece of seventeenth century ingenuity and engineering. Unfortunately, its beauty and symmetry were destroyed in 1984 when it was modified to allow Freycinet gauge grain barges to use it.

About a kilometre and a half after the lock we came to one of the more famous bridges along the canal. Le Pont des Trois Yeux, the Bridge with Three Eyes. This is one of the 126 stone bridges that had been built across the canal, many of which are still in use today. The bridge gracefully spans a wider section of the canal, providing one arch for navigation, one for the towpath and one to allow better water flow.

Four kilometres further along we came to a complex structure, l'ouvrages du Libron, the works of Libron. This unique structure allows the canal to traverse the Libron River. Where the canal and river intersect, the Libron is at sea-level and the Canal du Midi is very slightly above, so a traditional aqueduct was not an option.

The problem was further exacerbated by the Libron's propensity to flash flood up to twenty times a year. The problem was originally solved by the building of a pontoon aqueduct known as the Libron Raft which utilised a flush-decked barge to fill and protect the canal channel in times of flooding, allowing the river and its debris to flow over the barge.

Because the raft blocked the canal, navigation was closed until the flooding subsided and the barge could be removed. To replace the barge, in 1855 the works were built to better allow the two streams to coexist. The structure was designed to allow the river to be directed to enable a canal boat to safely pass, while limiting the mud and debris being deposited into the canal by the flooding river.

The works allows the river path nearest an approaching boat to be stopped for sufficient time to allow the boat to cross through that area and rest for a time in the "protected area" between the two paths. The river path behind the boat is then returned to flow and the path in front of the boat is halted, allowing the boat to cross this second path without interference. This *new* solution, which is over a hundred fifty years old, still elegantly solves the problem.

After we had uneventfully passed through écluse Portiragnes and under another graceful stone bridge, at 1640 we secured to bollards on a quai just downstream from the lock in Villeneuve-lès-Béziers. We had come 49.9 kilometres and had passed through three locks since leaving Frontignan mid-morning.

At 0855 on Wednesday morning we slipped and headed up toward the next lock a hundred metres along. There was no delay in passing through it, nor with the next one a kilometre further up canal. Four kilometres further along we pulled over and secured to bollards to wait for three boats to be cycled down in écluse Béziers.

While we were waiting we were entertained by the manoeuvring antics of a bangy boat attempting to moor on the canal bank astern of us.

We watched as it somehow got perpendicular to the Left Bank. With no wind and with negligible current in the canal, we could not figure out how the skipper managed to do it. Then, as if to reverse the error, the boat somehow managed to do a 180° turn and ended up with its bow on the Right Bank.

The boat was sufficiently far away from us that we thought it safe to divert our attention and watch the parade of boats coming out of the lock. They were all rentals, and offered a more immediate threat. By the time they were safely past us, the bangy astern had finally managed to secure parallel to the bank.

Écluse Béziers is a modern parallel-sided lock built to replace the original curved-sided double lock. Because it replaces two locks, it is double height. At 4.24 metres, it would be nearly impossible for an up-bound boat to loop the bollards at the top, so the chamber has mooring rods set in indents in the walls. The looped mooring lines slide up these as the water rises.

The next lock, écluse Orb is only 375 metres further along and appears to replace an original multiple set of locks that brought traffic back up from the River Orb. It has parallel sides, is 6.19 metres high and has the same system of mooring rods. Locking through both of these was easy.

From écluse Orb we motored across le pont-canal de Béziers. This superb bridge was built in 1856 to finally end the need for barges to descend to the Orb, cross it and lock back up the other side. It is 198 metres long supported on arches of seventeen metre span and it is considered to be by far the most attractive of the nineteenth century stone canal bridges.

A kilometre further along we came to the traffic light controlling the entrance to l'échelle de Fonserannes. We were delighted that it was green as we approached; passage is tightly scheduled. Up-bound traffic this time of year is between 1000 and 1215 and then 1600 and 1815 and the passage takes 30 to 45 minutes. L'écluse de Fonserannes, which is also called les neuf écluses, is a staircase lock that now consists of seven interconnected ovoid lock chambers. There were originally two additional locks that connected the bottom of the staircase to the river, but since the 1856 opening of the aqueduct over the river, these are no longer used and the entrance to the system is now in the side of the seventh chamber.

I dropped Edi off on the quai at the entrance and then continued into the chamber, while she climbed the stone steps to the top of the first lock to take the lines. We settled into place and the gates closed behind us.

The sluices were opened to flood our chamber from the next one up the ladder. When the next chamber above us had drained into our chamber, our upstream doors were opened and the sluices in the next set of doors were opened to continue filling the two chambers. The initial surge of water was a bit frightening to see coming down the chamber toward our bows like a tidal bore. However, Zonder Zorg was well-positioned and the brunt of the wall of water passed down our port side making it rather easy tending the bow line to keep us against the wall.

We had the stern line led aft to a bollard as a spring and I was motoring forward on it with the tiller lashed to keep the stern in. I monitored this from the bow and had to tie-off the bow line and go aft only once or twice per lock to adjust the stern line or the tiller.

Once the water level in our chamber had reached the proper depth above the gate cill, l'éclusier motioned us to move forward into the next chamber. As we waited for the next buttons to be pressed by l'éclusier, we had the entertainment of watching the bangy boat bounce its way in and bumble alongside the other wall. Finally the gates were closed astern of us and the process repeated.

Les neuf écluses, was built to take barges 21.5 metres up and down the steep embankment above the valley of the Orb. It was one of the major engineering accomplishments on the canal. Pierre-Paul Riquet had planned and engineered and lobbied for thirty years before King Louis XIV approved the construction of the canal in 1666. Among his planning was a visit to the Canal de Briare, where he studied les Septs Écluses, the flight of seven locks there. The Briare had been completed in 1642 to connect the Loire to the Seine, its 57 kilometre length and 36 locks taking barge traffic up 41 metres from the Loire Valley and then down 85 metres on the other side toward the Seine. It was the first summit level canal with pound locks in Europe and Riquet gathered ideas from it for the Canal du Midi.

The Canal du Midi

The building of a canal to link the Atlantic to the Mediterranean was an old idea with many projects being previously devised. Leaders such as Augustus, Nero, Charlemagne, François I, Charles IX and Henry IV had dreamed of it for its political and economic benefits. François I even brought Leonardo da Vinci to France in 1516 and commissioned him to survey a route for the section from Toulouse to Carcassonne.

We quickly progressed up the staircase of locks and a little over twenty-five minutes after we had entered the first chamber, we were looking into the final one. Then a couple of minutes later we were in it and rising to its top. L'échelle de Fonserannes is one of the three most popular tourist destinations in the Midi, along with Pont de Gard and Carcassonne and we were closely watched all the way up. Too closely! The tourists obliviously interfered with our attempts to get mooring lines onto bollards and they very dangerously and obviously ignorantly stood in line bights.

The view back down the flight is superb, and it is made all the more so by having just spent half an hour getting to know it intimately.

At 1211, thirty-one minutes after we had entered the first chamber, we exited the final one. L'éclusier could take an extra four minutes for lunch. From the top of l'échelle de Fonserannes begins the longest bief of the canal. For 54.2 kilometres there are no locks as the canal winds around the sides of low hills and ridges at 33 metres above sea level.

We motored out and soon came upon a logging scene, as crews worked at removing infected plane trees from the canal banks. At 1303 we came to the spud pole and a pin pounded in the bank just upstream of Pont de Colombiers. We had initially stopped for lunch, but found the place so pleasant and restful that we decided to stop for the day. We had come 14.2 kilometres and up 11 locks.

Astern of us was a bangy boat flying a huge Canadian flag. We chatted for a while with the two couples from Vancouver who were aboard; they were near the end of a two week rental. As we chatted, we watched a prime candidate for the ugliest boat design award motor past.

After a very peaceful night, at 1055 on Thursday we raised the spud pole, retrieved our mooring pin and continued up the canal. Within a kilometre and a half we came to the entrance to le tunnel de Malpas, which was the the first canal tunnel built in Europe. As we approached, we acknowledged the canalside signage by blowing our air horn to signal our presence to any oncoming traffic. We were hoping that any oncoming bangy boat skippers were aware of the meaning of our signal and knew to blow their own horn.

We received no response from our toot, so we continued into the one-way tunnel. The interior is lined with a vault of cut stone and there is a narrow towpath along the Left Bank side. About midway through, the cut stone vaulting gives way to a coarse spongelike surface, which some sources call sprayed concrete and others eroded limestone. It looks like a combination of the two.

Just beyond the exit were two bangy boats waiting for us to complete the passage. They had obviously heard our horn signal and stopped. Because they were down-bound vessels, they had priority, but this fact and the requirement to blow a sound signal seemed not to have been part of their few minutes of pre-rental instruction.

At 1252 we secured to bollards on the bank about twenty metres short of the new bridge in Capestang. We had come only 11.8 kilometres and it was still early, but this looked to be a very pleasant place, so we decided to pause for a while.

Beside us was the wifi antenna we had read about and there were also shore power and water outlets. A German man came from the boat moored 50 metres astern of us to take our lines, but we were secured by the time he arrived. We asked if he had the wifi code and he told us the system had never worked since he arrived two years previously. Rather than paying €18 per day for moorage with no wifi, we moved back down the canal to the first available free spot. There was no water or electricity there, but our battery was fully charged and we still had over 500 litres in our water tanks.

Nor were there any bollards, so we secured to a pair of mooring pins we pounded into the bank.

Moored ahead of us was Vrouwe Antje, a 1902 Dutch klipper with a youngish British couple, John and Jane and their 3-year-old daughter, Sophie. We spoke with them and learned they have been cruising and sailing their antique klipper through Europe since 2006. John is a pilot for British Airways, and he commutes from wherever the barge is to London to fly his monthly allotted schedule. Because he usually has this compressed into a single run of days, this arrangement leaves him with about two-thirds of every month to enjoy the barge.

The old town of Capestang lies on the slopes below the canal and from the banks beside us we had a fine view across the roofs to the massive bulk of thirteenth century Collegiate church of St Etienne.

The church, intended to be the largest in the region, was begun on the remains of an eleventh century church, but never completed. The nave and transept were built, but funding ran out before a choir, apse or ambulatory were started, thus the strange shape.

We spent a restful few days in Capestang. Among our daily walks was one down the hill for breakfast croissants or a multigrain baguette from the artisan baker in the square across from the church. There is a large Intermarché at the edge of town, less than half a kilometre from our mooring, so we had plenty of fresh provisions.

Sunday morning is market day in Capestang. The set-up is in the square by the church and a visit to the stalls provided us with some more fresh produce and some local cheese, and on the way back, bread from the baker.

The Canal du Midi

Shortly after 1100 on Sunday we recovered our mooring pins and motored the hundred metres or so to a set of bollards under the bridge and connected our hose to the water outlet there. With our water tanks again full, we slipped just after noon and continued up the canal. The banks were lined with moored boats, a few itinerants, but mostly boats and barges buttoned-up for the winter.

We shortly arrived at le Pont de Capestang, the infamous wrecker of boat and barge superstructures. While not the lowest bridge on the Canal du Midi, its somewhat irregular arch has low shoulders that have rearranged many barge wheelhouses.

As we neared, we could see the obvious chipped masonry where centuries of barges have inadvertently enlarged the opening. We saw no need to lower our mast and we watched as it easily cleared the centre of the arch. The chipping away of the edges of the masonry was even more advanced on the upstream side of the bridge.

For some reason there is a fable spread through the barging community that le Pont de Capestang is the lowest bridge on the Canal du Midi. It is common for purchasers of barges or designers of wheelhouses to base their decisions on the dimensions and profile of this bridge. Unfortunately for them, there are lower and more restrictive bridges further along the canal.

We followed the canal as it wound a crooked route along the sidehill. Several times we were able to look back at Capestang across a stream gully or from the end of a rib. The windings of the canal are such that the first nine kilometres of canal out of Capestang took us only 4.4 kilometres away from the town.

It was very pleasant motoring along all by ourselves under cloudless skies with the temperature in the mid-twenties. It was absolutely calm with our motion generating a gentle cooling breeze. Edi sat in her lounge chair on the foredeck reading or knitting and from time to time giving me an advanced look for oncoming boats around a bend or through a bridge hole.

Because we are traveling upstream, we need to give way to down-bound traffic at narrow spots. At this bridge we just made it before a large down-bound barge, which was not visible until we were within a few metres of, and committed to the hole.

We crossed several aqueducts across streams that flow down the slopes. These were a necessary part of the complex hydraulic works that were done over three centuries ago to build the canal.

At 1736 we approached the bank and dropped the spud pole just upstream of le Pont de Roubia. We had come 33.7 kilometres and we decided to stop for the day. To secure our stern, we used a pin pounded into the bank, ignoring the steel eyes that had been bolted to the tops of the infected plane tree stumps. Although mooring to trees is forbidden, it is very commonly done with no obvious policing.

Here the village of Roubia seems to encourage it by adding mooring eyes to the stumps, so that those who were accustomed to using the trees for moorage can still easily stop. It is now believed that the rapid spread of the canker stain has been exacerbated by mooring lines around infected trees picking-up spores and moving them along the canal to healthy trees. We were stunned to see the practice being so openly encouraged.

Monday was again a clear, warm morning when at 1038 we raised the spud pole, recovered our mooring pin and continued along the canal. As we motored, we continued to see derelict and sunken boats.

In less than two kilometres we arrived at the lock that marks the end of the 54.2 kilometre pound and we were joined in the chamber by a British sailor in his 7-metre boat. Over the next five kilometres we passed through double locks at both Pechlaurier and Ognon and then a single lock at Homps.

At 1240 we secured to rings on the quai just upstream of the first bridge in Homps. We had come another 9.3 kilometres and had passed through 6 locks. The place looked pleasant, so we decided to call it a day. There is a small supermarket about half a kilometre through town and out on its edge on the through road. A short distance along from our moorage were the rental bases for Le Boat and Les Canalous. We were relieved to see that most of their rental boats were moored in their ports, thus greatly reducing the number of them that would be out jeopardizing safe navigation.

At 1103 on Tuesday morning we slipped and continued up the canal. Two and a half kilometres along we arrived at the next lock, écluse Jouarres, where there was a down-bound boat locking through, so we moored on the wooden waiting quai, trusting our lines to some wobbly posts sitting in the canal bank and passing as bollards.

From the top of the lock we continued along through placid water and under another graceful stone bridge.

Across an aqueduct, around a bend and across another aqueduct, we came to Épanchoire d'Argentdouble, which was built in 1693 to allow excess water to spill out of the canal. The towpath crosses the spillway on a series of stone arches. At 1222 we arrived in La Redort to find a very tranquil spot, so after only 6.4 kilometres and one lock, we stopped for the day.

On the way to the Intermarché is the mouth of a tunnel across the street from Château La Redorte. The château was built at the beginning of the 18th century and in the 1840s it was acquired by Count Mathieu de la Redorte, who restored it and enlarged its gardens and their enclosing walls. He had been Ambassador of France to the Court of Madrid and was later active in politics. He was on the wrong side during Louis-Napoléon's coup in 1851, and he was banished to his château and forbidden to set foot in his gardens. To deal with this and to comply to the letter of the law, if not its spirit, he had a tunnel built allowing him to leave the château without setting foot in its gardens.

We spent a relaxing two days in La Redorte, where in the Tourist Office we finally found a working wifi with a working connection to the Internet.

The widespread lack of connectivity in France today was in large measure caused by the country continuing to use the Minitel far beyond its usefulness. France in the 1980s and 90s was well in advance of the rest of the world with computer connectivity. With the Minitel, users could make train reservations, do online purchases, search the telephone directory, check stock prices, have a mail box and chat in ways that are now possible with the Internet. At the end of the century, forty-two percent of the population was connected by Minitel. It was not expandable; it was dead-end technology. The French held on. The service was finally retired in the middle of 2012.

The Canal du Midi

Shortly past 1100 on Thursday we slipped our lines and continued up the canal. We passed a work site where the stumps and roots of cut-down plane trees were being grubbed out of the canal bank. Three kilometres along we came to écluse Puichéric, a double lock in which we could see three boats just beginning their descent. We pulled over to wait.

While waiting we walked up to the locks to watch the action. Edi stopped to talk with a British couple in a narrow boat.

When the locks were clear, we locked up and headed along the three kilometres to the next lock, écluse Aiguille, also a double one. It was closed for lunch, so we secured to the bank astern a bangy boat overflowing with a group of German men. There is a water tap symbol shown in the guides, and I had placed our bows next to the tap to make it easy to fill our tanks.

L'éclusier returned from lunch a comfortable time after the scheduled 1330 reopening and we headed into the lock to work our way up. A kilometre and a half further along we came to écluse St-Martin, another double one, where we waited for down-bound boats. Between the time we had first seen the Germans and the time we entered the lock, we had seen them down four rounds of steins of beer and glasses of wine. Possibly they thought the swerving from their alcohol impairment would counter their swerving from inept steering.

We worked our way through the single écluse St Martin and then the triple écluse Fonfile. As we approached écluse Marsiellette we saw a luxemotor with a large plane tree laying across it. At first we thought: "Oops!", but as we passed we realized the barge was being used as a temporary platform for felled trees, which were then cleared away by equipment on the bank.

We locked through the single écluse Marseillette and at the end of the tree-clearing operation we passed a very pretty skûtsje of about 12 metres length with a full sailing rig. Just beyond it at 1640 we dropped the spud pole and secured a stern line to a post in the recently denuded bank. We had come another 13.7 kilometres and had passed through ten more locks. Beside us was the road leading up into the village to the bakery.

On Friday morning We motored along the next bief past dead and dying plane trees. Then we came to a major logging operation, where an assortment of heavy equipment was on the bank cleaning-up from a very recent tree-cutting spree. Pulling up the rear was a man with a huge vacuum cleaner sucking up the wood chips and blowing them into a truck. We wondered whether the dust drifting away from the operation contained the spores of the canker stain fungus. A pair of white swans serenely surveyed their rapidly changing home.

We arrived at the triple lock, écluse Trèbes shortly after a boat had started down the series. We knew that we would have a long wait, so we wandered up to watch the operation of the lock. Even though the locks were cycling the boats down, it was valuable to watch the workings of the multiple chambers from a totally different and detached viewpoint. We gained additional dimensions to our understanding of the system and again saw how dramatically easier it is to cycle down than it is up.

We were finally allowed in and worked our way to the top to exit past an upscale restaurant installed in the former mill house. Upstream of Trèbes we passed under another graceful old stone bridge.

Then shortly thereafter, we spotted a short concrete quay with stout bollards in a quiet rural setting. Even though we had come just 11 kilometres and had worked our way through only a triple lock, we quickly decided to stop for the day.

Up the hill from us were vineyards of the Minervois and across the canal and down the hill were planted grapes for Corbières. I checked the Minervois vines for overlooked clusters, but because they had been mechanically harvested, there was not a single remaining grape.

Carefree Through 1001 French Locks

At 1013 on Saturday we slipped our lines and continued up canal. We moved along the two kilometres to the next lock at the canal speed limit of 8 kilometres per hour. Very soon we had a boat closing quickly from astern. I checked the GPS speedometer on my iPad to confirm I wasn't dawdling along and it confirmed we were making a point or two either side of 8. The boat continued to close rapidly at what must have been its hull speed, about double the speed limit. I mentioned the canal speed limit and bank damage to the skipper as he joined us in the next lock. He said he had no way of knowing the speed of the boat, so he just opened it up, thinking the rental company would have a speed governor on the engine.

With the bangy boat more slowly in our wake, over the next five kilometres we passed through two more single locks, then a double and another single. Then we came around a bend and through the gaps between the trees, we caught our first glimpses of La Cité, the medieval fortified city.

At 1245 on 19 October, four months to the day after we had left Harlingen, we arrived in le Port de Carcassonne and secured to the quai just below Pont de Marengo. Zonder Zorg had arrived in her winter home. We had come 2072 kilometres and had worked our way through 332 locks as we navigated 75 different canals, rivers and lakes and crossed three countries.

Chapter Twenty-Five

Wrapping-Up for the Year

Just forward of our temporary moorage was Pont de Marengo, the bridge that leads from the rail station to the la Ville Basse, the main shopping district. When we were in Southern France in mid-February on our way back to Friesland from Spain, we had stopped in Carcassonne to investigate possibilities for winter moorage and we were delighted to have secured a place for Zonder Zorg here.

Our assigned moorage spot for the winter was through the bridge, up the adjacent lock and across the basin, but with a hotel barge in our spot until Monday noon, we remained on the quai below the bridge and lock. Edi hauled out the sewing machine and began assembling waterproof nylon covers to protect Zonder Zorg's lier, coo coo, fries and roer. Meanwhile, I went off into the city to scout-out the supermarkets and other shops and services.

The Port is located on the north edge of Bastide St Louis or Ville Basse, which is referred to as the New Town. New because it dates only to 1247, when Louis IX founded it across the River Aude from La Cité, the fortified citadel on the hill. The first settlement in the area dates back about 5500 years, but it wasn't until about 2100 years ago that the Romans fortified the hilltop. In 462 they formally ceded the province of Septimania to the Visigoth King Theodoric II, who had captured Carcassonne in 453. Through the following thousand years and more, Carcassonne was the target of attack by seemingly anyone with any aspiration to power in Western Europe. Among these were the Goths, the Franks, the Saracens, the Burgundians, the Moors, the Albigensians, and the Cathars.

Carcassonne is home to about 50,000 people. It is about mid-way along the Canal du Midi from the Mediterranean to Toulouse, and is within a few kilometre from being the most southerly part of the French canal system. A major portion of its businesses is tourism, centered around La Cité and the Canal du Midi, both of which are UNESCO World Heritage Sites. Wine growing is also a major focus here, it being in the heart of the Languedoc-Roussillon wine region, with many famous areas, such as Minervois, Corbières and Limoux nearby.

While we waited for our moorage to be freed, we were entertained by bangy boats going past us and under the rail bridge. There was no wind or current, they were not sailboats, but they appeared to be tacking down the canal. The first one made it through unscathed. But the second one decided to take an adventurous route.

It crashed into the bridge abutment, bounced off and reversed to bang its port quarter on the other abutment. With lots of engine noise and only two more bounces, it made it through the bridge hole, only to crash into the bank.

It clawed itself off the bank and crossed the canal astern of an up-bound boat and was last seen crashing into boats moored on the opposite bank.

I have from time to time through this book vented my thoughts on rental boats and the habit of the rental companies allowing people with absolutely no boating experience, knowledge or skills to rent boats up to fifteen metres in length. The companies get away with this by giving a short briefing on the operation, handling and navigation of the boat. In a few minutes, a complete novice becomes a very dangerous captain. When I rented my first boat in France in 1984,

I was shown how to start and stop the engine, how to shift to forward and astern, how to use the galley stove and how to flush the heads. With electric heads these days, the training briefing has likely been pared down.

On a more professional level, we watched as Lady Sue approached Pont de Marengo, which at 3.33 metres is the lowest bridge on the entire Canal du Midi. They gracefully glided through and into the lock with a couple of centimetres to spare above their pilothouse roof.

Écluse Carcassonne is immediately beyond the bridge and lifts the canal 3.32 metres to the basin of the port. We went through late on Monday afternoon, after the skipper of the French hotel barge had finally been convinced that he was more than five hours beyond his promised departure from our contracted mooring. Our assigned spot was across the basin from the lock on the first piece of canal bank beyond. The finger floats in the basin are meant for boats of less than fifteen metres.

After a rather grey and cool few days, Tuesday was splendidly clear and warm, with the temperature passing through 25° before midday. While Edi continued sewing covers, I worked at little jobs to prepare Zonder Zorg for wrapping-up for the winter.

We began planning our trip back to Vancouver by looking at stand-by seats on Air Canada flights from Barcelona, Paris and London, which had one, two and six daily departures respectively. We chose to fly through London, primarily because it had the most departures, but also because it was the only one with a non-stop flight to Vancouver.

Getting to London began with catching a train, so I walked up to the train station, which is directly beside the port, to look at schedules. Of the six departures listed on the board, three were late: 15 minutes, 25 minutes and 1 hour 15 minutes. We needed to leave lots of slop time in planning the first leg of the trip.

From the Capitain de Port I got the contact information for a young man who had two years previously set-up his own boat service and repair business after years doing similar work for a boat rental company. He agreed to come at 1400 on Thursday to winterize Zonder Zorg. I could have done it myself; however, clauses in most insurance policies mention denying freezing damage claims unless preventive work had been done by a licensed professional.

We spent much of Thursday morning buying train tickets to Toulouse, checking bus schedules from the Toulouse train station to the airport, booking seats to London Gatwick and printing the boarding passes, booking a taxi from Gatwick to Heathrow, reserving a room at the Ibis Heathrow and printing boarding passes from London to Vancouver. It was more complex returning to Vancouver from Carcassonne than it had been two years previously from Patagonia. In our spare time, we cleaned and packed.

Loic arrived in French fashion, half an hour late for his appointment. Two hours and €155 later, we had completed winterizing the engine, generator, furnace, hot water system, dish washer, washing machine, sinks, shower, toilet and waste sumps. Edi and I then did a final clean-up, locked-up, lashed the cover over the doors and said farewell to Zonder Zorg for a season.

Wrapping-Up for the Year

We had only carry-on luggage, so the walk to the train station was pleasant. The train was on time, the bus connection was easy, as was the Easy Jet to Gatwick. Our taxi was waiting for us and the hotel reservation had worked. Friday morning the shuttle bus efficiently got us to Terminal 3, our seats were confirmed on the flight to Vancouver and it departed on time, a SkyTrain was waiting as we reached the platform in Vancouver and it wasn't raining for the block and a half walk from the station to our loft.

We had arrived in Carcassonne thirteen and a half months after we had taken possession of our skûtsje. During that time, Zonder Zorg had spent eight months in the shipyard preparing for and undergoing her refit and we had spent just over five months boating. In our boating before and after the refit, we travelled a total of 2370 kilometres on 102 waterways and passed through 342 locks and three countries. Among our observations are the following:

In the Netherlands:

- Pleasure boating is very common among the Dutch; there are about 150,000 cruising boats in the small nation.
- The waterways and their infrastructure are set-up to make pleasure boaters feel welcome.
- The arrangements in locks and in the waiting areas before them, appear to have been designed by people who understand the needs of the boater.
- The environment in and around the canals is pristine and well cared for.
- During our eleven weeks on the Dutch canals, we saw no broken facilities.

In France:

- Inland pleasure boating is much less common among the locals. Of the 20,000 private pleasure boats actively cruising inland in France, over fifty percent are foreign.
- The large locks on the Saône and Rhône function efficiently, and although they are very busy with huge commercial barges, they are welcoming to pleasure boaters.
- On the other hand, the Freycinet gauge locks are in poor condition. During our eight weeks on the French canals, we found over two dozen with disfunctional mechanisms that required us to wait for a roving éclusier to override or repair.
- Many bollards are missing in the locks, often in critical positions. Safety fencing has been installed to protect the public and to make it extremely difficult for lock users. In many places, sluices dump a heavy cross-current immediately before the entrance to the lock chamber.
- The entire system appears to have been designed and maintained by people with little or no knowledge of and even less concern for the needs of the pleasure boater.

- In many places, the canal banks are lined with derelict, abandoned and sunken boats and barges and except in one place, little appears to be happening to address this pollution.
- Seventy-five percent of the annual thirty to thirty-five thousand rental boat contracts are to foreigners, who need no boating qualifications or experience.
- The canal banks are being badly eroded by speeding boats, mostly rental boats, which have no speed limiters.
- Convenient and inexpensive connectivity to the Internet is sparse, but this situation is slowly improving now that the Minitel has finally been abandoned.

With all these negatives, barging in France is still a wonderful experience. If only a few of these shortcomings can be addressed and remedied, then barging on the small French canals can return to the sublime experiences I remember from previous decades. With the continued strength in pleasure boating in France, perhaps some of the revenue generated will be directed toward repairing and improving the infrastructure and making it more boater friendly. That way we'll likely dawdle a bit longer in France before heading back up to the superb boating we found in the Netherlands.

Chapter Twenty-Six

Return to France

After having spent the winter of 2013-14 in Vancouver relaxing and doing presentations for the various yacht clubs, cruising associations and at the Vancouver International Boat Show, on Monday 10 March we rolled our luggage train to the SkyTrain station to begin our commute back to Zonder Zorg in Carcassonne. We had seats on a flight from Vancouver to Toronto with a connection to Paris and we had a rental car reserved for a week. Everything went smoothly and shortly after 1100 on Tuesday morning we were on the ring-road around Paris and heading toward the Midi.

In the early afternoon we paused at an autoroute rest stop for a baguette and cheese and a two hour nap to help stave-off the nine hour jet lag. Mid-evening, after about 800 kilometres of driving, we stopped in Brive-la-Gaillarde for the night. On Wednesday we decided to take the small roads, rather than the autoroute and had a delightfully winding drive through the Périgord and upper Dordogne, with pauses in Lauzerte and Cahors.

Shortly before 1700 we arrived aboard Zonder Zorg in Carcassonne. The weather covers that Edi had sewn

together had stood-up to the winds of the winter. They had done their job well protecting the decorative woodwork and keeping the interior fresh. The cover over the skylight had allowed us to lock one side open a small crack, and with the portlight in the shower slightly ajar and adding cross-ventilation, the interior had remained wonderfully fresh.

I filled the water tanks, purged the water lines, purged and charged the hot water tank, reconnected the gaz bottle and we were comfortably back aboard. After a quick trip to the supermarket for supplies for dinner and breakfast, we had mid-evening dinner and fell into bed to continue adjusting to the nine-hour time change. For the next few days we made full use of the rental car to chase around the stores picking-up obscure and bulky items that would be difficult to source or to transport later by bicycle.

On Saturday our shopping took us into the wine country to replenish Zonder Zorg's cellars. Carcassonne is adjacent to three famous wine regions, Limoux, Corbières and Minervois. We began with Limoux, where the basic sparkling wine process appears to have been invented. By 1531 the monks at the abbey of St.-Hilaire, a few kilometres southeast of Carcassonne, were making a bottle-fermented sparkling wine. By the early 1660s the British were deliberately adding sugar to induce a secondary fermentation to add sparkle to bottled wine. Among these revivified wines were those they had imported from the Champagne region. By the late 1690s, because of the popularity of the fashion in England, the wine producers in the Champagne had begun using the process themselves to make the predecessor of what we now know as Champagne. Dom Perignon, who is famously reputed and ceaselessly marked in the Champagne region as having invented Champagne, had nothing to do with the process of adding bubbles. In Limoux we visited wineries and tasted many wines, but left the region empty-handed; nothing had excited us.

We continued along eastward into Corbières, pausing at small producers along the way to taste. Finally, in the village of Monze we found some wines we liked. Domaine La Luze was established about a decade ago by Christian and Thérèse Caverivière to exploit the vineyards that Christian's father and grandfather had worked for many decades. The harvests had previously all been sold to the local Cave Coopérative, but Christian wanted more quality control, so once he had taken over the family vineyards, he built a small, very modern winery.

Return to France

The vineyards total seven hectares and are planted to Grenache Noir, Syrah, Carignan, Cinsault, Grenache Gris, Viognier and Marsanne. To ferment and condition these grapes, Christian has installed a range of temperature-controlled stainless steel fermenters in sizes that allow him to keep small lots separate. He makes a white, two rosés and four reds. The reds, which are all AOP Corbières, were of most interest to us.

The first of the reds is made from younger vines and lesser plots and is packaged in five litre containers and we passed on it, but the other three reds all impressed us. The 2011 Marie de Sobransac is made with a blend of Syrah, Grenache Gris and Carignan and offers a delightful mouthful of fresh red fruit with a pleasant tannic edge and a long finish. The 2010 Guillaume de Sobransac comes from a blend of Syrah and Grenache Noir and is aged in small oak. It has a firm structure with overtones of vanilla and chocolate layered on complex dark fruit flavours that carry into the long finish.

The 2010 Monz' & Merveilles is also made from Syrah and Grenache Noir and it is aged for twelve months in small oak, about twenty-five percent new. On the nose it shows aromas of coffee and plum jam and these persist into the palate, combining with intense cassis flavours. The power of the fruit is well supported by strong, persistent and well-integrated tannins, which should allow the wine to evolve finely over a decade or more. To show the maturation style of the Monz' & Merveilles, Christian opened a bottle of the 2008 for us to taste. We were immediately hit by aromas of cassis syrup. The tannins were resolving wonderfully and the intensity of the fruit was almost overpowering. Delighted with the wines, we bought three cases, and to thank us, Christian threw in a bottle of the 2008 Monz' & Merveilles.

Domaine La Luze is midway between Carcassonne and Lagrasse, so after we had loaded our wines into the car, we continued along toward Lagrasse looking for more wines. Many of the vineyards had completed pruning, but along the way we saw several where it was still ongoing. While the temperature was warm, in the upper teens, the wind was very strong and the workers were well bundled-up against it.

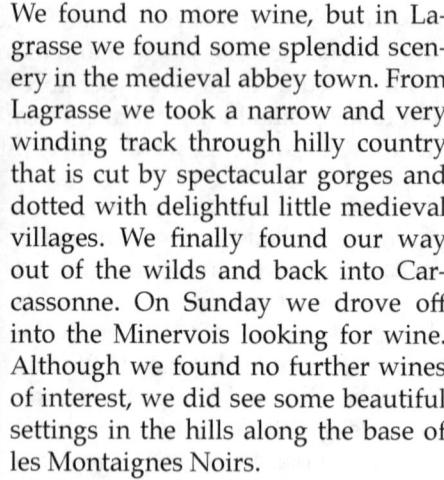We found no more wine, but in Lagrasse we found some splendid scenery in the medieval abbey town. From Lagrasse we took a narrow and very winding track through hilly country that is cut by spectacular gorges and dotted with delightful little medieval villages. We finally found our way out of the wilds and back into Carcassonne. On Sunday we drove off into the Minervois looking for wine. Although we found no further wines of interest, we did see some beautiful settings in the hills along the base of les Montaignes Noirs.

Back in Carcassonne, we took a walk through Ville Basse up to the fortified La Cité, which was inhabited by the Romans in the first century BC. Ville Basse, the Lower Town, which is also referred to as the New Town, since it dates to only 1247, was destroyed by the Black Prince in 1355 and then quickly reconstructed and fortified by the residents. We walked across the Vieux Pont, which had been built in the early thirteenth century to replace an older wooden bridge and give access between La Cité and Ville Basse.

We followed the roads to Place St-Gimer and the footpath that led up from there across the west slopes to Porte d'Aude. Back across the slopes, the view of the western ramparts and le Château Comtal was impressive. The hill has been fortified for at least two millennia, starting with the Romans, though possibly by others earlier.

The fortifications include a three-kilometre-long double wall with fifty-two towers. In 1849, because of their deteriorated state, the French government decided that the fortifications should be demolished. The locals strongly opposed this and a campaign was launched to preserve the fortress as an historical monument. Because of the pressure, the government reversed its decision, and in 1853 a restoration of La Cité was begun, led by architect Eugene Viollet-le-Duc. He continued until his death in 1879 and the work was largely completed by the end of the nineteenth century. Viollet-le-Duc was known for interpretive restorations, adding creative fantasy to historical fact. He had added a third tower to Notre Dame de Paris, and with La Cité, he added pointed roofs to the wall towers in a style typical of northern France, but unusual in the south.

Inside the walls are many shops and restaurants, two up-scale hotels and some guest houses. Although it was Sunday and all the businesses down in Ville Basse were closed, up in La Cité everything was open for business and they were mostly bustling with clientele.

After a final shopping spree Monday, on Tuesday we drove down to the Mediterranean coast and southward to Perpignan to return the rental car

and to catch a train to Barcelona. The newly completed rapid rail line linking the French SNCF and the Spanish RENFE systems means that passengers can now enjoy the high speed of the TGV to cross-border destinations. The 195-kilometre train trip took only an hour and a quarter, including two station stops.

We were in Barcelona to embark on a cruise to warmer climates to warm-up while we waited for the Canal du Midi to reopen for the season. Our ten-day itinerary would take us to Casablanca, Madeira, the Canaries, Malaga and back to Barcelona.

Chapter Twenty-Seven

Up Canal From Carcassonne

Mid afternoon on Friday, 28 March we returned to Zonder Zorg from our ten-day cruise. We settled-in and began preparing to continue up canal from Carcassonne. There is no large supermarket near the marina, but there are two small ones within a few blocks. Also, a market is set-up in Place Carnot on Tuesdays, Thursdays and Saturdays, so on Saturday morning we took our wheelie on a walk to begin provisioning Zonder Zorg's pantry and fridge. We paused at several stands tasting hams and sausages and selecting cuts of those that pleased us.

There was a wide selection of local and imported produce of varying quality, so after we had made a full round of the stalls, we chose potatoes from one, red and green peppers and a huge

cauliflower from another, avocados and mushrooms from a third and so on until our shopping cart was near full.

We topped-off the load with some multi-grain bread. After lunch we took a shopping bag to Monoprix, a mid-town department store with a small supermarket on its main floor. There we stocked up with meat, fish and a few grocery items to hold us until Zonder Zorg is closer to a large supermarket.

Up Canal From Carcassonne

On Monday morning I stopped by the Capitainerie to organise a fuel delivery for the following day. Before we had gone on our cruise, I had spoken with Stéphanie Bourgain, the Capitain du Port about getting diesel fuel for the boat. She said she could arrange for a truck to come to the boat, but the company needed a day's notice. While we were away she had polled the other boat owners in the basin and had compiled a list of seven boats wanting fuel. It was now simply a matter of her contacting the company, confirming a price per litre and getting a delivery time.

We were the first stop when the truck arrived at 0830 on Tuesday and we took on 300 litres at €1.325. This price was 12 cents per litre cheaper than at the filling station a few hundred metres up the street and only 3 cents more than for the potentially problematic high biodiesel blend at the Leclerc supermarket pump on the western edge of town. The fill took our gauge to about the 350 litre level, about 100 litres from full and sufficient to do us through the spring and into the summer.

Shortly after noon on Wednesday, after paying our moorage bill and saying farewell to Stéphanie, we slipped our lines and continued up the canal. Almost immediately the canal entered a deep cut with steep, complex masonry walls.

The cut was through a high ridge of land for about a kilometre and it appears to have required extensive excavation and masonry wall reinforcements. I looked at the Du Breil Guide

Fluvial chart and saw a double dashed line to the north of the city with the notation: "Ancien canal". The canal had been re-routed and I decided to dig further after we had paused for the day.

At 1305 we secured alongside the bank with the spud pole and a line to a pin pounded into the ground. We had come 2.3 kilometres and were in a delightfully quiet spot just outside the city and next to a track that led to a small road that led about 400 metres to the huge Leclerc supermarket. It had been cool, glum and rainy for several days and it was forecast to remain the same for a few more, so we decided to pause here for a while waiting for the weather to change.

With a good internet connection on my iPhone, I dug into the background of the re-routed canal. In planning the route of the canal, Riquet had used the easiest lines, the ones that required the least digging and the fewest works such as locks, bridges and aqueducts, so the natural route took the canal two kilometres to the north of Carcassonne. In 1669 Riquet's agents began negotiations with the city of Carcassonne to have the canal run past the city walls. The price asked for the additional work was 100,000 livres. A powerful faction in the city resisted the proposal, so in 1673 the canal was dug along the easiest path, next to the Fresquel River.

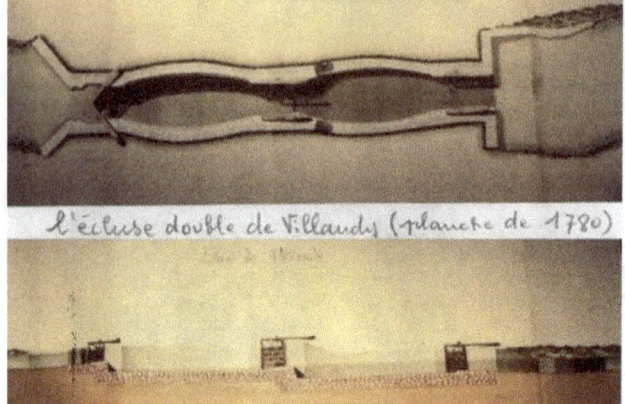

Along the way, a double lock had been built at Villaudy at a contracted cost of 16,061 Livres, so the 100,000 quote to Carcassonne for a longer route requiring extensive excavation and masonry work, two locks and several bridges does not seem out of line.

In 1787, after more than a century of watching the canal traffic bypass their city, Carcassonne finally decided to dig a seven-kilometre diversion to bring the canal under their city walls. The works were slowed by the Revolution, but were finally completed in 1810. Pont Marengo, where the canal abuts the north side of the city, was completed in 1800. It did not follow the template of the bridges built between 1667 and 1681, and at 3.3 metres, it is the lowest bridge on the Canal du Midi.

During a pause in the rain we walked along to the supermarket and picked-up some fresh supplies for dinner and breakfast. Back onboard, we enjoyed a celebratory meal to mark the beginning of the year's cruising: Sautéed dos de cabaillaud with roasted mini potatoes and steamed green beans accompanied by a bottle of Crèmant d'Alsace.

On Thursday we puttered around onboard, cleaning and organising, reading, writing and relaxing. In the evening I played in the galley. I had bought a fresh 550 gram foie gras de canard.

I sliced thick slabs of veinless pieces off its sides and stuffed this into Bresse chicken breast. Then I lightly coated them with some seasoned bread crumbs and placed them in a hot buttered pan to sear on both sides. They were then more slowly sautéed for a few more minutes before the pan was covered for the chicken to finish cooking.

The resulting suprême de volaille de Bresse au foie gras de canard was served with steamed asparagus spears with a mustard mayonnaise, basmati rice and sliced Roma tomatoes. It was accompanied superbly with a 2010 Monz' & Marveilles Corbières from Domaine La Luze. We were slowly getting back into cruising.

The glum weather persisted, with rainy spells, cool days and nights down in the 8º to 10º range. On Friday morning, about ten minutes after I had turned-up the central heating thermostat to warm the boat for breakfast, there was a loud bang and then a rapid running of the water pump. I quickly lunged to open the pump and furnace breakers, which brought quiet. Under the galley sink was steam and a growing pool of water. I removed the back panel and saw water flowing out of a dangling hose with a broken fitting on its end.

The fitting appeared to have blown off the side of the Webasto heater which supplies radiators in the heads and in the kick beneath the galley counter.

Looking to bypass the heater by connecting the line to its inlet with the line from its outlet, I unscrewed the clamp and took the broken fitting off the hose. The connecting barb showed gouges and scrapings from having been crudely modified, most likely by the installer trying to force the fitting into a cold hose, rather than heating the hose. Since everything in Zonder Zorg is new, I needed to determine why a ten-month-old fitting blew. I suspect it had been overstressed and weakened while the installer whittled down the hose barb.

I walked through the rain to the Bricomarche that is in the rear of the Leclerc and searched for a piece to connect the two sections of hose. After looking at many different bits and pieces, my best option seemed to be a 14 millimetre female-female copper water pipe solder junction. The straight junction had insufficient shoulder to hold the clamps, so I chose a 30° one.

I slipped the hose ends over the elbow, tightened the clamps and switched on the water pump. The bypass held against the water pressure. After I had refilled the heating system with water and had purged the air out of it, I closed the breaker for the furnace, closely monitoring the repair for leaks as the furnace began again warming the cabin. We then enjoyed a very late breakfast.

By Saturday morning the rain had stopped and the forecast for the following few days showed hope for an end to the prolonged glum. We continued to relax, and as the weather allowed, we spent more time out in the cockpit, relaxing and enjoying lunch.

With the activity and the smell of food, we soon attracted some of the local ducks, including this mother teaching her young ducklings how to flirt with boaters and beg for crumbs. As the weather improved, there was an increase in the number of locals out walking along the towpath. Many of them had bags of leftover bread and they lined-up on the banks overfeeding the ducks as if they were producing foie gras. This reminded me that we still had a very large of fresh foie gras de canard in the fridge that needed tending to.

On Sunday I prepared the remainder of it, de-veining the liver, seasoning it and pressing it into three small terrines. These were placed in a bain-marie and then into a 150° oven for half an hour. Out of the oven, they were allowed to stand for two hours before being placed in the fridge for three days to resolve.

Monday morning dawned clear and after some preliminary espressos inside, it was pleasant enough to continue with breakfast in the cockpit. Edi prepared a Zonder Zorg version of eggs Benedict: she basted eggs and served them napped with freshly made sauce Béchamel on ham slices on lightly toasted multigrain bread and garnished with avocado and tomato slices and fresh basil and served with more cups of espresso.

The selection of fresh seafood in the Leclerc is superb, and we have been spoiling ourselves with fish for dinner most evenings. On Monday there was a special on dos de cabillaud, so I bought a large piece, which I lightly breaded, seasoned and sautéed in butter and served with pomme de terre rissolées and steamed asparagus spears with a mustard mayonnaise.

The weather had stabilised, so on Tuesday afternoon we made one last shopping expedition to Leclerc to stock our fridge and pantry in preparation for continuing up the canal on Wednesday morning. We were comfortably back aboard Zonder Zorg, all her systems were working, we had recovered from ten days of eating American cruise ship style and had reacquainted our systems to proper food. We were ready to continue up canal.

Chapter Twenty-Eight

To Castelnaudary

We had spent the past week moored in a peaceful rural area just 2.3 kilometres from our winter moorage in the Bassin de Carcassonne. During our many years of cruising, we have often found it useful at the beginning of a voyage to leave, go a few nautical miles or kilometres and then pause to finish the departure preparations. This way the farewells are done and all the distractions of "home" are left behind.

Having finished our preparations, at 0840 on Wednesday, 8 April we recovered the mooring pin from the bank, raised the spud pole and continued up canal. Twenty minutes later we were in Écluse la Douce, our first lock of the year. We had to reacquaint ourselves with the routine of working our way up an ovoid Midi lock. Half an hour later we were in Écluse Herminis, our second lock of the season to reinforce the lessons.

Just 315 metres beyond Herminis we came to the double lock, Écluse Lalande, by which time we were fully retrained. The process we use is to drop Edi off on the landing quai just short of the lock. She walks up to the bollard at

entrance to the lock to receive the stern line that I toss up as I slow Zonder Zorg in the chamber. She

loops the bollard and drops me the end of line, which I snub around the stern cleat to stop the barge. Then I shift the engine into ahead and walk forward as the spring slowly moves the bow to alongside the chamber wall. I toss the bow line to Edi for her to turn around the bollard and tail it as the lock fills. I head back to the cockpit, shift the engine to neutral, shut it down and tail the stern line as we rise.

With multiple locks, Edi takes the bow line and walks it along as I motor the barge to the next chamber, where we repeat the process. During the last 50 or 60 centimetres of the rise, the turbulence of the incoming water has diminished and the lines need much less attention, allowing me the freedom to wander around and shoot some photos.

After leaving the double lock we passed under more of the old stone arch bridges that were built in the 1660s and 1670s. We are amazed that so many of these have survived for three and a half centuries.

What have not survived are the plane trees along the canal banks. We passed an area where many that have been stricken by the canker stain fungus were being removed. The works had churned-up the bottom of the canal, and as we passed through the disturbed water, our engine's raw water strainer clogged and we lost cooling water flow. There is a subtle sound change when this happens, which comes from the loss of exhaust cooling. At 1140 we pulled over to the canal bank to clean the strainer, and since we were stopped, we decided to remain there for lunch, knowing that we wouldn't be able to make it to Écluse Béteille before it closed for lunch at 1230.

To Castelnaudary

We timed our departure from our lunch stop to have us at the lock as it opened at 1330. Another five kilometres beyond Écluse Béteille we arrived in Bram and rafted onto a pair of Nichols rental boats; bow on one and stern on another. The port was full to overflowing with rental boats; the season was just beginning and very few had been rented thus far. We filled our water tanks from the rental base hose bib and then continued the hundred metres through the bridge hole where at 1550 we moored for the day on the abutment extension, using the spud pole and the stern line around a pin in the bank. We had come 21.5 kilometres and had passed through six locks.

We offloaded the bicycles and pedalled into the town of Bram and out the other side to the supermarket. Along the way we passed an asparagus farm advertising fresh asparagus at €5 per kilo. The local season was just beginning.

Wednesday we had a late breakfast in the cockpit and shortly after noon we raised the spud pole, hauled the pin and continued up canal. We arrived at Écluse de Bram at 1223 to find it closed for lunch ahead of schedule. Because we had quite recently finished breakfast, we could not spend the waiting time eating, so after we had secured to the small quai, we went exploring the area.

To occupy the remainder of the time, we harvested gravel and topsoil from the canal bank to use in bedding pots of flowers into our gunwale planter. The lock keeper returned from his 1230 to 1330 lunch at 1339, so using French logic, since he was nine minutes late we assumed he must have begun his break nine minutes early.

We continued through the lock and through two more before securing at 1455 with the spud pole forward and a stern line to a pin on the bank in a peaceful wilderness setting just upstream of l'Épanchoir de Villepinte. An épanchoir is an overflow to drain excess water from the canal, and many of these were designed into the canal when it was built. This one has an elegant eight arch bridge on which the towpath crosses the spillway.

After settling-in, we enjoyed a mid-afternoon lunch in the cockpit, centred on one of the terrines de foie gras I had prepared a few days previously. It paired splendidly with glasses of Duc de Castellac 2009 Monbazillac. I made a note to myself to add a few grinds more pepper and a splash more Armagnac in the next terrines.

At 0951 on Thursday morning we slipped from our moorings and continued up canal. Ahead of us only eleven kilometres away was the city of Castelnaudary, but in that distance are fifteen locks; six singles then a triple, a double and a quadruple. We set out with the intention of going until we were tired. At one of the singles, la Peyruque, the lock keeper's house had been transformed into a boutique offering locally made handicrafts, but it had not yet opened for the season.

At 1246 we arrived Écluse de Gay, which was closed for lunch, so we secured on the quai just below the lock to have lunch. We had already done the six single locks and the triple. We still had a double and a quadruple to do before Castelnaudary, but since the centre of the city was only two kilometres away, we decided to press on after lunch. We were joined in the double lock by a rental boat, which entertained us with its manoeuvring antics.

To Castelnaudary

The rental decided to pause between the double set and the quadruple, likely so it could have the set of locks to bash about in all by itself after we had gone through. Without the encumbrance of the badly-handled rental boat, we quickly worked our way up the flight, and at 1425 we were at the top of the upper chamber and preparing to head into le Grand Bassin toward central Castelnaudary.

Le Grand Bassin, the largest port on the entire canal, was dug in the 1670s as a reservoir to feed the double, triple and quadruple locks leading up to the city. Unlike in Carcassonne, the city fathers here had quickly agreed to pay the additional costs associated with re-routing of the canal into their city. For nearly three centuries the basin served as the major grain shipping port on the canal and Castelnaudary grew in prominence and wealth.

In beautiful sunny weather we motored across the basin. What was once a thriving commercial hub, buzzing with activity, is now devoid of any commercial activity except for a rather drab Le Boat rental base. We passed under the humpbacked bridge at the end of the basin and into the port, where at 1447 we secured to the wall directly in front of the Capitainerie.

We were greeted by la Capitaine, who directed us into the spot and then asked us to come to her office when we had settled-in. She was full of information and gave us local and regional maps and a very informative guidebook to the area. We decided to stay for four days, and the moorage bill came to €40.76, including water, electricity and a strong wifi signal aboard.

The city of Castelmaudary was first mentioned in 1118 as "Castellum Novum Arri", "Arius' New Castle". The castle was sited on the crest of a steep ridge, and around it the town was built and it became a strategic fortified town during the Crusade against the Cathars. In 1355, during the Hundred Years' War, it was razed by the Black Prince. In 1553 Catherine de Medici became the Countess of Lauragais and made Castelnaudary the seneschal seat of the region. We walked into the centre of the city and followed the maze of narrow, winding streets and lanes.

The Lauragais region, with Castelnaudary at its heart, was then and remains today a very important grain growing area. It straddles the broad and rather level saddle that forms the watershed between the Atlantic and the Mediterranean and is bordered to the south by the Pyrenees and to the north by the Massif Central.. With the coming of the canal in 1681, the city had a port from which to easily ship its grain and flour to both the Atlantic and Mediterranean. To handle the grain, by the late seventeenth century there were thirty-two windmills on the ridge at the edge of the city.

Of these, only Moulin de Cugarel remains today. It had operated until 1921 and somehow survived rather intact from then until it was fully restored in 1962. It retains all of its inner machinery, the spur wheel, the shaft and two millstone chambers. It was a very calm day with barely any movement in the air, except on the ridge near the windmill. It had been very well sited to take advantage of the thermal winds created by the ridge.

To Castelnaudary

We followed winding ways back down from the ridge top through the old town, looking for la Chapelle Notre-Dame de Pitié. During the French Revolution, in order to save the chapel and its contents from being sold as a national asset, thirty inhabitants of the neighbourhood got together and bought it. It remained the property of these thirty families until 1972, when it was sold for one Franc to the town council.

This tiny sixteenth century chapel is famous for its ten gilded carved wood panels depicting the Stations of the Cross. They had been exquisitely carved by three local artists during the eighteenth century and today the chapel and its contents are a listed national monument.

Four kilometres south of Castelnaudary, in the village of Villeneuve-la-Comptal, Clément Ader flew his first glider in 1873. After his first flights he continued to develop flying machines, driven by the idea of powered flight. His first powered flying machine, l'Éole was built in 1886 as a bat-like design with 14 metre wingspan and powered by a four cylinder steam engine he had designed. The engine weighed only 60 kilograms and developed 15 kW to drive a four-blade propeller. The all-up weight of the aircraft was 300 kilograms.

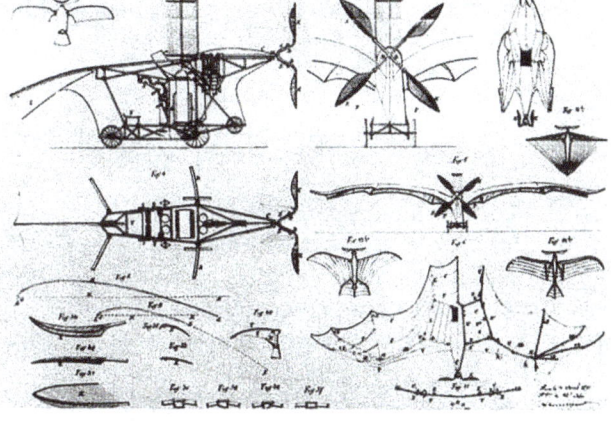

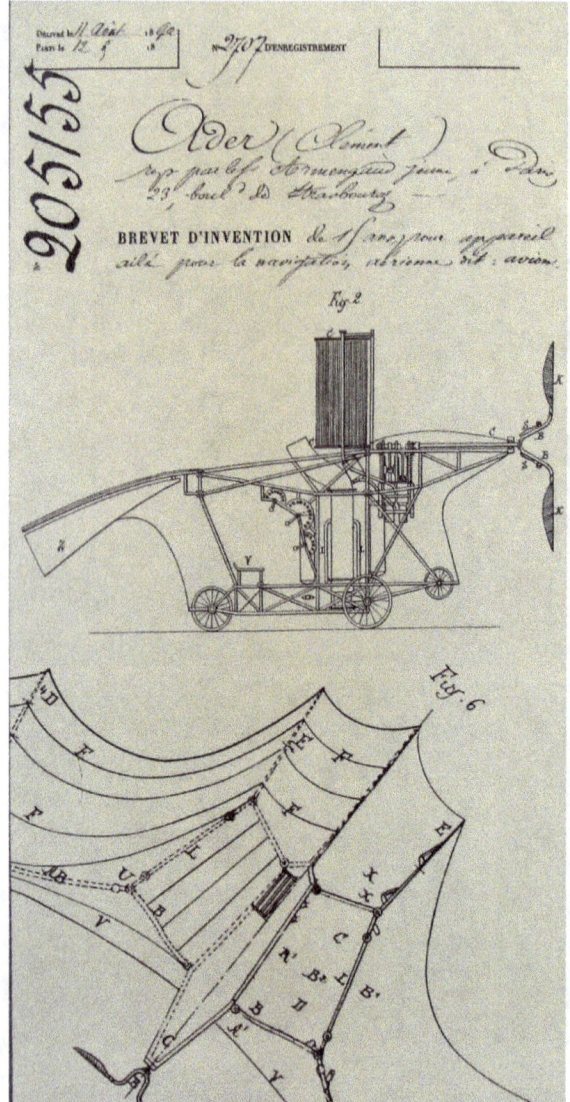

Clément Ader is known as the "Father of Aviation" as the first man to design, build and fly a powered aircraft. On 9 October 1890 he took off in l'Éole and flew for approximately 50 metres becoming the first heavier-than-air machine to take off under its own power. This was more than thirteen years before the Wright Brothers' famous flight at Kitty Hawk, which most people in the USA still erroneously believe was the world's first powered flight.

In August 1890 Clément Ader was issued a patent by the French government for "…appareil ailé pour la navigation aerienne, dit: avion", which translates to "winged device for air navigation, called: aircraft". He had coined the French word for aircraft: "avion".

He continued to refine his designs and in 1897 he had built his third aircraft, Avion III, a twin-propeller machine which is shown flying in this old photo. His designs lacked controls, so the flights were rather erratic.

Clément Ader's Avion III is now on display in Musée des Arts et Métiers in Paris.

The first controlled heavier-than-air flight was made in Fairfield, USA in August 1901 by Gustav Weißkopf (Whitehead), a German immigrant, in his Monoplane Number 21. This was twenty-eight months before the touted Wright Brother's "first" flight. I am constantly amazed with such examples of the recreation of history in the United States?

To Castelnaudary

We enjoyed our days in Castelnaudary, wandering the main streets and the many narrow, winding back lanes, enjoying the mix of old and new. Castlenaudary is famous for its cassoulet, a combination of white beans with pork, fat and sometimes duck confit, baked in a deep earthenware dish called a cassole. However; having grown up eating the poor people's food of baked beans and pork fat, we could not get our minds around it as a dining treat, so we passed.

On Monday 14 April we made a shopping excursion through the weekly market that sets-up in Place de la Republic and then augmented our purchases by a visit to the nearby supermarket. Back on board and properly resupplied, we then unplugged our shore power, slipped our lines and continued up the canal.

Chapter Twenty-Nine

Over the Hump to Toulouse

It was shortly after 1300 by the time we had slipped our lines from the quai and continued up canal from Castelnaudary. We had timed our departure to put us at the next lock, Écluse la Planque after it had reopened from its lunch break. The lock was ready for us as we arrived.

With our mast still up, we easily slid in under Pont la Planque, which at 3.42 metres is the second lowest bridge on the Canal du Midi. Its vertical clearance at 5 metres width is only 2.27 metres, making it the lowest shouldered bridge on the canal.

After we had worked our way up la Planque, we continued 1.2 kilometres to la Écluse Domergue and then 1.1 kilometres further along to the triple lock, Écluse Lourens. It took about twenty-five minutes to negotiate the three chambers and then after another 1.3 kilometres we arrived at the double lock, Écluse Roc, which like the previous locks, was ready for us as we arrived. There had been very little traffic thus far on our voyage, but we knew that would change as the rental season begins in earnest with the Easter weekend just a few days away.

At 1535 we were in Écluse Méditerranée, our last up-bound lock. Within ten minutes we had entered le bief du partage, the 5.1 kilometre summit pound at 190 metres above sea level. We continued along for three kilometres to le Ségala, where at 1608 we stopped for the day, secured along the bank with spud pole and a pin pounded into the ground. We had come 11.9 kilometres from Castelnaudary and had passed through our final eight up locks. It was all downhill from here.

On Tuesday morning we got an early start, slipping at 1005 and motoring along the summit pound. Near its end, at Col de Naurouze we passed the mouth of la Rigole de la Plaine, the feeder stream to the canal from the reservoir in les Montaignes Noir to the north. The key to the building the Canal du Midi had been in finding a reliable source of water higher than the summit pound to feed the canal. Riquet had solved this problem in 1660 with the idea of diverting streams in the Black Mountains and feeding them toward the summit pass, Col de Naurouze.

At about 600 metres up in the Black Mountains, Riquet had diverted two streams on the Mediterranean side of the pass and one on the Atlantic side and fed these through artificial channels to reservoirs. One of these reservoirs was held by a dam 780 metres long and 35 metres high, which at the time of its construction was the largest dam in the world. A short distance on, at a bend in the canal just before Écluse Océan, we tucked Zonder Zorg's stern into a nook, dropped the spud pole and snugged the stern up to a narrow two-metre-long quai.

We were at the beginning of a trail that led to the works at the summit and to l'Obélisque de Riquet, the monument erected to honour the builder of the canal. Water from the reservoirs led to a large octagonal settling pound alongside the canal next to the lock that begins the descent toward Toulouse. Unfortunately, after only a few years of use, the basin silted-up and a new canal was built to bypass it. We followed the signs to the obelisk, which led along a trail between rows of plane trees across the former bed of the octagonal pound. On a ridge in the distance we could see the obelisk sticking up through the trees.

After nearly a kilometre, the last part up a hill, we came to the monument. The twenty-metre-high obelisk had been erected in 1827 by the descendants of Riquet. It is sited on an exposed complex of tertiary conglomerate and quaternary sedimentary rock known as les Pierres de Naurouze that had been used in ancient times for ceremonies.

Over the Hump to Toulouse

Unfortunately, when we arrived at the monument, its gates were closed and locked. We walked around the encircling two and a half metre high wall looking for a breech in the defences; there was none. It wasn't a national holiday, it wasn't a Sunday or a Monday, it wasn't lunch time, it was simply closed. We walked back down the hill disappointed that we had been unable to see the bas relief allegorical figures that are carved around the pediment of the monument. On our way we passed at least half a dozen signs pointing back toward l'Obélisque.

It was here on Col de Naurouze in May of 1681 that the canal was inaugurated. Eight months before the opening ceremony, Riquet had died at the age of seventy-one. It was also here on 18 April 1814, that the English Duke of Wellington met with the French Maréchal Soult in the house of the canal engineer to sign an armistice between their two armies according to the terms of the week-old Treat of Fontainebleau. Two days later Napoleon left Fontainebleau for exile in Elba.

With our history lessons done, we returned to Zonder Zorg, raised the spud pole, slipped the stern line from the old stone bollard and headed the short distance into Écluse Océan to begin our descent. We were over the hump; this was our first down-bound lock since the one at Beaucaire on the Rhône in early October the previous year and we needed to retrain ourselves in descent techniques. We had never been down an ovoid Midi lock before, so we needed to experiment to find the best way for Zonder Zorg to do it.

A kilometre and a half beyond the lock is Port-Lauragais, which from the descriptions in the guides seemed like a good place to stop for the day. We nosed into the port and our first impression was of a highway truck stop, which in fact it is. The "store selling local produce" mentioned in the guide is one of the hugely overpriced rest stop convenience shops. On a peninsula in the lagoon there is what appears to be a hotel and a restaurant. The moorings around the periphery looked vapid and unwelcoming. We turned and headed back out.

265

Two and a half kilometres further along we came to a wooden rail on posts just short of Écluse Emborrel and at 1244 we secured for the day. As we lunched in the cockpit, we were entertained for twenty minutes by the manoeuvring and mooring antics of six people aboard a rental boat attempting to stop to wait for the lock to reopen after lunch.

Mid-afternoon we walked about a kilometre and a half along a road, across the autoroute, the railway tracks and the Route National to the village of Avignonet-Lauragais, which was built on a small hilltop in the thirteenth century. As we were descending the canal, we had been intrigued by its distinctive and imposing church tower.

During our walk up the hill into the centre of the old village we were impressed with the cleanliness of everything; the streets, the sidewalks, the houses were all very well maintained and cared for. It was a dramatic change from much we have seen the past few years in France; it reminded us of the Netherlands.

Near the centre of the village, just below the church is an old tower. A plaque on its side tells that it had been built in 1610 to reinforce the defence of Port de Cers, the principal gate to the community. It dominated and protected the lift bridge that gave access through the fortified wall and into village. The walls around the old town had long since disappeared, most likely dismantled for its building stones.. The statue on the tower's flank was added in 1893, and it is believed to represent Simon de Montfort, who had been the head of the Crusade against the Cathars.

Over the Hump to Toulouse

We continued up to the church that dominates the top of the hill. Notre-Dame des Miracles was built during the fourteenth to sixteenth centuries on the site of a much older church. Its tower soars forty metres and the bell balustrade is accessed by an interior stone staircase of 161 steps. This was closed, so we didn't go up.

From across the square, across the road and halfway across the facing garden, the church bell tower still looked huge. Its bulk dwarfs the church, which seems attached to it as an afterthought.

We walked down the slope to the Route National to town's boulangerie and épicerie. They were closed on Tuesdays. We walked along to the cave a vins to find it simply closed with no explanation. We walked back to Zonder Zorg to find her welcoming and comforting.

At 1130 on Wednesday we slipped our lines and continued down the canal. The lock is automatic, but there was no indication of this as we approached. Edi finally spotted a post by the side of the lock chamber with a set of remote control buttons and instructions in both French and English. The installation looks rather new, or at least is not worn-out and decrepit like much of the rest of the canal infrastructure. She pushed the down button and the lock began its automatic cycle, opening the upstream sluices to fill the chamber, then opening the upstream doors to allow Zonder Zorg to enter. Once we were secured in the chamber, Edi pushed the green button to continue the cycle, which closed the doors and then opened the downstream sluices to begin draining the lock chamber to the downstream level. With the lowering complete, the doors opened allowing us to continue down the canal. A sensor at the exit to the lock saw us pass and triggered the closing of the doors to await the next poke at the buttons.

A kilometre and a half further along we came to Écluse Encassan, a double lock. Likely because of its complexity, it was operated by a VNF éclusier. He pushed the appropriate buttons on his shoulder pack remote control, an ancient device that approached the size and appearance of a ghetto blaster radio without the speakers.

Just short of three kilometres further along, at 1226 we arrived at the automatic lock, Écluse Renneville. Its lights were out and it was closed. The lunch break for éclusiers on the Canal du Midi is 1230 to 1330 and it appears they close the locks five minutes early so as not to interfere with their time. It appears that the automatic locks follow the same routine, possibly to have a feed of electrons and hydraulic fluid. We secured to the stone wall just short of the entrance to the lock and relaxed.

When the automatic controls returned from lunch, we cycled through and continued the four kilometres to the next lock, Écluse Gardouch. There was a vacant spot on the quai, so we slipped into it and secured for the day. In response to my questions, l'éclusier at the double lock had told us of a grocery store and a bakery in Gardouche and a supermarket across the autoroute midway to Villefranche-de-Lauragais.

We decided to walk to the bakery in Gardouche to get some croissants for lunch and then later take our cart across to the supermarket for fresh supplies. It is three-quarters of a kilometre into the commercial centre of the village. We arrived at the bakery at 1500 to find a sign saying it was closed until 1600. We continued to the épicerie to find it simply closed with no explanatory sign. We turned around and walked the nearly three kilometres to the huge supermarket. It is open 0800 to 1930 and this explains very clearly why so few services remain open in the towns and villages.

Over the Hump to Toulouse

At the fish counter they were out of dos de cabaillaud, but the monger said he had some whole cod and could cut some. We also bought some thick albacore tuna steaks. The meats and produce departments tempted us to buy much more than we had planned, and since we had brought only one bag and no cart, we had to buy another large bag at the checkout.

It was nearly 1700 by the time we arrived back aboard with our booty and sat down in the cockpit to enjoy ham and brie croissants and cans of cold Heineken.

After a leisurely breakfast in the sun late on Thursday morning we slipped at 1120 and continued down the canal, through Écluse Gardouche and the double lock, Écluse Laval. We arrived at the automatic lock, Écluse Negra at 1223 to find it closed for its 1230 to 1330 lunch break. Maybe the locks close ten minutes before the start of the break.

We secured to bollards on the steel-faced wharf a hundred metres before the lock. There are electricity pylons and water bibs on the wharf with no sign of any fees. We plugged-in the shore power cable, hauled-out the hose and took on electricity and water as we waited for the automatic lock controls to return from lunch.

After the lock was re-energised we continued down, passing through it and then the double lock, Écluse Sanglier 3.7 kilometres further along. Looking at the chart in the Du Breil Guide Fluvial, I was curious to know why Écluse Sanglier with a total drop of 3.73 metres was a double lock, while Écluse Ayguesvives at 4.44 metres was a single.

269

We found out 1.5 kilometres further along when we arrived at Écluse Ayguesvives. It had originally been a double lock, but recently, the upstream lock gates were removed, the walls of the downstream chamber were heightened and higher downstream gates were installed. A drop of 4.44 metres is rather large, so we expected to see floating bollards in the chamber, or some other way to handle the lines.

At the downstream end is a pair of pipe bollards, one on each side of the chamber. The one on the right bank was missing its vertical pipe, and was therefore useless. There are no similar arrangements for the excessive height at the upstream end, where they are especially needed for up-bound barges. There is; however, on left bank of the chamber near the upstream end a bollard set into a recess midway up the wall. This appears to be another lock designed as an exercise by student engineers with absolutely no experience with canal boating nor any regard to the needs of boaters. We assumed that this desecration of the set of historic locks occurred before the Canal du Midi was named a UNESCO World Heritage Site in 1996.

To brighten-up the mood in the chamber, our planter of flowers on the gunwale received an automatic watering by one of the spouts that usually drain infiltrated water from the chamber walls on a descent.

During our planning for the day we had seen on the chart a supermarket symbol on the left bank a few hundred metres beyond Pont de Baziège, so we had decided to stop on the bank near the bridge. As we approached we heard repeated gunfire. At the bridge is the local gun club firing range; we continued along. Along this stretch of canal the cleared side with the path is the right bank. The left bank, where the supermarket is located, is dense bush. As we passed, we caught glimpses of an Intermarché, a Bricomarché, a Lidl and several other stores in a shopping complex about thirty metres through the bushes. We continued along looking for the landing. Where in the real world there would be easy access for passing boaters, here there was none. We wore around and headed back toward the bridge. The persistent gunfire caused us to turn around again, and finally at 1550 we came to the spud pole and a stern line around an old tree stump in the bushes on the left bank directly in front of the shopping centre.

The Easter weekend was beginning, and we didn't want to be caught short of food. This was our first Easter in France in many years, and we weren't sure on which days the shops would be open. Even without Easter, this is near unpredictable. We took the wheelie and bushwhacked to the road, crossed it and went shopping, returning with sufficient to see us through the long weekend.

Our bush mooring was very pleasant, and we remained there Friday so we could continue our shopping in some of the other stores. I walked to the car filling station at the entrance to the plaza and bought a second Butagaz Viseo tank of gaz for the galley. While we still had about a quarter tank in the other one, this was the easiest source for a spare that we had seen. Friday afternoon we walked the kilometre and a half past the village of Montgiscard to Écluse Montgiscard. We saw no landing along the way. We crossed the bridge at the lock and walked back along the right bank past Zonder Zorg, looking at the near unbroken tangle of bush along the opposite bank. Residential back yards ignored their waterfront. We were amazed; in the Netherlands there would be gardens, patios and private boat docks all along, plus a landing for the shopping centre.

At 1145 on Saturday morning we raised the spud pole, slipped our line and continued down the canal. Écluse Montgiscard is another former double lock, bastardised into a 3.82-metre-high single. The chamber was empty when we arrived, so we had to wait for it to fill. The automated locks drain and fill much more slowly than we have seen at the manned ones. This is likely because we appear competent and les éclusiers let us through the locks much more rapidly than they do for bumbling rental boaters; the automatic locks are likely set by default for the bumblers.

Downstream two and a half kilometres we passed under Pont de Donneville. It was built by Riquet of local Toulouse clay brick because there is a lack of stone quarries in the region. During the Battle for Toulouse in 1814, General Sault had blown-up this and many other bridges to protect his retreating troops. They were fully restored in 1821 and remain in fine condition today.

Another five kilometres brought us to the next lock, Écluse Vic, an original single. Seventeen hundred metres later we were at Écluse Castanet, another double lock bastardised into a single with a height of 4.98 metres. An up-bound boat had just activated the lock, which was full, so the chamber had to drain and then refill. The waiting arrangements before the locks are woefully inadequate, so to hold us in place during the half hour wait, I dropped the spud pole and looped a bollard with the stern line.

After we had locked through, we continued along about a kilometre, where at 1440 we came to the spud pole and the stern line to a pin pounded in the bank in a quiet rural setting about two kilometres before the beginning of urban sprawl of Toulouse. After lunch we harvested gravel, clay and topsoil and bedded three new basil plants, plus one we had been eating away at, into a planter we had bought at our bush stop.

The weather forecast for several days had been for three days of rain beginning on Friday. It then delayed to beginning on Saturday, then again to beginning on Sunday. We awoke Sunday morning to heavy rain and the forecast showed it continuing into the evening. We decided to stay put and continue on Monday, when there was predicted to be a brief clearing before the next spate of rains. We relaxed onboard all day to the sounds of rain on the cabin top and in the evening had a lovely Easter dinner of dos de cabaillaud seared with sautéed button mushrooms, pommes rissolées, steamed broccoli crowns and sliced tomatoes with shredded basil. This was accompanied splendidly by a bottle of Crémant d'Alsace Durenmeyer.

The rains stopped overnight and after a late breakfast on Monday, we raised the spud pole, hauled the pin and continued down canal. Within a kilometre we came to a near nonstop line of live-aboard barges along the left bank. Some were rather well converted and cared for, but most were clumsy conversions that were very short on care and maintenance.

The line of barges continued for over two kilometres and then as we entered Toulouse, the banks were again uncluttered. The plane trees here seem to have thus far avoided the canker stain and they all look healthy. During our three weeks from Carcassonne we have seen them go from their winter nakedness to now near fully leaved.

At 1250 we came to a T-head on a pontoon in Port St-Sauveur near the heart of Toulouse. We were only five kilometres and three locks from the end of the Canal du Midi and the beginning of Canal latéral à la Garonne. We decided to pause for a while to explore the city.

The port is located about 250 metres from le Grand Rond, which leads into Jardin des Plantes and Jardin Royal. To the north and west is the centre of the old city, with its many sights to see and pedestrian malls, squares and streets to wander. The banks of the Gironde river are only a kilometre to the west of the marina. Within a hundred metres outside the security gates of the marina are a boucherie, boulangerie and a fresh produce market plus two small supermarkets.

On Tuesday morning I dodged the raindrops as I dashed across to the boulangerie for fresh croissants for breakfast. In the late morning, seeing no letup in the rain, we took our umbrellas and went walking through the winding streets and lanes of the old city.

Among the things we were looking for was a chart book for le canal de Garonne and the waterways connected to it. We were within five kilometres of the end of our charts and we didn't want to fall over the edge without navigational and services information for the continuation of our route. Our preference is the Éditions du Breil series and we needed Volume 12 - Aquitaine and Volume 18 - Estuaire de la Gironde. The capitainerie was out of stock of both these, nor did they have any for the areas from the other guide publishers. Mme la Capitaine knew of no booksellers that would have any. I punched "librairie" into the maps app on my iPhone and off we went.

Over the Hump to Toulouse

We had no luck finding the guides at any of the book stores we visited, nor could any of the staff offer any suggestions of who might have some. We did see many wonderful old buildings along the maze of narrow, winding streets and alleys.

By the time we reached Place du Capitole, the produce vendors were packing-up their stalls in the market and lunch patrons were finding tables under the umbrellas on the patios of the bistros and restaurants lining the periphery of the square.

We continued north to Place Jeanne d'Arc where we had been told there was a shop selling computer supplies. After four years of very hard service, the cable between my iPad and its charger had finally failed. We found a replacement cable and then wound our way through the old streets back to Zonder Zorg, pausing in the midtown Géant Casino supermarket for fresh supplies.

After breakfast on Wednesday we took advantage of the break from the rain, hopped on our bicycles and continued our explorations of Toulouse. Our first stop was at Musée des Augustins, which is housed in a former monastery built during the fourteenth and fifteenth centuries. It was confiscated by the state during the Revolution and in 1793 it was converted into a museum.

Among its collections are thirteenth and fourteenth century church gargoyles, many of which are lined up on display in one of the courtyard ambulatories.

Leading up a grand staircase is a collection of sculptures, and at the top, there are three large rooms of paintings with French, Italian and Dutch masters. Many of the paintings on display are vary large, some monumental, and the room seems too small for some of them. Benjamin-Constant's 1876 depiction of The Entry of Sultan Mehmet II into Constantinople is one that would enjoy a larger room.

I was delighted to see among the paintings Alexandre Antigna's 1855 La Halte forcée, The Forced Halt. It brought back fond memories of my art classes in the late 1950s and early 60s when we studied this painting as an example of composition and the depiction of emotion.

The museum has a collection of Romanesque sculpture, which is classed as one of the richest in the world. Unfortunately, the huge display room was under renovation while we were there and it was temporarily closed.

We continued on to the displays of Gothic sculptures. Among the finer pieces is Nostre Dame de Grasse, a recently restored polychrome limestone carving from the fifteenth century.

There is a wonderful collection of carved epitaphs with about eighty pieces on display, dating back to the eleventh century. These took Edi back to her graphic arts studies and to her sign painting days, and me back to my days of studying and teaching calligraphy.

From the museum we pedalled across to the banks of La Garonne and followed the river northward to the beginning of Le canal de Brienne. This short canal leads fifteen hundred metres from La Garonne upstream of a weir to the end of Canal du Midi. We followed the canal to les Ponts Jumeaux, the Twin Bridges at the junction. Here a small basin called l'Embouchoure led through a lock down to La Garonne below the weir. With the construction of Le canal de Garonne, a third bridge was added in 1843 giving access to the new canal. The lock to La Garonne downstream of the weir was later closed.

We pedalled the five kilometres back to Zonder Zorg for lunch, and once re-energized, I hopped on my bike and chased off in search of the elusive canal guides. Finally, at Librairie Ombres Blanches in rue Gambetta, I found a copy of Guide Fluvial for Aquitaine. It was their only copy remaining of the only canal book they carry. We find it very strange that in a major city at the junction of two major inland waterways, Le Canal du Midi and Le canal de Garonne, it is so difficult to find a boating guidebook.

The Aquitaine guide has charts and directions for Le canal de Garonne, Le canal de Montech, La Baise and Le Lot. This will take us down to the beginning of the tidal Garonne, where we will need the guide for Estuaire de la Gironde. Fortunately, we have many weeks and many cities, towns and villages to search before we arrive at the first place we need it.

Chapter Thirty

Onward From Toulouse

At 1330 on Thursday the 24th of April we slipped our lines from the float at Port Saint Sauveur and continued northward through Toulouse on the final five kilometres of Canal du Midi. We had spent three days in the very modern marina with water, electricity, wifi and security for a total of €66.

Within fifteen hundred metres we came to Écluse Bayard, directly in front of the main train station. Edi twisted the lock wand as the bow passed under it, but got no response form the lights. I twisted it both clockwise and counterclockwise as it passed the stern, but still got no response from the lights to indicate the lock was preparing for us. I dropped Edi off at the bank, from where she had to fight her way over and around anti-vandalism grids, fences and other impediments to get into the lock area. She confirmed to me that the lock was cycling up for us; apparently the lights are dysfunctional.

Bayard had originally been a triple lock, but three and a half centuries of heritage were recently ripped out to convert it into a single lock 6.2 metres high. Fortunately it has been equipped with floating bollards that actually work.

Another fifteen hundred metres had us in Écluse Minimes, a former double lock bastardized into a 4.43 metre single.

A kilometer later, at 1440 we passed through Écluse Béarnais, our final lock on Canal du Midi. It had originally been a single and still is; however, its ovoid shape had been rebuilt into a straight-sided chamber. As we looked back on these locks, we thought of what a horrid representation they are of the great heritage of Canal du Midi. For those arriving from Canal de Garonne, this is their welcome and introduction to this UNESCO World Heritage site.

Ahead of us were les Ponts Jumeaux, the Twin Bridges marking the end of Canal du Midi. We slid under the arch and into le Port d'Embouchoure. At the end of the basin we could see the abandoned lock gate that for two centuries had let barges down onto La Garonne to head downstream toward Bordeaux.

As we emerged from under the bridge we saw the second twin, the bridge hole which led barges onto Canal de Brienne, the short branch to the upstream Garonne. On the masonry face between the two bridge arches is a high relief sculpture created in 1775 by local sculptor François Lucas. It represents the Occitan ordering the canal to receive the waters of La Garonne with two young men digging the canal. La Garonne is represented holding a cornucopia and at his side is a labourer driving oxen.

We wore around 255° to port and lined-up on the third bridge hole. The Twin Bridges are now Triplet Bridges; in 1843 with the construction of La canal de Garonne, a third arch was added to lead from l'Embouchoure to the end of the new canal. At 1446 we entered le canal de Garonne.

After three kilometres of industrial build-up along the canal banks, we came to the first lock, Écluse Lalande. We were pleased to see it is equipped with pipe bollards set into in recesses down the sides of the chamber. Locking through was easy, simply a matter of monitoring the bights of line we lead around the pipes from our cleats fore and aft. The industrial landscape continued after the lock, with a large number of homeless encampments along the banks and under the bridges. It does not appear a safe place to stop.

The main line of the railway joins the canal here and follows along its right bank for the next thirty-three kilometres. Two more locks in the next four kilometres took us to an easing of the industrial build-up. We carried on another three kilometres to just before Écluse Lespinasse, where at 1647 we dropped the spud pole and secured the stern to a bollard on a two metre wooden quai. We were almost within arm's reach of the lock's dangling actuator rod. The canal is built up between high retaining banks as it approaches the lock and this put us well above the train tracks and out of the direct line of the wheel noise of the frequent trains.

In the evening we celebrated having reached the end of Canal du Midi and having started down le canal de Garonne. For dinner we enjoyed pan-seared dos de cabaillaud topped with crisply sautéed pleurottes and accompanied by white asparagus with a mustard mayonnaise, pommes rissolées and sliced tomatoes with shredded basil. The Crémant d'Alsace from Arthur Metz added wonderfully to the meal and to the occasion.

It had been raining every day for the past week and the ten-day forecast showed only one day without rain. Knowing that it would be a long wait for a clear spell, shortly past noon on Friday we continued along under April showers. The countryside around the canal became increasingly pleasant as we worked our way down through six locks and along another 15.6 kilometres.

At 1550 we came to the spud pole and a stern line to a bollard on a solid wooden wharf just downstream of the bridge in Grisolles. We were at Kilometre 27 and far enough away from the unsavory outskirts of Toulouse to feel comfortable again. We took advantage of a respite from the rains to walk through the pleasant little town and out to the large supermarket on its other side. The rains recommenced just before we arrived, so we shopped long enough for the thunder shower to pass over.

The next morning we raised the spud pole, slipped our stern line at 0940 and we continued down the canal. Sixteen kilometres of pleasant countryside, several rain showers and one lock later we arrived in Montech shortly before noon. We drifted slowly past the halte nautique looking for an empty space. The arrangement is a modified Med mooring with stub pontoons jutting out about three metres every ten or so metres along a concrete wall.

We spotted an opening and I put the engine astern. For the next three or four minutes I gently played with the current and wind to drift gracefully in a single arc with only two engine moves and arrived perfectly stern-to into the tight space. At the last second I swung the tiller over hard to keep the rudder off the wall, dropped the spud pole to secure the bow and handed a bight of the stern line to the astonished bargee next door to drop over the cleat. This was one of those times when there should have been hundreds watching, but the hoards of onlookers seem only to be gathered when everything goes wrong.

We spent four rainy days in Montech for €28, including water and electricity. It is a pretty little town with many old half-timbered buildings along narrow streets. We went exploring during the brief interludes between the rain showers and also made visits to the in-town mini Carrefour supermarket and the large Intermarché a kilometre to the north of the canal.

Shortly past noon on Wednesday, after the day's showers had paused, we raised the spud pole, slipped our line and headed down canal. Within two hundred metres we turned to starboard into Canal de Montech. This 10.9 kilometre branch, also known as l'embranchement de Montauban, leads to the Tarn River at Montauban. It was opened in 1843 with the opening of le canal de Garonne, but fell into disuse and was abandoned after the middle of the twentieth century. It has recently been restored and re-opened.

Behind us we left the ruins of the abandoned paper mill that had operated alongside the canal de Garonne. It had been built in 1856 and once employed over three hundred people, producing originally newsprint, and after modernization half a century later, it changed to high quality papers. After the middle of the century its competitiveness declined and it was finally closed in 1968.

There are nine locks, which are open from 0930 through 1900 without interruption for lunch. Four and a half kilometres down the canal, at 1250 we came to the first lock, just before which was a sign giving a number to call to obtain a remote control to operate the systems. I called the number and spoke with the VNF roving éclusier, who told me he was at lunch and would be back at 1400. We pulled over, moored and waited. The locks are open, but not the lock keepers.

Shortly after 1400 the red light changed to red and green, the gates opened and we entered the chamber. L'éclusier issued us with a four-button télécommande and instructions for its use. The first eight locks are within three kilometres of each other and they are all equipped with ample bollards, including vertical pipes in recesses in the left bank of the chamber walls. The system works easily.

Onward From Toulouse

The lock houses appear all in good condition, though some are drably set, unoccupied and boarded-up. Others are attractively set and are occupied either by canal employees or are rented to locals.

As we worked our way down the canal a man and his young son paralleled our route, moving from lock to lock on an ATV. He carefully explained the workings to the five-year-old, pointing-out the various operations as Zonder Zorg made her way along.

After watching this intense interest for five locks, at the sixth I asked the man whether his son would like to come aboard for a couple of locks. The response was immediate and positive. We gained an additional crew member. He rode along with us through three locks in less than half a kilometre, quietly watching the action from a different perspective. We let him off at the eighth lock and we carried on the final two and a half kilometres and one lock to the end of the canal.

At 1613 we secure stern-to on a stub pontoon with our bows secured by the spud pole, directly in front of la Capitainerie in le Port Canal de Montauban.

The following day was the first of May, la Fête du Travail, the Festival of Work, when everything is closed and the people of France celebrate doing nothing. So that they don't feel discriminated against, the automatic locks are also given the day off, so we would be held captive until Friday. We paid the Capitain €14 for two days moorage, which included water, electricity and a strong wifi connection

The Port Canal is just above the locks that lead down to the Tarn. Our thoughts had been to moor down on the river on the new floats that had been installed by the city, but when speaking with l'éclusier at the beginning of our descent from Montech, we learned that there was no security and considerable vandalism there. This was confirmed by le Captain du Port Canal de Montauban, so we stayed above the river.

After a late breakfast the following morning we took our umbrellas and headed out into a brief lull between showers. We were well into our third week of rain every day, and combined with the unseasonably cold weather, it was becoming a tad depressing.

We passed through the narrow pedestrian tunnel under the railway tracks, the interior of which is now illuminated by strings of LED lights. The path led us past the lock chambers under the road bridge and out to the banks of the Tarn.

Down through the trees we could see the ruins of the old lock, Écluse Sapiacou, so we went down the slope to have a closer look. The river Tarn had been navigable upstream from its junction with the Garonne at Moissac for more than a hundred and forty kilometres through Montauban, Gaillac and a short distance past Albi. With the decline in barge transportation, the locks were abandoned.

Now all that remain accessible and navigable on the Tarn are two sections; this one for nine kilometres upstream of the weir from Montauban and the final twelve kilometres of the river before it enters La Garonne west of Moissac.

There has been discussion for several years on restoring the remaining twenty-six kilometre section between Montauban and Moissac. The river's drop is about twelve metres over this distance, and the restoration of the six abandoned locks between Montauban and Moissac would give a wonderful sixty kilometre circuit on la Canal de Garonne, le Canal de Montech and le Tarn through four historic cities: Montauban, Moissac, Castelsarrasin and Montech.

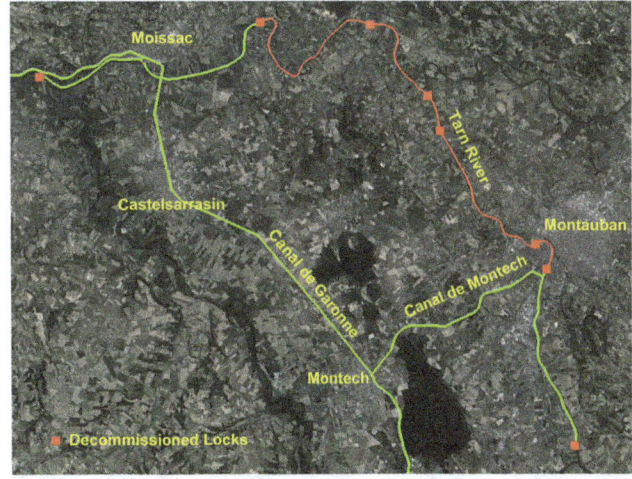

Beside the weir below the lock is an abandoned mill that would have used the abundant flow of the river to power its works.

Le Tarn is subject to flooding in the winter and spring, with levels three and four metres above the weirs not unusual. Any new works would need to be built to withstand being submerged during regular spates, so the restoration is not a simple undertaking. As evidence of the high water levels, we saw a large tree trunk lodged a good four metres above river level on a pediment of Pont Neuf.

Pont Neuf was built in 1913 to give a second crossing from Villebourbon to Montauban. Just downstream from it is Pont Vieux, which was built at the beginning of the fourteenth century. Built of brick with seven graceful arches, it has been sufficiently massive to have resisted the worst of the Tarn's floods for over seven hundred years.

We walked across Pont Neuf and into le jardin des plantes, a three-hectare park in a small stream bed that is planted with over four hundred species of shrubs and trees. Up a steep embankment above the park is the centre of the old city of Montauban, founded in 1144.

At the highest point of the city now stands the cathedral, a late addition, since the city had been one of the major bastions of Protestantism through the religious wars, and much Catholic property had been confiscated or destroyed. Catholic support seems tepid even today; the interior of the cathedral is in a very sad state of repair and it appears unloved.

More impressive in the cathedral square are the façades of some of the buildings, such as the former hôtel des Postes.

We continued along nearly deserted streets, weaving our way through the large grid of narrow pedestrian streets lined with upscale shops, all closed. We arrived in Place nationale to find it wonderfully photogenic with its lack of people. If the citizens weren't so seriously celebrating work, this place would be teeming with people.

Out from under the arches the square was deserted. The restaurants, bistros, bars and cafés under the arches around the entire periphery of the square were abandoned. The French take their celebration of doing nothing very seriously. We wondered whether we had missed the air raid warning and should have evacuated to shelter.

We continued down the slope past église Saint-Jacques, which is one of only two remaining medieval structures in the city, the other being Pont Vieux. The church dates to the thirteenth century and its polygonal tower was added in the fourteenth. Its façade is pocked with bullet and cannonball marks from a 1621 siege during the Wars of Religion, when it was severely damaged. With the Catholic reconquest in 1629, Cardinal Richelieu ordered its reconstruction and it served as the cathedral until the new one was built in 1739.

Onward From Toulouse

We continued past the former episcopal palace, which had been built on the river banks in 1664. It now houses le musée Ingres, an art gallery dedicated to the local artist, Jean-Auguste-Dominique Ingres. Of course, it was closed, so we continued across Pont Vieux with looks back at the deserted city. Our walk through Villebourbon on the way back to the port was equally deserted, but when we finally arrived at Zonder Zorg, we found her easily opened and very welcoming.

On Friday morning, with the electrons in our remote control having returned to work, we headed back up the canal. We waited for a break in the continuous series of rain showers so that we would be guaranteed at least one dry spell during the day's trip. The rain held off for the first two locks and we actually caught a few glimpses of blue sky, but always to the west were bands of ominous dark clouds.

Although we were now ascending the locks, they were easy to negotiate with their consistent and well-placed bollards, including two sliding ones on the starboard side. This made for an easy approach with our left-hand propeller and made line handling simple.

With only two heavy rain showers, we quickly made it up the first five locks in the series, then at Écluse Bretoille, the downstream doors refused to close. When they had begun closing, only the one on the right bank moved and then the automatic cycle aborted. The remote control device could not persuade it to either resume or start over. I phoned l'éclusier and got the standard reply: "J'arrive".

He did arrive after about twenty minutes and got the cycle going again. The final three locks went without a hitch, and at the last one we deposited the télécommande and its laminated instruction sheet into the dropbox.

Just short of four kilometres further along we moored for the day on the right bank with the spud pole and a pounded pin on a newly done steel embankment. We were a kilometre from the junction with le Canal de Garonne, but just along a narrow and thinly paved road was the Intermarché to which we had trekked during our stay in Montech. After lunch we wheeled our cart over to the supermarket and hauled back a load of fresh supplies.

While the roving éclusier was rebooting the recalcitrant lock gates, he had asked us our intentions once we reached le Canal de Garonne. We told him we wanted to continue down the canal the following day around noon. He recoiled in shock, telling us we can't do that; the passage commences with a series of five locks that, while automated, they need an éclusier to operate the controls and éclusiers need to eat. I suggested we begin at 1100 and he agreed that would be better. Slightly before 1100 on Saturday morning we passed under the final bridge on le Canal de Montech and re-entered le Canal de Garonne.

As we came around the corner, the light on the first lock turned green. The canal branches here; the channel to the left leads to the five lock series, while the channel to the right leads to la pente d'eau de Montech, the Montech Water Slope. The water slope was built in 1974 to speed the passage up and down the 13.3 metre elevation change of the canal. It is in basic terms a motorized squeegee that moves a wedge of water up or down a sloped concrete channel. It is restricted to use by boats of twenty metres or more and costs €58 to save about 45 minutes of transit time.

Onward From Toulouse

Between the trees as we left the third lock we saw the pair of diesel-electric engines that are joined across the water trench by a massive beam at the one end and by a huge squeegee blade at the other. The blade was in its raised position, allowing a descending barge to pass under it to continue down the canal or for an ascending one to pass into the wedge of water to be moved up the slope.

We continued down through the fourth and fifth locks with l'éclusier riding along on a moped down the towpath from lock to lock to work the controls. At 1144 we exited the final lock in the series, having spent less than the forty-five minutes that the water slope is supposed to save. I will assume that the time saving is calculated for a fully laden 40.5 by 6 metre péniche that would take a long time pumping its way into and out of the lock chambers.

After a couple of rain showers, three more locks and eleven more kilometres we arrived in Castelsarrasin. At 1335 we secured to bollards on a low wharf in Port Jacques-Yves Cousteau. The Capitainerie was closed, so I assumed it was for lunch. It was still closed mid afternoon, late afternoon and early evening, so I assumed it was closed for Saturday. On Sunday morning when I walked to the bakery for breakfast croissants, the Capitainerie was closed. It was still closed when shortly after 1000 we prepared to depart, so I assumed they did not want any money for the moorage, electricity and water.

We slipped at 1010 and motored along the canal in the first clear skies we had seen in over three weeks. The swans were out with us enjoying the warm, sunny day. We were back into the self-controlled locks, with dangling wands to twist to activate the cycle and the locks were all in our favour, so we moved along without any waiting for chambers to fill.

After five locks we came to le pont-canal de Cacour, a very graceful bridge that carries the canal nearly 400 metres across the Tarn. It was very solidly built of Toulouse brick and Quercy stone in 1845, designed to withstand the floods of the river.

After the devastating floods of 1930, which destroyed the adjacent railway bridge, one of the aqueduct's towpaths was used as a rail bed to carry the trains for two years until a new railway bridge could be built.

From the end of the aqueduct the canal steps down through three locks in a kilometre and a half. Shortly before noon we were overlooking le Bassin du Canal in Moissac. As we motored slowly through the crowded basin, le Capitain hailed us from the banks and told us of a short term spot at the far end of the basin.

At 1215 we secured to the solid quai on the left bank just short of the bridge into downtown. I tidied the lines, plugged into shore power and erected the cockpit umbrella while Edi prepared a nibbling tray our for lunch. We luxuriated in the sunny afternoon, dawdling over our first meal in the cockpit in over three weeks.

The wonderfully gentle atmosphere of Moissac quickly told us to pause for a week or so.

Chapter Thirty-One

Moissac to Buzet

Le Capitain told us the space that we were occupying was reserved from noon on Tuesday, by which time we would need to move. We told him we wanted to stay at least a week, likely two or three and we pointed to the long space across from us with a small boat moored directly in the middle of it, converting the two boat space into a single. We were told the space was reserved for a barge arriving Tuesday afternoon.

The near incessant rains of the previous three weeks had stopped and it had finally begun warming. so after lunch we walked the short distance into the town centre, past the market square and along to l'église Saint-Pierre, the church of l'Abbaye de Moissac.

The church is most famous for its sculpted south door, which was created between 1115 and 1130. The carvings on the tympanum depict Saint John's vision of the Apocalypse. They were exquisitely rendered and have remained in an excellent state of preservation, likely helped by the protection from the elements offered by the deep recess in which they are located. The work is considered one of the most consummate examples of narrative doorway and one of the great masterpieces of Romanesque architecture.

We passed through the doorway into the dimly lit porch and on into the nave with its graceful ribbed vaulting, which rises to a height of twenty-two metres. Most of the nave was built in the twelfth and thirteenth centuries and the apse was completed in the fifteenth to replace the structures at the eastern end of the church dating to the middle of the eleventh century. The walls, ceilings and arches are painted in repetitive decorative motifs in soft colours to give a sense of unity to the interior. These were restored in 1964.

Toward the altar there are rows of nicely carved seats for the monks and a large pipe organ hangs on the north wall.

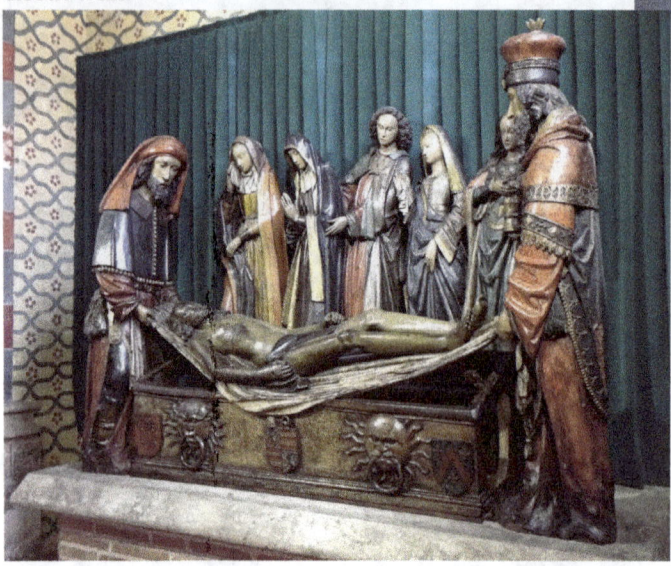

Several fifteenth century painted wooden statues are displayed around the interior. This larger than life size one depicting the entombment of Christ is particularly attractive. During its restoration in 1956, original paint was discovered and the colours were replicated.

There is a large wooden crucifix on which the style of the carving of Christ has led scholars to believe it was carved by the same artist who had carved the doorway between 1115 and 1130.

Moissac to Buzet

As we made our way back through the centre of town toward Zonder Zorg, we reflected upon the many churches we had recently visited. In most we had felt an emptiness. This one; however, feels like a spiritual place with a wonderful peacefulness and serenity.

On Monday we walked to the supermarket for fresh supplies, and while we were there we bought some geraniums to plant in a bucket and hang on the fokijser. Our little garden is slowly expanding.

Monday evening we took our folding chairs and joined about four dozen other bargees on the canal side. The first Monday of every month the Cod-en-Bleu fish and chip truck stops in Moissac for the evening. It is operated by a couple of expat Brits who offer genuine greasy British pub fare. Most of the crowd were Brits, but there were also Auzies, Kiwis, Canucks and Yanks. I put in our order for cod and chips, and just short of two and a half hours later, they were ready. It seems the chef can do only one order at a time and there were many ahead of ours. It was our first meal ashore in two months, so the heavy grease likely did little harm to our arteries.

On Tuesday morning we had to move from our mooring in the basin, but le Capitain had stories for the empty spaces. We still had much to see and do in Moissac and we weren't ready to move on. We had booked a lock time with l'éclusier to go down the double set of locks to the Tarn to do an excursion up and down the river, so we decided to moor down on the river until a space became available in the basin.

293

After locking through, we headed up river a few kilometres, passing under le Pont-canal de Cacor. We were impressed by its massiveness, recalling that for two years the aqueduct had carried both rail and barge traffic until a new rail bridge could be built.

We headed upstream about six kilometres to the end of navigation. wore around continued our excursion downstream past Moissac to the junction of le Tarn with la Garonne. Back up river, we paused for the day on the wall upstream from Pont Napoleon and Hotel le Moulin de Moissac, the upscale hotel now in the old water mill. The moorings include water and electricity, but le Capitain had told us they were not yet turned on for the season, so there would be no charge for mooring.

After lunch we went back over to the abbey to continue our visit. We had seen the abbey church, now it was time to visit the famous cloisters. There is reference to the abbey having been founded by Clovis I, King of France from 481 to 511, though modern scholars tend to attribute its founding to Dogbert's son, Clovis II, King of Burgundy and Neustria from 635 to 657. Whichever Clovis it was, the abbey goes back a long time.

In these drawings are the buildings that existed prior to the Toulouse-Bordeaux railway being built through Moissac in 1845. Destroyed for the rail line were #5 the refectory, #7 the Abbots residence, #8 the infirmary, #12 the chapel of the abbey, #13 the residence of the Canons, #15 the ovens, #16 the residential floors, stores and warehouses, #17 the underground aqueduct, #20 the wine cellars, #23 the Abbots gardens and #24 rue de l'abbaye. The cloisters very narrowly missed being destroyed.

Moissac to Buzet

Today the cloisters and galleries, which were completed in the year 1100, are a UNESCO World Heritage Site. They are magnificent. They surround a garth measuring twenty-seven by thirty-one metres, in one corner of which was once a large fountain fed by a spring in the adjacent hills. The gallery is supported by alternating single and double columns, each set with an intricately carved capital.

There is a total of seventy-six of these capitals, each different from the others.

At the corners and in the mid points of the sides are eight strengthening pillars. On the gallery side faces of each of these are carved reliefs in marble. Eleven of them depict saints: Peter, Paul, John, Matthew, Bartholomew and so on, and the twelfth is of Durand de Bretons, who from 1048 to 1072 was the first Cluniac Abbot of Moissac. From 1060 he was also the Bishop of Toulouse. He was responsible for re-energizing the abbey after many years of decline.

From the cloisters we climbed to the upper chapel, which is located above the porch in the tower of the church. About six metres above it is a roof supported by twelve radiating square arches, which meet in an annular keystone. Above that is the floor of the bell tower.

From the upper chapel we could look out into the church from a vantage point some fifteen metres above its floor. We were fascinated with the architecture from nine hundred years ago.

We slowly wound our way through the streets of the old town centre and back to Zonder Zorg on the river. In the evening, with our heads still filled with images of the cloister and the church, we enjoyed seared jumbo scallops with a butter sauté of pleurottes, criminis, shallots and garlic accompanied by steamed asparagus with a mustard mayonnaise, sliced Roma tomatoes with shaved salt and shredded basil and basmati rice. This was complemented superbly with a bottle of Alsatian gewurztraminer.

On Wednesday we unloaded our bicycles and pedalled back up the canal, across le Pont-canal de Cacor and continued along the towpath for six kilometres to a large shopping centre with two supermarkets and two building supply stores. The following day, 8 May was a national holiday, when some in France celebrate the 1945 end of the War in Europe. Most people in France; however, celebrate not working. We needed to stock-up for the Thursday holiday, and for all the closures on Friday as many traditionally 'fait le pont', 'make the bridge' to the weekend.

Besides food, we also needed a two metre piece of flexible hose with an outside diameter of 35 millimetres. Our engine's raw water intake continues to clog, and we determined that the easiest way to clean it is to remove the strainer basket and blow down the standpipe. I have used a water hose connected to a hose bib ashore, but these are seldom available when needed. In Mr. Brico we found some hose and other things we wanted after which we went across to the Intermarché for groceries. Among other things, we found albacore tuna steaks on special for €10 per kilo. With our bicycle panniers filled, we pedalled back down the canal and down to the river to Zonder Zorg.

One of the things we had bought in the hardware store was a can of semi-gloss black acrylic spray paint. We had a collection of off colour fenders that either came with the boat or we salvaged from the waters along the way. Two of these were particularly annoying to Edi; a large bright orange ball and a large white one, which hang prominently on

Zonder Zorg's bows. Edi's idea was to paint them black to match the remainder of the fenders. The plan worked and within an hour the off-colour fenders had been cleaned, degreased and painted. An hour later they were dry and much more pleasingly rehung.

On Thursday morning we passed the cenotaph on our way up to the canal basin. Wreaths were prepared for laying, soldiers were gathered and city officials were assembling for the 1100 ceremony. We continued to the basin to see if there were any open mooring spaces for us. There were three, and with moving a hogging boat, there were four large enough for Zonder Zorg. We talked with le Capitain, who told us that on Tuesday the city had

switched-on the electricity down on the river and that he needed to charge us for the moorage that he had earlier said was free.

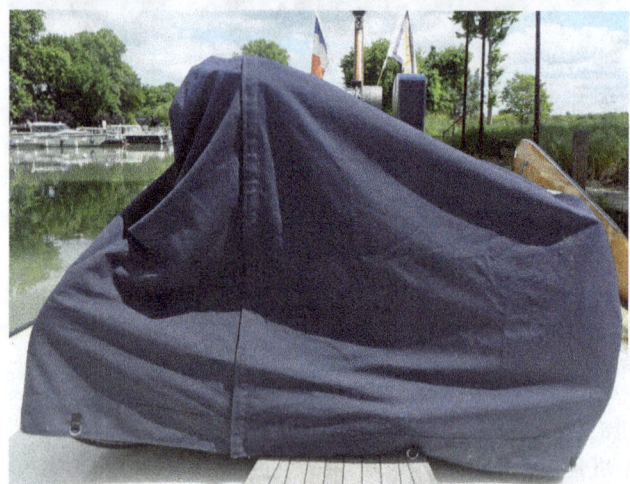

When we asked about mooring in the basin he told us to take the one across from where we had initially been. I asked about the empty spot within wifi signal range of the Capitainerie and was given a nebulous reply that inferred without clear explanation that he had other uses for that empty space.

We organized the VNF éclusier for 1130 and we moved back up to the basin. Spread out all along the right bank of the basin was a cross between a neighbourhood street sale and a flea market. We secured next to what appeared to be a professional vendor who had hauled a van and trailer load from his shop and enjoyed a brisk trade.

While the street sale progressed beside us, we pinned together a cover for the bicycles on the foredeck. We had some pieces of Sunbrella with zippers that were part of an old winter cockpit cover from our sailboat, Sequitur. We created a zipper-front design that draped over the bikes in their position locked to the mast tabernacle. Edi quickly stitched it up on the sewing machine and with only one additional fitting and minor adjustments, we had a fine waterproof bike cover.

After finishing the cover, we strolled around in the centre of the town, which was nearly deserted. It appears everyone here takes their holidays very seriously.

We could not get the wifi signal from our moorage, but by walking back toward the Capitainerie to the large empty mooring space, we found a full bar signal. On Friday the space was still empty and it was again on Saturday. The Capitainerie was closed until 1600, so we took the initiative and moved into the empty space and we finally had a strong wifi signal aboard. Unfortunately, the internet connection was down. Fortunately we had our iPhones with a strong cell connection and I was able to tether my computer to post a new addition to the blog. When the Capitainerie opened, we were told that they needed our space the next day. This was later reinforced by a reserved sign being hung beside us. I said we would move into one of the other empty spots in the morning. The internet connection remained dead, though we were paying for it as part of the moorage fees.

On Sunday morning we moved back to the place in which we had earlier been. Throughout the day we watched for the arrival of the boat for which the vacated space had been reserved. The space remained empty through the morning and much of the afternoon. It was nearly evening when a boat moved into the slot from its moorage across the basin. As wonderful as the city of Moissac is, after five mooring spots in eight days and a sense of our being less welcome than are the many boats that le Capitain has displayed for sale through his private brokerage company in this public marina, we decided to cut our stay short and move on.

At 1020 on Monday, 12 May we slipped our lines and continued downstream from Moissac. As we left the Bassin du canal and passed under the bridge, we reflected on our time in the city. Even with juggling through five mooring spots during the eight days, we had thoroughly enjoyed our stop.

Less than a hundred metres from the end of the basin is Pont St-Jacques, which has only 1.80 metres clearance under it, about 25 centimetres too low for Zonder Zorg. Fortunately it is also called le Pont Tournant; it is a swing bridge with an operator present from 0900 through 1900, except during lunch from 1230 to 1330 and except for four national holidays. We arrived at the bridge to find a double red light: "Out of Operation".

It was not quite 1030, still two hours before lunch, it was not 1 January, 1 May, 11 November nor 25 December. Then we saw legs among the mechanisms under the bridge; it appeared to be under repair or adjustment. We came to the spud pole and a stern line through a ring on the wall.

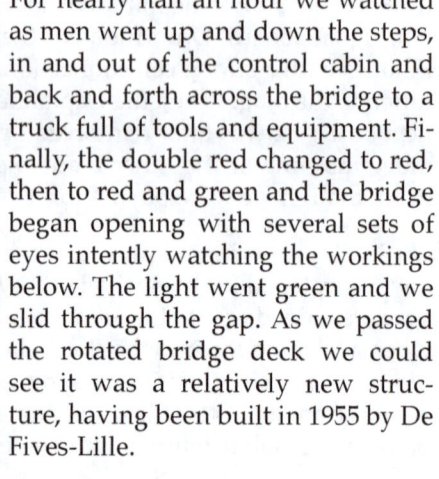

For nearly half an hour we watched as men went up and down the steps, in and out of the control cabin and back and forth across the bridge to a truck full of tools and equipment. Finally, the double red changed to red, then to red and green and the bridge began opening with several sets of eyes intently watching the workings below. The light went green and we slid through the gap. As we passed the rotated bridge deck we could see it was a relatively new structure, having been built in 1955 by De Fives-Lille.

We continued along the canal for another 17.5 kilometres, passing through six more locks and at 1343 we secured to a free mooring on a wooden wharf on the left bank. We were just upstream of a bridge that leads across the canal and into the town of Valence d'Agen. Across the canal from us was the town's halte nautique, with electrical and water points and fees for mooring. We needed no services, so we content not having to pay.

After lunch we walked across the bridge and along the two hundred metres to the edge of town. At the base of the hill on which the town was built is a wash house, which uses the water from a spring on the slope. It was built at the end of the eighteenth century and it is one of two in the area with this graceful circular design.

We arrived in the centre of town to find it almost completely deserted. It was past mid-afternoon, well after the end of lunch break, but all the shops were closed. Then we remembered it was Monday.

Moissac to Buzet

We continued through the wonderful old streets, which interlink four squares. La pharmacie de garde was open, as was an obligatory boulangerie. Following a zigzag route through the town centre, we came to the supermarket, which I had identified from the appearance of the parking lot on Google maps on my iPhone. I had given up using the search feature; the closest supermarché or supermarket or Carrefour or Géant Casino or Intermarché was generally the other side of France or in another country. It appears most French businesses have not bothered to enter their locations on the Internet maps; most are still stuck in the shock of the 2012 demise of Minitel.. The Casino supermarket was open… on a Monday. Its posted hours show it remains open through lunch and is open on Sunday morning. There is some progress in France.

On Tuesday morning we walked back into the centre of town and saw a dramatic change. The four squares and all the streets interconnecting them were crammed with market stalls. There was a bustling crowd and very lively business was being conducted.

Of all the markets we have seen in the past year in France, this was the largest and most active. It reminded me of those in Chalon-sur-Saône and Chagny in the Burgundy a decade and a half ago, spilling through the entire town centre.

As we were leaving, the lunch crowd was just beginning to gather at the several stalls that were dishing-up from huge cauldrons of cassolet, paella, moules and so on. We passed on the prepared foods and filled our cart with fresh produce, a good portion of it appearing to be directly from the producer.

On Wednesday we offloaded the bicycles and pedalled across a branch canal that supplies cooling water to the nuclear power plant downstream. We followed the road for five kilometres past some fine homes and two châteaux, one of them offering guest accommodation.

We crossed the bridge over la Garonne and came to a hill leading steeply up to the village of Auvillar.

We had come because the guides reports the village as one of the most picturesque in France and having among other attractions a museum devoted to barging on la Garonne. It was noon as we arrived at tour de l'horloge in the centre of the village and the place appeared abandoned.

We could find no bicycle racks, so we locked our bikes to a construction barricade under an arch of the galleries that surround the market square.

The streets were abandoned as we did a walking circuit along the narrow, winding lanes of the village. We saw a crêperie, a café, a bar, and a restaurant, but their menus and cartes didn't tempt us to part with the €30 or more that they were asking for lunch. Our circuit brought us back to the market square, which was still abandoned.

Moissac to Buzet

In the centre of the square is a circular grain market, which was built in the early nineteenth century and is noteworthy because it has both the old medieval measurements and the new metric ones. We continued our meander through the village, pausing to visit église St-Pierre, one of the historic stops on the pilgrimage to Santiago de Compostela.

Somewhat past 1300 we again passed the small epicerie/tabac across from tour de l'horloge and saw it had re-opened after lunch. Inside we had a quarter kilo piece sliced off a fresh brie, added a baguette and a 125 centilitre bottle of sparkling water and paid a total of €2.98. We walked along to a stone wall and enjoyed a delicious lunch for two. The village looks down on la Garonne and for centuries it was an important river port for the 3000 boats per year that went past.

We finished our lunch and headed across to le tour de l'horloge to visit le musée de la batellerie, the boating museum, which was just opening as we arrived at 1430. Inside we found a scattering of photocopied sheets taped to the walls of the clock tower. We went up to the next level and found the same, plus a model of a river barge. The next level up continued the theme, but rather than a boat model, there was a small telescope and a display of astronomy pictures. The top level was empty.

We left amazed and disappointed. To quench our thirst for river boating history, we saw an old anchor mounted on a pillar in the central square. The placard beside it tells that it was forged in Bordeaux in 1630. It was found complete with its chain in the gravel banks of the port below the village in 1950. One can only imagine the fate of the boat that had been attached to it.

At 1045 on Thursday, after a visit to the supermarket for fresh produce, we slipped from our mooring and continued down the canal. A short distance along we passed a public abattoir, which was very prettily dressed-up in the morning sun.

Four kilometres further along we cruised past the cooling towers of the local nuclear power plants. We had recently been remarking how France and the United States of America depend so much on nuclear and fossil fuels for generating their electrical power, while Germany, with much less sun produces more than a dozen times as much solar energy as does the United States and many fold the solar output of France. With the abundance of high-angle sun we wondered why more of its energy is not being harnessed. Surely a solar panel field is less costly to install and operate than are nuclear power plants.

As we left our second lock of the day we met Zelden Rust, a small Dutch flagged sailing tjalk, which had been below the lock waiting for us to descend. Edi greeted them in Dutch as we slowly passed.

At 1429 we secured with spud pole and a pounded pin on a dilapidated wooden wall in l'Hermitage, a basin on the north side of central Agen. The basin is large, but very unwelcoming, with no facilities for passing boaters. In a nook across the basin is a rental boat base and the remainder of that side is directly on a busy road beside the main railway line. After lunch I walked along the basin and then the continuation of the canal to see if Agen was more welcoming further along. They are not.

Further along, the road traffic turns away from the canal, but the bank is inaccessible from the towpath, making mooring in this quieter area impractical. A kilometre and a half along is the entrance to le Pont-canal d'Agen, the aqueduct over la Garonne. There is a small basin with mooring bollards immediately before the signal lights and their activating wand. Moored in the spot was Bees Knees, a British flagged rietaak and I saw room in front of it for Zonder Zorg. I walked back and we motored along to the spot.

There are only the two bollards in the spot, so we moved in ahead of Bees Knees, dropped the spud pole and secured the stern with a line to the available bollard. The spud pole landed on the sloping paved bottom of the canal leading to the aqueduct and our bow slowly drifted out.

After a few more tries, we finally accepted the invitation from Deirdre and Michael to raft alongside Bees Knees. They are a Canadian couple from Toronto, who after a few years on the canals had sold their 1912 Dutch barge and were giving-up possession in two days. As we had arrived, they told us that the bridge and the set of four locks the other end of it were closed. The traffic lights were out, they had received no response from the dangling wand and a passer-by suggested that there might be a strike.

A while later a fifteen metre tjalk, De Jonge Douwe, built in 1896 rafted outboard us. The two French men aboard had just taken possession of it two days previously and they were still trying to sort-out how it handles. We were now three old Dutch barges mid-afternoon wondering why the traffic lights were out.

We had a strong signal on our iPhones, and digging around on the internet we found that the canal bridge and lock closure was part of rotating strike by government employees protesting the new move by the Prime Minister to try to scrape the French economy off the basement floor. He had proposed a freeze in civil servant wage hikes for three years. We further learned the strike affecting us was for one day only. We were not yet ready to move on; we wanted to visit Agen, which despite its unwelcome face to visiting boaters is apparently a lovely city. However; with no water available to passing boats in Agen and not wanting to get stuck here if the labour action escalates, we decided to move on in the morning to the free water and electricity supplied by the village of Sérignac ten kilometres down canal. There we could wait out a prolonged strike.

The following morning, as 0900 approached we loosened our lines on Bees Knees and we motored forward about three metres carrying De Jonge Douwe with us until Edi could reach the dangling wand. The red light came on at 0900 and a second later Edi twisted the wand and the lights went red-green. Our end of the system had won the right to go first. Because of the length of the locks, we couldn't fit in with Bees Knees, so we let De Jonge Douwe go with them when the light turned green.

As soon as the light went back to red, Edi twisted the wand and a little over twenty minutes later we got a green allowing us to head across le Pont-canal d'Agen. The aqueduct is five-hundred-eighty metres long, supported on twenty-three arches each of twenty metres span. Its first stone was laid in 1839 and it was completed in 1843. At its downstream end are four locks within a kilometre that take the canal down about 12.4 metres to the level of the banks of la Garonne. The aqueduct and four locks act as an automatic set, each one triggering the next.

In the first of the locks we saw pipe bollards, which would make our descent easy. Unfortunately, the one at the downstream end of the chamber was missing its pipe and was therefore useless. Complicating this is the safety railing placed in such a way to make the lock more difficult and hazardous for boaters. It appears to be another design by a civil engineer with no knowledge of boating nor any concern for the needs of boaters.

Moissac to Buzet

We made it down the four locks and continued down a narrow, overgrown section of canal with many fallen trees well out into the channel. At 1128 we secured to large cleats on a solid quai in Sérignac-sur-Garonne. We plugged into the free shore power and filled our water tanks. With a butcher, a baker and a small épicerie in the village, were ready for a prolonged strike if one happened.

The history of Sérignac goes back to a Roman centurion named Serenus who gave the village its name in the first century. The locality was fought over through the middle ages and in 1273 it was fortified as a bastide and became part of the royal domaine of France during the reign of Henry IV. Several medieval houses remain in the village centre.

Much of the existing church was built from 1580 to 1600, but significant parts of it date back to the eleventh century. It was once remarkable for its helicoidal tower, which unfortunately was destroyed in a storm in 1921. Finally, after enduring the plain replacement tower for long enough, the citizens came together and in 1989 installed a new version of the old twist.

In the afternoon we bade farewell to Deirdre and Michael as they left on their final day of cruising, familiarizing the new owners with the workings of Bees Knees. Moored astern of us was the British narrowboat, Oxford Blue that we had first seen a few days previously in the moorage in Valence d'Agen. Like us, Barbara and Mike were heading to Buzet-sur-Baïse and the junctions with the rivers Lot and Baïse. We compared plans and thoughts.

The following day, Friday with our water tanks full and batteries at 100%, we decided to continue along the 17 kilometres to Buzet-sur-Baïse, where if needed we could go to the marina for water. The canal remained narrow with overgrown banks. The only clearing we saw was from the scattering of crude fishing shelters.

At the first lock there was a French-flagged Dutch aak waiting for the lock to prepare. We followed the aak into the chamber and rode down with it. After thirteen kilometres we crossed le Pont-canal sur Baïse, the aqueduct over the Baïse River. The towpath across the bridge has an attractive inlaid tile pattern. Out of the aqueduct is a pair of linked locks that take the canal down five metres to run parallel to and about three metres above la Baïse.

As we approached Buzet-sur-Baïse, we watched as a rental boat collided with the aak in front of us. We were less than two kilometres from the rental base and there is serious risk that the drivers of any approaching rental boats will have fewer than two kilometres of lifetime boating experience.

The bangy boat continued zigzagging along toward us, apparently oblivious of having caused the aak to end up in the bushes on the overgrown bank. By the time the rental weaved its way past us, we were stopped. They seemed unconcerned.

Soon we came to the short branch that leads to the double lock, which takes traffic down from le canal de Garonne to la Baïse and back up. We continued on past.

Two hundred metres further along, just upstream of the bridge leading into the centre of Buzet-sur-Baïse, we secured with the stern held on a rocky bank with a pounded pin and the bow held by the spud pole. We were at the point of departure for heading up two navigable tributaries of la Garonne: la Baïse and le Lot and we paused to await lock schedules and navigation conditions.

Chapter Thirty-Two

Heading Up the Lot

The organized moorage along the canal was completely filled when we arrived in Buzet-sur-Baïse mid-afternoon. It was mostly occupied by what appeared to be permanent or long-term moored boats. We secured on the wild bank just downstream of the junction with la Baïse. A hundred metres upstream and two hundred metres downstream of us were commercial marinas, but we needed neither water nor electricity nor moorage fees.

After a late lunch I harvested some topsoil from the bank and Edi reorganized the garden into larger pots. The flowers we had planted a few weeks previously were becoming root-bound and some of them were transferred to a larger hanging basket and others to a saddle planter on the gunwale.

On Sunday there was a street market in town, so we walked across the bridge to take a look. Spread along the street through the commercial centre was a mix of garage sale and flea market. Most of the tables were loaded with things the hopeful vendors no longer wanted. It appeared from the very full tables and the lack of interest, that nobody else wanted the stuff either.

Heading Up the Lot

We had enjoyed a few days without rain, so I assumed that the current in la Garonne would be down to a reasonable mid-spring speed and navigation on it downstream to le Lot would be open. Sunday afternoon I talked with l'éclusier at the lock leading down into la Baïse about our heading down the following morning. He said there was a car in the lock at St-Leger at the junction of la Baïse and la Garonne, but that it was being removed and there shouldn't be any problem for our going through on Monday. He then told us that we needed a pilot for the five kilometre descent of la Garonne to the entrance to le Lot.

At 0950 on Monday we slipped and headed down canal 200 metres to wear around in the widening at the marina and head back up to the entrance to the double lock leading down to la Baïse.

In the first chamber was the narrowboat, Oxford Blue. After they had locked through, we followed and caught up with them at the waiting wharves above Écluse St-Leger. It was 1130 and les éclusiers were not around.

This lock and the ones along le Lot are not run by Voies Navigables de France, the VNF who run about 7500 kilometres of navigable waterways in France. Instead, navigation on le Lot is run by le Conseil général du département de Lot et Garonne. There are about a hundred départements in France, each much smaller than an average county in a Canadian province, but they are incredibly more burdened with convoluted bureaucratic overload. We were not expecting an efficient and smooth operation, but we were open to being surprised.

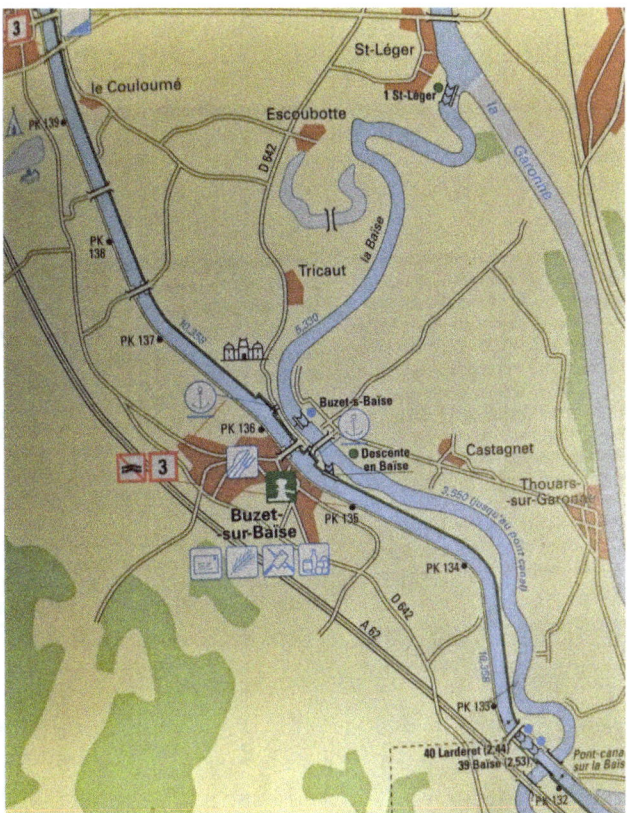

While we were waiting for the appearance of les éclusiers, we walked up to the lock chamber to see what we faced. The first thing we saw was a car just beyond the cill at the entrance to the chamber. It appeared to be nose down and laying back on its hood and roof. The rear tires and hints of the bumper were visible, one tire just breaking the muddy water's surface.

The story we heard was that the car had been pushed in by delinquents on Friday and was to be removed over the weekend. However; the weekend occurred on a weekend, so being France, nothing was done to remove the car. It was now Monday, so we could expect the continuation of the inactivity because it was Monday, when little happens. Fortunately, this lock fills about three metres, so there is ample clearance over the submerged car. At 1215 les éclusiers arrived with an up-bound boat and told us it was lunch time and that they would begin locking us through at 1300.

The lock door was open, so shortly before 1300 we motored into the chamber and secured to await the return of les éclusiers. They arrived at 1315 and we descended to la Garonne. For rental boats, a pilot comes aboard here and for a €16 fee, pilots the boat the five kilometres downstream through a buoyed channel that winds between isolated rocks, reefs and gravel bars. Private boats follow a pilot boat down the race.

The current appeared to be about six or seven kilometres per hour, except in Passage Monluc, where it was closer to ten. Mike and Barbara in Oxford Blue had asked us to precede them into the lock so that they could secure astern of us with spring and engine. Once at river level, they followed us out and motored in our wake as we followed the pilot.

Heading Up the Lot

The entrance to le Lot is through a lock at Nicole and it is a tad tricky. The swift current and the back eddies make the sharp turn back upstream complex, and I required two full asterns and two full aheads to make it into the lock without any bounces.

Les éclusiers said both Zonder Zorg and Oxford Blue could fit into Écluse Nicole together. The lock has a four metre rise and they told us they would handle our lines from the top; there are no mooring arrangements in the chamber. With both barges in diagonally, our fokijser jabbing into the upstream door and our quarter fenders removed, the downstream doors just grazed past the narrowboat's rudder. When the doors closed, we were both able to move back a metre or so and clear our jib sprit from under the ribs of the door.

I asked les éclusiers to pass our line ends down to us so we could control the boat during the ascent. They said they must control from the top. The back eddy of the flooding chamber pushed us forward and our sprit was again under a rib on the door. I told les éclusiers to haul us back and to properly tend the lines by using the bollards. They didn't understand what was needed nor how to do it. The eye at the end of our fokijser was now hooked behind the cross beam of the door and we couldn't be backed out. I asked that the flooding be stopped. By the time the flooding stopped, our fokijser was just about to gain a new angle. We needed the water level lowered so we could be pulled back. Finally, after some line tending explanation, we completed the locking process. I think we had just been served by some départemental office workers playing éclusiers on a rotating day outing. Either that, or it was their first day on the job and they hadn't yet done the training.

From the lock we headed up a narrow and very overgrown canal leading about three kilometres to the next lock, which would lift us to the river.

Écluse Aiguillon is fortunately automatic, so we wouldn't have to endure the services of the départemental éclusiers. I dropped Edi off at the tiny landing quai immediately through a bridge hole, so she could walk up to the control pylon at the lock chamber.

I then moved forward to come to rest along the pilings between which water spilled from the draining lock chamber. I needed no mooring lines; the water pressure held Zonder Zorg on the wooden beam on the pilings. I watched Oxford Blue manoeuvre through the hole astern of us.

After we had locked through, we secured to a dilapidated wooden wharf immediately upstream of the lock. Included in the free moorage are water and electricity, so we plugged-in and then went off to explore the area.

The old mills at either end of the weir are abandoned and derelict. In many areas of the world, these would have long since been transformed into upscale lofts and shops, but this is France. We walked across Pont Napoleon to the village of Aiguillon and among other things, we visited the bakery and supermarket to stock-up with fresh supplies.

A strangely arranged pedestrian sign at the crosswalk in town had me trying to figure out how to use the system.

After a leisurely morning and a late breakfast, on Tuesday we slipped and continued upstream. We were now on the river, which is about a hundred metres wide at this point. There was little current because of no recent heavy rains, but the river is subject to strong currents and rapid level rises with heavy rains.

Heading Up the Lot

Seven kilometres along the river we arrived at Écluse Clairac. I dropped Edi off at the waiting quai, which appears to have been designed for a five or six metre boat. I dropped the spud pole to stabilize Zonder Zorg against the turbulent current coming from the draining lock chamber as it cycled down for us.

The old mill building sits very close to the edge of the lock chamber, and from an ascending boat, there is access to only the downstream third of the chamber top. There are lines dangling down the walls of the chamber to assist in stabilizing the boat; however, unlike similar arrangements in the Netherlands, these are not fastened at the bottom, so they cannot serve as a sliding mooring, like a pipe bollard. The lines need to be tended individually as the boat rises, but with Edi on the chamber top, it would be awkward to do this.

We decided to have Edi handle the bow line from the top and I would do the stern with one of the dangling lines. This put Zonder Zorg too close to the downstream doors for them to close, so I opted to motor forward on springs from fore and aft and watch that I didn't pull the dangling line on the other wall into the propeller. Edi shouted a warning that the bow line was running over the sharp edge of the dangerous 'safety' fence. This is another example of a lock designed by theoretical engineers with neither knowledge of nor concern for the needs of the boaters.

We made it up without a rope wrapped around the screw, passed under the eastern arch of the bridge, turned in the stream, passed back under the bridge through its central arch and came to solid bollards on the concrete wharf in Clairac. On the wharf with us were Oxford Blue and Felix, a British cruiser. We walked across the bridge to Longueville to locate the supermarket that is marked in the guidebook. We found a large Intermarché and having replenished our fresh stock, we made it back to the boat just before the skies opened up in a cloudburst.

Through the series of thunderstorms and heavy showers, three more boats arrived on the wharf, all of them from upstream: Aquarelle, a steel cruiser from Ireland; Vertrouwen, a US-flagged klipper and Fennavera, a British-flagged steel cruiser. There are four service pylons on the wharf with a total of twelve electrical outlets rated at 16 amps each. When the fifth boat plugged into the shore power, the entire system on the wharf blew.

A town employee came to try to sort-out the problem. He discovered that someone in town is hooked into the circuit and is drawing much of the current. The electrical thief hadn't anticipated so many boats all wanting shore power at the same time.

Late afternoon, after the series of thunderstorms had finished rolling through, we took a walking tour through the medieval centre of the town. The narrow, winding streets are lined with old buildings, many rather attractive, but none outstanding, much like many we had already seen through the region.

Also like many towns and villages we had visited the past while, most of the businesses were closed and the streets were deserted.

Heading Up the Lot

Wednesday continued the spell of glum weather, working at becoming the coolest and wettest May on record. By mid-morning three of the other boats had already moved across to the lock to head back down river and Vertrouwen was preparing to do the same. Mike and Barbara in Oxford Blue had decided to turn back and follow the others down, so we bade them farewell as we slipped in the late morning to continue heading up the river during a brief break in the rain.

We cruised past river banks that are mostly wild, with occasional buildings along them. Some of these appear to be agricultural warehouses; this is a famous apple and prune plum growing area. There are a few chateaux, many of them shuttered and looking rather forlorn, though occasionally we passed more modern residences that appeared to be occupied and well cared for.

The most common structures along the river banks are the recreational fishing shelters, many of them furnished with tables and chairs, but since it was lunch time as we passed, they were unoccupied.

Seven kilometres along the river we passed a modern mooring float with a ramp leading to le musée du Pruneau, the Prune Museum. It was noontime as we passed and it was likely closed, so we didn't stop.

We had thought of stopping in Granges-sur-Lot, which the guide shows as having moorings with water and electricity and indicates a bakery and grocery store in the village. As we approached we saw that the floats were missing. The pylons were there, as were the orientation signs and the electrical and water bourne, but without the floats, all this was redundant.

We continued along, passing a barge that appears to have had its last voyage. A short distance upstream of the wreck we began weaving our way through a narrow channel marked with beacons and we wondered whether the demise of the barge was related to the rock-strewn channel.

We followed the winding marked channel for much of the next two kilometres until just before the town of Castelmoron, where shortly after 1300 we secured to the solid concrete wharf on the right bank, directly below the château.

We were intrigued by the Moorish appearance to the château and we walked up to find out its background. The building is now la Mairie, the town hall, and inside we were offered a surly greeting and dissinterested service by the woman behind the reception counter. She was the only person around and we likely interrupted her daydreaming. From the brochures she thrust at us, we unravelled some of the history of the building and of the town. The town and the château date to the thirteenth century and are named after the nearby marshes: Château des Marais or Mauron, later corrupted to Castelmoron.

Heading Up the Lot

Throughout the town are old buildings, some appearing to date to the fourteenth or fifteenth century. The church is relatively new, having been built to replace a thirteenth century one that had been demolished in 1845 to give access to a new bridge across the river. The château received its moorish embellishments around the same time. The new owner, born locally from an immigrant Portuguese family, had apparently been inspired by a visit to Spain's Alhambra and converted the Romanesque architecture into Moorish.

Just upstream of the town is the next lock, a ten metre high one to surmount the dam for the hydro-electric plant. The decline of the mining industry up the river, combined by the increasing use of rail transportation, had caused le Lot to be removed from the list of navigable waterways in 1926. Through the following decade large electric generating stations were built, with their dams flooding the old locks and weirs.

In 1991 an isolated section of the river up and downstream of Cahors was reopened. In 1995 the bottom 50 kilometres from Aiguillon to Villeneuve-sur-Lot were restored to navigation. In 2001 a further 15 kilometres upstream of Villeneuve was put back in service and in 2010 this was extended to kilometre 75 at Lamothe. Work is in progress on linking the downstream 75 kilometres to the upstream section. We walked up the river bank to take a look at Écluse Castelmoron.

We followed along the top of a concrete retaining wall for about half a kilometre to the waiting wharf below the lock. The gate giving access off the wharf was locked, so leaving Edi on the wharf, I climbed over the fencing and sidled along to the stairs leading up to the dam's control house.

I spoke with the watchkeeper about locking through and was told that we would be seen as we approached, stay aboard and wait for the green light. The floating bollards in the chamber meant there was no need to disembark. I told him that would likely lock through the next day.

We headed back to Zonder Zorg, making it aboard shortly before the start of a torrential downpour. The rain continued heavily for nearly three hours, during which time the current in the river increased to about 1.5 metres per second past our windows. There was strong turbulence in the water as it passed, but with our spud pole down we remained stable.

I lowered some more slack on the spud pole line and monitored our mooring lines as the level of the river rapidly rose. When we had moored, the top of the leeboard was below the level of the wharf, and we had used the leeboard as a ramp to disembark. When the rain stopped, the leeboard was nearly 30 centimetres higher up the face of the wharf. I then understood why the wharf is so high.

On Thursday morning the sun was breaking through the clouds and the current in the river was down to a more gentle flow and the water was back down toward its normal level. At 1020 we slipped and headed up the short distance to Écluse Castelmoron. By this time the clouds had mostly dissipated and we enjoyed a rare sunny day.

Heading Up the Lot

The lock was ready for us as we arrived. Although at 10 metres height, it is much lower than the high locks on la Rhône, such as Bollène at 22.5 metres. However; it is only 5 metres wide, compared to the 11.4 metres of Bollène, so there is a feeling of being confined when at the bottom of the chamber looking up. The ride up was gentle and was made easy by the floating bollards, which are well spaced along both sides of the chamber. It appears that the lock was designed with pleasure boaters in mind. The channel leading past the weir to the lock is well protected with stout concrete pylons and we emerged into a broad river valley.

After a couple of kilometres and an abandoned lock house, we began passing stately homes and châteaux gracefully set on the river banks. For the first time in a long while we saw homes utilizing their waterfront, rather than hiding it. There were even private mooring facilities in front of some.

Near St-Livrade-sur-Lot we passed a major renovation in progress. An old two-storey house was being expanded tenfold or more with huge wings and rooms. This ten kilometre section of the river upstream of Écluse Castelmoron is the most active we have seen along the two thousand kilometres of French waterways we have travelled in the past year. Here we saw fine homes new and old, yacht clubs, sailing schools riverside restaurants and resorts. During my six years on the canals a decade ago in northern and eastern France I recall nothing anywhere near similar.

Above St-Livrade the fine riverfront properties gradually diminished in frequency and then stopped. There are a few isolated châteaux, such as Château de Favols, but mostly the riverbanks are wild with occasional crude fishing shelters. Again the fishing shelters were vacant; it was lunch hour.

Twenty-five kilometres upstream of Écluse Castelmoron we negotiated a buoyed channel through a gap in the weir at a decommissioned lock. Instead of remaining abandoned, the lock house has been transformed into an upscale restaurant with a covered patio on a deck laid across the old lock chamber. As we went past we could see it was crowded with diners; it is only two kilometres from the city of Villeneuve-sur-Lot.

Villeneuve-sur-Lot appears very attractive when approached from the river. It is called New Town because it received its charter in 1264 to be built as a replacement for the chateau and buildings destroyed earlier in the century as a punishment for the residents' heresy.

The bridge is a seventeenth century rebuild of a 1282 fortified bridge that was washed away in floods. The original bridge had five arches, but the replacement was built with only four, one substantially larger to make it easier for the river barges to negotiate the currents between the piers.

We had planned on stopping here to explore the historic old city, but we had been told of repeated acts of vandalism to visiting boats, such as being stoned and having mooring lines cut.

As we slowly motored past the mooring places, we had an uneasy feeling, so we decided to continue upriver.

Heading Up the Lot

We were disappointed that the town was apparently allowing a few out-of-control youth to spoil the welcome to visiting boaters. There are few enough of us on the river; except for the crowd the first day in Clairac, we had seen no other boaters on the river. We motored under the bridge and sadly watched Villeneuve fade in our stern.

Two kilometres upstream is Écluse Villeneuve, where we secured to the concrete wharf to await its preparation for us. I walked along the wharf and up the steps to take a look at what lay ahead. This lock is thirteen metres high and was completed in 2001 to bypass the large hydro-electrical dam that had decades earlier been built across the river. The lock chamber was full when we arrived and it took a long time to drain the 2.6 million litres of water from it.

Inside the chamber are eight floating bollards, four on each side, so securing Zonder Zorg for the ascent was easily and quickly done.

L'éclusier told us that Écluse Lustrac, the next lock sixteen kilometres up river was closed until Monday. Fortunately, we had already decided to stop for a day or two in St-Sylvestre just seven kilometres up.

At 1516 we secured to floats on the right bank in St-Sylvestre, just downstream of Pont de Penne. About a hundred metres up the slope from the moorage is an Intermarché, just before it is a fuel station with diesel and a large selection of gaz tanks and across the street is a hardware store. An artisan bakery is half a block away and

on a hilltop across the river is a medieval village to explore. St-Sylvestre was well suited to our waiting four days for the lock to reopen.

On the mooring floats are pylons with both water and electrical hookups, but they were deactivated. Across the river in Port de Penne is a marina, which according to the guide is managed by Babou Marine of Cahors. From our mooring it looked full, so we walked across the bridge to see if there was any space. The mooring spaces were all filled with small runabouts and a float home. On the door of la Capitainerie was a sign indicating it was closed except for a few hours on Saturdays and Sundays. There was water and electricity at the pylons.

Friday continued our long spell of rain showers. In the early evening a series of thunderstorms rolled through and during one of them, as I was cooking dinner, our Butagaz tank ran out. I got drenched changing to the spare, but we were rewarded by a small crack in the clouds that generated a double rainbow.

Late morning on Saturday we walked across to la Capitainerie to find out about getting some water. Le Capitain told us he would have the technician from the town come to our float to turn on the system and to collect the mooring fees for our stop. We told him we were heading up the hill to look at Penne d'Agenais and would be back aboard Zonder Zorg to meet him at 1500.

We headed through the town and out the other side on a two-kilometre climb to the medieval ridge-top town. After a rather steady climb along the shoulder of the road – there are very few amenities here for pedestrians – we gained increasingly better views of the riverside communities.

Heading Up the Lot

We came to the lower gate, Porte de la Ville leading into the old quarter and we wandered the narrow streets that wind their way up the steep hillside. Leading off these are many very narrow lanes, many with steps to make the climb easier. Although the town dates to the early thirteenth century, the buildings appear to be mostly seventeenth and eighteenth century and they are finely restored and maintained. We continued to climb and wind our way up toward the top of the ridge past well cared for homes. At the top we came to the upper gate, Porte Ferracap, which looks both northward down over the river and southward down into the town. The view from the outside shows it to have been well fortified.

We passed back in through the gate and continued up the hill; we were heading to the silver-domed Sanctuary we had seen from the river.

We arrived at the summit to find a comical hodgepodge of Romanesque, Byzantine and Belle Epoque architecture that seems a failed attempt to combine too many styles. Fortunately its exterior is difficult to view because of the surrounding trees and the steep drop-off on three sides. The fourth side backs into rocky cliffs of a higher knoll to the east.

Construction on la Sanctuare Notre-Dame de Peyragude was begun in 1897 and according to the plaque inside, it was terminated in 1948. Whether the termination was upon completion is hard to determine. Also on the inside we learned that on this site sometime around the year 1000 a small chapel was built and it became a stop on the pilgrimage to Santiago de Compostela. The current Sanctuary is built on the ruins of a series of chapels destroyed during Hundred Years War, the War of Religion and the aftermath of the French Revolution.

We followed a sign indicating ruins. The mud path led us past a cave entrance, so we went in for a look. We saw large rooms with three entrances, and there are likely more, but the place was unlit except from the entrances and we hadn't brought our spelunking gear. We could see no signs or panels offering information on the caverns, but we later learned that they had been occupied during Paleolithic times. We wondered whether the lack of information was as a ploy to not distract tourists from seeing la Sanctuare.

Heading Up the Lot

We followed the path as it wound up to the summit of the knoll, where across a short lawn we saw some ruins identified only as royal. The remaining walls were very thick, indicating a large or strongly fortified structure once grew from them.

We climbed onto the south wall and scrambled along a sloping broken top to its apex at the corner with the west wall. We were looking down on roofs of the Sanctuary from which perspective it looked more Byzantine than anything. Below it in the distance we could see Zonder Zorg on the river.

We climbed down from our exposed perch and followed a mud path that led back to the Sanctuary. Buried in a photocopied sheet about the Sanctuary was mention of the problem of the former chapels being frequently destroyed because they were so close to the castle built in 1182 by le Duc de Gascogne. A century and a half later, during the Hundred Years War, Penne was a frequent battleground as the English lost and regained the strategically positioned castle four times, each time the battles causing serious damage to the town and the chapel.

As we walked back down the hill we ran our English history through our minds. The Duke of Gascony at the time the castle was built was Richard the Lionheart, the future King Richard I of England. His mother was Eleanor of Aquitaine and his father Henry Plantagenet, England's King Henry II. That this information is obscured and the castle ruins are nearly unmarked, amazes us. We later refreshed our knowledge that Richard was also Duke of Normandy, Duke of Aquitaine, Count of Poitiers, Count of Anjou, Count of Maine, Count of Nantes, and Overlord of Brittany, thus having control of a large portion of northern and western France. Yet he is almost ignored here by the French.

We were back down aboard Zonder Zorg by 1430, well in advance of the time when the technician was to come to switch on the water and electricity on the float. We waited aboard through the remainder of the afternoon and into the evening. We waited Sunday morning, then we walked across to la Capitainerie to find out what was happening, but found nobody there.

I tested the water spigot to ensure there was water, then we walked back over and motored Zonder Zorg to a temporary mooring with her nose under the low bridge end and her beam poised on a corner of the stone wharf. We filled our water tanks, found a rate schedule, but could not find anyone around to take our payment. We slipped and continued up river.

LONGUEUR DE BATEAU	PASSAGE		SEJOUR		
	JOURNEE	SEMAINE	MOIS	HIVERNAGE 1 Nov au 31 Mars	ANNEE
« 6.00 M	8 €	33 €	100 €	350 €	650 €
« 7.00 M	9 €	33.50 €	102.50 €	358.75 €	666.25 €
« 8.00 M	10 €	34 €	105 €	367.50 €	682.50 €
« 9.00 M	11 €	34.50 €	107.50 €	376.25 €	698.75 €
« 10.00 M	12 €	35 €	110 €	385 €	715 €
« 11.00 M	13 €	35.50 €	112.50 €	393.75 €	731.25 €
« 12.00 M	14 €	36 €	115 €	402.50 €	747.50 €
« 14.00 M	15 €	36.50 €	120 €	420 €	780 €
PENICHE	16 €	50 €	160 €	560 €	1040 €
Elec hors gel hiver Bateau non habité *			12 €	60 €	144 €
Elec année Bateau habité			39.50 €		475 €
Eau année Bateau habité					135 €
Eau 1m3 /J	2.35 €	16.50 €			
ELEC/jour	2.50 €	17.50 €			
SANITAIRES	2.35 €	16.50 €			

Four kilometres along we passed an old decommissioned lock, the top of which was just awash. The bollards and old iron hardware of the gates were still in place. The river had been removed from the list of navigable waterways in 1926, so the last maintenance on this lock would have been more than eighty-eight years ago.

A further five kilometres brought us to Écluse Lustrac, which was closed until Monday morning. Shortly after 1600 we secured to the small waiting float just downstream of the entrance to the lock.

It is a very pretty setting, and we were delighted to see that the old mill buildings had been converted into residences. We walked up to the lock to see if there was any indication why the lock had been closed for four days, but could see no signs of work, recent, present or future.

It rained heavily Sunday evening and overnight and it continued heavily on Monday, so we decided to remain on the float until the forecast clearing the following day. On Tuesday morning our card worked in the control slot and the lock prepared itself for us. As we left the lock we found a fresh current in the river from the rains.

Heading Up the Lot

We continued along for another three kilometres to Écluse Ondes, the final lock on the navigable portion of the lower Lot.

Two kilometres beyond the lock begins a 1.5 kilometre buoyed channel, which had been cut through solid rock in 2010 to the Saint-Vite weir, the current end of navigation. The guide says: "Do not move away from the centre as the rocks are sharp. On arriving at Saint-Vite, beware of the cross currents generated by the weir."

I looked at the narrow buoyed channel winding up the river and disappearing from view around the next bend; it appeared to be narrower than Zonder Zorg's length. With the current from the recent rains added to unknown eddies at the weir, turning around any further upstream than the beginning of the marked channel would be risky. We had come seventy-five kilometres up the river from la Garonne, but my mariner's prudence caused us to forgo the final fifteen hundred metres of the navigation along the river to Saint-Vite. We wore around just before the first set of buoys and started back down.

It was still heavily overcast, cool and threatening to rain as we retraced our way down through Écluses les Ondes and Lustrac. We were once again in down locks, so the process was simpler. There having been no traffic since our ascent, the locks were in our favour and we didn't have to stop to drop Edi off on the quai before the locks.

Midway between St-Sylvestre and Villeneuve-sur-Lot, as we passed Château Rogé, the current in the river moved in an undisturbed sheet over the bottom. The air was still and we were offered wonderful reflections of the passing scene.

A short distance downstream of Rogé we were challenged by a swan. As we approached, it swam across toward us with raised wings and lowered head and then turned parallel to our track and beat the water. It then turned and did the same in the opposite direction, moving quickly astern of us. We thought it had quit, buy shortly it began a low level kamikaze run toward our stern, veering at the last moment, just as I was ducking.

Edi shot a beautiful photo of it slowing to veer off. It did another challenging run, this time making foot stomps in the water on its approach. Just to make sure, it did a similar run down our other side.

Satisfied that he had scared us off, he headed back up river while we continued down. We shortly came to the waiting wharf for Écluse Villeneuve and secured to wait for the chamber to fill.

Heading Up the Lot

Within twenty minutes we had cycled down the thirteen metre lock and were heading out to continue down the river. By the time we reached Villeneuve-sur-Lot, the clouds had cleared to the southwest and it looked like we might have a rare sunny afternoon. We hadn't felt good with stopping on our way upriver and we thought we would give it another look on our way back down. As we passed under the old bridge, we still felt uneasy about the place. The stories of rocks thrown at visiting barges and of mooring lines being cut certainly influenced our thinking, but in the end, it was our gut feeling that told us to continue on past.

As we passed through the gap in the weir at the old lock, the current in the river seemed a bit quicker than it had been on our ascent. We arrived in St-Sylvestre at 1225, just five minutes before most of the supermarkets, grocery stores, butchers and bakeries close for lunch throughout France. We stopped anyway, just to check in case this town had some liberated shopkeepers that allow reason to defy tradition. It doesn't, so after watching all the hoardings come down and shutters close, we continued downriver.

At 1540 we secured to one of the two fifteen-metre wooden wharves in Casseneuil, where the €5 moorage includes water and electricity. The wharves are at the junction of la Lède, a tributary to the Lot. We strolled along the track that follows la Lède past the bases of ancient fortification walls. Among its early history, the town was a Roman settlement called Cassinogilum, from which it derives its present name.

Because of its strategic location, the town was a frequent target, having been sacked and partly destroyed by the Normans in 848 and repeatedly during the Cathar crusades, the Hundred Years War and the Wars of Religion. The walls look a patchwork of frequent repairs. In the town are many medieval buildings along the narrow, winding streets and hanging over the banks of la Lède. Église Saint-Pierre was destroyed in 1214 and was soon rebuilt as église Saint-Pierre et Saint-Paul, which still stands. It is reputed for its fine fifteenth and sixteenth century frescoes; however, the church was not open when we were there, so we didn't see them.

At 1015 on Wednesday morning we slipped and continued downstream. Along the banks we saw many pigeonniers, some restored and others converted as feature towers on homes. We descended the ten metre lock above Castelmoron and swept past the town on the rising current as rain came and went in frequent showers.

We continue to be amazed by the French fashion of shutting everything down for lunch. They even close the river to waterskiing at noon. We saw many signs along the river marking the start and end of watersport zones and giving the hours: 10H to 12H30 and 16H to 20H. Why they need to shut down waterskiing and fast motorboating for lunch makes as much sense to us as does shutting down nearly the entire country for a lunch break.

Heading Up the Lot

We motored past Clairac and into the narrow canal bypassing its weir. The 'lock-from-hell' at the mill in Clairac was easier on the descent, but its mooring arrangements still spoke loudly of having been designed by someone with absolutely no knowledge of boating.

At 1448 we secured to the dilapidated wooden wharf above Écluse Aiguillon. After we had settled-in, I phoned the listed number to arrange for lock keepers and pilot to lower us to la Garonne, guide us up the river and lift us into la Baïse the following morning. The person who answered the phone said: "C'est impossible, demain est un jour férié", Eet ees impossabeel, tomorreau ees a oliday". We quickly changed plans and re-organized the Lot-Garonne-Baïse transit for Friday.

It appears that on Thursday everyone in France will be sitting around thinking of Christ heading off into heaven, or something like that. With everything closed on Thursday, and many remaining closed on Friday to 'make the bridge' to the weekend and some continuing the closure through the weekend to keep the momentum going, we scrambled to restock our fresh provisions.

We walked across Pont Napoleon and through town to the Intermarché on its far side, returning in a heavy rain that increased to a downpour by the time we made it back to Zonder Zorg. Fortunately, the free moorage includes water and electricity, so we took long hot showers and enjoyed the dry cabin warmed by our electrical heater as the level and current of the river slowly rose.

On Thursday we crossed the bridge and walked through the town between rain showers, just to confirm that everything was closed. It was Edi's birthday and we thought it nice for all of France to take the day off to celebrate with her.

In the evening we added to the festive mood at dinner with a bottle of Champagne Veuve Clicquot, a birthday gift from elder daughter, Amy.

Chapter Thirty-Three

The River Baïse

It continued to rain through Thursday evening and was still raining when we went to bed at midnight. The clouds had begun breaking by the time we arose. We were just beginning our first cup of espresso when there was a knock on the hull. The pilots told us la Garonne was rising rapidly and is being closed to navigation. They could take us through if we left immediately. We quickly dressed and headed down through Écluse Aiguillon.

We followed the narrow winding canal that bypasses the final rapids on le Lot. With all the urgency that had been expressed by the pilots, we were surprised to see that the lock had not been prepared for our arrival. There seemed to be no sense of urgency among les éclusiers and we sat treading water in the approaches for nearly a quarter hour as the lock chamber was slowly filled.

Our first glimpse of la Garonne through the opening gates showed a swirl of turbulence near the exit from the lock, and beyond that the river was moving rather uniformly past. When the pilot came aboard, he said the river was running about twelve kilometres per hour with a swifter section through le Passage de Monluc. Zonder Zorg's theoretical hull speed is 17.5 kilometres per hour and I doubt that her 64 kilowatts of power could get her much past 17. This meant that at full power, the 4.8 kilometre passage up to the entrance to la Baïse could take an hour or more.

We headed out into the river and turned upstream to follow the pilot boat through the chocolate-brown water. At 2500 rpm we were making a big wake in the water, but the river bank seemed to be barely moving past.

We ground past buoys tugging at their mooring chains and in le Passage de Monluc, they seemed to pause beside us as we worked over the ledges.

We watched as Pont St-Léger slowly grew and we finally reached the turbulent 'lee' at the entrance to la Baïse at Écluse St-Léger. It had taken us an hour and eleven minutes; we had averaged 4.06 kilometres per hour at 2500 rpm. We could see much of the brown in the river was coming from la Baïse.

We locked through into la Baïse and started up a more placid river. The water here was even more muddy than in la Garonne, indicating a heavy runoff upstream.

The placidity of the river gradually changed as we approached Écluse Buzet, four and a half kilometres upstream. During the last kilometre before the lock we ran through increasing foam and turbulence and as we approached the weir we saw water flowing high over it. The wall beneath the weir, which separates the spillway current from the lock entrance was nearly awash.

From the top of the lock chamber we had a good look at the turbulence and the source of the foam through which we had been motoring.

At 1240 we paused on the stone quai a short distance upstream of the weir, just below Pont de Buzet. We were hungry, having had our breakfast plans cancelled by the early arrival of the pilots.

After enjoying our late brunch, we continued up la Baïse, passing the lock that leads up to le Canal de Garonne. Four kilometres further along we passed under the canal aqueduct as the skies attempted to clear.

The speed and turbulence of the river increased as we approached Écluse Vianne. The river was up with water flowing over the separating wall between the weir outflow and the lock entrance. The top of an abutment, a directional sign and some navigation beacons were all that slowed the water in the approaches to the lock.

I dropped Edi off on the finger below the closed lock gates and then sat in the turbulence on the bollards as she set the mechanism in motion to cycle the lock down for us. The back eddy below the lock doors made it a tad tricky entering the chamber, but our fenders were well placed and up to their assigned tasks.

From below the lock we had seen open mooring spaces along the quai in Vianne, so once we had locked through, we secured to the solid concrete wharf and called it a day. We had left the fast-moving Lot, ascended the faster moving Garonne and started our way up the raging Baïse.

It was 1440, well past time for the shopkeepers to be back from lunch, so we walked up into the town for fresh provisions. The only thing the town has going for itself is that its thirteenth century bastide walls are still standing. Inside the fortified walls it is nearly dead with most of the businesses closed. Of the two groceries, one was closed and the other may as well have been. Its shelves were near bare and its sparse produce selection was approaching compost. We picked among the overpriced baskets of strawberries and selected the one with the fewest culls on the top layer.

The River Baïse

At 0850 on Saturday we slipped our lines and continued up the river. The level had risen overnight and the current was noticeably stronger. Three kilometres along and around two sharp bends, we approached Écluse Lavardac, where there was a large churn of turbulence and spume the few hundred metres leading to the lock. I dropped Edi off on a two-metre wooden wharf sited in the middle of the eddies below the weir.

I dropped the spud pole and sat in the churn with a stern line to a bollard on the tiny quai, while Edi activated the lock. Her view from the lock made the conditions on the water appear much more sever than they seemed from onboard.

In the lock we found a continuation of the broken and dysfunctional infrastructure. Here the vertical pipes in the sliding bollards were missing. Missing bollards, badly placed bollards, broken bollards and other shortcomings do not allow a relaxation in the chambers, but force a constant adaptation and improvisation. The only predictable thing with the systems is that each lock will be differently setup and different things will be broken or dysfunctional in each.

At the next lock, the landing wharf was nearly awash, with its upstream end covered in foam. I stemmed the current and juggled with the eddies to land Edi so she could head up to the lock to activate the cycle.

The turbulence at the lock entrance made threading Zonder Zorg into the chamber a bit more exacting than normal. This was further complicated by the locks above Lavardac being only 4.15 metres wide, rather than the 5 metres on le Lot and the 6 metres on Canal de Garonne. This narrower width left us with just a couple of centimetres clearance on each side of our leeboard fenders.

We continued working our way upstream against the heavy current and erratic turbulence.

We arrived at Écluse Nérac after nearly three hours of very hard slogging against adverse conditions. We decided to stop for the day once we had cleared through the lock.

A few minutes before noon we secured to rings on the stone wall in Nérac. We were greeted with information that because of the flood conditions, the river had been closed to navigation for two days and would likely remain closed for another couple. We were unaware of the closure, since we had come directly from la Garonne and had bypassed the manned lock down from le Canal de Garonne into la Baïse where the closure is imposed for upstream-bound boats.

The River Baïse

We moored directly below the remains of Château Henri IV; much of it had been destroyed by the anti-royal fervor following the French Revolution. All that remains of the fifteenth and sixteenth century original is a portion of one wing of the quadrangle.

While we waited for the raging river to subside, we explored the area. Extending along the river banks for about a kilometre upstream of Nérac is a nature park, Parc de la Garenne, Among its attractions are several artesian fountains. In one of these, la fontaine de Fleurette, is a graceful sculpture of Fleurette, a maiden who wooed and won the fleeting attention of the young prince, soon to become King Henry IV. When the royal court refused her alliance with him, she threw herself into the river.

The attendant plaque tells that she was only sixteen, he three years older. Ravished by his love, she gave her life; he gave but a day. The word 'flirt' is said to derive from her brief royal tryst.

Across the river and back near the edge of town we passed the restoration of the royal bath house, its site showing the extent of the former gardens beneath the château. The elegant sixteenth century octagonal stone pavilion is now listed as an historic monument.

On Sunday afternoon the level of the water fell to 50 centimetres on the Nérac gauge and the river was again opened to navigation. Mid-afternoon Hoop op Zegen came through town, downbound in a rush to make it to Buzet before the locks closed; Nicholas and Marise were heading back to England from there on Monday for a couple of weeks and had been trapped upriver by the floods. With the level at 50 centimetres, the current in the river was still quite strong, making the entry to the locks still rather awkward.

We decided to wait until Monday before we continued upstream, not for river conditions, but primarily because we were almost out of diesel. We had burnt much more than usual churning against the swift currents in le Lot, la Garonne and la Baïse. Monday morning as we watched the level of the river drop, we scrambled to find a company to deliver diesel to the barge. None of the local companies offered unmarked fuel delivery.

After many phone calls with the assistance of the tour boat skipper and Capitain de Port we contacted a company to come with agricultural diesel. We decided it was an emergency and took on 200 litres of marked fuel at €0.94 per litre, about 45 cents per litre less expensive than transportation diesel. Besides the emergency, we justified the act by considering we had burnt well over 200 litres for heating and hot water during our year in France, and GNR, marked fuel is legal for those purposes.

It was 1540 by the time we finished fuelling and were able to continue upriver. As we passed the gauge on Pont Neuf, it was registering slightly below 40 centimetres, 10 centimetres below flood closure level.

The River Baïse

We continued along a winding and narrowing river with even narrower leads to the locks. From Nérac, we passed through seven locks. The first two were rather easy, but from the third lock onward there are narrow and very shallow bypass canals upstream from the locks. We dragged our skeg through the soft bottom as we pumped our way along at turns for six but making two or less.

At 1902 we secured to bollards on a stone wall beneath the bridge in Moncrabeau. The town calls itself the liars' capital of the world, but whether they are telling the truth about this, we didn't find out.

We left early on Tuesday morning and motored another eleven kilometres up the river, passing through three more locks to arrive in the city of Condom. We continued on through without stopping, not that we weren't interested, but we had decided to stop there on our way back down.

A kilometre and a half beyond Condom we secured at 1145 to the wooden wharf just upstream of Écluse Gauge. The chart book shows a supermarket 600 or 700 metres along a road above the lock, so we scurried to get there and do our fresh provisioning before all of France shut down for lunch.

We returned to Zonder Zorg with a fine selection of fresh fruit and vegetables, meat and fish bread and pastries and, of course cheese. We are constantly reminded of the stupidity of the Canadian dairy marketing board that rewards a few farmers by making Canadians pay nearly three times the world price for cheese and other dairy products.

For less than €16 we bought a selection of cheese that would have cost $75 or more in Canada.

The next lock along the river is a double one and it is operated by éclusiers. We arrived and tooted the horn as the sign requires and immediately two éclusiers came running. We were confused, we hadn't previously seen any pace above a slow saunter among the lock employees. As we approached, we saw two very fit young men, obviously eager to serve us and most likely university students freshly arrived for summer employment.

The workings of the lock have been kept totally manual; we watched as they cranked closed the doors and dropped the sluices.

They than ran up the stone staircase to the central doors and began opening the sluices to flood our chamber. By the end of the summer they will either have slowed down to the snail's pace of the standard French worker, or they will be very fit and extremely satisfied.

The next lock, Écluse Flaran, is 3.3 metres high and has a 2 metre flood wall above the full chamber, making it a long climb up the slimy ladder to the rim to push the button to trigger the cycle.

The River Baïse

At 1450 we arrived in the basin at Valence-sur-Baïse, the end of the navigable portion of the river. Alongside were two small cruisers in long-term storage and two rental boats waiting for clients.

After we had secured we walked up the steep hill beneath the cliffs on which the walls of the fortified town were built in 1274. It was mid-afternoon and the town was nearly deserted. All the shops were closed, most of them permanently. We paused at la Mairie and its tourist information office, where we found a very talkative woman. She told us of the new supermarket and other shops a kilometre and a half outside the town and of the declining population and dying businesses in the old town. Of the remaining businesses, they open only for a few scattered hours a few days of the week

Beside the town hall is a Romanesque church, which was begun when the town was fortified and it was completed and dedicated in 1303. The interior is simple and graceful with beautifully rendered reliefs and sculptures in wood and marble backing the main altar.

Late on Wednesday morning we headed back down the river, passing through two single locks and one double in the ten kilometres to Condom. Exactly two hours after having left Valence, we found a safe mooring place on the low masonry wall on the left bank of the basin in the centre of town. Condom is the capital of the Armagnac region and the Armagnac being with the Cognac, one of the two great brandy regions of France. We were between the two bridges, either of which lead into the historic centre of the town.

Besides Armagnac, the city is also famous as the home of d'Artignan et les Trois Mousquetaires and a newly erected sculpture of the foursome stands next to the cathedral. Alexandre Dumas had based his famous nineteenth century novel on a real historical figure: Charles Ogier de Batz de Castelmore, Count of Armagnac. As a young man in the 1630s the count had travelled to Paris to make his fortune with the musketeers. After a long and glorious career in the military, he died in 1673 during Franco-Dutch war at the siege of Maastricht. It was upon his memoirs that Dumas based his novel.

Also beside the cathedral are the wonderful sixteenth century cloisters. These are now a public space and above them are the offices of la Mairie.

Cathédrale Saint-Pierre dates from the fourteenth and fifteenth centuries built on foundations of an eleventh century abbey church. It has superb gothic vaulting in its nave. Among the more impressive aspects of the interior is the intricately carved masonry in the new choir, which dates to 1844.

The River Baïse

At 1115 on Thursday we slipped and continued downriver. The water level was down and we dragged our skeg through the ooze of the bottom in the canals leading past the weirs to the locks. Some of these narrow canals are short, but a few of them are near 500 metres long; progress was painfully slow. Above Écluse Lapierre we came upon a work barge poised across the river clearing a fallen tree. We waited while it swung itself out of the way and then we passed.

After we had pumped our way along the shallow canal leading to Écluse Lapierre, our bows were just entering the open lock doors when they began to close. A woman from an upbound rental boat had scurried up to the lock activation button and without looking, pressed it. The system shut down.

I backed clear of the doors and dropped the spud pole, then I phoned the number listed in the guide to report the incident and received the familiar response: "J'arrive". About thirty minutes later the work barge that we had passed upstream arrived astern of us and put its bow into the bank to land two workers to reset the system. It took another quarter hour for the men to sort out the reset before we were able to enter and lock down.

Finally, nearly an hour after we had begun entering the lock, we made it through. The rental boat had to wait for the lock to cycle back up to bring the work boat down before it could continue, so it was delayed nearly an hour and a half for not spending two or three seconds to see if it was clear to push the button. At 1659 we arrived in Nérac and secured on the left bank between the old and new bridges.

On Friday morning, after a trip to the supermarket to restock, we continued downstream at 1045. The water level was still a bit high, though dramatically below the levels we had been in on our ascent the previous week. The landing pier that had been nearly awash was now a good 30 centimetres above water, though the buoys downstream of it showed there was still a strong current running below the weirs.

At 1325 we passed under the aqueduct of Canal de Garonne. Four kilometres later we passed the sign indicating the junction with the canal and a short while later we were at the entrance to the double lock waiting to be taken up out of the Baïse. At 1428 we entered le Canal de Garonne and turned to starboard to continue downstream. We had spent just short of three weeks on three rivers: la Garonne, le Lot and la Baïse. Much of this time was in unstable rainy weather and we had been in high water and in flood closures. It was good to be back in the more stable and predictable canal.

Chapter Thirty-Four

Down la Garonne to Bordeaux

We had arrived back on the canal on Friday 6 June and continued downstream through Buzet-sur-Baïse. The banks on both sides of the canal through Buzet were still filled with moored boats and we were delighted to see all the idle rental boats in the marina. Half a kilometre further along, the marina at Coustet was also overflowing with idle rental boats. This meant that fewer would be out jeopardizing our safe and peaceful navigation. Five kilometres further along we paused for the night on a quiet wilderness bank downstream of Damazan.

On Saturday we continued down the canal, working our way through the locks. Edi got back into the swing of twisting the dangling wand to initiate the automatic cycle.

After three locks and fourteen kilometres, we arrived in Mas d'Agenais shortly before noon. As Edi prepared lunch, I trotted up the steep hill and through the old wall to the village to buy bread and fresh produce from the épicerie that I had seen mentioned in the guide.

Also mentioned in the guide is the eleventh century Église Saint-Vincent, in which is a depiction of the dying Christ painted in 1631 by Rembrandt. After a leisurely lunch, giving sufficient time for any lunch closure to be completed, we climbed up to the village and walked along to the church. The doors were locked. We walked around the building looking for the access and finally saw a sign indicating that the church is open only on weekdays and that it is closed on weekends. Here was a new twist to the French closure game, the church was closed on Saturdays and Sundays.

Not wanting to wait in the tiny three shop community until Monday, we decided to take a quick look around and schedule a weekday stop on our way back up from Bordeaux. There is a splendid view over the canal and la Garonne from a small park at the rim of the cliff. We followed the switchback path and stairs down the escarpment below the park enjoying the fine view of Zonder Zorg on her mooring just up the canal.

We continued down the canal for another dozen kilometres and two locks and at 1540 we secured for the day on a wilderness bank two kilometres downstream of the village of la Grande Route.

We slipped shortly before 1000 on Sunday and worked our way down another six locks and 27.4 kilometres to the basin in Castets-en-Dourthe. The water in the basin is rather shallow and it is very overgrown with underwater plants. There was no clear path through the weeds, so we had to cut our own route through, and in the process, the propeller became entangled and progress was very slow.

Down la Garonne to Bordeaux

Mid-afternoon we slowly plowed our way backwards through the dense weeds and secured on a wharf, the bow held by the spud pole. The mooring balls are set-up for boats half Zonder Zorg's length and simply got in the way of our maneuvering. We had chosen a spot next to the wifi antenna, from which we received a very strong signal. Unfortunately, like many other places in France, the wifi is not connected to the internet.

We were at kilometre 192.5 on the canal, half a kilometre from its end. At the end of the basin is Écluse les Gares and just beyond it is Écluse Castets, the lock down into la Garonne. On the wall of the Castets lock house is a flood gauge, which measures the level of the river above normal. The lip of the lock chamber is 7 metres above the river and the gauge goes up to 13 metres. Notations of floods border the scale, with dates recorded at more severe floods. Of note are the 12.5 metres in 1875 and the 13.0 metres in 1900, both floods just below the top floor landing of the double staircase.

We walked over to la Capitainerie to get information on descending to the river. We learned that access to the final two locks of the canal is scheduled to coincide with high water on la Garonne, as long as it falls within the working hours of les éclusiers. High water at Castets is two hours after the high tide at Bordeaux and the next high water was at 0815 the following day, but the lock keepers would be off duty then. Vive la France! We booked an appointment for Monday at 1830 and after ordering bread and croissants for delivery in the morning, we went back to Zonder Zorg to relax.

Since our scheduled locking through wouldn't be until Monday evening, there wouldn't be sufficient daylight remaining to make the four hour passage down the river to Bordeaux and

there are no safe stops on the river until Bordeaux. We would have to wait overnight Monday on the float in the river below the lock until Tuesday morning's ebb.

Le Capitain had told us there was an épicerie and a boucherie in the village, so after breakfast on Monday we walked up to buy some fresh things for dinner. Our fresh supplies were near exhausted, since we had anticipated being able to replenish in Bordeaux. We arrived at the grocery shop to find a crude note taped to the security roll-down. It translated to: Store Closed Saturday 7 June to Monday 9 June inclusive. Reopening Tuesday 10/6 at 0700.

We rethought the evening's dinner; we had end bits of a variety of vegetables, so a stir-fry made sense. We walked along to the butcher to select a piece of pork or turkey breast to add to the wok. The sign on its door told us that it was closed on Mondays. Vive la France!

We continued along to the twelfth century Église St-Romain for a look. I was intrigued that the church is oriented North-South rather than East-West and that it has its bell tower at the back, rather than at the front. We walked around it looking for an open door; there was none; it was closed. Vive la France!

On Monday evening we descended through the two final locks to the river, taking an inordinate amount of time waiting for a fiberglass boat to bounce its way into the two locks and pass their lines. We later found out the owners had just bought it and were still uneasy with its handling.

We entered la Garonne, swept downstream under the Eiffel bridge in the building current, turned and stemmed the current back up to secure on the dilapidated float, which had only two unbroken bollards and one ring for the two boats. With the current overnight expected to be up to ten kilometres per hour with the ebb and six or seven in the opposite direction with the flood, we had to be very creative with our mooring lines. Not long after we had settled in, a heavy thunderstorm rolled through. I dug some baby scallops out of the freezer and cleaned-out the fridge's vegetable drawer; we enjoyed a splendid stir-fry.

We had been told by le Capitain to leave the float at 0800 with the strengthening ebb. He said the passage to Bordeaux should take about four hours, and this timing would put us at Pont de Pierre at slack water just before noon. I looked at the 54 kilometre distance and saw we would need to average 13.5 kilometres per hour in an increasing and then a slowing current. We left the float at 0730 to give us half an hour in hand.

Two kilometres along, the first buoys on the river showed that the current was already rather swift. Running at 1400 rpm, which in still water gives Zonder Zorg 9 kilometres per hour, we were already making 13.8.

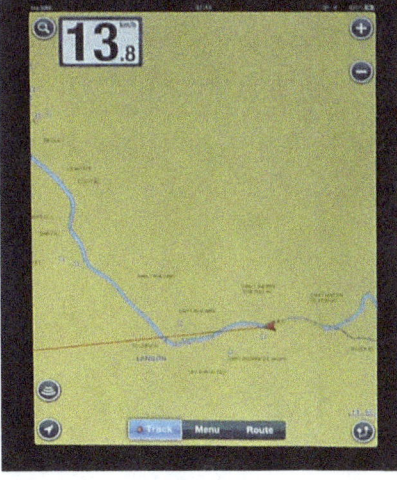

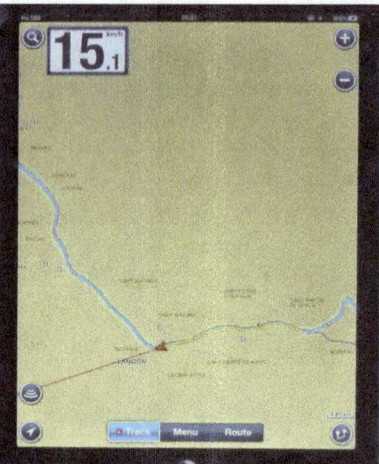

Four kilometres further along we passed Port Airbus, the terminal for the barges that bring upstream from Bordeaux huge fuselage and wing sections for the Airbus assembly plants. We were making over 15 kilometres per hour, the river current kicking our speed up by more than 6 kilometres per hour.

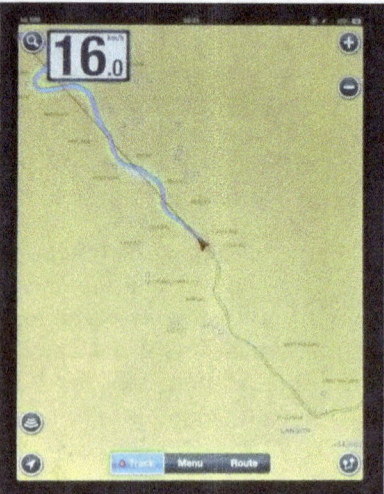

The beacons in the river had lengthening current trails and twelve kilometres further along we were hitting 16 kilometres per hour.

With increasing frequency we passed bank-side fishing installations. These are common on the lower Garonne and along the estuary of the Gironde, and they typically employ drop nets. Many of the installations are shacks on poles above the river, but some are on floats on pilings. They vary from extremely rustic to rather luxurious.

Down la Garonne to Bordeaux

Also along the banks we passed an increasing number of chateaux. Many of these were wine properties as we passed the dessert wine regions of Ste-Croix-du-Mont, Loupiac, Barsac and Sauternes.

There are very few riverside communities, most being built back in away from the banks and above the historic flood levels. Langorian is an exception to this, being built at river's edge on a bank only about four metres above the river. Its only moorings are overgrown sloping stone walls. There are few safe mooring possibilities in the more than 50 kilometres between Castets and Bordeaux and a power failure or steering gear problem would place a boat at risk over much of the passage.

At 1110 we arrived in Bordeaux. Five minutes later we passed through a heavy ebb current under Pont de Pierre and washed downstream as we turned to stem the current back up to our reserved spot on Ponton d'Honneur. We secured at 1123, having made the 54 kilometre passage in seven minutes short of four hours. We were alone on the float until the arrival an hour later of the plastic boat, with which we had shared the lock and float in Castets.

Up the ramp and across the street from us was Port St-Caihau, through which begins the heart of the old city. I had been to Bordeaux dozens of times in the 1970s, 80s and 90s searching for wines to import and conducting wine tours. My last trip had been in 2000 just before I shut down the wine company and bought my first Dutch canal boat. I was looking forward to seeing the city again.

Two hours after we had secured on the float, the barge Brion arrived beside us on its way up the river with AirBus fuselage sections. It sat stemming the current below Pont de Pierre, waiting for slack water so it could pass through and continue up to Langon. There is a strong current and turbulence around the arch bases until shortly before slack water. We had squirted through when we arrived.

Finally, after thirty-five minutes of treading water, it moved through the specially prepared centre arch of the bridge. It appears we had passed under the bridge more than two and a half hours before slack; no wonder it was rough.

Our moorage on the float was also rough, contending with tides of over five metres and a twice daily tidal bore meeting the heavy seaward flow of la Garonne. There is a pair of mooring basins, les bassins à flot, downstream three kilometres and through a tidal lock, but we had received no replies from our repeated emails and phone messages. We were stuck on the float in the currents and tides, where moorage cost €42 per night including water and electricity.

Down la Garonne to Bordeaux

We had been thinking of heading up la Dordogne, but since it was nearing full moon, the heights of the tides and the mascarets, the tidal bores was increasing. This would make navigation more difficult, less safe and the moorings would be even more tenuous. Waiting on the expensive and uncomfortable float for more than a week for tides of lower amplitudes made little sense to us, so after an afternoon, an evening and a morning seeing the sights of Bordeaux, we decided to head back up river.

At 1400 on Wednesday we slipped from the float and headed back up toward the arches of Pont de Pierre. It was still a good half hour before slack, but I wanted to buck the slowing current for the first while to put us further upstream to use the push of the flooding tide before the river current overcame it. We selected the arch with the best looking current and punched through at 2400 rpm, having the constricted flow slow us to a near stop under the bridge.

Upstream a kilometre is a set of three bridges, two road and one rail, with a navigational channel indicated by the red/white diamonds. Coming downstream was Brion, heading to fetch more fuselage pieces. I stayed on the outer edge of the marked channel, leaving seventy-five of the eighty metre width for Brion. She inexplicably veered toward us and crowded us out of the channel.

After nearly an hour we had lost the ebb current and began to feel the flood. By the time we were passing Langorian, twenty-two kilometres upstream from Bordeaux, we were moving along at better than 13 kilometres per hour.

Upstream we passed the tailored corduroy slopes of vineyards contrasted above the wild and ragged river banks. The sun was over our shoulders as we arrived in Langon and I called les éclusiers at Castets to advise them of our arrival at the lock. We were told the high water occurred after they closed for the day and that we would need to wait on the float until the morning tide. Both the paper and electronic tide tables we had showed this was not so; high water at the lock was within their working hours. However, this meant that locking us through would have entailed two or three minutes overtime. No further discussion was warranted.

At the height of the tide an hour later, we secured on the float below the lock as les éclusiers spent their last twenty minutes idle before quitting time.

After a leisurely breakfast the following morning we slipped from the float, headed under the Eiffel bridge and into the lock. Within a quarter hour we were out of the river and back into le canal de Garonne.

We decided not to stop in Castets, but to continue up canal to somewhere with a better work ethic. We joked that this might mean having to continue all the way across France and into Flanders or the Netherlands before we found such a place.

Chapter Thirty-Five

Up Canal de Garonne to Toulouse

It was Thursday, 12 June when we had left la Garonne and re-entered le canal de Garonne. Ahead of us was a climb of 132 metres through 52 locks spread over 193 kilometres that would take us back to Toulouse. The transit of the basin above the second lock was much easier than on our descent a few days earlier. A weed cutter had been operating in the pound for two days and had already cut a clear channel through the very heavy bottom growth.

We quickly got into the routine of up-bound locks and by 1700 we had passed through eleven locks and travelled 39.4 kilometres to Mas d'Agenais, where we secured to a very fine wharf with solid bollards in a clean, attractive setting with no moorage fee.

We had pushed onward without pause because we wanted to visit église Saint-Vincent and see its Rembrandt painting. The eleventh and twelfth century church a listed national monument and for some unfathomable reason it is closed on weekends. We had missed it on our way down the canal the previous Saturday.

After breakfast on Friday morning we walked up the steep hill to the church and found the door unlocked. Inside, hanging on a plain wall behind armoured glass, was the painting. Besides a cleaning lady, we were the only people in the church.

The painting was done in 1631 as part of a series of eight commissioned by Dutch Prince Frederik Hendrik of Orange Nassau to decorate his daughter's chapel. Six of the series are now in the Pinocotheque in Munich and the remaining piece is known only by a copy in the museum in Brunswick. The Crucifixion arrived in the church as a donation from the Dufour family a generation after they had left their native Mas d'Agenais for Dunkerque. They had bought the painting in a public auction in 1804 and gave it to the church in thanks. Rembrandt's depiction revolutionized the traditional representation of the Christ crucifixion. It was classified as an Historic Monument in 1918, and in 1959 after much debate on its authenticity, it was sent to the Louvre, authenticated and restored.

The cleaning lady appeared to also be the church custodian and she obviously had great interest in the contents of the church. She pointed-out sculptures and carvings, including a carved stone capital displayed high on a wall. It is thought to be by the same sculptors who had done the ones in the cloister in Moissac.

Up Canal de Garonne to Toulouse

Seeing our interest, she eagerly showed us around the interior of the church to point out other treasures. She pulled back heavy drapes and moved a barrier aside to allow us to enter the beautifully carved choir. The stalls had been created for the Benedictine Abbey of La Réole in the seventeenth century and had been acquired and installed here in the early nineteenth century. They are among the many superb works of art that someone with exquisite taste had collected for the church at that time.

After a thorough visit to the church, we paused at the bakery for fresh bread and croissants and then continued up the canal. A few locks along we almost had a pipe bollard collapse onto us as Edi looped the bow line onto it. We scrambled for other mooring arrangements and as we rose to the top of the lock, we had a clear view of l'éclusier doing the only maintenance that seems to be done: mowing the lawns.

The entire pipe bollard assembly at the downstream end of the lock was missing, hopefully not having injured anyone as it collapsed into the chamber. The lawns; however, were well groomed. We continued along the canal through six locks and 36.6 kilometres to Serignac, where in the late afternoon we moored on the pleasantly sited wharf.

We spent Saturday resting alongside on the peaceful wharf and. adding to our enjoyment was the free moorage with its free water and electricity. We topped-up our fresh supplies at the town's butcher, baker and small supermarket.

On Sunday morning we continued up the canal toward Agen. Just over eight kilometres along we came to the start of the flight of four locks that lead up to le pont-canal d'Agen, the aqueduct across la Garonne. We waited as the first chamber drained and prepared for us.

The first thing we saw in the chamber was broken and dysfunctional infrastructure. Three of the four pipe bollards were broken or missing, a 'safety fence' was in the way of easily looping one of the remaining bollards and the life ring was missing and its line appears to have been commandeered by passing boaters to assist in hauling mooring lines up to the few bollards from their boats.

Remembering from our descent that all the locks in this series were impaired, we decided that the easiest way to handle the series was to have Edi walk the three or four hundred metres between the locks and take the mooring lines from the rim as Zonder Zorg arrived. So as I waited for the final few centimetres of water to fill the chamber, Edi began walking along to the next lock.

We worked our way up the three remaining locks this way and then crossed the river on the 580-metre-long aqueduct. In the early afternoon we secured in the Hermitage basin in Agen. Moorage is on a grassy bank crudely 'improved' with driven wooden posts tied across their tops by and planks. There are no bollards, nor is there any water or electricity. The only moorage with any facilities appears completely taken over by a boat rental company.

Up Canal de Garonne to Toulouse

On Monday morning we walked the bridges across the canal and the railway yards to the centre of the city. All the shops were closed, but we were hoping that the market hall would be open. It wasn't, but we did find a Petit Casino, the French version of our Seven-Eleven with similar poor selection and rip-off prices. We noted its location and continued on to le musée des Beaux-Arts. We had been drawn to the museum by the description in the guide, which mentions some Gallo-Roman treasures, including the superb Venus de Mas and paintings by Goya, Sisely, Picabia and Dufy. We quickly found Venus de Mas d'Agenais, a first century BC Roman copy of a Greek sculpture rendered in marble from Asia Minor. Its placard tells that it was discovered in 1877 in Mas d'Agenais. This is around the time the Abbot and the Count were stopped by the Prefect from selling the Rembrandt from the church to the Louvre.

We continued through displays of Paleolithic tools and artifacts, as well as large selections of Bronze Age and Iron Age pieces, before entering the art galleries.

We were very disappointed by the poor lighting in the galleries, with badly placed lights reflecting glaringly off the surfaces of most of the paintings, making viewing very difficult and photography near impossible. A portrait by Anton Van Dijk was barely visible in the glare.

I shot many pictures at oblique angles in an attempt to prevent glare. With some creative distorting and skewing in Photoshop, these became much easier to appreciate on the computer screen than they are in the gallery.

This 1904 impressionistic scene along the Loing River at Moret is by Francis Picabia, who is more famous for the creation of Dada and Surrealism. He had encountered Impressionism in 1897 through Sisley and Pissarro.

On our way back to the barge we stopped in at the rental boat base to inquire about the price of diesel fuel. Agen is one of the very few places along the canal where it is available, and the company is holding their rental clients and the boating public to ransom with a price of €1.75 per litre, more than 40 cents above street prices. We passed on the fuel and in the early afternoon, we continued up the canal, stopping four hours, twenty-four kilometres and three locks later in Valence d'Agen. The next morning's market and the large Intermarché enabled us to properly refill our supplies.

Up Canal de Garonne to Toulouse

We continued to find broken and dysfunctional canal infrastructure at nearly every lock. This mooring post on a waiting quai was the only one left, and it came free and floated away as we used it.

Not wanting to endure another long series of moorage shuffles at the hands of le Capitain du Port in Moissac, we paused there this time long enough only for the swing bridge to be opened for us.

Even though there appeared to be plenty of open spaces in the port, instead of stopping there, we opted to continue across the aqueduct and along another six kilometres to moor on the quiet embankment close to the hardware stores and supermarkets on the outskirts of Castelsarrasin.

We found wonderful supplies, and in the evening we enjoyed a splendid dinner of seared albacore tuna steaks topped with crisply sautéed pleurottes, and served with steamed cauliflower with sauce Béchamel and nutmeg, saffron basmati rice and sliced Roma tomatoes with fresh basil. The accompanying Alsace Gewürztraminer complemented splendidly.

On Wednesday morning, after we had done more browsing and shopping, we continued along, ascending the series of locks past the water slope at Montech, before heading down le canal de Montech four kilometres to the peaceful moorage in Lacour St-Pierre.

After a very restful day and a delicious dinner, we enjoyed a superb sunset as we sipped our Armagnacs in the cockpit.

The next morning we moved back up the canal three kilometres to moor on the bank behind the Intermarché. We wheeled two 20-litre jugs of diesel fuel from the supermarket's filling station, bought at €1.295 the litre, rather than the €1.75 we had seen at the rental boat base in Agen. We then continued along, stopping for the night in Grisolles. On Friday we continued along to a wilderness mooring a short distance above Écluse Lespinasse, a safe distance from the outskirts of Toulouse.

We slipped at 0945 on Saturday morning and shortly began passing the industrial sprawl and derelict havens of Toulouse. At 1150 we entered le canal du Midi and began back up it, arriving within two kilometres at the first lock. The three locks through Toulouse are run from a central control station, and all we needed to do was twist the dangling wand before the first one to set the series in motion.

We quickly passed through the first and second locks and arrived at the third, an old triple lock bastardized into a 6.2 metre single canyon in front of the main train station. The traffic lights were turned off at 1222 as we approached the lock and the gates began to close. It was lunch time; the lock is scheduled to close at 1230, but our transit would have taken l'éclusier a minute or two into his own time, so we were forced to stop, back down the canal and scramble for a mooring spot.

Up Canal de Garonne to Toulouse

The only mooring place was occupied by two addicts melting heroin and shooting-up. I backed to within a line toss of the last bollard and dropped the spud pole to moor in the tangle of weeds just out of reach of the pair.

Welcome to Toulouse!.

Chapter Thirty-Six

Back Across the Midi

With French logic, because he had shut-down the system for lunch eight minutes early, l'éclusier re-opened it several minutes beyond the 1330 posted time. We passed through the lock and decided to continue through Toulouse without stopping. We had seen all we wanted during our four-day stop on our way down in April.

We worked our way up through six more locks and twenty-three more kilometres to Montgiscard, where at 1840 we secured to the wild bank across the road from the shopping plaza where we had stopped at on our way down. We resupplied with fresh produce at the InterMarché before it closed, and on Sunday morning we carted two jugs of diesel fuel from the service station. Our fuel had fallen below the bottom of the sight gauge on the tank and the additional forty litres gave us sufficient fuel to make the eighty or so kilometres to Carcassonne, where we could organize a truck delivery.

At noon on Monday, after shopping in the hardware store and supermarket, we slipped and continued up the canal.

Ten locks and twenty-six kilometres along, we stopped for the night just below Écluse Ocean, the final up lock before the summit pound. In the evening we enjoyed seared tuna steaks with butter sautéed mushrooms, steamed asparagus with mustard mayonnaise and sliced roma tomatoes with shredded fresh basil. The Alsatian gewürztraminer accompanied splendidly.

Tuesday we passed through the summit pound and began our descent toward the Mediterranean, pausing after fifteen kilometres and nine locks in Castelnaudary.

We spent two nights alongside in the basin relaxing and doing chores. Among the chores was the manufacture of insect screens for the portlights. We had bought some plastic strip and bug screen at the previous brico and Edi made custom-fitted hoops for each of the holes.

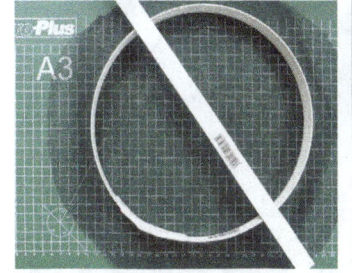 

At 1030 on Thursday, 26 June we continued down the canal, beginning with the four lock flight just downstream of le grand Bassin de Castelnaudary. We were fortunate to find the light green as we arrived and made a very quick descent. At 1520, after seventeen locks and fifteen kilometres, we secured for the night in a wilderness setting just downstream of Écluse Sauzens.

At 1020 on Friday we slipped the line from the roots on the bank, raised the spud pole and continued down ca-

nal. The water was busy with zigzagging bangy boats since we were between two major rental bases. Because no license, qualifications or experience are required to operate these boats, we are always extremely wary when any approach. The operators most likely have no awareness of regulations. We were the downbound vessel, so we had priority, but were the oncoming renters aware of this?

At a bridge hole, a rental was trying to force us to stop when we were already committed. The skipper likely didn't understand the navigational meaning of the Five Short Blasts on our air horn, but he very quickly slowed and veered. We continued along unscathed and after seven locks and 25 kilometres we secured on the bank on northern outskirts of Carcassonne shortly before 1600.

I phoned Stephanie, la Capitaine du Port and asked her to organize a delivery of diesel fuel to us in the port on Saturday morning. She said the fuel companies don't work the weekends, so delivery would have to be Monday. She phoned back a short while later, saying the fuel office was closed for the day and it would remain closed for the weekend, so she said she would call on Monday morning to arrange a delivery on Tuesday morning; the company requires a day's notice. While I had her on the phone, I reserved mooring in the Carcassonne basin for Monday night to be in position for the fuel delivery.

We were next to the road that led half a kilometre to the huge Leclerc supermarket, so we wouldn't starve over the weekend.

At noon on Monday we moved along the two kilometres to the basin and the following morning the truck arrived as we were sitting down to breakfast. Between bites I assisted the operator in adding 385 litres to our tanks. After shopping for fresh produce in the Tuesday market, at 1130 we slipped and headed down through the lock and under Pont Marengo, the lowest bridge on the Canal du Midi.

Back Across the Midi

After nine locks and seventeen kilometres, at 1715 we secured for the day to the bank with spud pole and pin just through the bridge in Millepetit.

We slipped at 1050 on Wednesday morning, heading to a rendezvous we had organized with Christian of Domaine La Lauze. By telephone on Tuesday evening, we had ordered four more cases of his wine and arranged for a noon delivery by the lock in Marsiellette, only a dozen kilometres from his winery. Because he works for himself, he was on time and everything was in order.

After loading the wine and catching-up on events since we had seen him in March, we continued through the lock and along very pleasant stretches of plane tree lined banks.

These were very frequently interspersed with sections where many of the trees had been removed or were dying from canker stain fungus. All along the canal work continues in its many phases of identifying stricken trees, girdling then and removing them. In many places, fallen branches have impaled themselves into the canal bottom and present navigational hazards.

Since 2010 more than 10,000 diseased trees have been identified and so far, no solution has been found to stop the spread, other than removing and burning the diseased trees. The average cost of felling and replanting each contaminated tree is €3,000.

Over the next twenty years more than €200 million will be needed in the effort to fight the canker stain, replace the affected trees and preserve the beauty of le Canal du Midi.

Also along the banks is the wonderful variety of engineering works from the seventeenth century. This spillway with a bridge of a dozen arches was built in 1693 to allow the haulage horses and their drovers to keep their feet dry as they crossed the overflow from the canal.

Also of interest to us were the masons' marks on the stones in the walls of some of the lock chambers. After three and a half centuries, many of these are still sharp.

On a dramatically less aesthetic note, we met a French cruiser using for fenders a collection of construction site pylons dressed in what appeared to be old socks.

We passed through fourteen locks over the following twenty-two kilometres. During the last half of this we were harassed by the very belligerent and obviously drunk crew of a rental boat, as we shared the locks with them.

They were apparently trying to make the final lock leading to the 54-kilometre bief to Beziers before it closed for the day. We were making the eight kilometre per hour speed limit and they seemed to be making close to double that, with a huge bank-destroying wake as they caught up to us. We all had to wait for an up-bound hotel barge to pump its way into the next lock and then slowly pump its way out of it. We watched as the rental barge crew became increasingly impatient and increasingly drunk.

They seemed bitter that they had to wait for another rental boat and Zonder Zorg at each of the frequent locks. A few glasses of shared wine allowed them to recruit the other rental boat to team up against us and they combined to hamper our way and try to block our entry and exit of the locks and harassed us in the pounds. We listened as they told successive éclusiers that we were a big problem for their progress along the canal. Eventually, communication between successive éclusiers pointed to the real problem, and after half a dozen locks of beligerence, some justice.

Below the third last lock is a flood gate and as we came around the corner toward it, it began to close. It remained partly closed for more than twenty minutes while we knowingly waited along the bank on the spud pole and the crews of the two rental boats fumed and cursed as they bumbled from bank to bank. Finally, after sufficient time had passed that the belligerent boaters could not make it through the remaining locks to the long pound before the locks closed for the day, l'éclusier opened the gates. We smiled. The rentals rushed off in a flurry, heavily banging both walls of the gate on their way through.

Once the turbulence and venom of their passing had subsided, we waved a thankful acknowledgement to l'éclusier and slowly continued along looking for a peaceful stop for the night. At 1845 we came to the spud pole and a bank-side post downstream of Bassenel. The following morning, ensuring we would leave sufficient time for the horrible examples of boater to pull far ahead, we had a leisurely breakfast in the cockpit before continuing along.

At 1020 we raised the spud pole and continued down through the remaining locks and into the long pound. After seventeen kilometres, at 1335 we secured for the day on the bank in Le Somail.

In the chart book is shown a supermarket about a kilometre south of the village, so as it began to cool at the end of the afternoon, I walked along for some fresh produce and fish for dinner. On the way back I paused at a winery I had seen on my way out and was offered a tasting of their line.

The first white was thin and lackluster. The next bowled me over; it reminded me very much of better Premiers Crus from Chassagne-Montrachet and Meursault, and even of lesser producers' Corton-Charlemagne. It was a rich, barrel-fermented, stirred lees Chardonnay with great balance and finesse. I continued with great anticipation into the reds. They all disappointed, including their star €24 Syrah. I went back to the Chardonnay and was even more impressed. At €6 Domaine du Somail Festiu drinks not far below a €40 or €50 Burgundy. I told the winemaker I was on foot from a barge, but when he said they will deliver, I bought two cases.

The following morning, with the wine delivered and stowed in Zonder Zorg's cellar, shortly after 1000 we slipped and continued down Canal du Midi. Within three kilometres we came to the junction of Canal de Jonction, which leads to Canal de la Robine, which leads to Narbonne. We made the turn and began the descent of seven locks in five kilometres. All the way along were rental boats; there are several rental bases between Le Somail and Narbonne and the area was full of boats with fresh skippers trying to learn boat handling. As we left one lock we were met by one bangy boat deciding to head up the wrong side of the canal and another appearing ready to mount the left bank.

By the time we had met the sideways boater, he had straightened up, and as we passed, he continued to turn and shortly rammed into the right bank.

After five kilometres the canal empties into the Aude River and at the seventh lock is a large panel with instructions and a chart showing the navigation along the short stretch of river leading to Canal de la Robine.

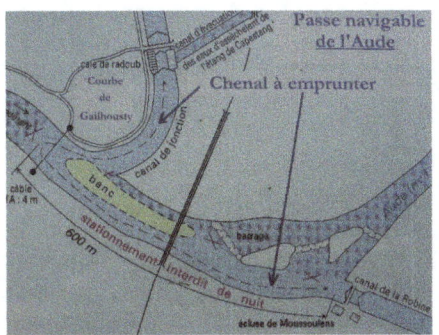

The instructions in French, English and German read: "River can be difficult depending on the water level. A moving sand and gravel bank obstructs the middle of the river. Follow the banks as shown on the chart." We had an easy crossing.

After four more locks, at 1550 we passed under Pont des Marchants in the centre of old Narbonne. Within five minutes we were secured to the wall across from Les Halles, the Public Market.

In the evening we relaxed with a dinner of seared albacore tuna topped with sautéed crimini mushrooms and served with rissolées, haricots verts almandine and sliced roma tomatoes with shredded fresh basil. The Willm Crémant d'Alsace complemented both the meal and the occasion wonderfully.

On Saturday morning I began some more serious troubleshooting of an electrical problem. On our way down Canal de Jonction the previous day, our house battery had quickly died and refused to take a charge, leaving us with no domestic electricity unless we ran the generator or plugged into shore power. Fortunately, there are electrical pylons in the Narbonne moorages, so we had plugged-in on arrival and I did some preliminary troubleshooting without much success. Our 855 Amp-hour battery would not take a charge from shore power, from generator, nor from our portable 30 Amp charger. I suspected a dead cell or a loose connection.

We were in a wonderful place to be broken-down; Narbonne has a long and complex history dating to its founding in 118 BC by the Romans. The site was at the crossing of the Aude River and at the crossroads of Via Domitia, which joined Rome to Spain and of Via Aquitania, which led across to the Atlantic at Bordeaux. Narbonne was the most important Roman city outside of Italy. To clear my mind after a few hours of blind-end troubleshooting, we headed off for a look.

A short distance back along the canal from Zonder Zorg was a portion of Via Domitia excavated in 1997 in front of l'Hotel du Ville. It crossed the Aude on a bridge that is now le Pont des Marchants.

The City Hall is located in the former Archbishop's Residence, which with other surrounding ecclesiastic buildings, has been recently restored and renovated. I recall wine buying trips through here in the 1990s and seeing huge works underway on the site. Housed in the buildings, besides the municipal offices are museums and galleries. We went exploring.

In the former Archbishop's Residence is also located la Musée d'Art et d'Historie. At the reception desk we were offered an option of single entry for €4 or entry pass to seven sites throughout the city for €9. We bought the passes and entered the museum.

In the galleries is a wide variety of paintings from the sixteenth century to the early twentieth. Among these are several of nautical topics, since Narbonne was once an important Mediterranean port before the Aude shifted its course and the port silted up. Narbonne is now a dozen kilometres from the Mediterranean, across salt water marshes and ponds.

Also broadly represented are paintings of religious nature, such as this exquisite early sixteenth century rendering of the adoration of the Magi by Jan de Beer.

We were delighted with the collection of works by Hippolyte Lazerges, a native of Narbonne. As an adolescent his family moved to the new frontier, Algeria. His talent for painting expressed itself primarily with religious subjects and these made him the toast of Parisian art public. However, his portraits and scenes of Algerian life showed his true virtuosity.

Satiated with art, we went next door to la Musée Archélogique, which occupies an even larger portion of the former Archbishop's Residence. While the museum has extensive displays of Paleolithic, Bronze Age and Iron Age artifacts, we were most interested in the

Back Across the Midi

Roman collection from the third century BC onward. The frescoes taken from the walls of nearby Clos de la Lombarde are among the richest of the Gallo-Roman remains.

There are also extraordinary mosaics, including a large intact floor with an intricate geometric pattern.

There are also mosaics with more graphic designs.

Marble sculpture is well represented with both individual pieces, and groups.

No display of Gallo-Roman artifacts is complete without some amphorae. Many corners throughout the museum are stacked with a variety of these.

Also, there are many large dolia on display. An outstanding large intact dolium is from the third century BC and it predates the Roman settlement in Narbonne. It was found as it had used, buried up to its shoulders in an oppidium four kilometres north of Narbonne.

We found a large Roman ship's anchor, but decided we were satisfied with the rather smaller one in Zonder Zorg.

From the Archeological Museum we went across to le Donjon Gilles Aycelin. This forty-two metre high fortified tower was built between 1295 and 1306 by the archbishop. A spiral staircase of 142 steps leads up past the treasury, the king's room and the defence post. Glimpses of the city below are offered along the ascent through the embrasures, or arrow slits.

Above the defence post is the terrace from which there is a splendid view over Narbonne. It also gives a fine perspective of cahtédrale Saint-Just et Saint-Pasteur, showing wonderfully its foreshortened shape. Built between 1272 and 1332, only the choir was completed before the parishioners refused to support any further construction. At 41 metres, the gothic choir is the highest in southern France and is among the highest fifteen in the world. It is higher than the 35 metres of Notre-Dame de Paris and the 38 metres of Notre-Dame de Reims.

In the other direction we had a fine view over the canal and down on Zonder Zorg. This reminded us that there was still some problem solving to sort through with our battery, so we descended from our perch and headed back to the barge.

The battery is composed of twelve 2.1-volt gel cells wired for a nominal 24 volts at 855 Amp-hours and it was assembled in April 2013. It is designed to give at least ten years of service, so something was definitely amiss, and since it weighs well over half a tonne, it is not simply a matter of popping it out for a closer look.

After I had done some preliminary troubleshooting and tried main engine, generator, shore power and a portable charger to add to the battery's charge, I had emailed a report to SRF in Harlingen with the sequence of events leading to the failure and added a variety of meter readings. While we were off museum visiting, an email reply came with suggested remedies, tests and a variety of questions. I redid all the remedies and tests and answered all the questions and emailed the results back to Friesland.

The following day, while waiting for a response, we went off to continue our museum circuit. We went to the cathedral for a look at its interior. It is amazingly disharmonious; with its great height and truncated length it is very much out of proportion. More harmonious are the cloisters, which were begun in the early fourteenth century when work had stopped on the cathedral. These joined the cathedral through the gardens to the Archbishop's residence.

Adjacent to the cloisters is the Annonciade Chapel and above it is another museum, le Trésor de la Cathédrale. It is in a vast room under a brick dome, which offers a wondrous acoustic effect, transmitting sound from corner to corner.

On display are a few of the luxuries the archbishops treated themselves to at the expense of the peasants. Among these are huge tapestries, such as a fine rendition of the creation woven in Flanders around 1500.

Other extravagances on display include portions of huge ninth century high relief panels of carved ivory that stand about two metres high and span many metres of width.

Filled with the excesses of the clergy, we headed along to the Horreum, which dates to the first century BC. This complex of subterranean galleries about five metres below ground is believed to have been storage rooms for ground-level markets and merchants' warehouses.

First rediscovered in 1838, they were partly explored beginning in 1935. While they are still being explored, charted and cleared, they were opened to the public in 1976. Thus far two adjoining wings, one of fifty metres and another of thirty-eight metres length, have been cleared and opened and work is progressing on a third wing. The work is slow because the site extends beneath occupied residences and businesses.

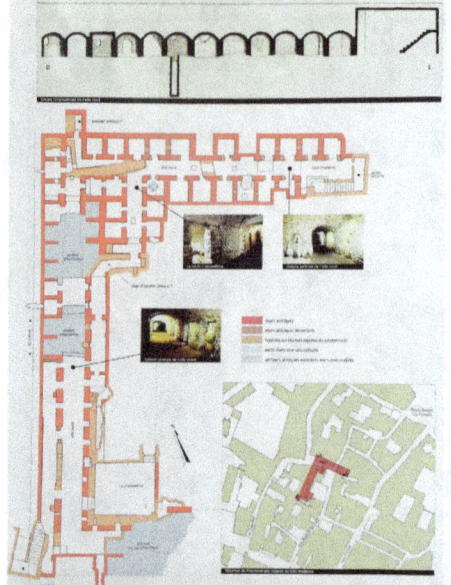

Three dozen storage rooms have been cleared, and displayed in many of these are capitals, friezes, reliefs, steles, inscriptions and fragments of tombstones gathered from Roman sites through the city. Also, amphorae from a huge 1990 Narbonne discovery are displayed.

From the Horreum we went to la Musée lapidaire, housed in a meridional gothic building, which was completed in 1299. It served as a Catholic church until it was confiscated during the French revolution. Inside are some 2000 pieces of carved stone from the Roman period of the city. The pieces are mainly stacked in a series of rows two to three metres high that run the length of the former church. Among these is a wide variety of capitals, friezes, cornices, figures, scrolls and inscriptions that once adorned the buildings of the ancient Roman city of Narbo-Martius.

Back Across the Midi

In five days we had seen a complex cross-section of Narbonne's past twenty-one centuries, but we had not been able to unravel the current mystery of Zonder Zorg's house battery. The partners at Scheepsbouw & Reparatie Friesland were in the process of organizing some engineers to drive south from Harlingen to meet us. We had decided to continue along to Capestang to await them there.

At 1320 on Wednesday, 9 July we slipped our moorings and headed down the canal a couple of kilometres to the winding basin, where we wore around to head back up. We passed again under Pont des Marchants, which we had learned was built on one of the arches of the old Roman Via Domitia across the Aude.

We worked our way up the locks of Canal de la Robine, wound along the short stretch of l'Aude and began negotiating the locks of Canal de Jonction. We were in up-bound locks again and these required that Edi disembark before the gates to receive our mooring lines from the top.

Having maneuvered into and out of over a hundred and fifty of these oval locks of the Midi, I was rather expert at getting Zonder Zorg in with a minimum of fuss. The correct angle of passage through the gates, the swing of the tiller and the timing of the astern kick are all coordinated to bring us to a stop alongside the curved wall below the bollards.

At many of the locks the locals and passers-by pause to watch and some even take the time to come over and offer greetings.

In the late afternoon we stopped for the day in the peaceful village of Sallèles-d'Aude. With no house battery and no shore power, we had to run the generator whenever we needed water or lights. This made sense for the dishwasher, but seemed overkill to brush teeth or flush the toilet. On Thursday morning we continued up the canal to the junction with Canal du Midi, where we turned to starboard to continue down it.

Along the way we passed some questionable examples of boat design. Shortly before Capestang we passed some dredging barges being relieved of their loads. Among the dredgings was a large pile of bicycles and other gear that likely fell off rental boats as they slammed into the banks, locks and bridges along the canal.

We passed under Pont de Capestang, one of the more restrictive bridge holes on the Canal du Midi. It is often named as the lowest bridge on the canal, but it is not. The lowest is Pont Marengo in Carcassonne and there are several other bridges lower than this one at Capestang. Its reputation likely comes from the axis of its arch being aligned about fifteen degrees from the axis of the canal. Barges not properly lined-up add to their own battle scars, as well as adding to the scars the masonry bears from many unintended encounters over the years.

Mid afternoon we secured to bollards along the bank between the bridges in Capestang and plugged into shore power. We again had running water and other conveniences without having to run the generator. We relaxed with a spread of food and a bottle of rosé.

The centre of town is down the hill from the canal, and in its old centre is Collégiale Saint-Étienne de Capestang. It was begun in the late thirteenth century, designed by the same architect who had designed the cathedral in Narbonne. The construction began with the choir and progressed arch by arch into the early fourteenth century, destroying as it went the eleventh century arches of the Romanesque church it was replacing.

Like the cathedral we had just visited in Narbonne, money and parishioner support ran out well before completion, leaving the church with an apse and a choir of two spans of Gothic vaults soaring to twenty-seven metres, butting against the much lower remainder of the Romanesque church. The roof of the older church is so low that it is hidden from view by the surrounding buildings. Visible though, are the jutting stones on the western corners ready to receive the continuation of construction.

Inside, we saw the high blanking wall built above the low Romanesque arch that was inserted to close in the incomplete structure. The very short church is missing its transept and nave and the old Romanesque portion serves as its porch. The altar is crammed into its apse. It continues in use as the parish church.

On Friday we stocked-up at the InterMarché, a short distance through town, laying in enough for the long Bastille Day weekend. On Sunday morning we wandered through the weekly market in the town square, but found nothing of interest, so we walked along to the InterMarché for fresh fish and a few vegetables. Many supermarkets in France are now opening on Sunday morning, but other businesses, like the big BricoMarché hardware and building supplies store beside the InterMarché remain closed. To fill the need there are now large mobile shops offering Sunday counter service and scheduled roadside pick-up of online purchases. The line-up attests to its popularity.

Shortly before 0800 on Bastille Day there was a knock on the hull as Wychard and his engineer, Falco arrived from Harlingen with a wagon load of tools, spare battery cells and parts to repair or replace a few little arisings from our first year of cruising since our refit. They had driven in shifts for fourteen hours and were a bit road-weary.

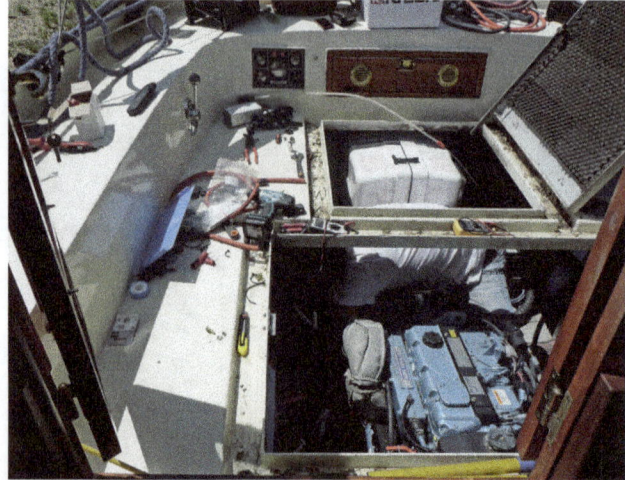

We greeted them with coffees and they soon set to work trying to sort out the problem with the battery. They redid all the tests that I had done and also came-up stumped.

After more than three hours of troubleshooting and eliminating the Victron inverter/charger as the culprit, they began the task of dissembling and removing the battery. This entailed first removing the exhaust cooling system for the engine and generator to make room to remove each of the twelve cells.

With each cell weighing 48 kilograms, removal from the confined space was not a simple task.

As one of the final cells was being disconnected, Wychard found a faulty connection. It appears that the installer had failed to properly tighten the connection and it had arced and led to the death of the battery.

By the time the last cell was ashore, it was well into the afternoon and the temperature was pushing past 35°, so we paused for lunch. Edi made a huge platter of ham, cheese and avocado croissants and the men devoured them. After the pause they continued lugging twelve new cells up the ramp, lowering them into the engine bay and moving them into place in the battery box. The temporary replacement they had brought from Harlingen was only 550 Amp-hours at 24 volts, less than two-thirds the capacity of the dead battery, so each cell was that much lighter and easier to manoeuvre into place. The two dozen internal connections were remade, as well as the connections to the inverter/charger, the generator, the engine alternator and the monitoring instruments.

Late afternoon, I threw the switch, the battery took over domestic chores through the inverter and the shore power connection began topping-up the battery. The system was again working properly.

Attention was then directed to replacing the Webasto heater in the galley. It had blown a connection in April and I had bypassed it. Based on photos of the ruptured piece that I had sent him, Wychard had brought an entire new heater. It was close to dusk by the time the two men had completed adjusting the water heater and central heating furnace and had repaired the air leak in the domestic water system and packed-up their tools and equipment in preparation for their fourteen-hour drive back to Friesland.

Not long after they left, the celebratory fireworks began. The French viewed them as celebrating the tumultuous beginning of their socialistic system, while we viewed them as celebrating the restoration of our carefree independence and freedom of movement.

At noon on Tuesday, after having paid our moorage account with le Capitain du Port, we slipped and continued along the fifty-four kilometre pound between locks, timing our departure to put us at the seven-lock staircase at Béziers toward the end of its afternoon down-bound opening. We figured this should have cleared the long line of boats waiting to head down.

As we approached Tunnel de Malpas, we sounded our horn, hoping any up-bound rental boat understood the meaning of the signal and that we, as a down-bound vessel had priority. Fortunately, there was no opposing traffic as we passed unscathed through the tunnel, which had been completed in 1680 as the first canal tunnel in Europe.

Further along we did see some very creative manoeuvring by the rental boat skippers as they banged from bank to bank on their first day out. We were fortunate that they hadn't yet found the nerve to venture into the tunnel.

Our earlier calculations and assumptions proved correct as we arrived at l'échelle de Fonserannes with the traffic light green and no line-up. We were joined in the staircase by a rental boat crowded with South Africans, so Edi had another opportunity to keep her Afrikaans in practice.

Back Across the Midi

The chamber tops were all crowded with tourists, many of them obliviously in the way of our safe use of the locks. Nonetheless, we made a very quick descent of the seven locks and then motored out the bottom past a long line of rental boats waiting for the up-bound cycle to begin at 1600.

We counted sixteen boats waiting and as we were nearing the end of the line-up, we met two up-bound commercial boats, which have priority in the locks. Their individual passages would delay the line-up by nearly an hour and we saw that many of the waiting boats would not make it up the flight of locks until late the following morning. The summer Midi madness was on and it was a great time to be leaving the canal.

We continued along across the 198 metre aqueduct over l'Orb with its views across the river at Béziers and Cathédrale St-Nazaire. We would love to have stopped for a couple of days to explore the city, but our delays with the battery had put us here in the crush of rental boaters who swarm the canal from Bastille Day through to the end of August.

From Béziers we worked our way down through four more locks and along another eleven kilometres. We passed through more of the wonderful hump-back bridges across the canal that were built in the late 1670s and early 1680s. That so many of these have survived in such fine condition both amazes and delights us.

At 1920 we secured to bollards for the night in a peaceful spot on the embankment just short of Écluse Portiragnes.

The following morning we waited in the lock as an old dory crewed by seven senior citizens rowed into the chamber. Rowboats, canoes and kayaks are not allowed in the locks, but this one was given an exemption because of its historic aspect and the difficulty of portaging it around the lock, as other paddlers must.

Along the way we passed a very creative use of hundred litre drums, seconded to serve as fenders sufficiently large for a boat ten times the size.

Also along the banks, we saw Pastis, one of the tjalken we had looked at online in the spring of 2012, early in our search for a barge. We had gone as far as contacting the broker to get further information and to discuss price. Seeing her now in the flesh, we were delighted that we had decided to pass on her and instead of chasing around all over France to look at Dutch barges, to go directly to the source: The Netherlands.

As we motored, we passed many boats that were dramatically less attractive, including a number that were sinking or sunk at their moorings. We were getting closer to easy access from the Mediterranean and the canal banks increasingly become final resting places for failed dreams. We continued to be amazed that the system allows boats to be abandoned, to fall derelict and sink. We have been told that a boat has to be demonstrated to have been abandoned for seven years before it can be removed.

Back Across the Midi

We passed under the graceful Pont des Trois Yeux, the Three Eyed Bridge, which was an elegant 1670s solution to cross a broader section of the canal.

Shortly beyond the bridge we arrived in Agde. Le Bassin Ronde d'Agde was built in 1680 with three exits to allow barges to go from le Canal du Midi to l'Hérault either above or below the weir. Below the weir leads to the Mediterranean, while above the weir leads after a kilometre to the continuation down Canal du Midi.

When we arrived, l'éclusier informed us there was traffic heading down to the lower Hérault, then traffic coming up and out to the upper Hérault, then traffic coming up le Canal du Midi before it was our turn in the lock. We secured alongside and watched the action for more than an hour as the bangy boats bumbled into and out of the lock.

When our turn finally came, we locked through into the upper Hérault, which we followed upstream to the flood gates that lead to the continuation of le Canal du Midi.

We had one final lock before the end of the canal and it closed for lunch as we arrived. After lunch we locked through and as we motored the final three kilometres to the end of the canal, the abandoned derelicts were in the majority, with many of these sunk to the bottom. With the French love of liberty, there are no regulations against this freedom to moor and the authorities apparently can do nothing to remove the abandoned boats, which in many places line the banks in unbroken rows, preventing passing boaters from mooring.

Shortly before 1400 we came to the end of le Canal du Midi.

Chapter Thirty-Seven

Up the Rhône

At 1400 on 16 July we nosed out into Étang de Thau, relieved that the winds were still light. Ahead of us was a 17.5 kilometre crossing of a very shallow brackish lagoon that is notorious for its steep chop in anything but light winds. Navigation is strongly discouraged in winds above Force 2. We were out of the 8 kilometre per hour speed limit of le Canal du Midi, which is imposed to prevent bank erosion, but is near universally ignored by the rental boaters who are apparently not informed of the speed limit. Once on the étang I upped the engine to 1800 and we moved along at 12 kilometres per hour.

The winds gradually rose to light Force 4 as we made our way across, but as they rose we increasingly gained protection of land to windward and the waves had little fetch. Within an hour and a half we left the étang and entered Canal du Rhône à Sète. At 1635 we secured to the wall on solid bollards just short of the lift bridge in Frontignan. It is a peaceful setting except for the fishermen who all run through the mooring at full speed in their outboard skiffs, pulling huge wakes, seeming to make a sport of disrupting the moored boats.

The town fathers long ago invented a story that the lift bridge mechanism is delicate and should be operated a maximum of twice daily to prevent premature failure. It opens for ten minutes at 0830 and 1600 during high season and once per day at 1600 during the low season. To add to the reason to stop in Frontignan, the first three days of moorage with water and electricity are free. At 0830 the following morning we watched the boats scramble to change sides with the bridge. The services are below the bridge, so we stayed put.

We visited the weekly Thursday market, which spreads through the town centre and we restocked our fresh supplies. In the afternoon we spoke with a British couple aboard Quo Vadis, a nicely converted tjalk. They told us that they had been in Frontignan for two weeks waiting for proper conditions on the Rhône for their ascent. They told us it still looked bad, so I began digging up information online to assess it for ourselves. We enjoyed the peacefulness of the mooring in the lulls between the high-speed passings of the belligerent fishermen.

On Friday morning we joined the 0830 parade of boats passing under the raised bridge, but instead of continuing with the herd, we stopped on the wharf just beyond the bridge. I walked across to the boulangerie for fresh croissants for breakfast, which we enjoyed before continuing along in peace.

The route onward was in a canal built through a chain of shallow, brackish lagoons just in beyond the low dunes from the Mediterranean coast. The banks of the canal are low, narrow

and fragile and there are prominently posted speed limit signs all along the canal to reduce wake and help prevent bank erosion. The local fishermen are the main users of this stretch of canal, and it is in their interest to keep the banks intact. In many places the narrow bank between the canal and the étangs had collapsed and much of the remainder was close to doing the same. We ensured that our gentle wake was doing no damage.

Less vigilant were the many fishermen we met who all pulled huge wakes as they sped along the canal. It seems to us that they were all in a big rush to get in all the fishing possible before their canal was destroyed. They seem totally oblivious to the connection between their wakes and the erosion and collapse of the banks.

Midday we passed the British couple in Quo Vadis. They had gone through the bridge the previous evening to move a bit closer to the start of the Rhône and had stopped for the night after a couple of hours. We continued along.

A few kilometres later we left the chain of étangs and entered the Camargue, the delta lands of the Rhône, where frequently along the banks we passed grazing Camargue horses.

Shortly before 1400 we arrived in Aigues-Mortes, a fortified town built in the thirteenth century by King Louis to serve as a Mediterranean port for France. Prominent in the town is la tour de Constance, which once was used to hold political prisoners. Below the tower is a railway swing bridge, which is normally open, but it closes ten minutes before a scheduled train.

It was closed when we caught first sight of it around the bend. We didn't know if it had just closed or whether a train had just passed and it was about to reopen. However; using our experience with French train schedules, we hedged our bets, and rather than remaining adrift at the canal junction, we put our starboard flank along a concrete wall with no mooring bollards, pins or rings and dropped the spud pole to wait.

After fifteen minutes, we knew the train was following the normal French local train schedule and was late. After nearly thirty minutes, the train rolled slowly across the bridge. After it had passed, it took a further eight minutes for the bridge keeper to start slowly opening the bridge.

At 1455 we secured to a rubble bank with our bow held off by the spud pole and our stern fender held against a large rock by a line to a pin pounded in the bank. We were a kilometre along from the centre of Aigues-Mortes and within a hundred metres of a large supermarket. The Rhône was only twenty-five kilometres away and this was our last chance to stock up before starting up the river.

We began trying to make sense of the data on the CNR website. La Compagnie Nationale du Rhône was set-up in 1933 with three main objectives: improvement of navigation conditions; creation of irrigation canals and the installation of hydroelectric plants to utilize the strong flow of the river. It appears that their main objective is now to generate electricity and that navigation has been relegated to an ongoing expense they must endure. A prime example of this are the flow rate charts on their site. These are given as water debits in cubic metres per second past a number of selected points along the river. To get

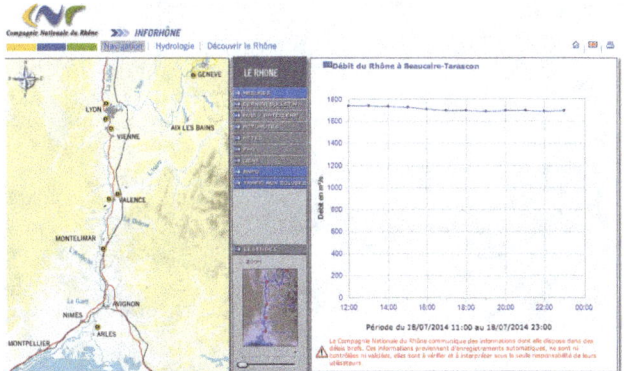

these flow rates, they must measure the river current at each point and convert the speeds into volumes using the cross-sectional areas of the river at those points. This is very useful for hydroelectric plant operators, but is near useless to navigators unless they know each of the cross-sectional areas of the measurement points along the river. Nowhere does CNR publish their original current speed from which they calculate their volume flow.

We were told that by multiplying the flow at Beaucaire-Tarascon by 0.0036 we would get the speed of the current at that lock in kilometres per hour. The table on the CNR site shows the volumes for the past twelve hours only, with no easy access to an archive and no forecast. With the flow rather steady at about 1700 cubic metres per second through Friday afternoon and evening, we calculated the current to be just above 6 kilometres per hour at Écluse Beaucaire-Tarascon.

Shortly after 1600 on Saturday we slipped and headed toward Écluse St-Gilles about 23 kilometres along. This lock takes boats from le canal du Rhône à Sète to le Petit-Rhône. From there it is 300 kilometres against the current to Lyon, the first 20 in le Petit-Rhône, the remaining 280 in le Rhône. We arrived at the lock shortly after its 1900 scheduled closing time and looked for a place to moor for the night to await its 0700 opening. The area was in the midst of reconstruction and there were no mooring facilities, so we turned around and retraced the last kilometre back to canal du Rhône à Sète and headed up it toward the town of St-Gilles.

Less than a kilometre up the branch we saw three short wooden mooring wharves, each about three metres long and each with a pair of bollards. We turned about again, placed our stern against the first wharf, dropped the spud pole and settled in for the night. To give us some forecasting of the river current, I had decided to watch the weather forecasts for areas where rains would feed the Rhône.

On my iPhone I began following the forecasts for Geneva and Lyon for the upper Rhône, Macon and Chalon-sur-Saône for la Saône, which dumps into le Rhône at Lyon and I also watched Grenoble for water into l'Isère, which dumps into the Rhône upstream of Valence. It had been rather dry in all those places for a few days, but two

and three days of heavy rains were forecast beginning on Sunday. From this, I expected a slug of water to enter the upper Rhône and take a couple of days for the resulting higher currents to reach us. We had a very weak connection in our mooring, so I could not update the weather forecasts nor the volumes in le Rhône.

We were up early on Sunday, timing our departure to have us arrive at Écluse St-Gilles shortly after it opened and we watched the light turn green as we approached. I spoke with l'éclusier as we passed through the lock, and he confirmed my thoughts that the river current was normal at the moment.

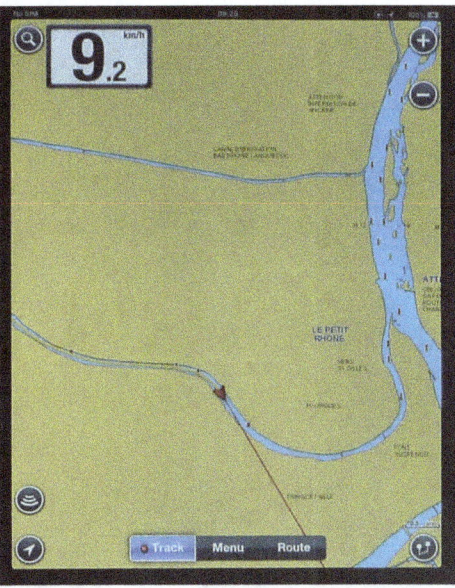

We were delighted to find such a gentle current in le Petit-Rhône. I upped the engine to 1600 rpm, which in still water gives Zonder Zorg 10 kilometres per hour, and in the light current we moved along at a bit above 9 kilometres per hour. I expected the current to increase as we approached the main body of the Rhône, but sixteen kilometres along at PK 184, just four kilometres from le Rhône the current was still light.

We were still making above 9 kilometres per hour with turns for 10. As we passed under Pont suspendu de Fourques and the flimsy wooden jetty that serves as Petit-Rhône moorage for Arles, the current began increasing.

I upped the engine speed to 1800 and we continued to move along at above 9 kilometres per hour as we neared the junction with le Rhône. The current appeared to be about 2.5 kilometres per hour with very little turbulence.

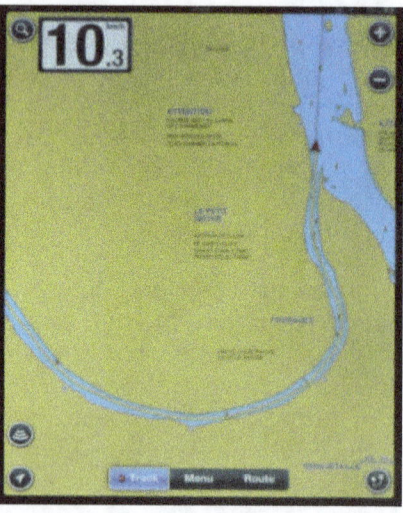

At 0950 we entered le Rhône and I ran for a while in the slower water along its right bank, making over 10 kilometres per hour with turns for 12. We shortly had to work our way across the river to meet a down-bound hotel barge and we lost the lee of the bend and slowed to just under 8 kilometres per hour.

We continued up the river using where safe the reduced current in lee of its bends. We moved along through the water at 12 kilometres per hour and making between 7 and 10.5 over the ground.

Just below Beaucaire we entered the bypass canal leading to Écluse Vallabrègues and the strength of the current increased. We were in a narrow branch carrying all of the water that was being discharged from the hydroelectric plant beside the lock. We passed through heavy turbulence under the bridges linking Beaucaire and Tarascon and were slowed to under 5 kilometres per hour. As we passed Château du Roi René we were back up to about 6.

Up the Rhône

At 1145 we arrived at Écluse Vallabrègues, having taken one hour and fifty-five minutes to do the 15 kilometres from our entry onto le Rhône. Our average was a little over 7.8 kilometres per hour. Thus far our theoretical calculations of the expected conditions on the river seemed to be acceptably close.

Above the lock the turbulence and current were much reduced and once we had left the narrowness of the diversion canal and re-entered the river, we moved along at above 11 kilometres per hour.

An hour above the lock we stopped at a new marina on the river at Aramon. We were low on water and we pulled in to see if we could get some. We were greeted by Olivier Pallier, a recently retired French fighter pilot. He had just three months previously opened the new marina complex, which offers a full suite of services, including short and long-term moorage, dining and local region tours. We told him all we wanted was water and he said it was at no charge. As we filled our tanks, we chatted about our common backgrounds as Air Force pilots and career officers, though he had stayed Air Force while I had switched to Navy.

After topping our tanks, we thanked Olivier and wishing him success with his venture, we slipped and continued up le Rhône. The weather ahead looked ominous, but we decided to press on. Edi dug-out my foul weather gear. As we approached the mouth of la Durance we were hit by a severe squall line with a drenching downpour and winds of Force 10 and above. Edi retreated below.

The wind was directly abeam so I put my back to it and glanced upriver from time-to-time through rain-splattered glasses. The frontal passage took more than ten minutes, though the severe winds lasted less than five before beginning to abate. Fifteen minutes after it had begun, the rain eased and the winds nearly died as we passed under the arches of Pont TGV Méditerranée. My understanding of weather systems and my previous experiences of boating through similar situations certainly helped me keep a steady hand on the tiller here.

Our progress up the river was very good, so we decided to press on beyond Avignon; we had spent a few days there on our way down river last year and both Edi and I had visited the city several times over the years. We passed through Écluse Avignon in the mid afternoon and continued up the river, passing along the way the ruins of Château de l'Hers on the riverbank and those of Chateauneuf du Pape across the fields and vineyards.

At 1800 we were at a fork two kilometres below the next lock, Écluse Caderousse. One branch was the canal leading to the lock, the other was the continuation of the river to the weir. We opted to follow the river to find a mooring for the night. Five kilometres up, near the village of l'Ardoise is a marina, but we were not headed there. Instead, our destination was to a pair of chart symbols about 1500 metres up the river that looked to be moorings. We slowly went past the overgrown sloping stone walls for a look and decided to turn about to use our favoured starboard side for a closer approach. The upstream wall had partially collapsed and we saw random stone rubble fouling the bottom alongside. We moved slowly along to the downstream wall, which appeared intact. At 1825 we secured alongside with the spud pole holding the bow and a line to a tree stump holding the stern. We had made 81.6 kilometres.

It is a very peaceful spot, far enough away from the wakes of the passing barges and out of the current. We slept very soundly and long. With steaming cups of espresso, at 0930 we slipped and headed back downstream in clearing weather to join the canal leading to the lock.

In response to my VHF call, l'éclusier said there was a down-bound commercial in the lock and we would have a short wait. We had barely secured on the waiting float when the lock door opened and a large hotel barge came out.

We soon got a green, entered and had the entire chamber to ourselves for the ride up the 8.6 metres. Once we had cleared the lock, Edi went below and prepared a platter of open-face sandwiches and more cups of espresso for our breakfast, which we enjoyed as we moved along the canal back to the river.

As we regained the broad river two kilometres above the lock we saw our speed increase appreciably in the weaker current. I played with reduced currents in the lee of river bends and at times we were making in excess of 11 kilometres per hour with turns for 12. Midway along the 25 kilometre pound to the next lock we were still making above 8 kilometres per hour in the main current.

The canal leading to Écluse Bollène is over 10 kilometres long and carries the full flow from the hydroelectric plant down its relatively small cross-section. The current is noticeably stronger and we were slowed to below 6 kilometres per hour. We were steadily being overtaken by a large unladen barge.

It was alongside and moving at more than double our speed as we passed beneath the bridge 2 kilometres from the lock. With our reduced speed in the increasing current as we neared the lock, it would take us another half hour to reach the lock and we would fall too far behind the commercial to lock through with it. We resigned ourselves to a long wait for the lock to cycle up, discharge the traffic, reload and cycle back down. Luck was with us; we came around the bend half an hour later to see the commercial still waiting for the lock to empty a group of down-bound vessels.

We had to slow only when we entered the slack water behind the breakwater at the mouth of the lock, and we followed the commercial into Écluse Bollène. At 22.5 metres, it is not only in the highest lock on le Rhône, but the highest lock in the European network.

Out of the lock we watched as the commercial quickly pulled away from us as we worked our way up the 20 kilometre canal above the lock that continues Bollène's long bypass of some of the greatest navigational hazards of the former untamed Rhône. The rocky shoals, the winding course and the strong current made boaters respect places like le Robinet de Donzère, la passage de Pradelle, le Banc Rouge, and le Désirade. After nearly three hours we finally left the stronger current in the confines of the canal, regained the river and passed beneath the cliffs of Donzère.

Five kilometres further along we came again to a fork. Straight ahead was the canal that led two kilometres to Écluse Châteauneuf, to the left was the continuation of the river leading to below the weir. We followed the river for half a kilometre to the marina in Viviers, where at 1608 we backed in on a small rickety finger and dropped the spud pole to hold the bow.

Up the Rhône

We had come another 49.9 kilometres and had 166 remaining to Lyon. The river current had been increasing through the day and we were beginning to see the slug of water I had predicted from the two days of heavy rain in Geneva, Lyon and Grenoble. We expected to see the current increase further, so we decided to stay put in Viviers for two or three days until the current dropped to more reasonable levels.

Across the fields behind us, an old flood plain of the river, was the town of Viviers, perched on a hilltop with a line of sheer cliffs. Through the evening we were hit by a series of severe thunderstorms, but Zonder Zorg's spud pole held her bow securely. We hunkered-down inside and studied at the graphs on the CNR website, trying to make sense of the convoluted information.

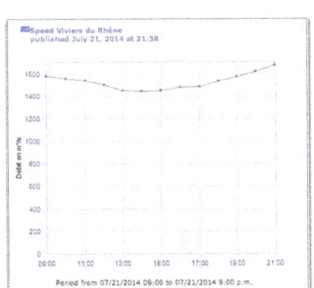

After decreasing until 1300, the reported debit of water at Viviers began steadily increasing through the afternoon and evening. I assumed that since water debit is the volume of water per second past a point, that it is directly related to the speed of the water's flow, which is the river current.

Thirty-five kilometres upstream at Valence, the current had begun increasing at least six hours earlier in the day, which made sense based on a five or six kilometres per hour current over that distance. I saw that we could expect a similar 30% to 35% increase in the current at Viviers by midnight as the result of upstream rains in Geneva and Lyon made their way down.

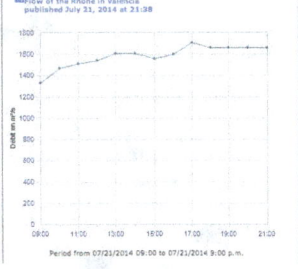

L'Isère drains the basin on the western end of the Alps, and there had been heavy rains in the areas around Grenoble. The river, which dumps into le Rhône a short distance above Valence, showed a near doubling in its flow volume over the corresponding period. With two days of heavy rains upstream of us, we settled-in for a two or three day wait for the current to reduce to a manageable rate.

After breakfast on Tuesday we walked toward the base of the cliffs to visit the weekly market. It was mainly filled with trinkets for the hotel barge passengers and other tourists. There were a few produce stalls with prices considerably above reality and at the fish truck we saw dos de cabaillaud at €37.50 per kilo. In the supermarkets these are generally between €15 and €20. We walked away empty-handed and continued up the narrow, winding streets into the town.

The town dates to before the Roman conquest. It became the capital of the Gaulish Helvii tribe and in the late Roman period it grew in importance as a bishopric and the capital of Pays Vivarais.

On the highest point is cathédrale Saint-Vincent, the seat of the Bishop of Viviers. Its tower dates from the eleventh century and the remainder of the building mainly from the twelfth century; it was consecrated in 1119. Because of the narrow streets, steep hill and abrupt cliffs, it is difficult to get a photo of the exterior.

Inside, the apse is hung with a selection of very large tapestries with religious themes. These are difficult to view because of the poor light and the glare coming from the windows above them.

Behind the altar is an elevated platform with stairs, which allows a better viewing angle, so I climbed up the stairs to get a better look, but the glare from the windows still made them very difficult to fully appreciate.

Back outside there is a large courtyard, likely once the bishop's private gardens. It is surrounded on three sides by sheer cliffs and from it are views down over the town to the south and west. To the north the river can be seen near the end of the line of trees that follow the road across the plain to the port.

Mid-afternoon, I pasted together portions of three previous screen captures of the water flow at Viviers to see the trend over the previous thirty hours. This showed the rate to have increased from slightly above 1400 when we arrived to slightly above 2200, an increase of nearly 60% in the current.

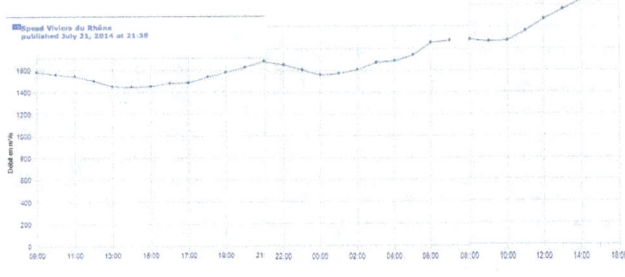

The following morning, the graph showed that the volume had paused slightly in the evening and then steadily increased overnight. It was now above 2400, an increase of about 70% from the time of our arrival.

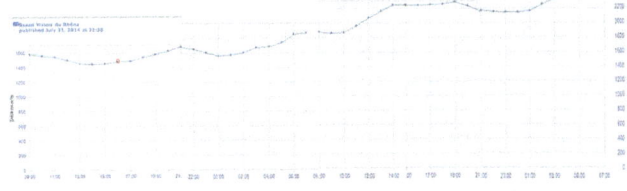

The heavy rains had ended upstream and the past couple of days had been dry, so we expected to see an end of the spate of water coming down the river. On Wednesday afternoon we went to the supermarket to replenish our fresh supplies in preparation for a Thursday morning departure. The midnight graph showed a substantial drop in the flow, confirming our predictions.

The online graph on Thursday morning showed the flow rate had continued its downward trend, and was sitting just above 1800 cubic metres per second. The pause in the decrease around midnight and the small rise through the small hours was most likely from flow control measures by the CNR when hydroelectric power demand dropped overnight. With industrial power demands increasing, I expected that the water draw-down would continue.

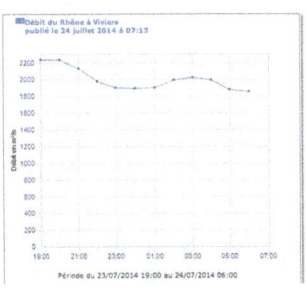

Sitting on a masonry wall and on a series of pylons just off our port bow were two large hotel barges, each with 150 to 200 passengers. Each day since we had arrived, two or three had moored here, either for the night or for several hours for the passengers to visit the town.

At 0725 we slipped, headed out of the marina, ran down the river half a kilometre and turned up the two kilometre canal leading to Écluse Châteauneuf. Not long after we slipped, the two river cruisers followed and before long, one was overtaking us. The lock is 190 metres long and each of the cruise ships was well in excess of 100 metres, so I knew only one of them would head in.

I called l'éclusier on VHF and told him in a slightly questioning tone that we would follow Arosa Stella into the lock. He confirmed. Arosa Stella is within a few centimetres of the 11.4 metre width of the lock, and she took a long time to pump herself into the chamber. There was plenty of room for the two of us inside for the 16.5 metre lift.

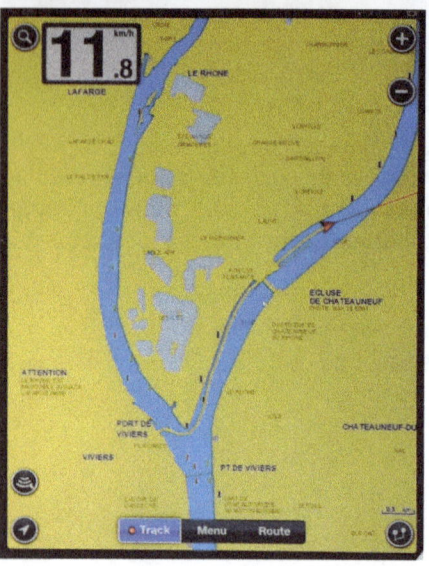

We were soon out of the lock and into the eleven kilometre canal leading back to the river above the weir. The current was rather light above the lock, but nonetheless, I took advantage of the lee below the curves to gain some speed. Doing this, we often made speeds over the ground that were close to our speed through the water. We had to be careful with this, since there was down-bound traffic that would be embarrassing to surprise. Fortunately, the banks are mostly sufficiently low to see over the curves and we were never surprised by any traffic.

We were still in the canal as we passed Montélimar and found the current to be around 5.5 kilometres per hour, slowing us from 12 to about 6.5. We were pleased with our progress and pressed on.

Once we broke out of the confines of the canal and were back into the broad and deep river, the current decreased considerably. Our speed increased by three kilometres per hour. At 1000, as we approached le centrale nucleaire de Cruas, we were overtaken by Van Gogh, the other cruise ship that had left Viviers with us two and a half hours earlier.

In an obvious attempt to show how friendly and benign the nuclear power installation is, one of cooling towers has a mural painted on its upstream face. It appears to depict an innocently nude child playing with sand on a beach.

Two kilometres upstream of the power plant, we passed le Port de Cruas. As we looked in at the floats, we could see two of the T-heads were unoccupied, so there was space for us if we wished to stop. We were making such good progress that we saw no need to cut the day short.

Half an hour later we arrived at Écluse le Logis-Neuf to find Van Gogh just beginning to enter. She had had to wait for down-bound traffic, so we slid in behind her.

At 1245 we crossed the mouth of la Drôme and saw negligible current flowing from it. We didn't expect any change in our speed over the ground. We were making either side of 7 kilometres per hour, showing the river current to be about 5. At 1320 as we were passing la Voute, we met an empty down-bound barge and saw a large river cruiser alongside the wall in front of the historic old town.

Three kilometres further along is Écluse Beauchastel, and by 1430 we were through it and continuing up its feeder canal. Edi brought up a board of open-face sandwiches to share as we motored.

At 1600 we were arriving at le Port de Plaisance de Valence, a huge marina just on the south side of the city with berths for 420 boats. We were delighted with our continued progress and decided to continue upriver. Our speed over the ground was in excess of 8 kilometres per hour

Half an hour later, just before we entered Canal de Fuit, the arm leading to Écluse Bourg-les-Valence, we were overtaken by Swiss Corona, a large river cruise ship. It appeared as if our luck with the locks was over; we would be too long arriving at the lock for the ship to wait.

We came around the bend immediately below the lock and saw Swiss Corona waiting for the lock to finish discharging a group of down-bound pleasure boats.

We followed her in through the tunnel beneath the guillotine door and settled in astern of her for the ride up the 11.7 metres.

Out the top of the lock we were in the continuation of the bypass canal that runs another eight kilometres to take traffic back to the river. Half way along this is the confluence with l'Isère, which brings into le Rhône the major drainage from the western end of the Alps. As we approached the mouth of l'Isère we were slowed to below 5 kilometres per hour.

Up the Rhône

After we had passed upstream of the confluence, our speed over the ground increased to nearly 7 kilometres per hour. We were still in the constricted canal and we expected to see another increase in speed once we re-entered the river. Our predictions were confirmed as we left the canal and regained the river. As we passed le Roche de Glun, our speed blipped between 9.3 and 9.5 kilometres per hour on the GPS, indicating the current was down to about 2.5. There is a moorage listed here, but we had decided to press on to Tournon, another eight kilometres along the river.

At 1915 we secured to a ring and a bollard on a masonry wall just upstream of Pasarelle de Tournon. We had come 77.0 kilometres on the day and were more than two-thirds of the way up the river to Lyon. Across the river from us rose the hill of Hermitage with la Chapelle on its crest above some of the finest Syrah vineyards in France. Blocking the view of Tain-l'Hermitage were five hotel barges lining the banks. These did not diminish the view in the foreground as we enjoyed a celebratory bottle of bubbly for our progress thus far.

It had taken us three days of travel to make 209 kilometres against the current up the rivers Petit Rhône and Rhône from the Midi. If we had not paused for the three days in Viviers to await the passing of a spate of fast current, but had opted instead to run against the heavier current, we would likely have reached Tournon at the same time. This though, would have meant finding three additional moorings, would have burnt much more fuel and would have required refuelling along the way. As it was, our fuel tanks were getting low, showing about 80 litres remaining.

At 0710 on Friday we slipped and continued. Our goal was to reach the fuel dock in the marina in les Roches-de-Cordrieu 50 kilometres upstream before running out of fuel.

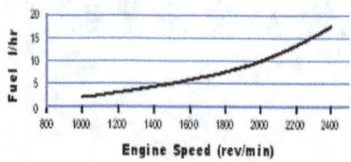

Zonder Zorg burns about 3 litres per hour at 1200 rpm, which in still water gives a speed of 8 kilometres per hour. She burns about 7.5 litres per hour at 1800 at a speed of 12 kilometres per hour. With the expected current running around 6 kilometres per hour, 1200 rpm would take us 25 hours and 75 litres to make the fuel pump. Upping the engine speed to 1800 rpm would give us a speed over the ground of about 6 kilometres per hour, which would require 8 hours and 20 minutes and burn 62.5 litres. At 2000 rpm our speed over the ground would increase to a little over 7, taking us a little under 7 hours of motoring and burning just shy of 70 litres. I calculated that the sweet spot for time and fuel with the prevailing current was around 1900 rpm, so that's where I set it.

Écluse Gervans was ready for us shortly after we arrived and we made a quick passage up it. The bypass canal is short and we were soon out again into the broad river and making good speed. As we passed the ruins of Tour d'Aras, we met a downbound klipper tjalk. We continued to use the slower current in the lee below the bends to increase our speed. At times we were making close to 11 kilometres per hour.

As we approached Andance we found a stronger current, which slowed us to just above 6 kilometres per hour. Upstream of Andance, we were foiled in our desire to find slower currents at the edges of the river and in the lees of the bends. A six kilometre stretch here has groins not far underwater extending well into the river to train the channel down its centre. We were forced to remain in the full force of the current.

Shortly after the end of the section of groins, we entered the canal leading the next lock, Écluse Sablons and the fast current continued. Fortunately, the diversion canal below the lock is short and the lock was being readied for us as we arrived. Out of the lock there is a nine kilometre canal, where the current allowed us to make 7.5 kilometres per hour. As we broke out of the canal into the wide river, our speed increased to near 10.

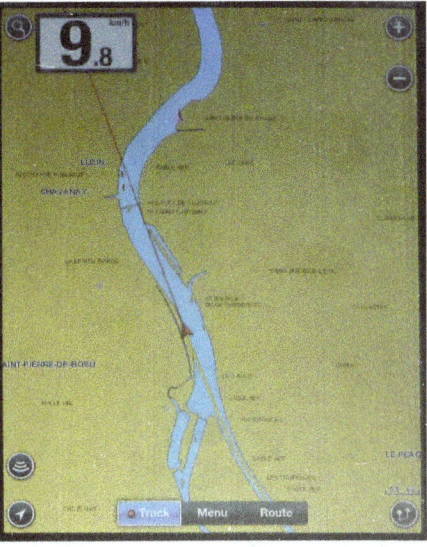

We continued to meet down-bound traffic, this one a deeply laden bulk carrier as we approached the vineyards of Condrieu.

Condrieu and Château Grillet are planted with Viognier and winemakers around world have long used the wines produced here as their benchmark for that variety. Fortunately for consumers, Viognier wines of equal and better quality are now regularly produced in vineyards around the world at substantially lower prices, leaving these Rhône examples to the label drinkers.

At 1516 we arrived in the marina in les Roches-de-Cordrieu with no diesel left visible in the sight glass on our fuel tanks. We had motored a total of 7 hours 35 minutes and had spent 31 minutes with the engine off in the locks. I figured we had burnt about 67 litres and had about 13 remaining. We refuelled at €1.45 per litre, about 12 cents above the road station price, but the cheapest waterside pump we had seen since a tanker truck delivery the beginning of April in Carcassonne.

It was still early when we had finished fuelling, but there were still 41 kilometres and two locks remaining to make it to Lyon. The final lock was 37 kilometres along and it would have been a very tight race trying to make it through before it closed for the day at 2100. Instead of searching for a place to stop further up the river, we decided to take a mooring in the marina. Le capitain directed us to a T-head and, while I walked to la boulangerie for a fresh baguette, Edi prepared a nibbling tray for a relaxing late lunch in the cockpit.

After a leisurely breakfast, we slipped at 0925 and continued up le Rhône. Above Condrieu, the river bends so that the slopes above the Right Bank face south-east. The vines changed from Viognier to Syrah as we approached the town of Ampuis, where we were looking up at the vineyards of Côte Rotie, the Roasted Slope. Like with the Viognier in Condrieu, winemakers around the world have long looked on the wines of Cote Rotie and Hermitage as the holy grail of Syrah wines.

Again, fortunately for the wine lover, Australia, California and British Columbia, among other New World wine regions are now making as fine or finer Syrah wines at lower prices, leaving these Old World examples for those who prefer to drink fame and reputation instead of quality and appropriate pricing.

Our luck with the locks continued as we arrived at Écluse Vaugris just as a large commercial began entering. We followed it in and rode up the 6.7 metres. Out of the lock we quickly rejoined the broad river and were making good speed against the current, which seemed to be running about 5 kilometres per hour.

Upstream of Vienne we were quickly overtaken by a full lock length convoy of two barges and a pusher. Fortunately the next lock was sufficiently far upstream that the convoy would be well through and clear before we were anywhere near.

As we approached the final lock, Écluse Bénite, we were overtaken by another commercial. This barge was only 130 metres long, so there was still a good 60 metres left for us to settle in astern of it. We had been extremely fortunate with the Rhône locks, having had to wait briefly for only one of the twelve.

At 1540 we passed the bifurcation beacons marking the confluence of le Rhône and la Saône. We bent our course slightly to port and started up la Saône into Lyon.

The creative, colourful and highly unconventional architecture along the banks of the mouth of la Saône was as startling when we viewed it heading upstream as it had been when we first saw it on our way downstream the previous year.

I had emailed Place Nautique de le Confluence and had received no response. I had been phoning as we came up river and as we neared, I called them on VHF on their listed CH18, all with no response. We arrived, entered the port and saw one open spot. We backed into it, dropped the spud pole and at 1600 secured the stern to a five metre finger. I went looking for le Capitain, but the office was closed.

Remembering the huge Carrefour supermarket in the shopping centre off our bow, we grabbed our shopping bags and headed across the passarelle. In the wine department we saw some representative bottles from the vineyards we had been passing. Remembering the days when I drank these for a tenth of the price, we decided to continue passing. We stocked-up instead on more perishable items for our planned continuation up la Saône then went back to Zonder Zorg.

In the evening we enjoyed a fine dinner of seared dos de cabaillaud with a sprinkle of cumin, crisply sautéed pleurottes, steamed green beans and sliced roma tomatoes with shredded fresh basil. Accompanying this and to celebrate our very easy ascent of le Rhône, we had a bottle of Rebmann Crémant d'Alsace Brut Rosé.

Chapter Thirty-Eight

Up the Saône

On Saturday night, 26 July I had pasted together three CNR website screen shots of graphs showing water flow at the next lock up la Saône. These showed the data from 0700 on Friday through 2300 on Saturday and I figured the graph represented the 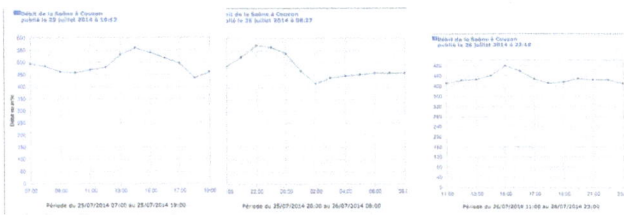 volumes we would have experienced on Saturday as we ascended le Rhône from les Roches-de-Condrieu to Lyon. There were three days of heavy rains forecast upstream in Macon and Chalan-sur-Saône that would increase the flow in the coming days, so we decided to move on up river before the current increased too much.

On Sunday morning, I walked over to le Capitainerie to find it still closed. I continued along to la boulangerie for breakfast croissants, but it was also closed, so I continued along to the Carrefour supermarket, which I knew was open. After breakfast, le Capitainerie was still vacant, so we couldn't pay. We slipped and headed out of the port and up la Saône. A couple of kilometres along we saw the flashing blue lights of two police patrol cars and then some armed police descending the stairs to the river bank ahead of us. We were hugely relieved to watch an apparent drug deal being busted.

A short while later we passed in front of le Palais de Justice and beneath Basilique Notre-Dame de Fourvières. As we passed through the narrows at Ile Barbe we were slowed to about 7.5 kilometres per hour, down from the 9 we had been making in the hour since our departure. The sky was clear, there was no wind and the day was warming rapidly.

Four kilometres along we passed peek-a-boo glances through the trees at the Michelin three star Restaurant Paul Bocuse. Then as we passed the mooring float for the restaurant, I fondly remembered visits in the 1970s, 80s and early 90s before the dramatic price distortions took places like this out of the reach of food lovers and placed them predominantly in the domain of wealthy status seekers. We carried on past, not only because we had just finished breakfast.

We continued upstream, passing two locks and making a total of 70.4 kilometres before stopping for the night on a float in front of a campground. After we had secured I went up to a small building that looked to be the port office. It appeared to be abandoned and the faded sign on its front still listed moorage and services fees in Francs. The only other building near the float is a restaurant-bar and the staff there knew nothing about the floats. We enjoyed our free mooring, but had to put up with the thump-thump music of the summer night bar scene.

On Monday morning we were away shortly before 0700 with the heavy overnight dew wetting the decks and the house. An hour later we were passing through Macon with the river current allowing us to make around 8 kilometres per hour with turns for 12. Alongside the banks in front of Hotel de Ville were two large river cruise ships.

Up the Saône

Shortly after 1000 we left la Saône and entered the lock leading to Canal de Pont-de-Vaux. The safety geeks had placed a chain fence at lock side directly over the bollards to make it near impossible for boaters to toss a loop over the bollards from the deck of an up-bound boat in the chamber.

Within half an hour of our passing through the lock, we came to the end of the canal in Pont-de-Vaux. There appeared to be no visitor moorage and we could see no open spots among the congested crowd of boats in the small basin. Nor could we see anything of the town to entice us to even think of creating a mooring spot along the bank. That it was raining didn't help matters, so we wore around and headed back down the canal.

It was raining heavily as we reached la Saône and an hour later as we turned into the mouth of la Seille, it was raining very heavily. Edi had long since gone below to stay dry.

La Seille, a river that rises in the Jura Mountains, had its lower thirty-nine kilometres canalised at the beginning of the nineteenth century to give access to and from the city of Louhans. We arrived at the access float before the first lock, and because it was too small and flimsy to moor Zonder Zorg on, I put the stern against it and dropped the spud pole.

The locks here are all manual and self-operated. We left Zonder Zorg against the float and went up to drain the chamber and open the gates in preparation for our entry. While Edi stood by to take the mooring lines, I went back down and brought Zonder Zorg in.

Once in we had to close the downstream gates and close their sluices before opening the upstream sluices to flood the chamber. While the chamber was filling, there was some important work to do; there were plum trees with perfectly ripe fruit waiting to be plucked. We filled a large bowl.

The rain continued unabated as we worked our way up the river. After thirteen kilometres and two locks, we decided to call it a day at the foot of the hill leading up to the town of Cuisery. We moored and hunkered-down inside as the rain continued.

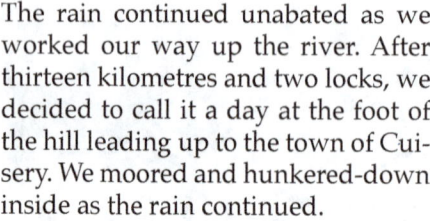

Mid-morning on Tuesday we slipped and continued up la Seille. We were thirteen kilometres and two locks up from la Saône and had twenty-six kilometres and two locks remaining to Louhans. The current in the river had increased and there were many large pieces of wood floating down.

The steady rain had let-up and we were alternating between fine drizzle and intermittent showers.

We were in the heart of the Bresse region of the Burgundy, a region famous for its chicken and in many places along the river we saw flocks ranging freely.

We arrived in Louhans and looked for a place to moor between the gap-toothed array of small cruisers that were Med-moored on the floats in the marina. All seemed to have left about three metres space and we needed four. Finally we found a four-metre gap between two ex-rental boats, backed in and dropped the spud pole. The river is too deep at that spot for the pole to find bottom, so we had nothing to hold the bow.

The skipper of the boat alongside said there was a mooring wall through the bridges and we looked low enough to pass under them. We dropped the mast and slid under the railway bridge and the road bridge with nearly half metre to spare. We came alongside a sloping stone wall, dropped the spud pole, looped a bollard and pulled the stern in against a large round fender. The top of the sloping wall was about thirty centimetres above our deck and was an easy step up. Above the sloping wall is a narrow walkway and above it, a two and a half metre high vertical wall. Beyond the vertical wall is the main parking lot for the downtown core and directly across the lot is a large Spar supermarket. It is a great location to moor.

The next morning the top of the sloping wall was nearly half a metre below our deck. The river had risen about eighty centimetres overnight and the bridges were now too low for us to get back under. Fortunately, I had foreseen a rise in the river and had let out plenty of slack in the spud pole chain and the pole simply slid down in its caisson as the barge rose.

Even although the rains had stopped, the river continued to rise through the day from run-offs upstream. Late afternoon on Wednesday the level was about five centimetres below the top of the sloping wall.

The sky remained rather clear and we basked in our first hot day in a long time. It was still warm in the late evening as we dined in the cockpit. We enjoyed dos de cabaillaud with gnocchi in a mushroom, shallot and garlic cream sauce, green beans and sliced tomatoes, accompanied by a wonderful brut rosé.

The river continued to rise, and before going to bed at midnight, it was lapping over the top of the wall. The fenders had lost their effectiveness, so I placed the point of a boathook into a crevice in the stone wall, put the butt end against the side of Zonder Zorg's house and lashed the pole to the gunwale. With the tiller lashed over against the river flow to keep the rig in place, this kept our stern stable and outside the line of wall top.

On Thursday morning the spud pole continued to hold the bow and the boathook continued its fending effect with the stern.

Up the Saône

The downstream portion of the walkway was well under water, with only the upper portion of the bollards showing. Upstream there was a short patch of dry disconnected from the broader continuation of the walkway, so I lowered our boarding ramp onto it to enable me to make it ashore to the bakery for a fresh baguette for breakfast.

One of the dogs from the barge moored on the float ahead of us waded along the flooded walkway for a closer look at the situation, and from the expression on its face, it didn't look good. I sidled along the narrow patch of dry and had to do a standing jump across the watery gap of flooded walkway. Getting bread had become more difficult than simply searching the town for a boulangerie that was open.

On my way back, I sighted across Zonder Zorg's highest point at the bridges and figured were now nearly half a metre too high to make it back under.

We were interrupted during breakfast preparations by a uniformed police officer who hailed us from the top of the wall and told us to move on; we had been on the moorage for more than two days and it was meant for temporary mooring only. We pointed to the water level and to the bridges and suggested that he either lower the river or raise the bridges.

Also, we pointed out there were no signs to indicate mooring restrictions along the wall. It was then that he realized he had been sent to remove the two boats from the restricted float ahead of us. Both of the boats, a luxemotor and a steel kruiser were more than half a metre higher than Zonder Zorg and he was greeted by them with similar requests to lower the river or raise the bridges. After amiable chats and a sense of duty done, the police officer retreated to his car.

We went back to having breakfast and waiting for the flood to crest and then recede. Mid-morning the river appeared to have crested, and this was confirmed in the early afternoon as the walkway on the top of the sloping wall began to become more visible as the depth of water over it shallowed. By mid-afternoon the walkway began drying and we were finally able to get ashore again. With our new-found freedom, we headed off into the town to explore.

We took advantage of Zonder Zorg's continuing captivity by wandering along the streets of the old town exploring the shops under the arcades. It appears that heavy, prolonged rain is not unusual here; the town boasts 157 arcades to keep shoppers dry. Our explorations had been put on hold when we arrived on Tuesday because the town is closed on Tuesdays. We had wandered then as if through a ghost town. Now the shops were open, except during two or three hours for lunch and except for those shops that had decided to start a little early with their August vacation shut-downs.

Louhans is the capitol of the Burgundy's Bresse region, famous for its chicken: Poulet de Bresse, renowned for its gamey depth of flavour and fine, tender flesh. In 1957 it was granted an Appellation d'origine contrôlée, the first livestock given such protection in Europe. The meat is highly prized throughout France and sells at double the price of other chicken. Guarding the bridges that had trapped us was a huge model of the bird with its patriotic red comb, white feathers and blue feet.

On Friday morning as I headed out for breakfast croissants, I saw that the river was down to a level that would allow us to squeak under the bridge. After breakfast we slipped and wore around in the river current to slide under the bridges with a few centimetres to spare. At 0940 we began retracing our route back down la Seille toward la Saône.

The river current was strong, sweeping us along at between 13 and 15 kilometres per hour with the engine at 1200 rpm, turns for eight. We had some tricky manoeuvring in the tight bends and in the approaches to the locks where the current wanted to take us places we didn't want to go.

Within three hours we had descended twenty-eight kilometres and were in the fourth and final lock before la Saône. Unlike on our way up, the lock was now manned. L'éclusier warned us of very high levels on la Saône, approaching two metres above normal. We locked through with barely any drop in level, and as we passed the landing float below the lock, we saw it was an island. The ramp giving access to the bank led down at a steep angle and was totally submerged about three metres short of dry land and useless.

Along the winding half kilometre arm leading to la Saône we had to guess where the river bank was. Fortunately, there were a few lines of trees to give clues, and the GPS marker on our Navionics screen in the iPad gave even better ones. At 1358 we entered la Saône and turned upstream into the current.

Six kilometres along we arrived in Tournus, where we saw the river was about two metres above normal. It had flooded over the tops of the mooring walls that we had planned to use and the level appeared to be a metre above the top of the broad quai we had moored along during our visit in October 2013. As shown in this photo, the quai had been at deck level when we stepped ashore in 2013, but now it was about a metre up the sloping upper wall. We had been looking forward to another stop in Tournus, but prudence caused us to continue on.

Seven kilometres along is the next lock, Écluse Ormes. Its rise is listed as only 2.53 metres, but when we entered the chamber, we were already nearly to its top. We rose less than half a metre and continued out into la Saône, which appeared to be at normal levels above the lock. At 1615 we secured in l'Ancienne écluse de Gigny, one of the old locks that were replaced by new works. It is now being used as a marina.

At 0930 on Saturday we slipped, backed out of the old lock and headed upstream into the current. As we approached Chalon-sur-Saône I phoned the marina, requesting permission to pause on their floats for half an hour to visit the adjacent Carrefour. After restocking our fresh supplies in the huge supermarket, we slipped and continued three kilometres upstream to the junction of le Canal du Centre.

Up the Saône

In the first lock we watched the guillotine door behind us close on the currents of the river.

Out of the top of the lock we rejoiced that after more than 500 kilometres and fourteen days in river currents, we were finally back in the placidity of the canals.

Chapter Thirty-Nine

Canal du Centre

Early on Saturday afternoon, 2 August we had entered le Canal du Centre, passed through the first lock and stopped for the day in Fragnes. We had expected the halte nautique to be crowded; we were on one of the most popular canals in France. Also, it was the beginning of August when most of France heads out on vacation and the rental boats are notoriously as thick as files in a cow barn. In the marina with us was one other barge and through the remainder of the day, one private cruiser and a rental boat came through and stopped.

This was my sixth trip along the Canal du Centre and each time I have been delighted with the improvement in the moorage facilities in Fragnes. It is one of the nicest small town moorages in the region. The rain that had been our rather steady companion the past two weeks seemed to have taken a pause, so we enjoyed breakfast in the cockpit on Sunday morning. It was so pleasant that we decided to spend the remainder of the day relaxing and told l'éclusier we would continue up canal at 0900 on Monday.

Canal du Centre

With our internet connection I caught-up on the results the skûtsje racing in Friesland, which had concluded its annual eleven race series on Saturday. This is an extremely serious sport, which traces its beginnings to a century and a half ago. The skûtsje is a very nimble barge with its length relatively long for its narrow beam and shallow draft. Its curved bilges, gracefully curved bow and stern and its pronounced sheer make it more agile and capable than other tjalken.

These barges were built to haul peat, sand, gravel, compost and other cargoes across the shallow Friesian lakes and through the narrow canals. During quiet times when few cargoes were available, the skippers would gather together, move their furniture and belongings ashore and race with each other. These impromptu and casual races continued. As motorized barges replaced the old skûtsjes, many were scrapped as the skippers moved on to more efficient vessels. The better skûtsjes survived by being converted to jachten or woonbooten, yachts or residential boats, as was our Zonder Zorg.

In 1945 the SKS, the skûtsje racing association was founded and began an annual series among the skûtsjes representing fourteen Friesland towns and cities. Among the regulations is that the skipper of each barge must be a direct descendant of a commercial skûtsje skipper. Our skûtsje, Zonder Zorg was owned and operated by Albert Douwe Visser, and this week his grandchildren made him proud again, with Douwe, Douwe and Albert Visser placing first, second and fourth among the fourteen skûtsjes. This is the eleventh time one of the Visser grandsons have won the annual championship in the past twenty years. Above, Douwe Visser in the Grouw skûtsje pulls ahead of his cousin Douwe Visser in the skûtsje from Sneek.

At 0850 on Monday we slipped and headed to the next lock, which was ready for us as we arrived. A couple of locks along we had to wait for a down-bound rental boat to lock through. The landing float was too small to moor on, so I looped a bollard for the stern and dropped the spud pole to stabilize the bow while we waited. Another boat came up astern of us as we were waiting and we invited it into the lock with us.

The eleven locks between Fragnes and Chagny are a mixed bag of heights, ranging from 1.83 to 5.80 metres, with five of them above 5 metres high. These high ones have floating bollards, so it is easy to secure in the chamber. However, to activate the lock, a blue cord dangling from a chamber-side pole has to be pulled. This worked well until we arrived at a lock in which the cord was broken and didn't reach below the lip of the chamber. Edi had to climb the slimy ladder to reach the end of the cord.

There had been warnings that the canal would be closed before the end of August, possibly as soon as mid-month because of a lack of water in the reservoirs feeding the upper locks and the summit pound. We spoke with l'éclusier for confirmation, who told us the traffic had been light and the rains heavy; there was sufficient water that if properly managed would last to the end of August. In the pounds between the locks we saw many spillways where excess water was being bled from the canal in large, steady flows, so the water shortage story seemed a bit like mismanagement.

At the eleventh and final lock in the series to Chagny, the numbering on the lock house was rather confusing. Chiseled in the door lintel is XXXI and on the sign above the door is Écluse N° 31. Above that is a 24 plaque and in the guidebook, it is referred to as Écluse 24. The rebuilding of some of the original multiple locks into high singles has caused the renumbering.

Out of the top of the lock is a narrow stretch for the final few hundred metres to the basin in Chagny. As we approached, I saw a large péniche entering from the far end. We pulled into a notch and waited as the hotel barge, Hirondelle made her way through.

Canal du Centre

I had previously visited Chagny five times by canal, the first being in 1984 and the last in 2005. On each of the previous trips I been disappointed with the very poor facilities and the uninviting aspect of the moorage. I had forced myself to put up with the ugliness of the basin twice to stop and dine at one of my favourite Michelin three-star restaurants, Lameloise a short walk away. I was hoping that in the nine years since my last visit that the town had taken a hint from Fragnes and improved the moorage. It had not. Since we had no reservations at Lameloise, we continued along the canal with a fine view of the vineyards of Cassagne-Montrachet and Santenay.

As we approached Santenay I responded to a horn blast and shortly we saw the hotel barge Adrienne coming around a tight bend. We paused before a second bend for her to slide past.

We continued along through four more locks and paused for the night in St-Léger-sur-Dheune, where I remembered a convenient supermarket from previous trips. We lowered the spud pole and looped a stern line around a pin pounded in the grassy top of a rough rubble embankment. Moored astern of us was Owlpen, a skûtsje we had looked at online when we were first searching for a barge. Seeing her here in the flesh, we were delighted that we had searched further.

With our fresh supplies again replenished, we relaxed and in the evening we enjoyed another delicious dinner in the cockpit.

On Tuesday morning we continued up the canal. The series of locks we were now in had security fencing adjacent to the upstream gates that made it impossible to loop the bollard. We had another boat in the locks with us, so there was no choice but to use the front bollard.

Edi had to climb the ladder and struggle to manipulate the bow line onto the bollard through the security fencing. We continued to be saddened by the deteriorated state of the canal infrastructure and amazed that the only apparent work on the locks is to make them more difficult and dangerous to use by boaters. Edi was in great danger of falling into the chamber while trying to use the only available bollard.

We continued up, passing a couple of moorings I had used in the past. We still felt fresh, so we kept going through the remaining nineteen locks to the summit pound, hoping that the mooring facilities had been improved in Montchanin in the nine years since my last visit.

We motored along the 3.5 kilometre cut across the summit and arrived at a grubby mooring dock next to the VNF works yard. The former port across the way, instead of being upgraded, had fallen further into dereliction.

We decided to continue and began descending toward the Loire looking for a suitable place to moor. A rather busy road follows directly beside the canal and there are no peaceful settings, so after a few locks we figured that we may as well press on to Monceau-les-Mines.

Canal du Centre

On the outskirts of town we passed the huge Leclerc shopping centre and were pleased to see that someone in their marketing department had the sense to provide not only a mooring pontoon for passing boaters, but also a huge sign indicating its presence. At 1700, after a day's total of twenty-seven locks and 31.6 kilometres, we moored on a T-head in the port de plaisance in the centre of Monceau-les-Mines.

We took Wednesday off to explore the town and to refresh our pantry from the supermarket just back along the canal. At 0850 on Thursday morning we slipped and fell-in astern of two other boats that had left the marina a few minutes earlier. We all had to await the opening of the first two of three lift bridges leading out of town.

The first of the series is a wonderful bascule lift with colourfully painted parts, clearly showing the massive counterbalance and the complex of semicircular tracks and rollers connected by huge bicycle-type chains and enormous turnbuckles to the bridge deck. The second bridge is triggered at the same time, and is a much simpler affair, hinged to the left bank and lifted by hydraulic rams. After the third bridge we had to wait as the two boats ahead of us passed down through the lock and then as the lock cycled back up for us.

A few locks along we were stopped as VNF technicians worked at resolving a problem with the automatic lock operating mechanisms.

On our way again after a half hour pause, we continued to work our way down through the locks and soon came to a deep masonry lined cut through a ridge.

After our seventh lock of the day, at 1236 we secured in the wonderfully peaceful basin in Genelard. This was the fifth time I had stopped here in thirty years, and it has never failed to please. Across the basin from us were several barges, both converted and new-built that appeared to have found home.

At 0840 on Friday we slipped, timing our departure to put us at the next lock several minutes before it opened and ahead of the three or four other boats we had identified as intending continuing down canal at the same time. We won the lottery and were first in line of four. Next in line was Sancerre, a plastic cruiser we had first shared a lock with descending into la Garonne at Castets-en-Dourthe in June. We had subsequently shared many other locks as well as moorings in Bordeaux, Castelnaudary, Narbonne, Frontignan, Fragnes and Genelard. We each jokingly accused the other of following.

The early morning light along this section of the canal makes ordinary things look extraordinary and makes beautiful scenes exquisite.

At 1255 we secured to a wood-faced wall in the centre of Paray-le-Monial and after refreshing, went to see the basilica. It was as disappointing as it had been on my previous visits. The Romanesque building had been elevated to a basilica as a place to worship a nun who had visions of a bleeding heart in the nineteenth century in a nearby chapel.

In the twelfth century the Romanesque church was begun on the site of a tenth century monastery. It was completed in the fourteenth century with additions in the eighteenth and nineteenth. Its greatest merit is its age and the interior is also rather bland and lifeless.

One thing had changed in the town since my last visit. Except for a small one, les épiceries were all shut-down. The remaining little grocery store had a very short selection of produce, none of it fresh. We returned to Zonder Zorg empty handed.

We decided to move on to Digoin Saturday morning and stop in front of the InterMarché beside the canal. I had used this convenient supermarket several times in the past, and with the weekend coming, we needed to re-stock. We slowly motored along the stretch where I remembered the InterMarché to be, but all we saw was

a BricoMarché. We rounded a bend and were committed to the approaches to le Pont Canal de Digoin, the aqueduct across the Loire.

I then realized that the InterMarché must have been converted to a BricoMarché and the new supermarket was nowhere to be seen. Since we couldn't turn around, we decided we had finished with le Canal du Centre and kept going.

Chapter Forty

Roanne and the Loire

We had committed ourselves to the single-lane traffic of the aqueduct across the Loire. The next turning basin was in le canal latéral à la Loire, beyond the lock at the far end of the aqueduct, so we decided to keep going and find food for the weekend somewhere along the way.

Just over a kilometre past the lock we turned left off le canal latéral à la Loire and into le canal de Roanne à Digoin. This 56 kilometre canal was completed in 1838 further upstream along the Loire to link the city of Roanne to the rest of the canal network.

The map panel on the door of the first lock house showed us where we were and gave the names of communities around us, but there was no indication of any shops or services that might be in any of them. L'éclusier told us the closest supermarket upstream was a few kilometres east of Chambilly, across la Loire in the town of Marcigny. Our map showed Chambilly to be twenty-two kilometres and two more locks along the canal. It was just after 1400 and we saw there was sufficient time to make it before the store closed for the weekend, so we decided to press on.

The next two locks are rather high at five and seven metres and neither of them was ready when we arrived, so we had a long wait as les éclusiers drained the chamber and cranked open the gates for us.

The second of these locks is so high that the downstream doors close against an upper jamb and the water fills to more than three metres above the door tops. Because of the massiveness of the gates, they had high reduction gearing in their operating mechanism and their movement was slow.

At 1745 we secured in the small basin below the fourth lock in Chambilly. I unloaded my bike and pedalled across the Loire river and found the Atac supermarket to restock our fresh supplies before France ground to a halt. Shortly after I had returned to Zonder Zorg, the skies opened up with a torrential downpour. As the evening progressed, the lightning became closer, many thunder claps occurring within a few seconds of the flashes, some almost simultaneously. For a long while we seemed to be directly under the heart of the storm and we had the clatter of hail on the cabin top. We remained very comfortable and dry and the storm moved on before we went to bed.

After a sleep-in on Sunday morning, I walked the kilometre to the boulangerie for a breakfast baguette and paused at the lock on the way back to inform l'éclusier that we wished to pass through his locks at 1100. There are three locks in quick succession and he began preparing them for us. When we slipped forty minutes later he was opening the gate for us.

We passed through the three locks, and just before the system closed for lunch at noon, we entered an eighteen kilometre pound to the next lock. In many places the banks of the canal have collapsed and there are shallow shoals well out into the channel. The grazing cattle take advantage of the easy access to drinking water.

Along the banks we also saw signs of lightning strikes from the previous evening's storm. At one place there was a mature tree fallen into the canal, nearly blocking it. We coasted over the upper branches of the tree and I shifted the transmission to neutral as the stern approached the tree. We slid over, our momentum depressing the limbs and the trailing propeller preventing its snarling and possible damage.

Roanne and the Loire

Below and to the east of us flowed the Loire with Charolais cattle grazing beneath riverside farms and hilltop villages. We were on a very tranquil canal that follows the general course of the Loire River, which is the longest river in France.

The Loire rises in the Massif Central less than 150 kilometres from the Mediterranean and then flows in a grand arc for 1012 kilometres across France, draining over one fifth of the land area of the country before emptying into the Bay of Biscay at Saint-Nazaire.

Roanne is about 150 kilometres downstream of the source of the Loire and it is sited at the head of navigation along the river. Upstream of the city begins a series of narrow gorges that make further river navigation impossible. Freight was offloaded here and was taken on an overland route to Lyon on the Rhône, making Roanne an important transshipping centre between Paris and the Mediterranean from the twelfth century onward.

We continued along the canal, passing through idyllic pastoral settings and a few small peaceful villages, than at 1645, after negotiating another three locks, we entered the basin in Roanne and secured to rings on the low masonry quai.

We were at the blind end of the canal and the basin here is a very popular over-wintering place for barges. Moored along the quai were many barges, a couple of them named rather like the words we use to describe some of the horrible and a-hole boaters that we meet along the way.

Moored astern of us was a small tjalk, which we recognized immediately. It was the online brokerage listing of this particular barge that moved our thinking from another modern cruiser to a converted tjalk. On our way up the Brazilian coast from Cape Horn we had pored over her photos and details and had exchanged many emails with the broker. The photos showed the interior to be chaotic and in need of gutting, but the broker remained firm with the price. Seeing her here in Roanne, still listed by the same broker but with the price now slashed dramatically to well below what we had been willing to offer two and a half years previously, we were delighted that the broker had been firm with the price. It was a great lesson in patience.

The following morning we walked across to les Halles, the covered the market to find it closed on Mondays. We asked a passer-by where we could find a supermarket and were directed around the corner. Inside we found a wonderful selection of fresh produce, including the finest pleurottes we have seen in France. Here was more evidence that the supermarkets are bettering the traditional markets, not only with selection and price, but also with freshness and quality. Also, they are open.

We explored, but could find very little of interest in the city to keep us longer, so shortly before 1300 we slipped to head back through the lock and down the canal out of the urban environment and back into the pastoral.

Along the way we passed fine agricultural estates that had obviously been built, expanded and operated by many generations.

Here and there we passed more modern seventeenth and eighteenth century homes that had been built to add to the farmers' comforts. There is a look of long-standing general prosperity through the area.

At 1820 we paused for the night just above the series of locks leading down to Chambilly. L'éclusier came by to ask what time we wished to continue down through the locks; we said 0900. On Tuesday we waited for the lock to be prepared and after half an hour of no action, I went looking for l'éclusier. He had drained all three locks in preparation for an up-bound Zonder Zorg. It took him a while to realized he had misread the instructions in the lock book. He had to refill all three locks and we were delayed well over an hour.

We met very little traffic on the canal, both coming and going. When we did meet the few boats, we had to be very careful not to run aground on the shoals that often extended far out from the banks. In many places the banks have collapsed and the wakes from passing boats wash the soil further out into the channel. There is no apparent recent remedial work.

Shortly after 1500 we came to the mouth of the canal and re-entered le canal latéral à la Loire. We turned to port and continued downstream. The unsettled weather of the previous few days appeared to be settling on being cold and wet. We were still in the first half of August and we were wearing multiple layers of fleece. Shortly after 1700 we secured to bollards on a quai in Pierrefitte-sur-Loire and scurried below as the first wave of thunderstorms hit.

On Wednesday, after a leisurely breakfast we continued down the canal another nineteen kilometres and five locks to Beaulon, where we secured for the day at 1410.

We walked the long kilometre into the town to find les boulangeries and l'épicerie that are advertised on the canal-side signs. The town appeared abandoned. Both bakeries were closed, one for August holidays and the other for its regular Wednesday closure. We strolled around the very pretty village looking at the sights and searching for the épicerie. It is also closed on Wednesdays.

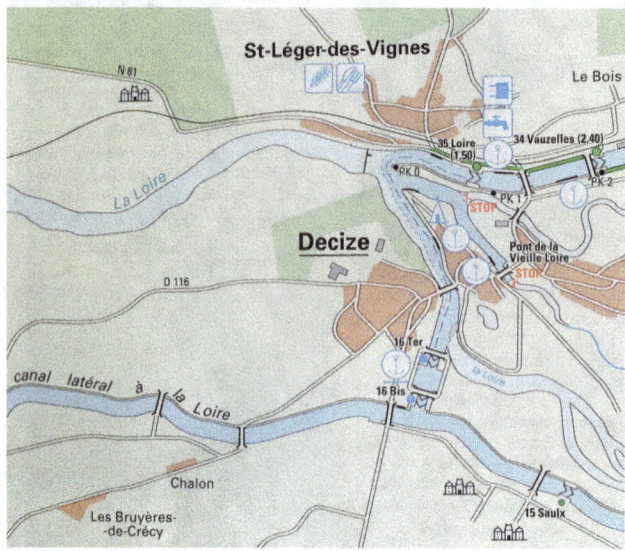

Thursday morning we slipped at 0945 and continued through another seven locks before leaving le canal latéral à la Loire and passing down through two locks into la Loire at Decize. As we left the second lock the skies opened in a torrential downpour. Edi scurried below while I navigated down the swiftly flowing river, under the bridge and found an eddy on the right bank to help me make it alongside a concrete quai in the blinding rain. At 1520 I secured to bollards, shut down and went below out of the storm. Less than a kilometre ahead was the beginning of le canal du Nivernais.

Chapter Forty-One

Canal du Nivernais

It was Wednesday, 13 August when we had arrived in Decize and within half an hour the line of thunderstorms had moved through. After it had cleared in the late afternoon, we walked across the bridge and along a little over a kilometre to the supermarket to stock-up. Ahead of us we had a long distance to cover with very few large communities along the way.

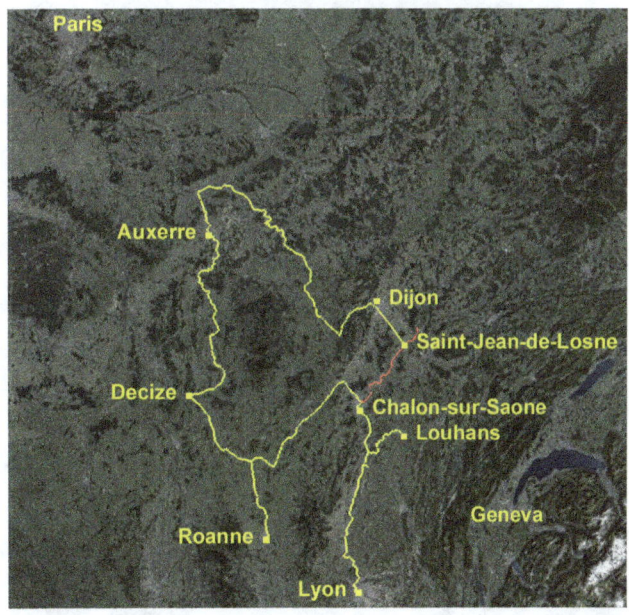

When we had arrived in Chalon-sur-Saône at the beginning of August, we were just two locks and about sixty kilometres from our winter moorage in Saint-Jean-de-Losne. With two good months of cruising weather remaining, we had opted to wander west across Canal du Centre, then along le Loire to Decize and from there follow Canal du Nivernais and Canal du Bourgogne back to la Saône. This loop added over 800 kilometres and more than 400 locks to the 1831 kilometres and 375 locks we had done since leaving Carcassonne at the beginning of April. The Nivernais and the Bourgogne are rated as two of the prettiest canals in France and we were looking forward to them.

The construction of the Canal du Nivernais was begun in 1786, initially as an outgrowth of an earlier idea to dig channels for the floating of timber from the forests of the Morvan. Transportation advocates overruled the timber floating faction and the summit level canal was designed. Work was interrupted during the Revolution, recommenced in 1809 and the canal was finally completed and opened for navigation in 1843.

Carefree Through 1001 French Locks

Our unsettled weather continued with a cold, bleak day on Thursday. There was a light drizzle as we slipped at 1000 and headed the short distance down the Loire to the beginning of the canal.

After a rather bland section, the canal became more attractive and we passed under an interesting variety of graceful stone bridges. Intermittent rain showers continued, so at 1255 after three locks and 16.9 kilometres, we decided to stop for the day on a float in the river at Cercy-la-Tour.

On Friday morning we awoke to dead calm with the early sun just beginning to burn through the fog. The decks were thick with dew.

Our fuel gauge was showing about 100 litres remaining in the tanks, so we had about 30 hours of motoring remaining before running dry. The next fuel facility shown in the guide is 180 kilometres and 111 locks further along the canal in Auxerre. With very careful engine use we could just make it, so as we continued along the canal we kept our eyes out for earlier possibilities.

The locks are manually operated along much of the canal, and many of the lock keepers live in the lock houses. Most of these have been tastefully restored and enhanced by the lock keepers as an obvious demonstration of their pride in the locks.

Canal du Nivernais

In the more remote areas, the lock keepers are students on summer employment and they operate a series of two or three locks, traveling between them on motor scooters or by bicycle. Along one section, the locks all needed to be drained before we could use them so with the short distances between locks, we generally had to wait five or ten minutes while the lock keeper scooted ahead, closed the upstream gates, closed their sluices, opened the downstream sluices, drained the chamber and then opened the gates.

Where possible I moored next to a ladder so I could climb out to close one of the downstream doors. Also, I always opened one of the upstream doors once the chamber had filled. This made the job much easier and quicker for the keepers and seemed to encourage them to pass a good word along the canal as they contacted the next lock keeper to announce our coming. We enjoyed very good receptions all the way along.

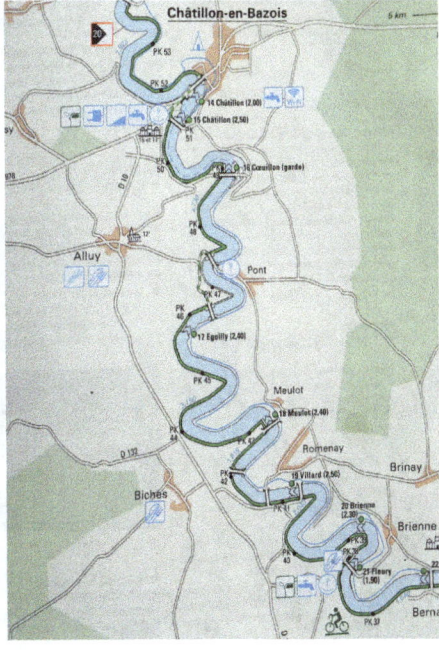

Mid-afternoon we stopped for the day along a low wall with bollards just short of Écluse Fleury. We had come 30.1 kilometres and had worked our way up nine locks. Ahead of us was a very winding course, which takes fourteen kilometres and eight locks to make only five kilometres up the slope.

After a leisurely breakfast, at 1015 on Sunday morning we slipped our lines and continued along the canal. We continued past beautifully restored and maintained lock houses, most of them built in the early 1840s as the canal was being completed.

I assisted les éclusiers with the gates at each of the locks and we moved along quickly. After pausing for an hour at noon for the lunch closure of the canal, at 1405 we arrived in Châtillon-en-Bazois and secured for the day below the imposing château.

On Monday morning we explored the area and restocked our fresh provisions from the mid-town supermarket. Shortly before 1300 we slipped and continued up the canal, timing our departure to have us at the next lock as l'éclusier returned from lunch. Along the way we passed wonderful old structures, we wound our way through tranquil pastoral settings and beneath peaceful villages on the hills, both in the foreground and in the distance.

Canal du Nivernais

After sixteen kilometres and fourteen locks, at 1830 we secured to a stone wall on bollards and rings. The wall is the dike between the canal and Étang de Baye, the main reservoir of Canal du Nivernais. We had reached the beginning of the summit pound.

To enhance the supply of water to the upper reaches of the canal on both sides of the summit, in the late eighteenth century engineers had harnessed the drainage above the summit into three reservoirs with a surface area of 220 hectares and a volume of 6.63 million cubic meters. Now Étang de Baye, Étang de Vaux and Étang de Perchette supply sufficient water to keep the canal open in all but the driest years.

To celebrate our arrival at the summit, in the evening I seared some huge noix de St-Jacques and served them with girolles sautéed in butter with minced garlic and shallots and a liberal splash of Bas Armagnac. Saffron basmati rice, steamed green beans and sliced roma tomatoes with shredded fresh basil completed the plates. The accompanying Rebmann Crémant d'Alsace rosé complemented splendidly. We love wilderness boating.

The following morning, As Edi prepared pain perdu for breakfast, I walked the 200 metres back to the previous lock to inform l'éclusier that we wished to continue across the summit, through the tunnels and begin descending the other side. He told us he would give us a green light at 1015, so at 1010 we slipped and motored along the canal beside the reservoir toward the beginning of the 3800-metre one-way section that passes through deep cuts and three tunnels. The light was green for us as we arrived.

We approached the first of the tunnels and found it unlit. Fortunately, it is only 758 metres long and we could easily see the light at its far end. We were quickly through and heading into a deep, narrow cut.

This led to the next tunnel, a shorter one at only 268 metres. The third tunnel quickly followed and we passed through its 212 metre length and out into a long and wondrously serene trench through overhanging trees.

The trench is close to 1500 metres long, interrupted only by an abandoned bridge. As we neared the end of the trench, faint images of an arched bridge began to emerge in the glare of the light at the end of the vegetative tunnel.

As we approached, these slowly resolved into a graceful stone bridge. Through the arch we could see the beginning of the regular width canal, the end of the wilderness of the summit pound and the beginning of the descent.

Ahead of us were eleven locks in the next two kilometres before the slope began to ease. At the first lock we were received by a young couple that obviously had the hots for each other. As they efficiently worked the gates and sluices, their body language was a mutual seduction and we watched in fascination as they shared a telepathic foreplay.

They opened the gates for us and indicated they would see us at the next lock. The romancing continued through the next two locks. Not only did they give excellent service, but they were also a joy to watch. We left them warmed by memories of our own unfettered youth.

We continued down the series of locks with a new éclusier, who was very efficient, though not anywhere near as entertaining. I assisted him by closing the second upstream gate and opening the second downstream one so he didn't need to walk around the lock to do them. This sped the locking procedure by four or five minutes and we moved through quickly. However, because the next lock was empty, we had to wait as l'éclusier filled it for us.

Our luck turned with the following lock; an up-bound barge was coming out as we arrived, so it and the subsequent locks would be in our favour. We waved at Rival as she passed and we happily slipped into the prepared lock as she equally happily headed to the waiting one we had just left. With each lock now ready for us, we made very good time down through the series.

The lock houses along this section are rather small, but very pleasingly designed. Many are well maintained and tastefully decorated. Some residents have extended their maintenance and decoration to the locks, often with gate railings matching the paint scheme of the house.

We arrived at a series of locks where the gates are the original style oak with massive balance beams. The weight of these beams balance the non-floating weight of the gates, relieving the hinges.

Canal du Nivernais

Operating the beam was akin to handling Zonder Zorg with her massive oak tiller. Many of these basic gate operating mechanisms have now been replaced with rack-and-pinion, chain windlass, bell-crank or even hydraulic ram systems, much like with the trend in barge steering gear. I continued to close a door behind us to assist l'éclusier and help speed our passage. As each of the chambers drained, I climbed up out of the chamber to open a downstream door, often closely supervised by passing tourists. Closing the sluice on the door also eased the load of the lock keeper.

At 1455 we secured to bollards on a low wall in a basin near Sardy-les-Épiry. There are no facilities at the moorage, nor in the village a kilometre away. Although we had come only 8.3 kilometres, we had passed through three tunnels, sixteen locks and one lift bridge and it was good place to stop for the day.

Shortly before 1000 on Wednesday we slipped and continued along the canal. Though we were in the Burgundy, we had left its Charolais region, but the Charolais cattle did not seem to pay heed to the boundaries; they grazed all along the canal. This is another of the main Burgundy beef areas and we saw many future cauldrons of boeuf Bourgignon à la pied.

A couple of locks further along we were picked-up by another hot couple. Though they were not as hot as the couple near the canal summit, they still exuded a rather fluent erotic body language. As we locked down, an up-bound boat approached the locks, so again, we would have prepared locks ahead.

A few locks further along we met two more up-bound boats, bringing the count to four in two days in what is one of the most popular canals in France in the middle of its high season. Besides the moving boats, we had seen only five moored boats. We wondered where everyone was.

Some of the lock houses along this section are too simple to lease out, so they have been restored to a basic waiting station for the roving lock keepers.

Shortly before 1400 we secured to bollards on a wood-faced quai in Citry-les-Mines. We had come another 8.8 kilometres and had passed through twelve more locks. Our fuel gauge was down to about 65 litres. When le Capitain came by, I asked about the fuel pump beside the quai. He said it was no longer in use; the next canal-side fuelling facility is in Auxerre, 86 kilometres and 52 locks along. We had a little over fifteen hours of motoring remaining before we ran out of fuel, so I eagerly paid the mooring fees and plugged into the port's shore power so we didn't need to run the generator to top up the house bank.

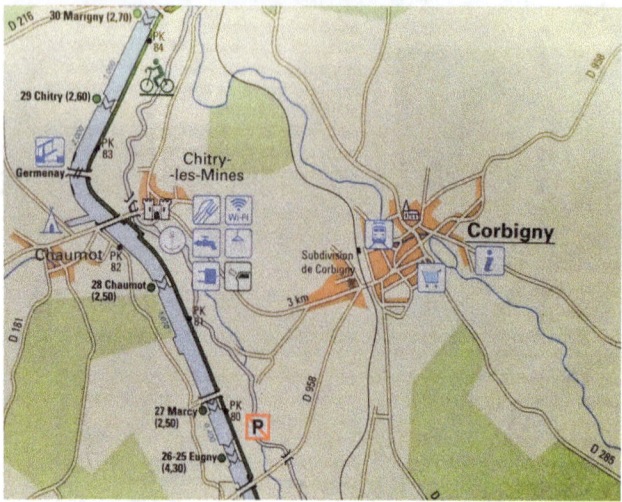

Three kilometres east of the canal is the town of Corbigny, which the guide shows as having a supermarket. I asked the port captain about it and he said it is uphill both ways, across a ridge and down into the next valley. We were out of bread and our fresh supplies were near gone, so I unlocked my bike and ground up the hill. I found the supermarket on the far side of the town. Fortunately, the return hill with the loaded pannier is slightly less steep.

At 1040 on Thursday we slipped and continued down the canal. Just around the first bend is the first of a series of lift bridges. We secured to the short quai and I got off to crank open the span. We were fortunate to be followed by the first down-bound boat we had seen this side of the summit. After we had both passed through, they stopped to close the bridge.

The first lock was not ready when we arrived and we had to wait for l'éclusier to prepare it for us.

In the next pound, the water was covered with freshly cut grass, which clogged our cooling water intake. It appears that the only canal maintenance that is done is grass and weed cutting, and the clippings are dumped into the canal to foul lock gates and sluices and clog boat engines. The rotting mass that sinks to the bottom accelerates the pace of lessening the depth of the canals. We wondered when the VNF would finally relate cause to effect.

We were fortunate a couple of locks later to meet an up-bound boat in the chamber, which meant the following locks should be in our favour.

We came to a pair of lift bridges in quick succession, and since the boat that had been following us had stopped, Edi got off to crank open the bridge and close it after I had passed. Instead of picking her up at the landing, I continued along while she walked the 300 metres to the next bridge.

Both these bridges have electrical winches operated by push buttons, so other than the walk, it took little effort. By the time she had lowered the bridge, I was waiting on the landing downstream and we continued through more locks and lift bridges and past cute little lock houses, and regularly passing herds of Charolais cattle grazing and watering along the banks.

The next lift bridge spans what appears to be an old lock. There is no landing upstream of it and the water along the banks is shallow and foul. There is no ladder in the chamber and its walls are high. This presents no problem for most boats, but for me to climb up from our very low skütsje, I needed my to draw on mountaineering background. It was too much for Edi to do.

After I opened the bridge, climbed back down to Zonder Zorg and motored her through the bridge hole, we saw that the newly-installed landing quai is stupidly placed across the canal from the bridge operating mechanism. With the bridge open there is no way to access the mechanism from that side. We had to pick our way onto the rubble-strewn left bank, drop the spud pole and land the boarding plank for Edi to get ashore to close the bridge. This is another prime example of the disconnectedness of the design engineers from the reality of the use of their designs.

We came to our last lift bridge of the day to see it being opened for us. We thanked the man who had opened it and seeing that he remained at the crank, we continued the last 300 metres to our intended mooring for the night in Flez-Cuzy. We secured at 1615, having come 13.7 kilometres through eight locks and six lift bridges.

In the small basin is a Le Boat hire base – it is called Tannay after a nearby town, likely because Flez-Cuzy doesn't market well. I spotted a fuel pump on their wharf, so the next morning I walked across the bridge to the office to inquire about fuelling. I was told it is for the rental boats only.

Canal du Nivernais

Shortly past 1100 we slipped and continued down canal and were delighted to see the next series of locks ready for us. We worked our way down past attractive lock houses, one using nothing but overgrown ivy as its decor. We passed under many bridges, only two we needed to open and close.

A few minutes before 1700, after 18.1 kilometres and nine locks, we secured along a masonry wall at the edge of Clamecy. The last three locks were operated by a friendly éclusier. When I had asked him about fuel had said there were two supermarket filling stations about two kilometres from the moorage in Clamecy and he immediately offered to drive me to one after he finished his shift at 1900. He arrived at Zonder Zorg shortly before 1900 and we took our two gerry cans across town for filling. The forty litres took our fuel gauge from below the sight glass to about the 75 litre level. We had been down to about ten hours remaining.

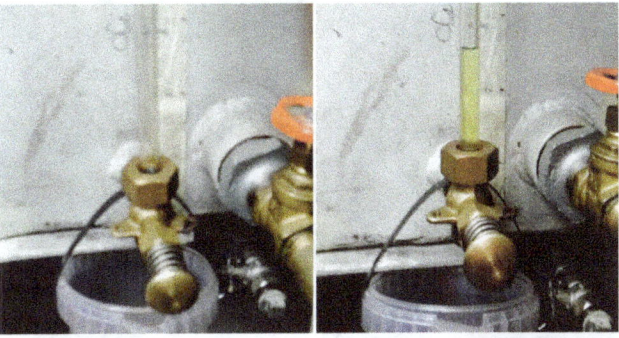

We spent two days in Clemacy wandering the winding, narrow streets.

Many thirteen to sixteenth century buildings still stand on the slopes leading up to the church and the market hall and the entire area is a French protected historical sector.

Église Saint Martin dominates the town from the hilltop. It dates to the thirteenth century and after years of neglect following the Revolution, it was named an historic monument in 1840 and restored. We searched in vain for attractive fresh produce in the weekly market that fills the square and streets around the church on Saturdays. Finally, we pedalled the two kilometres to the supermarket for a wonderful selection.

At 1000 we slipped and motored the 100 metres into the next lock, which we had arranged to have ready for us.

It was a wonderfully warm and clear Sunday morning as we headed down a section of the Yonne between locks. Sharing the water with us were a few rowers.

Along the river and canal banks were imposing old châteaux, many of them more like tiny fortified farming communities from an era when such protection was needed.

Effectively there were home, farm buildings, staff quarters and stables for an extended family; a centuries-old beginning of a village.

As we worked our way down the locks, we increasingly came to rock outcrops along the canal, which here follows the course of the Yonne River, mostly beside it, but sometimes in it. As we progressed, the size of the river increased and we were into the beginnings of its meander through a limestone cap.

After twenty-three kilometres we passed through our thirteenth lock of the day and entered a stretch of the Yonne. This led us for two kilometres and around a sharp bend with sheer limestone cliffs soaring from the banks. At 1600 we secured to a new float beneath the centre of the escarpment. Edi prepared a nibbling tray as a late lunch and we enjoyed it with a bottle of Rebmann Brut Rosé Crémant d'Alsace. We stuffed our faces as we watched the climbers working-out routes on the faces above us.

We had a leisurely breakfast on Monday and it was nearly 1030 by the time we slipped and continued down the river. Within a kilometre and a half we were back in the canal and working our way down the locks. It was an absolutely calm day with a low overcast, which rendered wonderful reflections in the water.

At 1634, after thirteen locks and 23.4 kilometres we secured to large iron rings set into a concrete wall along the banks of l'Yonne in Vincelles.

We were greeted by a cat, which after an appropriate pause, hopped aboard to do a mouse inspection.

As Edi prepared a tray for our late lunch, I headed off for a baguette. I followed the quayside sign to la boulangerie and was pleased to see on its placard that its closing day is Wednesday, so I pushed the door. It wouldn't open.. I again checked the sign for its afternoon hours: 16h00 à 19h00. It was 16h45.

I tried the door again and noticed a small note indicating the bakery was taking a break and would close on Monday at 13h00 and remain closed this week from Tuesday to Saturday.

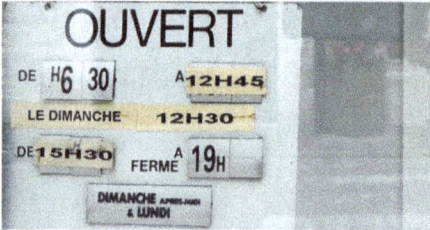

I walked along to the town's other bakery to find the sign on its door indicating it was closed Mondays. I then walked along to where I had remembered the guidebook indicated there is a supermarket. Because it had the only bread in town, it was nearly out of stock, but I did manage to score the second last baguette.

In my wanderings through the small town, I spotted a BP sign in a courtyard and a sign indicating heating and motor fuel delivery. On Tuesday I went to their office to ask whether they delivered to boats. I was delighted with the positive response, and even more delighted with the quoted price; it was 4 cents lower than at the supermarkets. The office clerk said the earliest possible delivery was late afternoon on Thursday and I quickly agreed. It was barely past lunch break on Wednesday when the fuel truck rolled-up astern of Zonder Zorg. We filled our tanks to the bottoms of their fill pipes.

At 1015 on Thursday, 28 August we slipped and continued downstream. From Vincelles, much of the remainder of the Canal du Nivernais is actually in the course of the Yonne, with short canals leading to and from locks bypassing the weirs. In some places, the canal runs along beside the weir, the other side of which runs the river two or so metres below.

Shortly before 1400 we were in the 116th and final lock of Canal du Nivernais, looking at the skyline of Auxerre. We motored past the congested marina and spotted an empty mooring space at the extreme downstream end of the left bank quai. We turned and stemmed the current back up to it and secured. We were directly below cathédrale St-Étienne, in the heart of the ancient city. In two weeks we had worked our way up 35 locks and down 81 as we navigated along 175 kilometres of spectacular river and canal scenery. We were ready for a bit of a break.

The next morning we went exploring the old city. Off our stern was Place Saint-Nicolas with a large statue of the patron saint of mariners set into the wall of a fine old building that once housed Yonne boatmen. We paid our respects and continued up the hill to the cathedral.

A Christian church has stood in this site on the crest of the slopes above the Yonne River since St-Pelerin built a basilica here at the end of the third century. Around the year 400 the basilica was replaced by a cathedral, which lasted 500 years until it was destroyed by fire. The replacement lasted only until 950, when it also was destroyed in a fire. The cathedral was again reconstructed, only to be destroyed by the great fire of Auxerre in 1023.

A new Romanesque cathedral was immediately begun and completed in 1059. In 1215 construction began on the current gothic cathedral, replacing the Romanesque structure gradually. By 1478 the vaults of the nave were completed and all that remained of the former buildings were the crypts.

From the cathedral we wound through a maze of streets and arrived at la Tour de l'Horloge, one of the old city gates dating to the fifteenth century. In its base are vestiges of third and fourth century Roman structures.

We walked through the old gate and into Place des Cordeliers, the fashionable shopping street, where we browsed the shops and relaxed.

Chapter Forty-Two

Canal de Bourgogne

On Saturday morning, suitably rested and with our larders fully restocked from the nearby supermarket, we slipped and headed down the Yonne and into the first lock.

From Auxerre navigation is along the Yonne River, which flows another 108 kilometres to its junction with la Seine. It was a clear, calm day as we made our way downstream.

We were following the river for only twenty-three kilometres of this, heading to the junction of le canal de Bourgogne. There are nine locks along this course and we soon came to one with a vertical wall and a sloping one. I chose to secure on the vertical side, but as the basin began to drain and we handled the lines, we watched floats on the other side gently slide down rails on the slope. I made a note to myself to select one of the floats the next time we came to a similar lock.

We continued down the river through mostly very pretty scenery. We did come to another slope-sided lock and used one of its convenient floats.

Canal de Bourgogne

Approaching the bridge hole leading into Derivation de Gurgy, we saw an up-bound that appeared to be ignoring our down-bound priority. Not only were we the priority vessel, but we were also closer to the narrow passage. I sounded five short blasts on Zonder Zorg's air horn: "You are standing into danger" or "I don't understand your intentions". Whether the skipper of the boat understood the meaning of the signal or was just cowered by the authoritative, deep toots, he immediately took action, slowed and veered toward the bank.

At 1530 we arrived at the junction and turned to starboard into the basin below the first lock of le canal de Bourgogne. We waited for down-bound boats to clear the lock and then we started up the first of 113 locks toward the summit. At 1610 we secured in the basin in Migennes and called it quits for the day. Ahead of us were another 188 locks and 242 kilometres back to le Saône. We reflected that in the Netherlands there are only 186 locks in its entire 6000 kilometres of canals.

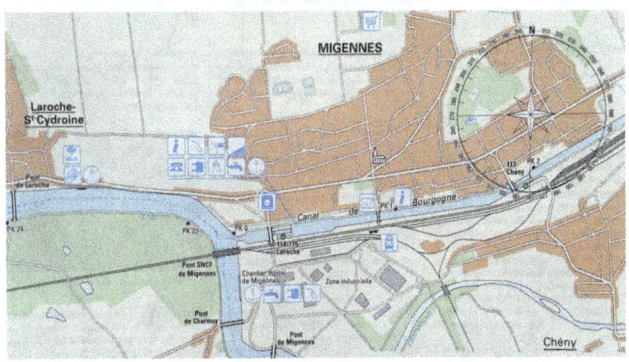

Tom Sommers' summary charts in his EuroCanals Guides offer an excellent overview for planning. Beginning at the junction with the Yonne at 79 metres above sea level, the canal rises through 113 locks to an elevation of 378 metres at its summit pound, making it the highest canal in France. From there it descends through 76 more locks to reach the Saône at Saint-Jean-de-Losne at an elevation of 179 metres.

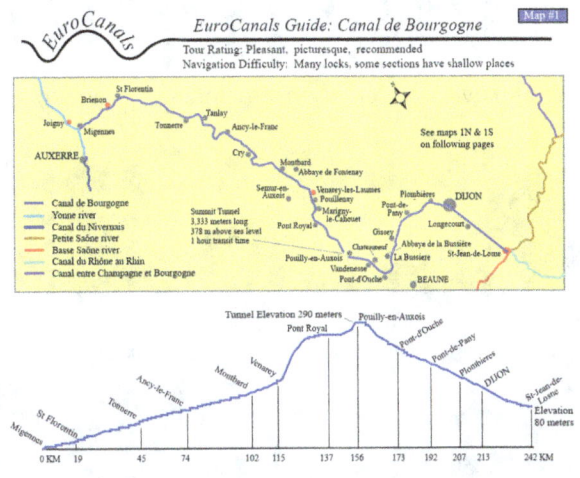

After plans that go back as far as 1604, construction on the canal began in 1774 with an edict signed by King Louis XV. The Crown would finance the portion from the Yonne to the summit and the portion down from there to the Saône would be paid for by les États de Bourgogne. After a pause for the horrors of the Revolution, construction resumed in 1808 under a decree from Napoleon. The canal was officially opened to navigation in 1832.

After a relaxing breakfast on Friday morning, we headed up canal. We were delighted to see so many rental boats in the LeBoat base as we left Migennes and even more thrilled to see a full house at the Nichols base in Brienon two locks further along. This meant that there would be few bangy boats endangering us along the way.

The lock houses along this section of the canal are small, but very attractively designed and many of them are well-maintained and decorated by their occupants. We got into a relaxed routine passing through the locks, tossing our lines up to loop the bollards, stopping beside the ladder so I could climb up to help close the gates, then help to open the gates before motoring out toward the next lock.

We were looking forward to some pleasant locks on our own, but when we arrived at the next lock, l'éclusier told us we had to wait for another up-bound boat. After a few minutes, a rental boat came zigzagging toward the lock. The skipper continued to oversteer and over-correct all the way into the chamber, managing to hit both corners on the way by.

We resigned ourselves to an extra five to ten minutes per lock as our new companions bumbled in, fumbled with their lines and then scraped their way out of the chamber. Waiting for them caused us to be too late to make Écluse St-Florenitn before the lunch break.

We had planned on stopping for the day in St-Florentin, but the moorage is through the lock and we had to wait an hour for l'éclusier to return from the noon break. We locked through and at 1320 stopped beneath some large trees just upstream of the lock.

We heard what we thought was a stone hitting the boat and looked around to find the responsible kids. There was another thunk on the deck, but no stone throwers were in sight. Then we saw a walnut bounce off the roef and onto the side deck. We were sitting beneath walnut trees with heavy, mature loads. We grabbed the boathook and assisted the nut fall.

On the hilltop of St-Florentin sits a beautiful gothic church, which was begun in 1500. On its doors are signs indicating that a door key is available in l'office de tourisme. Saturday morning we got the key, let ourselves in and as instructed, locked the door behind us. We were the only people in the church and we wandered in awe of the spectacular interior.

Église St-Florentin is famous for its many stained glass windows of the Troyenne school, which was reaching its high point as the church was being completed. Very few intact examples remain outside this church. The Troyenne school is notable for its use of bright reds, rich blues, emerald greens and silvery yellows. It also used subtle grey tones to add shadows and dimension to bodies, clothing and objects.

The organ dates to 1620. It has been restored and is still in use with its cabinet and many of the pipes being original. For more than an hour we were totally alone and undisturbed as we absorbed the beauty of the church interior.

We then locked-up, returned the key to the tourist office and went shopping for fresh provisions in the Saturday morning street market. We arrived back at Zonder Zorg shortly before noon and taking advantage of the locks being closed for the lunch hour, we harvested more walnuts from the ground and the trees.

At 1240 we slipped and continued up canal, timing our departure to have us at the next lock as it reopened. Écluse St-Germigny is a former double lock reworked to a single one over five metres high. Fortunately, l'éclusier was on the chamber top to take our lines; it would be nearly impossible to loop lines around unseen bollards from Zonder Zorg's low decks.

Among the things we have noticed with canal de Bourgogne is that there are very few waiting facilities at the locks. The banks are usually very shallow and rubble-strewn. There is seldom a safe place to moor to await the preparation of a lock. Fortunately, Zonder Zorg has a spud pole, which allows us to be very creative in mooring and in waiting for locks. We'll often drop the spud and grab overhanging branches to stabilize the stern.

Where there are no branches, a pole lashed to the aft gunwale and jabbed into the bank will keep the stern off the crud. Here, a bollard is nearly all that remains of a collapsed wall and the pole keeps our stern off the rubble with a line holding us in while we anticipate a huge wake to be set-up by a down-bound rental boat. We have no idea why they all seem to want to get up to hull speed immediately out of the lock.

While we were waiting for a down-bound boat to lock through, a zig-zagging rental caught up to us. It was the same one we had shared locks with the previous day. After slowing us again for a few locks with their incompetence, while exiting Écluse Flogny they appear to have collided with two down-bound boats that were waiting to enter the lock. We didn't see the collision happen; we just heard the loud crashes and turned to see the jumble of the aftermath.

With the lock keeper close at hand there was no need to call authorities, with the shallow water a holed boat wouldn't sink very much and even the injured could step ashore. We decided there was nothing we could offer that wouldn't add to the confusion, so we continued along under the Flogny-la-Chapelle bridge. We had the remaining five locks of the day to ourselves. In the late afternoon, after twelve locks and twenty-five kilometres, we secured to bollards in the basin in Tonnerre.

Tonnerre is on the Armaçnon River and was established at a crossroads: north-south between Auxerre and Troyes and east-west between Dijon and Paris. The settlement was enhanced by the Romans and called Tornodurum. The following morning we headed into the centre of town, about a kilometre south of the canal across bridges over two branches of the river and the railway tracks. The focal point of the old town is the imposing structure of l'Hôtel Dieu de Tonnerre, the largest civil monument in the Burgundy.

Construction on the charitable hospital began in 1293 and in 1295 the first patients were admitted. La Grande Salle des Pauvres housed the poor in forty wooden cubicles along its walls, and at 96 metres it is the longest medieval hospital in Europe. Its immense roof is supported by massive curved wooden arches.

For several centuries the immense building served as a model for architects and in 1443 the famous Hospices de Beaune, south of Dijon was based on its design.

The hospital was built by Marguerite de Bourgogne, Countess of Tonnerre and Queen of Sicily, Naples and Jerusalem. She was the widow of Charles of Anjou who was the brother of the King of France, Saint Louis. Besides donating the hospital to the community, she became a servant of the poor, nursing invalids and comforting the dying and she was broadly respected and admired.

Her original tomb was destroyed during the Revolution, along with other vestiges of royalty and the revolutionaries used the building to store hay and straw. In 1826 a fine white marble replacement of Queen Marguerite's tomb was erected.

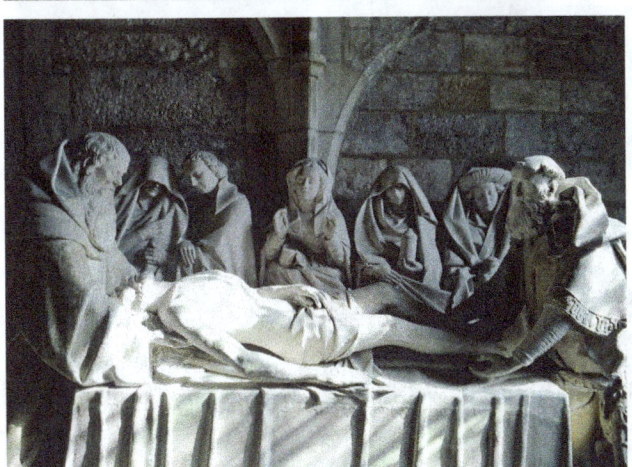

Fortunately, the fifteenth century sculpture of the entombment of Christ was not destroyed. This magnificent piece, completed in 1454 by Georges et Michiel de la Sonnette, is considered one of the most beautiful and best preserved representatives of mid-millennium Burgundian art.

Carved into the floor stones and spanning the width of the great hall is a set of meridian lines. The light of the sun projected through a pinhole in the south wall falls on the meridian line at noon. There are also indications of 1145 and 1215, and the instrument is a very precise measure of time. The position of the spot of light along the curves across the floor, which represent the sun's progression along its ecliptic, indicates the day of the year.

In display rooms at the western end of the building is a wonderful scale model of the roof structure.

The model is about four metres long and is set on trestles to allow close examination not only of the exterior, but also of the interior.

We did an extended examination of displays that include Queen Marguerite's favourite ring, the golden cross she wore around her neck containing a piece of wood from Christ's crucifixion cross, the original 1293 charter for the hospital and her last will and testament. Also on display are dioramas of historical scenes from within the hospital.

It was late morning by the time we had left the hospice and headed along narrow, winding streets toward l'église Saint-Pierre, perched on a rock high above the medieval centre of the town.

Along the way we passed la Fosse Dionne, a circular basin that is fed by a spring in the hillside behind it and once served as a public wash place. Above it we climbed a steep trail with winding stone stairs leading to the church and were soon well above the rooftops. We had a splendid view across town to the huge roof of l'Hospice de Tonnerre.

We arrived at the church at the top of the hill to find it closed. Parts of the church date to the ninth century, but the literature reports that it has undergone many modifications and restorations over the centuries. We wound our way down the hill, through a park and along the streets back to Zonder Zorg. Our minds and souls were so filled with images and information that we decided to take the remainder of the day off.

On Tuesday, 3 September we continued up le canal de Bourgogne. Eight kilometres along we came to Écluse Tanlay, our sixth lock of the day. A couple hundred metres beyond the lock we moored and walked through the town of Tanlay to visit its famous château.

We arrived at the gate to see a sign indicating that the grounds and the château are closed on Tuesdays. We have adopted a new meaning of the expression WTF: What The France? We decided to remain in the mooring overnight and try again on Wednesday.

Château de Tanlay was commenced in 1550 on the foundations of a thirteenth century château-fort. During the Wars of Religion, Tanlay became a centre of Huguenot resistance and the construction project slowed. It was finally completed in 1649. In 1704 the property was acquired by the Thévenin family and still remains in their hands. It sits in vast lands on the northern edge of the town.

After a sleep-in and a leisurely breakfast, mid-morning we again walked across the bridge and through the town to the gates. This time they were open.

Inside the outer gate is a large courtyard. Off to the right is an arched gateway leading into a large quadrangle 65 by 85 metres in size. The buildings that form the quadrangle are the former stables and workers' quarters. These all appear well-maintained and several of the larger rooms are now used as art galleries for visiting exhibits.

After a look around the buildings of the quadrangle, we walked back out into the courtyard and headed through the stone arch of "le petit château". On approaching the château grounds from the town it is easy to assume that this building is itself Château de Tanlay. What in many other settings would be considered a large three storey chateau is only the triumphal arch leading to the inner courtyard.

Through the arch we caught our first look at Château de Tanlay. It is a wondrous mix of architectural styles and elements that the designer, Pierre Le Muet managed to blend into a harmonious whole.

We walked across the broad inner courtyard to the high stone wall on its far side to get views from different angles. Through an arch lie the gardens.

Looking back past one of the corner towers of the main château we could see "le petit château" and it was easy to see why it is so often confused for the real château by those peering through the bars of the locked gates on Tuesdays. If I hadn't researched beforehand, we wouldn't have waited overnight.

The main structure of Château de Tanlay is set in a broad rectilinear moat and access to it is across a bridge. We headed to the bridge over the moat and walked up to the entrance booth. It was closed for lunch; it was 1135 and the sign said the next visit will commence at 1415.

As we walked back across the moat bridge, which is flanked by a pair of intricately carved obelisks, like a pair of thrusting fingers, we muttered another WTF. Having spent more than two thousand days cumulatively in France since my first visit in 1966, I am still trying to comprehend its mysterious opening and closing rituals.

We slowly walked back through the small town of Tanlay, watching the shops prepare to close for lunch. Back aboard we relaxed until 1240 before slipping to continued up the canal, timing our departure to have us at the next lock as it reopened after lunch. A couple of locks along we had to wait for the arrival of l'éclusier to prepare it for us. As we waited, we wandered up to the lock to see it full of grass clippings. We continue to be amazed that the only apparent maintenance along the canals is the grass cutting, which fouls the canals and their works with clippings.

At Écluse d'Argentenay we asked l'éclusier about nice mooring recommendations in the next few kilometres. He told us there was little to choose from; the moorage four kilometres along at Lézines is close to a railway bridge and noisy. Ten kilometres further is a quiet stop at Ancy-le-Franc. We decided to head there.

Shortly after 1800 we moored with the spud pole and a pin pounded into the grassy bank. The local ducks came over to guard our mooring line as we relaxed with a bottle of rosé to revive us from the day's twenty-one kilometres and eight locks. We plugged into the electrical pylon and filled our tanks with water, but nobody came to collect moorage fees, and we could see no port office nor any indication of where one might be. The ducks settled-in with their guard duty.

Canal de Bourgogne

On Thursday morning we worked our way through seven locks on our own to Cry-sur-Armançon before we picked-up another boat to accompany us in the chambers. Fortunately, it was a private boat, and the owners were experienced, so we moved through the locks efficiently.

We wound along through very pleasant rural settings with a scattering of small villages. Most of these are many centuries older than the canal and had been built along the course of the Armançon River. The canal followed beside its course when it was built in the 1820s and 30s. After a total of twenty-nine kilometres and sixteen locks, in the late afternoon we moored in Montbard.

Our moorage was in the centre of the town, across from some huge metal fabricating plants. It is difficult to find a place in the town that is not close to a metals plant, since there are nine large metallurgical companies here. From the canal, their grimy and bleak façades hide vast building complexes that are best seen from space.

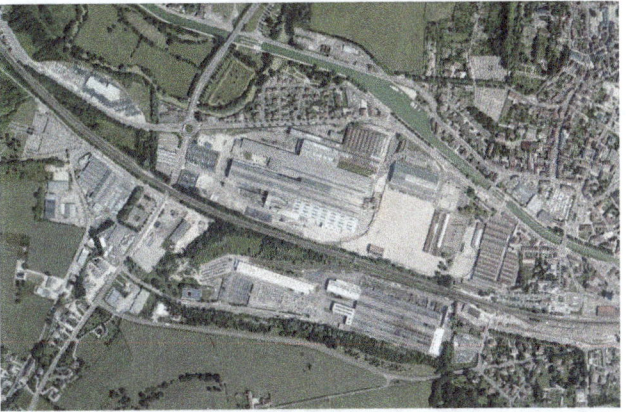

Fortunately, there are also some other sights, such as the Parc Buffon around the remnants of a tenth century château, once the property of the Dukes of Burgundy, on the hill above the centre of town. In 1733 the king authorized the young Georges-Louis Leclerc, naturalist and future Count Buffon to be the custodian of Château de Montbard and for the next two decades Buffon systematically demolished most of the château to

make terraces and gardens for his exotic trees and plants. Buffon became a world renowned naturalist, mathematician, biologist, cosmologist, philosopher and writer. He is equally famous as a pioneering metallurgist and built a state-of-the-art industrial forge in the nearby village of Buffon, where he conducted experiments.

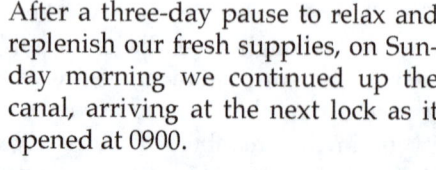

After a three-day pause to relax and replenish our fresh supplies, on Sunday morning we continued up the canal, arriving at the next lock as it opened at 0900.

Unhampered by any other boat traffic in either direction, we moved along through placid waters in the early slanting sun. The countryside was becoming more hilly as we moved up the valley. Charolais cattle grazed in the sloping pastures, which were occasionally punctuated with ancient farm buildings and tiny villages.

While moat of the lock houses were occupied, a few smaller ones, like this one at Écluse Venarey were unoccupied and rather forlorn looking.

After nine locks and just short of fourteen kilometres, we decided to stop for the day in Venarey-les-Laumes. In a basin off the canal is a Nichols rental base and we were pleased to see so many boats in port. We moored along the canal bank, plugged into the shore power and topped off our water tanks.

A lock-filling hotel barge arrived and moored ahead of us. In the late afternoon it appeared to be storing ship in preparation for departure the next morning. I walked over and asked the skipper his intentions and was relieved to hear they were heading back downstream at 0830 Monday. As we

enjoyed breakfast in the cockpit the following morning, we watched the barge slip and begin to wear around in the turning hole.

Canal de Bourgogne

We had not been visited by the port captain to collect for our mooring and services. We could see no activity around the office, nor any indication of hours or contact phone numbers, nor could anyone we asked offer any information. At 0850 we slipped and headed toward the next lock. We were about to begin a steep ascent with forty locks in twenty-four kilometres. Fortunately we were on our own again in the locks, and the passages all went smoothly and quickly.

Shortly before noon, after only 4.6 kilometres we arrived at Écluse Pont de Pouillenay, our tenth lock of the day. It was also our one-thousandth lock in Zonder Zorg. In just short of two years since we left Aalsmeer, Noord Holland on our maiden voyage, we had been through one thousand locks.

There is a good mooring basin just above Écluse Pont de Pouillenay, which is a standard stopping place before the next steep series of locks and we had decided to stop there for the day. The guide shows an épicerie and a boulangerie in the village of Pouillenay the better part of a kilometre east of the canal. We waited aboard giving the shops ample time to reopen after the noon break and then walked across the bridge overlooking the lock and continued into town.

We arrived at the épicerie to see signs in its window indicating Fermature Hebodomadaire le Lundi, its weekly closing day is Monday. We wandered the tiny hamlet looking for the bakery and finally found a man who directed us back to l'épicerie, saying it was the only retail business remaining in the village; it does everything. We walked back to Zonder Zorg empty-handed.

In the late afternoon we began taking on a list to port. After the reflex action of checking the bilges, I realized we were aground; the water level in the pound had quickly receded and was nearly forty centimetres down by the time we noticed it. Our starboard side was on the sloping bottom.

I phoned the VNF office to report the water level problem and then we set about getting Zonder Zorg off the bottom and away from the sloping bank. We loosened the mooring lines and I gave the bow thruster a long kick to port. The bow slowly swung out away from the bank. When it was a good metre off, I dropped the spud pole. We rocked the barge and slowly pushed the stern off the crud and away from the bank. Once we were in deeper water, we held the stern off with the gang plank tensioned against a line to a bollard.

With the excitement of a spontaneous grounding over, we relaxed. In the evening to celebrate our thousandth lock, I prepared scallops sautéed in Normandy butter with diced shallots, garlic and red peppers with a liberal splash of Bas Armagnac served on a bed of saffron basmati rice. The Wolfenberg Crémant d'Alsace complemented the meal splendidly and the full moon added to the occasion. We were looking forward to our next thousand.

Canal de Bourgogne

On Tuesday morning we slipped and headed up the canal to begin our second thousand. Immediately ahead of us was a quick start; there are thirty locks in the next ten kilometres. Again we were alone in the locks, so passage through them was routine, uncomplicated and rather quick. By mid-afternoon we had made 6.1 kilometres and had ascended nineteen locks to arrive in Marigny-le-Cahouët, the recommended stopping place.

We secured for the day and walked into the village to the l'épicerie and boulangerie for fresh supplies to discover that Tuesday is their closing day. We walked back to Zonder Zorg empty-handed.

On Wednesday we slipped at 0855 and were followed into the first lock by a small sailboat owned by a French couple. We fell into a routine with the sailboat skipper closing one of the downstream gates and me opening one of the upstream ones, making the job of l'éclusier lighter and speeding our progress.

The first series of locks was in our favour from down-bound traffic on the previous afternoon and by noon we had made four kilometres and twelve locks to reach the end of the steep climb. As l'éclusier took his lunch break, we dawdled along toward the next lock in our first long bief in forty locks.

Ahead of us was a gentle climb through four locks in twenty kilometres before the final steep eleven lock rise to the summit of the pass. We were well above the river and the countryside had become noticeably less steep.

After the lock keeper rejoined us, we passed through the next lock and entered a narrow section. Here the trees hang far out from the banks, and in places constrict the passage to little more than a boat width. This half kilometre narrowing is a trench across a ridge that allowed the canal to make its way more directly through gently undulating terrain.

After passing through two more locks we entered another narrow section. This one is 1130 metres long and is cut so deeply into the ridge that the sides needed to be reinforced with masonry walls. This cut allows the pound between locks to be over ten kilometres long.

Along the way we were challenged by a swan. With its wings raised to make it look bigger, it swam aggressively toward Zonder Zorg as we approached. Maybe it thought that we had our leeboards raised in challenge. After we had passed it chased us furiously to ensure we kept going. At 1725, after a final steep climb of ten locks in four kilometres, we arrived at Écluse Pouilly, the 113th and final lock up to the summit. As we were locking through, l'éclusier gave us a sheet of paper with instructions on passing through the tunnel and asked if we intended to continue in the morning.

We had done twenty-six locks and nearly thirty kilometres since morning, so we told him that we wanted to take a day off and that we would probably leave early Friday morning. We motored into the summit pound and took a mooring on the stone quai next to an electrical and water pylon. In the evening, to celebrate our reaching the summit, I prepared dos de cabaillaud with sautéed mushrooms, potatoes rissolé and steamed green beans and we enjoyed it with a bottle of Crémant d'Alsace.

We were at the highest point in the French canal system, having ascended through one-hundred-thirteen locks since leaving the Yonne. Forty-five of these locks had been in the past two days and we wanted a break before starting the steep descent toward Dijon. On Thursday morning we watched as the French sailboat, which had accompanied us in the locks the previous day, slipped and headed toward the tunnel, leaving only Zonder Zorg and Milo in the basin.

We continued to be amazed at the lack of traffic as we watched the sailboat motor into the early morning mists. Forward of us at the northern end of the basin was an arched structure. We walked over to take a look.

Inside is a decommissioned electric chain tug. It was used from 1893 to 1987 to haul barges through the 3350 metre tunnel. This tug had replaced steam tugs, which had operated since 1867 and they in turn had replaced a team of six men that had done the haulage from the time of the canal's opening in 1832.

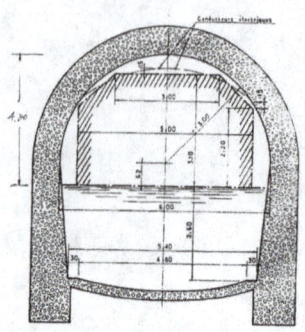 

The profile drawing of the tunnel shows its restrictive dimensions. At 5.0 metres width, the vault is only 2.2 metres high, making it too low for unladen barges. However, the depth at 2.4 metres is well in excess of the 1.8 metre depth of the rest of the canal. In 1910 a solution for unladen barges was devised. To take advantage of the extra depth, a floating dry-dock was constructed, which when partially sunk, lowered an unladen barge sufficiently to allow it to fit into the tunnel. This old postcard shows the electric tug coming out of the tunnel towing the partly sunk scow containing an unladen barge.

After looking at the old chain tug we walked the three hundred metres to the Atac supermarket to stock-up on fresh supplies. By the time we had returned, Milo had departed and Zonder Zorg was the only boat remaining in the large basin. We relaxed on board and did small chores. Among these was shelling the walnuts we had harvested along the way.

I reviewed the sheet that l'éclusier had given us. It detailed the procedure for the transit of the tunnel. This was our tenth tunnel in France with Zonder Zorg and the only one we had seen with any instructions. In addition to the requirements on the paper handout, l'éclusier told us that he must inspect our spotlight to ensure it is sufficient for the tunnel. Fortunately, I had earlier read about equipment inspections here, so the previous week we had bought a light. Rather than an expensive marine light, for €7 we got a basic home outdoor security floodlight and I wired a plug to its leads.

At 0845 on Friday I walked back to the lock house to ask l'éclusier for a transit ticket. He gave me a radio specially set-up to allow communications from within the tunnel and then he accompanied me back to Zonder Zorg to inspect our life jackets and tunnel light. I hung our new light on the mast step, plugged it into the 220v outlet and we passed the inspection. At 0910 we slipped and motored past the old chain tug and past the green traffic light toward the narrow cut leading to the tunnel.

The six-metre-wide cut is about eight-hundred metres long, the first half of which is a gentle 45° curve toward the mouth of the tunnel.

The stone walls supporting its sides gradually increase in height as the cut advances into the ridge. There are a few places with crumbling masonry, some of these shored-up by temporary scaffolding. From the appearance of the vegetation that is well established in the rubble, the wall looks like it has been waiting a long time for repairs.

When we finished the bend and caught our first sight of the tunnel entrance, we were pleased to see the traffic light was still green. It would have been an awkward reverse back around the narrow bend.

It was still early as we approached the tunnel mouth and most of the tourists were likely too busy with breakfast to watch the day's boat traffic head into the tunnel; there was only one spectator on the viewing plaza near the middle of the town of Pouilly.

We were surprised to see lights down the centre of the tunnel. Of all the tunnels we have been in, this is the best lit. We turned on our navigation lights, but we saw no need to turn on our new floodlight.

Zonder Zorg tracked perfectly down the centre of the tunnel, well balanced by the pressure waves from the sides. She needed very few adjustments to the tiller. We could see the light at the end of the tunnel more than 3.3 kilometres ahead.

We glanced back from time to time to watch the light astern gradually diminish in size.

Slowly, the ellipse of light ahead grew. I kept myself from being hypnotized by watching and keeping track of the numbers along the side walls marking the distance from the end.

At 0958, thirty-four minutes after we had entered the tunnel, we came to its eastern end. Of the long tunnels we have been through, this was by far the easiest; it was easier than the 4820-metre Balesmes and easier than the 2302-metre Mont-de-Billy.

Just over a kilometre along from the exit is the first lock, Écluse Escommes, where we were met by l'éclusier, who had prepared the lock for us. He told us there were three hotel barges on their way up that we should meet by early afternoon. Only one was continuing past Vandenesse-en-Auxois, and after the fourth lock we met it.

L'Impressionniste shuttles between Fleurey-sur-Ouche and Escommes on week-long luxury cruises, Sunday to Saturday and then reversing the itinerary with new clients the following week. We slid past each other and continued along.

Through four more locks past beautiful scenery, we arrived in Vandenesse-en-Auxois at 1130, where l'éclusier told us to stop and wait for the locks to reopen at 1300 after lunch. The second hotel barge was approaching from downstream and les éclusiers would be busy handling it.

About half an hour after we had moored in the small basin in Vandenesse, hotel barge Wine & Water arrived and began to wind around in the basin beside us. She was completing her Dijon to Vandenesse itinerary and would disembark her clients here the following morning. On Sunday she would embark new clients for the six-day itinerary back to Dijon.

At 1250 we slipped to arrive at the next lock as it reopened after the lunch break. On the way we passed Wine & Water, which by this time had stabilized alongside at the end of the basin and was beginning to disgorge passengers for sightseeing. Around the bend as we approached the lock, we had the beginning of a long series of views of Châteauneuf-en-Auxois on the hill above the canal.

We were followed into the lock by a strange-looking vessel that was handled even more strangely than it looked. The skipper seemed to have given-up using the engine and rudder and resorted to excessive and ineffective use of the bow thruster. He then relied on the four crew who manhandling the boat into the chamber.

At the first lock we thought he had simply blown the approach, but when he repeated the same incompetence at the next lock and the one after that, we realized we were in for a very slow passage. In the beginning l'éclusiers assisted with the manhandling, but after a couple of locks, they rolled their eyes and gave-up.

We were delayed fifteen to twenty minutes per lock and we thought of stopping for the day to let the boat continue its bumbling on its own, but with the crumbling canal banks with generally foul bottom alongside, there were no suitable places to moor. We resigned ourselves to watching the passing scenery.

A couple of locks along we met the third hotel barge, Prospérité and we waited as it cleared the chamber and then the bridge hole.

The lock keepers along this stretch operate several locks and travel between them on motor scooters. After they had finished with the gates and sluices of the previous lock, they often passed us before we got to the next lock.

To control the water levels in the pounds, the lock keepers often leave sluices in the locks opened a crack to bleed water downstream through a series of locks. We arrived at one lock to find the downstream sluice had apparently been left a bit too widely opened and the chamber had partly drained. L'éclusier was still refilling it as we arrived.

Canal de Bourgogne

We had intended stopping for the day in Pont d'Ouche. L'éclusiere told us it was very crowded with permanently moored boats, itinerant hotel barges and visiting cruisers. She phoned the port to see if there was space for us and was told there was none. Fortunately, there was space for the smaller boat that had been delaying us in the locks through the afternoon, so we decided to continue along through three more locks before stopping for the night. We told l'éclusier that the following morning we would be at the next lock when it reopened, ensuring we wouldn't have to share chambers with the bumbling boat.

At 1750 we secured in the village of Veuvey-sur-Ouche. The water along the quai was rather shallow and rubble-strewn, so we held ourselves off a metre with our spud pole and boarding plank. We had come 21.8 kilometres and through twenty-two locks, eleven of them with frustrating company.

There is no bakery in the tiny village and we had no croissants, so we slipped without breakfast. Just beyond the bridge we passed a new quai that appears to have good depth alongside, though no mooring rings or bollards.

At the first lock l'éclusier replied to my query with information about a bakery a hundred metres from Écluse la Bussière, three kilometres and four locks along. While Zonder Zorg was lowered in the chamber, I walked into town, made my purchase and made it back in time to assist with opening the gates. By the time we had arrived in the next lock half a kilometre along, Edi had prepared ham and cheese croissants and had pulled fresh cups of espresso.

As we rounded a bend a few locks further down the canal we saw three barges. At first I thought we were meeting a parade, but soon saw they were all moored.

The first appears to be a long-abandoned dream of converting an old péniche into a hotel barge. As we passed, we could see that work on the project had long since ceased. This was likely when the owner grew aware of how huge the conversion task was, both in time and expense. There are many broken dreams like this moored along the canals.

The next barge in the line-up appears to have completed its conversion, but apparently has no business. Vaillant is not listed among the more than fifty active hotel barges in France, so either it is new or simply has failed in its marketing.

The third barge in the line-up appeared to be another incomplete conversion project that has run out of energy.

Shortly before noon we arrived at a lock filled with a luxemotor, so we moored on the bank to wait. We recognized L'Escapade as she began rising in the chamber and chatted with David and Evey as they locked through. We had moored next to them in May in Sérignac-sur-Garonne and in July in Carcassonne.

Even though it was striking noon, l'éclusiere invited us into the lock and worked us down a few minutes into her lunch break. She then told us we could wait in the next lock during lunch. Just above the lock were two boats, including Xenia, and as we were waiting in the lock, Charles came over and chatted. We had been sharing experiences on the DBA Facebook group for many weeks, and it was fun to finally meet face-to-face.

Canal de Bourgogne

After l'éclusiere returned from her lunch break, we continued through the lock and down the canal. We were following the valley of l'Ouche, in places rather steep-sided and desolate. A few of the lock houses were unoccupied or were being used only as waiting places by the roving lock keepers. Most, though were occupied by tenants who obviously took pride in maintaining the buildings and their surroundings.

Limestone crags appeared more frequently above us as we neared the end of the ridges coming down from the back side of the Burgundy's Côte d'Or. Just across these ridges, near the bases of their eastern slopes are such famed vineyards as Morey-St-Denis, Chambertin, Musigny, Vougeot, Vosne-Romanée and Corton, home to some of the world's greatest Pinot Noir. I fondly remembered my many years of barrel tasting there and selecting wines to import. I salivated at memories of the spectacular old bottles and the exquisite cuisine the wine growers so proudly and generously shared.

Around a few more bends we passed the junction of the road that was beginning its wind up and over the ridge toward Grvrey-Chambertin and the great Burgundy vineyards, but Zonder Zorg insisted on staying in the canal.

A little further along we passed under a large divided highway, which then began to follow closely beside the canal. After working our way down through seven more locks, we stopped for the day just below the lock in Velars-sur-Ouche.

It is not a pretty place, but it is along a short loop away from the noise of the highway that leads into Dijon. Beside the solid concrete quai is a supermarket so new that its exterior sheathing was not yet completed. During the day we had passed through twenty-three locks and had made just over twenty-seven kilometres.

In the evening to celebrate my seventieth birthday I prepared dos de cabaillaud with a cumin crust served with gnocchi in a mushroom, garlic, shallot and armagnac sauce and fresh Peruvian asparagus. The bottle of Champagne Canard-Duchêne complemented superbly.

When I had gone to the bakery for breakfast croissants, I had also bought an entremet glacé de mousse à la framboise. With a candle on its top, it made an appropriate birthday cake. We enjoyed it with some wonderful Alsatian Gewurztraminer Vendange Tardive.

Around 0845 on Sunday, as I was pulling our second cups of espresso and Edi was nearly ready to plate our pain perdu au lardons, madame l'éclusiere knocked on the hull to ask when we wished to continue through the next lock. I told her we would depart directly after breakfast and be there by 0915. We had nine locks to Dijon and the last one was a little over eight kilometres along from the first, so to make the last lock before it closed for lunch, we needed about two and a half hours.

Canal de Bourgogne

.Shortly after 0900 we slipped and were in the first lock seven minutes later. The locks were all in our favour and we moved along steadily until the ninth and final lock, where we had to wait for an up-bound boat.

At noon we were motoring past the collection of moored hotel barges on the northern rim of the D-shaped basin of Dijon. A semicircular island sits in the middle of a larger semicircular basin and moorage is around the basin's periphery. The basin is within a kilometre of the medieval heart of the city.

We motored past the crowded moorage in the port. We had no reservation and we were hoping that there were still open spaces along the quai along the right bank. The previous day both David and Charles had reported there being plenty of open spaces when they were there a couple of days earlier.

We were in luck. There were several open slots. At 1210 we secured to solid bollards on the stone wall a couple of boat lengths short of the next lock. There are no water or electricity points on the quai, but neither are there charges for mooring. We were within a hundred metres of the tram line to downtown and a block away from a large supermarket.

We still had twenty locks to pass through to the end of le canal de Bourgogne at Saint-Jean-de-Losne. The onward route runs across the flat expanse of the Saône plains for thirty kilometres with only one slight bend, less than 5°, to relieve the monotony. Rather than rushing off to complete the canal, we decided to pause in Dijon for a while.

Chapter Forty-Three

Dijon - Capital of the Burgundy

We were in Dijon, the capital of the Burgundy. We had entered the region in July as we ascended the Saône and we had explored along the Seille, crossed the Canal du Centre, ascended the Loire to its end of navigation, then descend it to the Canal du Nivernais, which we followed to the Canal du Bourgogne and then along it. We had done a clockwise circuit near the boundaries of the Burgundy. The boundaries of present-day Burgundy, one of the twenty-two Regions of Metropolitan France, have changed dramatically over the years.

Human settlement in the area is traced to Lower Paleolithic times and one of the final phases of the Upper Paleolithic is named Solutrean for evidence of the 20,000-year occupation of the site at Roche de Solutré near Macon. The Burgundy region was later an important crossroads during the Bronze Age and waves of migration from Central Europe swelled the population. Predominantly Celtic in origin, these people were called Gauls by the Romans. The Gauls of the area around present-day Burgundy were the last hold-out against the Roman conquest. With the surrender of Vercingetorix to Julius Caesar in 52 BC, the area finally fell to the Romans after 150 years.

In the fifth century, as the Roman Empire was collapsing, the Burgundians moved in from the northeast. Thought to be originally Scandinavian, these peoples had long before settled along the Baltic coast of what is now Germany before migrating and occupying the area that is now eastern France and western Switzerland.

Dijon - Capital of the Burgundy

In the sixth century another Germanic tribe, the Francs conquered the Kingdom of Burgundy, but finding the Burgundians strong, they allowed them to continue to occupy and govern the area. The Frankish territory continued to expand for another three centuries, reaching its greatest extent in 814 as the Carolingian Empire under Charlemagne. After inheritance disputes among Emperor Charlemagne's three sons following his death, the Treaty of Verdun in 843 divided the empire into three portions: East Francia, Middle Francia and West Francia.

West Francia was a major portion of what would become the present-day France. The dividing line between West and Middle Francia ran through the Burgundy, dividing it in two.

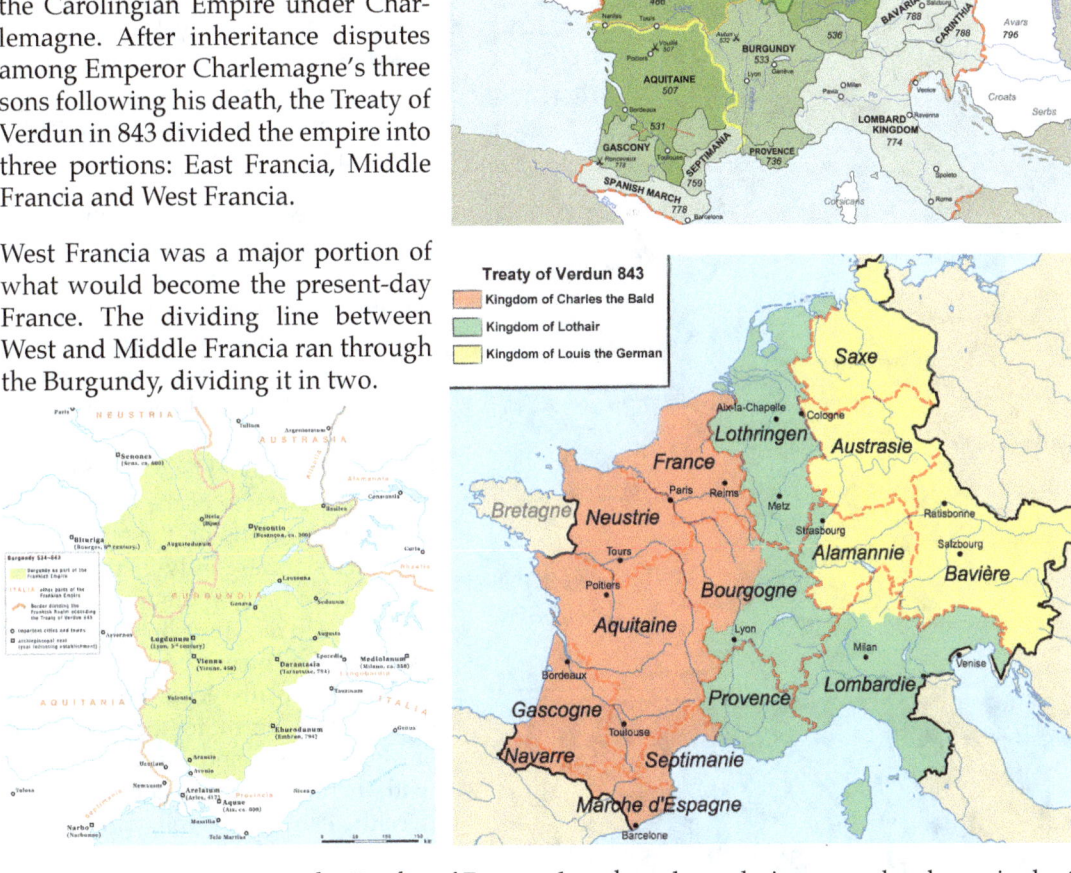

The smaller piece became the Duchy of Burgundy, whose boundaries were closely equivalent to today's Burgundy Region.

Starting from this small base, successive Dukes of Burgundy expanded their holdings through battle, negotiation, marriage and inheritance. By 1477 their possessions included much of modern-day Netherlands, all of Belgium and Luxembourg and portions of Germany and Switzerland. This was nearly the size of the lands of the Kingdom of France, but it contained the biggest cities and the richest territories of the region. The court in Dijon outshone the French court both economically and culturally. In Belgium and in the south of the Netherlands, a 'Burgundian lifestyle' still means 'enjoyment of life, good food, and extravagant spectacle'. During the Hundred Years War and the years following it, the Dukes of Burgundy tried on many oc-

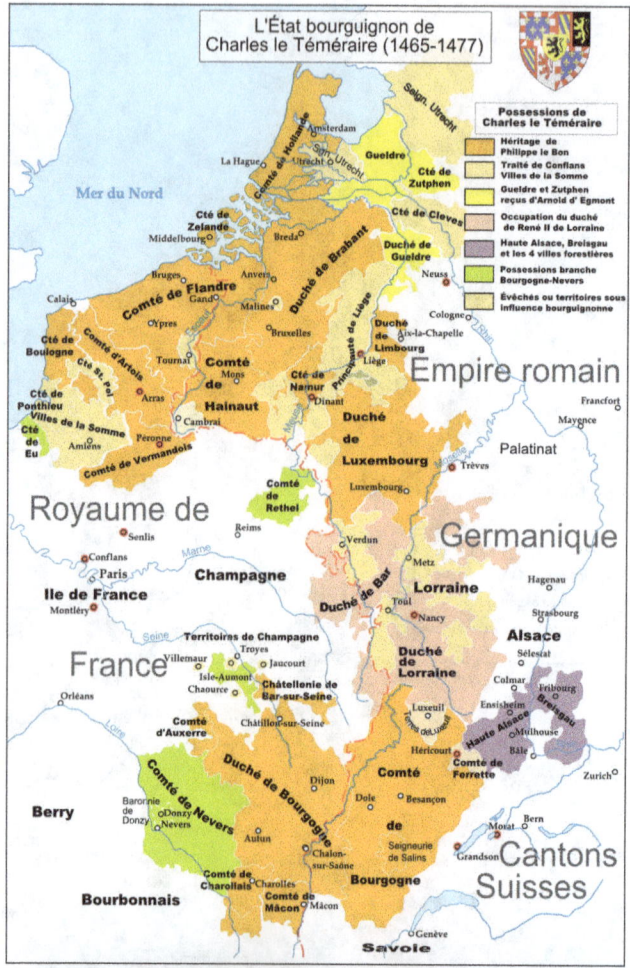

casions and through various means to take the crown of France. Duke Charles the Bold was killed in battle in January 1477 in one such attempt.

Louis XI of France quickly moved to seize Burgundy by presenting himself as the protector of Charles' daughter and heir, Mary. By the beginning of February 1477, the royal army had entered Dijon and occupied the Burgundy. Duchess Mary refused the King's offer of marriage to his son and heir, the Dauphin Charles.

She retreated north to the Low Countries, and then a few months later she married Maximilian of Austria, the future Hapsburg Emperor, Maximilian I. By 1520 their grandson, Charles V wore many crowns, among them as the King of Spain and the Archduke of Austria and he had been elected Holy Roman Emperor. He was the most powerful monarch of the time, but he focused his efforts on the recovery of the duchy of Burgundy that had been seized from his grandmother. He was successful in regaining only the Charolais, the southwest portion. By this time, many of the loyal Burgundians had moved north to the Burgundian Netherlands and Flanders, taking with them their arts, talents, skills and wealth. During the following century, the Netherlands dramatically increased its standing as the most prosperous region in Europe.

With this background in mind, on Monday we headed into the medieval heart of Dijon. There were dramatic changes since my last visit in 2006. A new light rail transit system opened in 2012 and much of the downtown core had been converted to pedestrian-only streets. What was once a noisy and exhaust fuming traffic gridlock is now peaceful and inviting. There are many medieval buildings and classic Burgundian glazed tile roofs throughout the city centre. We continued through very pleasant streets to place de la Libération and past l'Hôtel de Ville, the City Hall, which is housed in the former royal and ducal palace buildings.

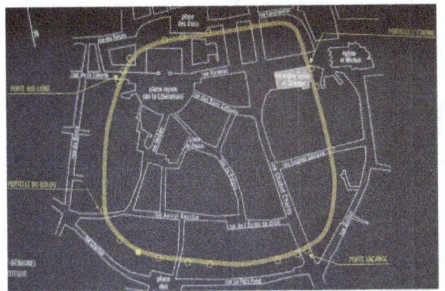

Through a side entrance toward the rear of a building, we entered the transept of the former église Saint-Étienne. The church was confiscated after the Revolution and became the wheat market, later the stock exchange. During renovations in the choir, remains of the Roman castrum were found beneath the floors. This city wall was built between 270 and 275 by Roman Emperor Aurelian. It was ten metres high, four-and-a-half metres thick and with its thirty-three towers and four gates, it circled twelve hundred metres around the city. Very few traces of it remain.

We walked back to place de la Libération and its imposing buildings. The palace of the Dukes of Burgundy and later the Kingdom of France is comprised of several interconnecting parts. Among the oldest parts are the Gothic style ducal palace from the fourteenth and fifteenth centuries, most prominent being la Tour Philippe le Bon. The remainder of the buildings date from the seventeenth and eighteenth centuries and are in the Classic style.

Inside the courtyard we approached la Tour de Bar, which had been built by Philip the Bold in 1365. Beside it is the entrance to le musée des beaux-arts de Dijon, which was established here in 1787. It is one of the finest, most important and oldest museums in France. We were delighted to find that admission is free.

Carefree Through 1001 French Locks

Among the more important pieces in the museum are the tombs of the Dukes of Burgundy. Philip the Bold built a Carthusian monastery at Champmol, just outside Dijon to house the tombs of his dynasty. He gathered many artists to decorate and embellish its interior and to work on his tomb featuring his recumbent effigy. Work on it began in 1381.

Three successive sculptors, Jean de Marville, Claus Sluter and Claus de Werve worked on it for thirty years, completing the project in 1410, six years after Philip's death. The polychrome and gilt decoration was done by Jean Malouel, official painter to the Duke. The tomb is considered to be one of the finest works of Burgundian sculpture.

Philip's heir, John the Fearless expressed a wish for his own tomb to resemble that of his father, but as a double one with his Duchess Margaret of Bavaria by his side. Work had not been commenced when he died in 1419. Finally in 1443 a Spaniard, Jean de La Huerta, was contracted and was sent drawings for the effigies. He completed most elements, but not the effigies, before leaving Dijon in 1456. Another master was brought in, and in 1470 the monument was finally installed.

Among other remarkable aspects of the two tombs are the mourners standing in alcoves surrounding the bases of the biers. There were originally eighty-two of these free-standing alabaster sculptures, each about forty centimetres high and each an individual character. Following the Revolution, the tombs were badly damaged by anti-royal vandals and a dozen of the mourners were stolen. Some ended up in Dijon homes, others were marketed to museums and private collectors.

In 1819 the tombs were restored and placed in the Dijon museum. In 1945 an English collector returned a choirboy sculpture to Dijon. Soon after, the Louvre donated its mourner and the Cluny Museum returned two mourners. An American collector had bought four mourners from French collectors and after his death, his estate sold the sculptures to the Cleveland Museum of Art, where they remain today. In 1959 the museum gave replicas of its

mourners to the Dijon Museum. Only two of the niches remain empty and it is presumed the missing sculptures had been destroyed during the vandalism of the Revolution.

Among the museum's vast collection of medieval art are many pieces from the monastery at Champmol, including a splendid folding altar piece. It was carved by Jacques de Baerze and gilded and painted by Melchior Broederlam. The quality of the 1398 painting on the exterior of the door panels is superb, particularly the scene of the Annunciation.

In addition to the medieval collections, the museum has large collections of ancient Egyptian, Greek, Etruscan, Gallic, Roman and Byzantine material. We were more interested in Burgundian art, so we passed quickly through to the galleries of Renaissance and later paintings.

Peter Paul Rubens is well represented among the Burgundian Netherlands and Flemish artists, and the pieces in the museum arrived under interesting circumstances. During the conquest of the Austrian Netherlands by the armies of the French Revolution, paintings were seized in 1793 and 1794 from churches in Flanders. Two of these were transferred from the Louvre to the Dijon museum in 1803. These were the side panels from a triptych altar piece commissioned in 1632 by a parishioner to honour her dead father. One was "The Entry of Christ Into Jerusalem".

The other is "Christ Washing the Feet of the Apostles". These two side panels were painted on wood and appear quickly rendered. Once thought to have been produced by Rubens' workshop, experts now believe they came directly from the master's hand.

The central panel of the triptych is "The Last Supper", which measures three metres by two-and-a-half, is oil on canvas and is decidedly more meticulously painted. It is now in Pinacoteca di Brera in Milan.

A brochure on Rubens produced by the Dijon museum reads: "One of the greatest masters of Flemish painting, has always been admired in France. The three tables in this room, seized by French troops in 1794, are representative of that interest. They also reflect the final changes in the seventeenth century on the medieval form of hinged altarpieces."

This altar, composed of a central panel flanked by two movable flaps, was commissioned in 1618 by the tailors' Guild in Lierre, near Antwerp for their chapel. It was seized in 1794 by French troops, and arrived at the Louvre. It was dismembered in 1809 and the central panel was sent to Dijon. The side flaps remained at the Louvre and were claimed in 1815 and had to be returned to Lierre. We wondered why these other stolen Rubens pieces are still displayed in Dijon.

We slowly made our way through the galleries, looking at seventeenth and eighteenth century European works and then we wandered through a grand assortment of paintings by the nineteenth century French painters, including Delacroix, Courbet, Tissot, Monet, Manet, Sisley and Pissarro. We found "la Japonaise au bain" painted in 1864 by James Tissot a refreshing departure from the heavy religious themes we had been viewing.

As we made our way out from the museum and back toward Zonder Zorg, we reflected that this region was once the principal centre of art and culture in Europe. Since the departure of the last of the Dukes of Burgundy, the region has lost this reputation, but through their continuing refinement of viticulture and viniculture, when the word Burgundy is spoken today, most think of great wine. Many of the world's finest Pinot Noirs and Chardonnays come from here, but that's another story.

On Tuesday morning I walked over to the lock keeper's office to ask to be locked through after the lunch break. We had twenty more locks to the end of le canal de Bourgogne at le Saône. Initially the canal passes through several kilometres of industrial area and alongside the rail yards before it passes close by the end of the airport runways. The first ten kilometres is not an exciting stretch. After ten locks and eleven kilometres, we had finally left the bland and industrial behind and stopped for the night under a grove of walnut trees in Epoisse.

The following morning, after a final scour of the grounds for fresh walnuts, we slipped and continued. At the first lock, l'éclusier told us there were three hotel barges coming upstream. Two locks along we met the first of these, S.Antonius as she was entering the chamber, so we had a rather long wait.

Two locks along, the second barge, Le Haricot Noir was part way up, so the wait was not as long. Just before noon we met the third barge, Adrienne just entering the lock chamber. It would be well past the beginning of the lunch break by the time she exited, so we resigned ourselves to waiting until 1300. As Adrienne churned past us at 1225, we saw l'éclusier waving at us, motioning us to come into the lock. We recovered the gangplank and headed in, thinking he was going to let us wait his lunch break in the chamber. He surprised us by locking us through and then telling us to continue along to the next lock. The next lock was his home and since there was no more traffic expected on the canal, he could then take a long break.

At 1430 we arrived at the final lock of the canal and passed down into la Saône. We motored out onto the river for a look at St-Jean-de-Losne and to my surprise and delight, we had our choice of three open moorings on quai National. In some twenty passings of the quai since 1984, this was only the second time I have found a moorage spot here. We enjoyed a relaxing late lunch in the cockpit.

It wasn't very long before some of the locals came by to share our lunch.

In the evening we celebrated our passage along the 242 kilometres and through the 189 locks of le canal de Bourgogne and of our arrival in Saint-Jean-de-Losne, Zonder Zorg's planned wintering place.

Chapter Forty-Four

Preparing for Winter

We remained moored alongside Quai National on the bank of the Saône enjoying the superb location and the wonderful late summer weather. The town of Saint-Jean-de-Losne is quite legitimately referred to as the centre of the French canal system, since from this hub it is possible to head in six different directions along the canals and navigable rivers.

From here boaters can choose to head southward along the Saône and Rhône to the Provence, the Mediterranean and the Midi, they can head southwestward across canal du Centre to the Loire, northwestward along canal de Bourgogne to the Yonne and Seine, northward on canal entre Champagne et Bourgogne to the Champagne, northeastward along canal des Vosges into Loraine or slightly more eastward across the Vosges on canal du Rhône au Rhin into the Alsace.

With its location at this important barging hub, a marine infrastructure grew in the years following the 1832 opening of canal de Bourgogne. Much of this was immediately upstream of the first lock on canal de Bourgogne in the wide basin that was designed to serve both as a port and as a repair and maintenance facility.

Preparing for Winter

Just below the lock, at the level of the river is a large excavated basin, la Gare d'Eau on the north side of the town. This was dug as a place of refuge for barges when the Saône was in flood. This Water Station was also used to store rafts of logs that had been floated down from the forests of the Vosges and Jura.

With the dwindling barge trade, by the mid-1970s the repair and maintenance services had little business, so many of the skilled marine tradesmen had moved to other jobs or to other regions. As commercial use of the canals declined, pleasure use was slowly increasing. In 1972 Blue Line began offering self-drive rental boats in France and they had expanded to more than 300 boats by the mid-1980s. They had set-up in Saint-Jean-de-Losne in 1980 and I rented a canal boat from them in 1984 from their base in la Gare d'Eau. Blue Line is the small cluster of boats on the right side of the 1987 photo and the lines of floats in the bottom right of that photo belonged to the then newly established business named H2O. The 2012 photo shows the expansion of the marine facilities since then.

We had based our Dutch kruiser, Lady Jane in Saint-Jean-de-Losne in 2000, 01, 02 and 03 and again in 2005 and 06. Not only is it a great hub from which to explore the waterways, but it also has a dry dock, crane haul-out facilities and a marine railway. There are marine chandleries, a thriving community of marine trades and a strong boating atmosphere.

Two large marinas in le Gare d'eau, Blanquart and H2O offer moorage, both long-term and itinerant and there are three supermarkets within a kilometre of the basin. In addition, long-term moorage is available from H2O in l'Ancienne Écluse three kilometres downstream from town and from Bourgogne Marine between the final locks of canal du Rhône au Rhin four kilometres upstream at Saint-Symphorien.

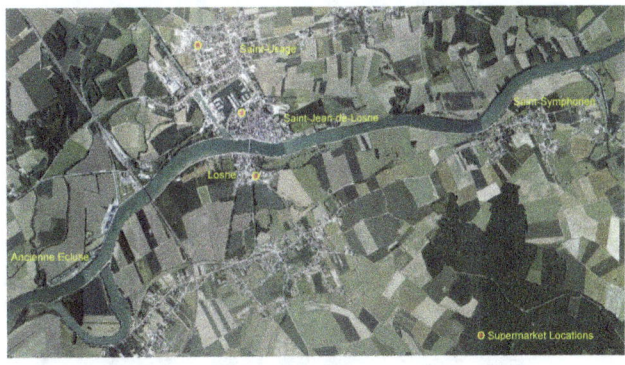

Saint-Jean-de-Losne dates back to at least 675 and it grew to prominence with its bridge across la Saône. Taxes and tolls from the crossing allowed it to fortify itself and it is famous for its

defences. In 1636 the townspeople held-off the sieges of a much superior Austrian army that had already wiped-out most of the surrounding towns and villages. King Louis XIII rewarded the town's courage by exempting taxes. During the Battle of France in 1814, the town again foiled the attacking Austrians, this time as they chased-down Napoleon. The townspeople crossed the bridge, and after attacking the Austrian camp, they retreated, destroying the bridge behind them. The following year, after Napoleon had returned from exile in Elba, he conferred la Légion d'honneur on the town.

On Thursday morning we awoke to heavy fog, which hid the left bank of the Saône and nearly obscured the bridge. By the time we had finished breakfast the fog had begun lifting, the bridge gradually appeared and navigation was slowly resuming on the river. The bridge had been replaced after the Napoleonic wars, but was destroyed again by the Germans in September 1944 to cover their retreat eastward.

At the entrance to the bridge is a square, place de la Libération and in the centre of the square is a monument commemorating la Belle Defense de 1636. Bordering the square is Église Saint Jean Baptiste, begun in the fourteenth century, now nicely restored and maintained and resplendent under its classic Burgundian glazed tile roof. Among the embellishments on its interior is a large model boat suspended in the transept since before the opening of the canal de Bourgogne. It is dedicated to Saint Nicolas, patron saint of boatmen, to honour the boating tradition of Saint-Jean-de Losne.

Preparing for Winter

We walked through town to the offices of H2O to find out what moorage had been assigned to Zonder Zorg. Months before, while we were in the Midi I had emailed Charles, H2O's founder and director to ask about winter moorage and had received a reply confirming a place for us in le Gare d'Eau. We arrived in the Capitainerie to find that we were not listed for moorage; Charles had apparently omitted to inform le Capitain.

I tracked-down Charles, whom I first met in 1984 and had done brokerage, repair, maintenance and moorage business with him many times since. I found him overseeing excavation work for the footings of a new building. He seemed a bit flustered when I mentioned that le Capitain said there was no more space available to fit Zonder Zorg and that he knew nothing about my request for winter moorage. Charles countered with: "We have plenty of room for you at Port Royal".

Port Royal is a new marina with 150 moorings that was created in 2011 by H2O in cooperation with the city of Auxonne, eighteen kilometres up la Saône from Saint-Jean-de-Losne. We had moored along the wall on the river at Auxonne many times over the past thirty years and had enjoyed the town. Last year on our way down la Saône we had moored in the port and were impressed with the location and the layout. Although I had my mind set on wintering in Saint-Jean-de-Losne, I told Charles that we would take the moorage in Port Royal and that we would head up to Auxonne after a few more days enjoying our spot on Quai National.

After four days relaxing in Saint-Jean-de-Losne, on Sunday 21 September we slipped from our moorings on the quai and headed up la Saône. It was pleasant to be on the river again; it is broad enough for relaxed steering and we faced negligible current. Auxonne, eighteen kilometres upstream is a small city on the eastern edge of the Burgundy region, just three kilometres from the boundary with the Franche-Comté. Being on the Franche-Comté side of the river made it very a strategic location in battles for many centuries.

A pleasant two hours later we arrived at Écluse Auxonne, the only lock on the route. Two down-bound boats were just beginning to enter the chamber, so we had to wait. After manoeuvring Zonder Zorg's bow for Edi to twist the dangling rod to signal the automatic lock system, I backed down toward a duc d'Albe, tossed a loop around its bollard and secured the bow with the spud pole. The waiting station is set-up for 38-metre barges and there is no provision for anything shorter.

Out of the lock we motored along the bypass canal back to the river and then past the centre of Auxonne to Port Royal. Four years previously the marina had been dug out of an unused copse on the northwestern corner of the city close under its centuries-old fortification walls.

It was midday when we entered the port and nobody was around, so we took a mooring on the T-head at the end of the second wing of floats and settled-in to wait for an assigned mooring.

Preparing for Winter

Beside us, across the city walls were the military barracks that occupy much of the northwest quadrant of the city. In the late afternoon there was a knock on the hull. Roy, the Capitain was looking for mooring fees and asked how long we would be staying. When I said we were here for the winter, he told us that he had no notice of our coming. I told him Charles had said there was plenty of room for us, but it became apparently that he had done nothing beyond that.

Roy told us there was no more room for anything above fifteen metres. I then told Roy of my email exchange with Charles many months before and of the confirmation of winter space in Saint-Jean-de-Losne and of how that ball had also been dropped. Roy took sympathy and said he would juggle a space for us.

Relieved to have a place for Zonder Zorg to winter, we began to acquaint ourselves with her new winter quarters. We also reflected on where we had been. In the twenty-five months since we took possession of our skûtsje, she had taken us a total of 4968 kilometres through 1120 locks on 121 waterways in three countries. During all this travelling, we had experienced only four problems with the barge: fluid leaks on the new engine, a connection in the central heating had failed, the automatic igniter for the water heater/furnace required frequent resetting and the house battery had suddenly died and refused to take a charge. We counted ourselves extremely fortunate.

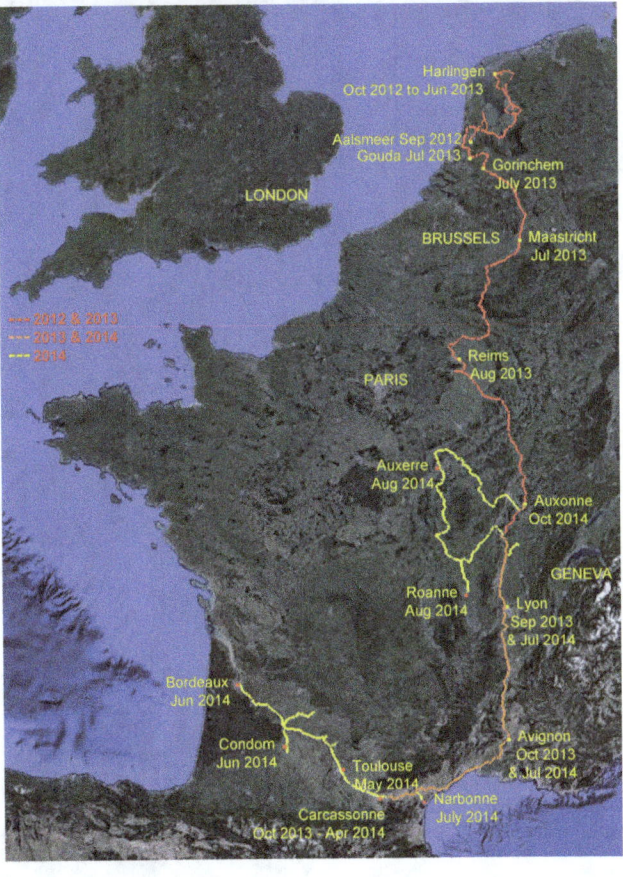

We were waiting for the arrival of the engineers from SRF in Friesland with our 22.6 kWh battery. In July they had removed it and installed a temporary 14.6 kWh bank. Even though they suspected the problem was from a faulty connection, they wanted to ensure the battery was not damaged, so they had

taken it back to Harlingen. Now their drive was considerably shorter than it had been to Capestang in July. To make their job easier, the evening before their scheduled arrival we moved Zonder Zorg to the float alongside the haul-out ramp. This would give them easier unloading and reloading.

From their 0400 start in Harlingen, Wychard and Lourens arrived mid-morning after an 875 kilometre drive, and after greetings and cups of espresso, they set to work. While Wychard began disconnecting the temporary battery, Lourens loaded the more than half a tonne of battery cells from the van to the float and moved them along to the barge.

They manhandled the cells of the temporary battery out of the engine compartment, after which they paused to devour a large platter of ham and cheese croissants. Then they began lowering the cells of our original battery back into place and connecting them. The manufacturer had bench tested each cell and all were deemed to be in top condition.

By mid-afternoon, after moving a tonne and a half of battery cells, the electrical system was reconnected and with crossed fingers, we threw the switch. We had power. We load-tested the system using the oven switched to microwave-convection and Edi's hairdryer on set to high, giving a total load of 2950 Watts. The system passed.

The men then set to work on the faulty igniter system on the water heater/furnace. By 1830, after many blind leads, they had it repaired and we shared some welcomed Heinekens after their very long day.

Preparing for Winter

Later in the evening, after Wychard and Lourens had gone off to their hotel for a well-earned rest, Edi and I celebrated the completion of repairs with some Arthur Metz Cuvée Speciale 1904 Crémant d'Alsace. To go with it, I prepared cumin-crusted dos de cabaillaud topped with butter sautéed pleurottes and accompanied by pommes rissolées and steamed broccoli.

After dinner we heard the water heater start to cycle and we listened closely for it to ignite. It aborted the cycle. I emailed Wychard, hoping he would see the message before starting north.

There was a knock on the hull as I was pulling our first espresso the following morning. Wychard and Lourens spent over an hour in the engine compartment before Lourens finally noticed that a sensor had been installed in an inverted position, which caused it to often give wrong information on the flame, which led it to abort the start-up. He turned it over and successfully ran the system through several start-up cycles. They left mid-morning, hopeful that the last of our short list of problems from the refit had been resolved. All this service was without charge; rather it was a follow-through on SRF's commitment to ensuring their work is satisfactory. Wychard and Lourens, both professional engineers are two of the four partners in the company. Wonderful!

Over the next several days we explored Auxonne. The first settlements in the area date to the ninth century as fishing camps along the river. The 843 Treaty of Verdun to divide Charlemagne's empire had used the Saône in this area as the boundary between East Francia and Middle Francia. This placed the site of Auxonne on the Germanic side of the line.

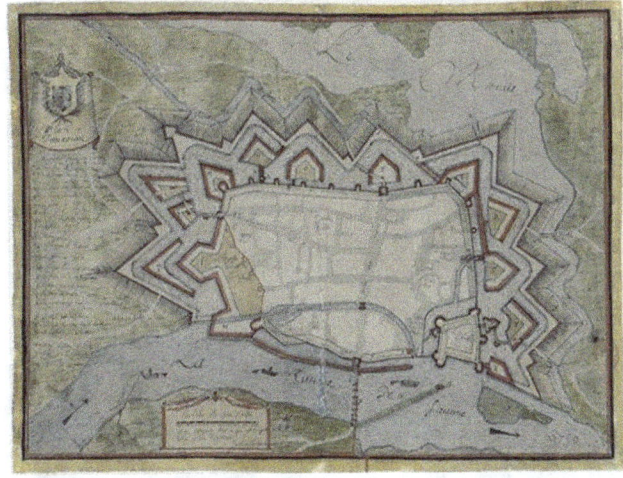

In 1197 Stephen II, Count of Auxonne moved his allegiance across the river to the Duke of Burgundy, and forty years layer all the possessions of the Count of Auxonne in the basin of the Saône were transferred to the Duke of Burgundy. With this, Auxonne became a bridgehead of the Duchy on the eastern bank of the Saône, ostensibly on Holy Roman Empire soil. This position was progressively fortified, culminating by the middle of the fourteenth century in a wall with twenty-three towers and four gates girding the small city.

The walls were strengthened and the gates were rebuilt. One of these, Porte de Comté was built in the fifteenth century on the eastern side of the city, leading out into la Franche-Comté. It is preserved today as a national monument. Another surviving gate is la Porte Royale built in 1775 to replace an older one on the northern side of the city, near the new marina, Port Royal. This makes a splendid play on words and on history.

In the centre of the city stands Notre Dame, which was begun in 1200. Among other notable features of the church is the helicoidal spire. This rises thirty-three metres above top of stone tower. In the square beside the church stands a statue of the young Lieutenant Bonaparte. Napoleon trained as an artillery officer for a total of twenty months during his two postings here between 1788 and 1791.

Among the treasures inside the church is a statue named la Vierge au Raisin, the Virgin of the Grapes. It is attributed to Claus de Werve, who had completed the magnificent sculptures for the tomb of Duke Philip the Bold in 1410, which we had seen in Dijon.

Besides exploring Auxonne, we began preparing Zonder Zorg for winter. Among the tasks was tending to the paint scrapes from more than eleven hundred locks and two hundred moorings. There were surpris-

Preparing for Winter

ingly few, mostly on the black rubbing strake, which had been designed to take the punishment. After some wire-brushing, cleaning and a few dabs of paint, she looked like she had never left port. I moved the bicycles down into the fo'c'sle along with other attractive portable items. The bicycle cover was repositioned to wrap the stricken mast, the mast step, the spud pole winch and the gaz lockers protecting them from both eyes and winter.

We had been steadily reducing the contents of our pantry, fridge and freezer, aimed at having nothing perishable remaining by the time we closed-up Zonder Zorg for the winter. On the morning of 7 October we took the train to Dijon and drove back in a rental car from Hertz, which is conveniently across from the train station. In the evening we had our season's last dinner aboard. I seared some jumbo coquilles St-Jacques, smothered them with crisply sautéed pleurottes and served this with saffroned basmati rice and sliced tomatoes with fresh basil. It was deliciously accompanied by an Alsatian Gewürztraminer.

The next morning the mechanics from H2O came to do the winterising of our systems. When they had completed their work, we closed-up Zonder Zorg, placed the protective covers over the tiller and klik, the skylight and the friese and rear of the roof. We then loaded our luggage and drove to a hotel near CDG airport to be ready for our flights to Canada the following morning. Zonder Zorg was ready for winter, but we weren't.

Concluding Comments

Our observations after a second year cruising the French canals in Zonder Zorg are similar to those after our first:

- The French interest in inland pleasure boating is slowly growing, though of the 20,000 private pleasure boats actively cruising inland in France, over fifty percent are foreign.
- The large locks on the Saône and Rhône function efficiently, and although they are very busy with huge commercial barges, they are welcoming to pleasure boaters.
- On the other hand, the Freycinet gauge and smaller locks are in poor condition. On the automatic sections we still found many locks with disfunctional mechanisms that required us to wait for a roving éclusier to override or repair.
- Many bollards are missing in the locks, often in critical positions. Safety fencing has been installed to protect the public and to make it extremely difficult for lock users. In many places, sluices dump a heavy cross-current immediately before the entrance to the lock chamber.
- The entire system appears to have been designed and maintained by people with little or no knowledge of and even less concern for the needs of the pleasure boater.
- The most obvious maintenance along the canals is crass cutting and weed trimming. Unfortunately, the trimmings are frequently dumped into the canals, fouling the locks and clogging engine cooling water intakes.
- In many places, the canal banks are lined with derelict, abandoned and sunken boats and barges and except in one place, little appears to be happening to address this pollution.
- Seventy-five percent of the annual thirty to thirty-five thousand rental boat contracts are to foreigners, who need no boating qualifications or experience.
- The canal banks are being badly eroded by speeding boats, mostly rental boats, which have no speed limiters.
- Wifi connectivity to the internet continues to be a problem.
- In most small communities, the businesses have closed or have limited opening hours, making it increasingly difficult to not to use the supermarkets. Most often the supermarkets have far fresher produce and much lower prices than can be found in the local stores and street markets.

With all these negatives, barging in France is still a wonderful experience. If only a few of these shortcomings can be addressed and remedied, then barging on the small French canals can return to the sublime experiences I remember from previous decades.

Our intention for the coming boating season is to head up le canal du Rhône au Rhin into the Alsace, explore from Basel to Strasbourg and then head back across the Vosges on canal de la Marne au Rhin, hoping that the repairs have completed on the Arzviller inclined plane. From there we want to explore the waterways across the north of France.

www.ingramcontent.com/pod-product-compliance
Lightning Source LLC
Chambersburg PA
CBHW081352290426
44110CB00018B/2351

NOEL

An Authorised Memoir

For: Geoffrey, Peter, Tania, Natalie Kathryn
and their families

Dedicated

to

The Men & Women of Australia's Tractor and
Machinery Industry

§§

NOEL

An Authorised Memoir

as told to

Terry L Probert

PREFACE

"If your actions inspire others to dream more, learn more, do more and become more, you are a leader."

*John Quincy Adams Sixth President of the United States of America

In writing his memoir with him, I have tried to help Noel record the important moments of his life in the tractor and machinery industry as best he remembered them. I believe he understood just how much time he had left and did not want to leave this world without leaving a record of this part of his working life behind.

The cancer he had been fighting over the last four or five years had begun exacting its toll and, often answers to my questions became interrupted by another important fact he wanted me to describe. A keen participant in the process, he would review and correct every chapter I sent him. Correcting a statement here, or changing the wording of a sentence to better represent his sentiment.

Where possible, I have researched facts and double-checked statements with his colleagues. As in everything he did, Noel wanted this to be his story, and every opinion expressed, his own.

My personal journey throughout this process has had its moments of enlightenment, joyful memory and extreme sadness. When Noel passed away in 2021, it was like losing an older brother, and I'm extremely grateful he allowed me to become that close.

Terry

NOEL   An Authorised Memoir

FORWARD

"One of the most positive people I ever met."

I first met Noel Howard when, both in our early 30s, we were sitting by a hotel swimming pool in Michigan, USA not far from Detroit watching actor Tony Curtis and his family enjoying themselves in the water. Tony Curtis was playing in a stage production in Detroit and we were there to see our respective bosses at the Michigan HQ of Ford Tractor Operations.

Growing up in post war Australia, Noel had to leave school early to run his father's farm, volunteered as an ambulance medic, went to night school to qualify as an engineer, found time to get married to Joan at age twenty-one. Joined Massey Ferguson then Ford and became the youngest country manager in any of Ford's affiliate companies worldwide. Twenty-six years, four children and five successful farms later Noel spent a short time running J.I. Case, before retiring from the equipment business in 1994 to become an extremely successful entrepreneur.

Always seemingly able to fix the "unfixable", I recall the look on the faces of Qantas ground staff in San Francisco Airport who regretfully refused to allow our Sales & Marketing VP to check in for a flight with me to Melbourne. The VP's secretary had forgotten to arrange for an Australian visa! One phone call to Noel and 30 minutes later we were on that Qantas flight.

An engineer and farmer, Noel always fought hard to defend Australian farmers' interests, particularly when future product design was involved. Sometimes those battles and the strength of his convictions did not always win Noel friends back at corporate HQ. Nevertheless, as I know this account will show, he generally persevered...."

Bernard Sarfas
Staff Director
Ford New Holland

Tact is the ability to step on a man's toes without messing up the shine on his shoes.'
Harry S. Truman

CHAPTER ONE

Seventeen minutes into United Airlines Honolulu to Auckland Flight 811, on the 24th of February 1989 is a defining day in the life of Rhoden Noel Howard, a time when the Gods decided there was much more for him yet to do.

Completing a scheduled round of meetings with Ford New Holland's North American Dealers and another round in Honolulu, the past week has been a heavy haul. However, Noel is looking forward to the next two days in New Zealand, meeting with Norwood's management team in Palmerston North and doing it all again. Tired and keen to be standing on Australian soil again, he checks the time, it's just after 1.00 a.m.

Flying has been in his blood since childhood and remembering the first time he sat in the front seat of a Tiger Moth brings a smile. It only seems like yesterday he had wind from its propeller tugging at his hair under a borrowed flying helmet. The smell of hot oil floods back. Instantly, he is nine years old, his senses alive with each vibration and every response as the tiny plane responds to incremental movements of the joystick coming up from the floor beneath his feet.

Warmed by the memory and looking at his watch, he sets it to New Zealand time and thinks about calling home before boarding. Making a quick calculation he dismisses the idea; Joan would be in bed.

Thinking tonight's flight was a bit of a milk run for United Airlines Boeing 747, N4713U. Noel knew the aircraft had left New York for Los Angeles and this was a refuelling stop over in Honolulu. Only another two days before he can leave Auckland and then on to Sydney's Kingsford Smith Airport, catch his connecting flight to Albury where he can find Joan, pick up his luggage and drive home.

The boarding call cuts above the chatter in the room and breaks his reminiscing. Finishing his beer, he checks his boarding pass. Standing, he shoulders his jacket, grabs the handle of his briefcase and begins making his way to gate lounge seven.

Again, he finds his mind wandering and tries to break free of any memory of FORD's New Holland buyout and the resulting unpleasantness suffered by so many of his friends. However, no matter which way he looked at it, it seemed his treacherous Ford colleague had been duped, by what Noel perceived to be an incompetent New Holland Australia's management team. This revolving door of thought was getting him nowhere, but he would have time to go over every line of Gor Cowl's offer to relocate to Dallas when he was on board.

NOEL   An Authorised Memoir

Flying to the United States for business had become routine and boarding was just another ritual he didn't have to think about.

Dodging his way past a throng of people farewelling loved ones, he thinks about his own children, wondering about their careers. Others are shuffling forward, all eager to find their seat, hoping to sleep their way to New Zealand.

An earlier thunderstorm had washed everything in its path and the airport still wet from rain, sparkled. Through the windows he could see the colours of his United Airlines Flight 811 reflecting in pools along the tarmac. Hazard lights from the numerous airport tenders flashed amber spears into the night as they scurried in and around waiting aircraft. The headlights of a tender flashed along the portside wheelset of the 747. He had always marvelled at the advances in aviation and his mind flashed back to those Boeing executives he had met during his Harvard days. One of them could even be captaining this flight, but the chance of that would be remote, and he dismissed the idea.

Tossing his jacket into the crook of his arm, he showed his boarding pass to the gate steward and started toward the forward door. Looking at the details and memorising his seat position again was just another habit. He knew where he was sitting, but having done this so many times it had become a reflex action. Smiling at the flight attendant and thanking her as she took his jacket, she pointed him toward his seat.

As a senior executive within Ford, international flying had become no different to the daily slog of any other commuter on their way to work. Even now with Ford Asia Pacific, the pressure of travel remained.

Settling into his seat and noticing his reflection in the window, he wondered if the silver around his temples had come with turning fifty, or was it from the constant demands of work. Chastising himself for being vain, his priority was to become comfortable, complete meeting reports and settle in for the long flight to Auckland.

Flying around the globe started when he first joined Ford Tractor Operations Australia in February 1964. This flight, like those before it, had become just another part of life. He couldn't remember the first time he started using a pilots' case as a briefcase, but it was ideal and could hold all the files he would require on trips like these. As was another habit whenever he could, tonight he strapped it into the empty seat alongside him. Another habit he had maintained from the first of his Massey Ferguson days was to complete his expenses claim, opening the case he took out a blank sheet and began collating his receipts just as he had with hundreds of similar forms before it. Stapling everything together and tucking the file into its pocket, he started thinking about reports he had yet to write.

Believing he had been lied to by Gor Cowl, Ford Tractor's architect for the smooth transfer of personnel and assignment of corporate responsibilities of staff from both New Holland and Ford Tractors, Noel began listing the facts that would become the backbone of his report. However, even though this was crucial to smoothing the transition of both Australian operations working together as Ford New Holland, he had another more personal set of facts to consider. Cowl had made him an offer to relocate, taking up a position based in Dallas Texas.

Hearing the familiar thump of the front cargo door closing, he flipped to a new page in a spiral bound note book and looked at his watch. They were already two minutes late. The plane moved back and stopped; he tapped his pen against his teeth thinking things through and listening for the tug to uncouple. Hearing the engines spin up he put his pen down, knowing take off wouldn't be too far away.

Noel waited while the big jet turned onto the runway and once airborne, waited a few minutes before beginning to resolve the strengths and weaknesses of his two employment opportunities. Ford had made a very generous offer that would require him taking a position based in Dallas.

Case IH however, had been searching for someone to create order out of the mess their Australian division had found itself in. They knew of the high regard Ford's inner circle had for him and, Noel Howard was their preferred candidate.

Honolulu Time: -- 02:08.00

The beep of the fasten seatbelt light going out told him that ten minutes had passed since take off, he paid no attention to it and went back to concentrating on the task at hand. His seat belt was still in place and he thought about releasing the clasp, but wanted to finish his point in the against column. Case IH was an unknown.

Believing he had been betrayed by Gor Cowl clouded his thinking and Noel doubted they could ever resolve things enough to reach a point where he could be civil to him. As senior managers within Ford New Holland, their paths would cross often and it would take more than an apology from Cowl, for any middle ground to be found.

He put his pen down, closed his eyes and pushed those thoughts to one side. He loved working for Ford and be damned if the treachery of three New Holland men should change that.

Leaning his head against the rest and thinking a move now would be the first change of employer in over twenty-five years, he knew weighing Case IH's offer would be difficult. Managing a major Ford New Holland territory on the west coast of USA would be great for his career within Ford, but it too had risks. A move like this required close examination.

Honolulu Time: -- 02:09:09

His ears began pulsing with pain, all around him air rushed past tugging at anything not bolted in place. The noise consuming everything, the cabin air thinning and making it difficult to breathe. Believing a bomb had exploded, he turned back to see a hole the size of a bus missing from the starboard side of the fuselage. A tremor buckled along the floor and an orange glow filled the cabin. Focussing through squinting eyes, he could see three rows of seats in Business Class were missing. Screams coming from terrified passengers cut through the roar, displacing any ringing he had had seconds earlier. Wondering what it was he could see in the amber glow from the burning engine he realised the forward flight attendant had fallen and was now sliding along the aisle, her hands flailing, trying to grab onto anything. Then as her hand latched around the metal of a seat base it stopped her slide and being sucked toward the hole in the side of the plane. Slipping out of his belt and steadying himself he moved toward her. Bracing against the pull of wind swirling between the forward bulkhead and the void, he managed to grab her. Dragging her into an empty seat, he clicked her belt into place.

Insulation debris wafted in clouds and the decreasing cabin pressure had him struggling for breath. Seeing the oxygen mask waving in front of him he snatched at it, tugging it to release clean air. The mask smelled of aging plastic and without air flow, he discarded it. Figuring they would now be falling through a bit over twenty thousand feet. If this plane was going down, these would be the last four minutes of his life.

Over the screams coming from terrified passengers, he fought his own fears, and searched for an inner calm. He looked over at the injured Flight Attendant, Mae Sapolu, and knowing he had done all he could for her, pulled his own seatbelt tighter. To him it had been a logical move, if they found his body at least the seat would identify the occupant. Cowl and his cohorts could go to hell. Now, with pictures and memories from his childhood tumbling through his mind, Noel only had thoughts for Joan and their family.

CHAPTER TWO

Honolulu Time: -- 02:10:00
   On the flight deck, Captain David Cronin declares an emergency, begins procedures to descend to an altitude where air is more breathable, and sets the aircraft on a return course.

Despite the chaos around him, Noel's mind was rolling through his memories stopping at a story and the photos his mother had shown him from the 5th of January 1939.

The Howard home bustled, filling with sounds of a baby coming into the world. On meeting their third child, Finlay and Lillian Howard could never have imagined the heights their son would climb during his career. All they could do for now, was to give him a name.

After much thought, Rhoden Noel was chosen. However, it was not long before they realised his siblings were struggling to enunciate his given name. Lillian discussed their difficulty with her husband and together they decided calling him Noel would be easier for everyone.

Finlay and Lillian Howard were married in Kiewa's Anglican church during the spring of 1933, and to much joy, their first child Graeme came along the following year. As Fin had qualified as a baker and pastry cook, his father-in-law

Walter Quonoey suggested Kiewa would be a good place to set up a business. This appealed to the newly-weds, as the same man had offered them land alongside the bakery on which they could build their home.

Hearing that houses were being demolished to make way for the spillway at the Hume Weir, Fin made enquiries of the authority and secured a suitable house. With help from family and friends he relocated their new home to Kiewa. Now the town had a bakery and before long, Fin Howard's bread, pies and pastries were being delivered to farms and homes along both sides of the river and throughout the greater Kiewa Valley.

As a small child, Noel remembers sitting on the bench alongside rows of yeast buns. His mother allowing him to help, always encouraging him. Lillian explained, that by painting the dough with the sticky sweet sugar syrup would make them gleam golden-brown after baking. Something that would make them more tempting to people's taste buds.

Whenever a customer came in only to buy a loaf of bread and just because Lillian had asked, often they decided to add a couple of buns to their order. In this way she was showing her children how important it was for people to be drawn to the sight and smell of everything in their bakery. By achieving that, everything they made would be easier to sell. Each product was responsible for making a profit and by doing so, they were securing their family's future.

A farmer's daughter, Lillian knew the need for cashflow and it was something she made sure her children understood.

From an early age, every Howard child learnt the importance of, "glazing the bun".

Noel knew Kiewa as a small village of around two hundred and forty people. The surrounding farms were a mix of graziers, tobacco growers, dairy farmers growing pigs, along with a good number of mixed farmers. Everyone getting by on small holdings. He was never bothered by class or race distinction and everyone in Kiewa was doing their best to assist with the war effort. Even though it was part of the hydroelectricity plan, Kiewa was yet to be hooked to the Victorian electricity grid. So, growing up with the smell of kerosene lamps, was something kids like Noel thought of as normal. Feeding the bakery's wood fired ovens was a never-ending chore that would later become a metaphor for his working life. If ever he felt drained and needing a break from the troubles he faced, a vision of his father swinging an axe would flash into Noel's mind. Tired and weak from long hours as a baker, there would always be a furnace for Fin to fuel. That image and others like it, always reminded Noel to press on.

The Howard home, with a young boisterous family brought up on the smell of fresh bread and pastries coming from the bakery, made a perfect place for Fin's young family. Lillian was ever diligent, completing her household chores and raising her brood around the demands of their business. However, having family members serving in the armed forces, she enjoyed spending a few minutes each day reading the Melbourne Sun aloud to her children. Believing it important for their future by keeping them abreast of local news and Australia's commitment during the War.

These were times when skills outside those of the bakery, made getting by easier. Fin was an accomplished gardener and although his vegetable patch provided greens for both bakery and family, it was his skill as a fisherman and hunter, that provided much needed protein.

Having a body of fresh water to work with, Fin would direct Noel where to swim to in the Hume Weir to fix his gill nets. The family always had fish swimming in a cage in the Kiewa River and not all the fish came from the weir. Fin had often set drum nets and set lines along the Kiewa River. It was a time of rationing and a man needed to be resourceful to feed his family.

Starting in the one roomed Kiewa School before his fifth birthday, Noel was one of the smallest there. Schoolwork did not come easy, but with encouragement from his mother and siblings, Graeme and Yvonne, Noel's grades were above average. The teacher, like many of his colleagues in small schools, was responsible for all the grades from one to eight. For Noel, it exposed him to many lessons he would not have had in a bigger school and allowed him to see pathways available to someone with a good education. It is hard to remember the moment he first set his heart on becoming a doctor, but knows it was from an early age. For the next few years Noel toiled at study, always trying to improve his chances of getting into university and studying medicine.

Noel discussed his prospects with his mother and she counselled him about the amount of work involved. Telling him that only by getting the best grades possible, would he be able to make it happen. Driven by those thoughts and her encouragement, he ensured his homework was done on time.

Outside of school hours, with his homework and chores completed, Noel was free to visit his grandparent's farm. The farm offered an escape where he could pretend to be Hop-a-long Cassidy, or any other radio cowboy when he was riding Daphne, the farm's horse, and bringing the cows in for milking. It was always somewhere he could learn something new. Whether it was fixing a fence with his grandfather, or driving the tractor, being in the open air and responsible for something bigger than himself made his heart sing. Being someone more than just another Kiewa school kid, was a great feeling.

Spring had its hazards and as part of his chores, Noel was responsible for walking his father's racing dogs after school. Scanning the skyline and the trees that lined the road for nests, he was always on alert for attack from the air. Kiewa's magpies swooped from behind and in silence, only at the last minute could you hear them pecking. That last second and with their beaks rattling like a machine gun as they snatched at the hair of their victim, was capable of driving fear into the bravest of souls. However, magpies were only one risk from the air, plovers nesting in clumps of native grasses on the flats seemed, only too ready to swoop anyone nearing their territories too.

As a kid he had to be diligent, keeping his grip on the leads, his mind on the path and all the time watching the skies, preparing to defend from attack. To add another element of excitement to the task, Noel had to make sure he found a place to walk without treading on a red bellied blacksnake.

School holidays however, were a time for adventure and riding a horse trailing a couple of packhorses was one of those. A time when opportunity, combined with responsibility would forever imprint a sense of resourcefulness within him.

His Uncle Des had been talking to Fin and Lillian, asking if it would be okay with them if Noel joined him on a ride from Tawonga to the Wallace Hut on the Bogong Plains in Victoria's High Country. This was an annual event where tough and experienced cattlemen rode out to muster, yard and mark their calves, prior to bringing the cattle down to pasture for the winter.

Before dawn, with Daphne saddled and two horses packed with provisions for twelve men, a young Noel Howard checked his gear before setting off. His grandfather had shown him how to saddle a horse a few years ago. All the while talking to the animal and instilling in Noel the importance of checking his own equipment, a habit he has followed throughout life. This morning's adventure beckoned and with a boost from his grandfather, Noel dropped into the saddle, his right foot fishing for the stirrup. Weeks of thinking about the trust Uncle Des was placing in him had made his stomach churn. Sometimes daydreaming with excitement and sometimes fear, planning forward to today. From that first tentative discussion when Des suggested the idea, Noel understood the importance of this event; knowing it would shape not only how others would see him, but how he would see himself.

As first light broke sending spears of orange and red across the sky, he noticed himself sitting a little taller in the saddle than he had on those occasions when he and Daphne brought the cows in. With his back a little straighter and looking down at his mother, who had walked from the bakery to see him off, he smiled.

With dawn casting long shadows, it was time to move out and knowing they were riding east, Noel pulled on the rim of his hat to shade his eyes. Taking the lead rope tethered to the packhorses from his grandfather, he nudged Daphne forward with his heels and called for the pack to follow.

Hearing his mother calling to be careful as he rode to join the others, he was unaware that this adventure would become a metaphor for his life, a time where through example, he was shown the path to becoming a man. By watching and learning, this band of weathered men would teach him where to look when crossing a stream, how to find a trail and that preparation was as important as the task itself.

The group had gone on ahead, leaving him with a rough map to follow, assuring him that if he made sure of his land marks and assessed every crossing, he would make the hut before dark. Often double checking his position, Noel, Daphne and the pack horses followed on. Nervousness and self-doubt found a way into his thoughts whenever he had to choose where to cross a stream, or where to follow a trail.

After hours in the saddle, he was sore and tired, but hearing cheers from his uncle and grandfather as he arrived, his spirits lifted. Somewhere inside him, he knew today was another life lesson learned. He had brought the provisions through as planned.

At night the campfire provided an opportunity to listen to stories of the high country and he marvelled at the joy his companions found in the satisfaction of an honest day's work. Every rider had a story, some were variations on the same theme, but none of these men could be seen as a bludger and all of them possessed a positive outlook.

Even the fire had a life, built high at twilight, it would flame and crackle until dropping back to coals, making way for cooking duties. At dinner time, apart from the clatter of utensils on steel plates the camp would drop silent, staying that way until the meal was finished. Exhausted, Noel crawled into his swag and lay there listening, as stories of fishing, breaking brumbies and working on the Kiewa Hydroelectric Scheme swirled in his mind, filling his night with dreams.

When the ride was over his body had changed, he had become stronger and being exposed to the elements, every muscle and sinew had tightened to the task.

Hugging him hello, Lillian squeezed his biceps and pushing him away she looked into his eyes telling him how much she had missed him.

Brushing a tear from her cheek, he noticed his back straighten as he heard her say she had seen him off as a boy and was now welcoming him home a man.

Even as much as he found something special about riding with a band of men into the High Country. Noel, like a lot of young Australians knew he wanted to carve out a life less ordinary, he had aspirations of becoming a doctor. However, life as the baker's son in Kiewa, made studying medicine at university a dream for later life.

CHAPTER THREE

School work was important and being an eager student, Noel tried to complete his lessons and homework on time. However, as much as he loved winning accolades from his teacher Mr. McKenzie, it was spending time on the Quonoey dairy farm that excited him most. It was there as a child he met George Reid, the proud owner of a Tiger Moth.

Why George took an interest in Noel he can't be sure. Maybe it was because he had identified the boy's inquisitive nature, forever asking questions about machinery, why it was needed, or how it worked? Whatever the reasons, George took a shine to Noel and became a close friend and mentor up until his death.

The small biplane appeared fragile and birdlike, George encouraged Noel to run his hands along the fabric, feeling the ribs of the wings. Explaining their shape and how it was aerodynamics that gave the craft lift. He told Noel to move the aileron, and using his hands showed him how the movement of air passing over the wing would be interrupted, causing a pressure shift above and below it, making the plane change its yaw. George impressed upon him how important it was for a pilot to know how his machine would work, understand how it was designed and what it could do.

They tested the rudder and looked for any damage to the flaps. All the time Noel was asking questions and in his patient manner George was answering them. They looked at the tyres and landing gear, as George stressed the importance of looking ahead, planning a strategy for every disaster and being aware of your surroundings. It was important for a pilot to always remember that if you were up, you were always coming down. Therefore, it was best to make sure you planned.

He called it, "looking over the horizon".

Explaining that a flyer should be weighing up every option along the flight path, he stressed that no-one could rely on every landing to be a soft one.

At the front of the Tiger, George stood behind the propeller and ran his hand along its leading edge, questioning the boy as to why a pilot might do that, scruffing his hair with approval, when he answered that they would be looking for damage.

That night, Fin told Lillian about George offering to take Noel with him in the Tiger Moth the next time he was flying to inspect their property at Balldale near Corowa in New South Wales. Although she was cautious and thought the idea crazy, she wanted her son to experience all life could offer him.

To George flying was a hobby. However, he would smile and often justified it by saying the aeroplane was a tool the same as a tractor, or any other piece of farm equipment, although a bit of prompting would have him agreeing it was more fun than most of the other tasks around the farm.

Hearing the four-cylinder engine splutter into life found him tingling with excitement. Even today it's a sound that takes him back to his nine-year-old self.

Taxiing downwind to the end of the strip, Noel heard George's voice crackling through the speakers in his flying helmet, explaining to him how pushing up into the breeze would make take off shorter and safer. Every stone and bump seemed to travel through the frame and into the seat. The fabric over the spars in the wings buzzed and the wires whistled, he felt the change in vibration and the tail wheel free spinning as they built speed. Within seconds the seat pushed up as they lifted off. Free from the runway the plane came to life, as if it were a dragon woken from its sleep and free of its chains. Then as George banked to the right, he began shivering with excitement, elation tickling its way into every sense of his body. The landing wheels slowed to a stop and the Tiger found a new rhythm as it climbed above the trees.

All the while George's voice coming through the helmet explaining every movement of the controls and urging Noel to look at the changes to the wings and feel the movement they made. On the return journey, George, confidence building in his protégé, suggested Noel take the controls. Instructing him how to move the stick and asking him to repeat the instruction was George's way of making sure Noel was taking mental notes of what he was doing. Experiencing how those changes to the controls altered the aircraft's movement added even more joy to their flight. Too soon Noel recognised the valley with Kiewa's houses in the distance, praying George would want to go around and do it all again.

When they buzzed over the house to announce their return, everything seemed new. It was as if he was seeing Kiewa for the first time. Looking down on his mother bringing in washing and his father crush out a cigarette while he waited by the baker's van, has become a memory to savour.

Standing beside his father listening to George telling Fin Noel had a natural appreciation for flight and saying, if he and Lillian were agreeable, he would like to teach him to fly, making it a moment that filled him with pride.

Pushing the aircraft back into its hangar that afternoon became the beginning of a love affair with flying that would stay with Noel for life.

Honolulu Time: 2.20.00

The screaming subsides, the mood of the passengers going from panic to sombre acceptance.

In a flash he remembered riding in the baker's van. The smell of its gas producer would again become familiar when he started learning to weld during his apprenticeship as a motor mechanic with Charlie Milthorpe.

For Noel or any of his siblings, riding with Fin while he made his deliveries throughout the valley was a treat, each child guarding their turn until the novelty wore off.

The Howards owned a dairy farm near the Bonegilla Migrant Camp. However, it was not for Lillian. She was unsettled, wary of the great number of men women and children flooding into the disused army barracks. A world's people displaced and with strange accents coming to settle in her part of Australia until they moved away. She had become frightened of them when some of them began raiding the farm's vegetable gardens, or pilfering milk and cream from the dairy.

The culture Lillian knew was changing, and under no circumstances would she move to the dairy farm. In Kiewa she had everything that made her happy, they were close to her parents and it was somewhere safe for her young family to flourish.

Noel's grandfather Walter ran a Stock and Station business, alongside the Quonoey farm. Over the years their business changed, harness giving way to spark arrestors and driving lights. It was an exciting time, changing farming practices were driving advancements to the type of farm machinery being offered and rendering some stock obsolete. Heavier baler twine was replacing binder twine. Horses no longer pulled sheep feeders and fertilizer spreaders, but tractor mounted Power-Take-Offs provided auxiliary power.

Witnessing the improving mechanization of dairy and horticulture, Noel observed that wherever there is change, there is opportunity and it became a skill to set him apart from his peers in later life.

Ever present were small complimentary notebooks sitting in neat stacks on top of the wide red gum counter and these were always a source of interest. Crammed with information that was alien to someone from the city, they became an ever-useful constant in the breast pocket to the man on the land. Holding gems of information like tables to calculate bushels to bags, how to age sheep and cattle by looking at their teeth. Formulas for estimating wheat, or barley to the nearest hundredweight, or ton. Best of all for a Quonoey grandchild, it was full of blank pages and complete with a pencil that slid into the spine pocket, it was a gift that could entertain them for hours.

The Quonoey business represented several livestock and land-sales agencies. Visiting commercial travellers calling to a regular itinerary were always welcome as they would bring news of new opportunities products and farming trends. One time an insurance agent calling as expected had driven to Kiewa in a new 1948 Holden and it didn't take long before a crowd began gathering around it.

Prime Minister Ben Chifley, when unveiling it to the Nation, billed it as Australia's car for the people and demand soared, making spotting a Holden in a country town a rarity. However, such was Walter Quonoey's standing with the National Mutual Insurance Company, that their representative offered Noel and his siblings a ride in the new machine and after some cajoling, everyone had been offered a drive.

When it was his turn, Noel slid across the glossy front bench seat and breathing in the new leather, thought this car very different to that of the baker's van. Reaching down, his fingers searching for the chrome plated knob to release the seat adjuster, he pressed down and using his heels pulled the seat forward so he could reach the controls. The clutch was smooth and light, the gear lever glided into first and easing the clutch out they were away. Driving a new car for the first time is a memory he has never forgotten. Comparing it to that of the machinery on the Quonoey farm, became a reference he would return to often during his career in the farm machinery business.

Adventure was everywhere and Noel was often found helping on their neighbour's farm. While there was always a fence to mend and cows to milk on the Quonoey farm, it was helping with mowing, raking and baling on the Bartell farm that brought an opportunity to earn a decent wage. Driving a World War 2 Jeep towing a rake during spring, brought its own version of flying, only this time the wind in his face was slower and what was filling his nostrils was fresh mown hay.

Mr. Bartell believed in offering equal pay to whoever was doing the work, so if a man earned two pounds for a day of baling, then his neighbour's kids would be paid the same amount.

Walter Quonoey, Noel's grandfather had been the same, encouraging his grandchildren to challenge themselves and showing them how to improve with each repetition of a task.

Without them knowing it, he was demonstrating how showing trust in someone, would help them to understand responsibility. Instilling a belief that sometimes when mistakes are made, you own up to them, fix the problem and get back to work.

CHAPTER FOUR

Although his brother Graeme attended Scott's School at Albury and his older cousins had attended Wangaratta Technical School and lived as boarders; for a twelve-year-old boy living at the Anglican Hostel was still daunting and yet at the same time, exciting. The thought of being left to figure things out for himself was a challenge worth looking forward to and having been regaled for hours by Graeme's stories at Scot's, he knew how the hostel worked. Besides, he knew his aunt's brother Ted Barton had started at the hostel a couple of years before and when Lillian had asked, Ted promised that until Noel had settled, he and his mates would look out for him.

Lillian had enrolled him well before his twelfth birthday and the day before he started, the family were driving him to his accommodation. Yvonne started cracking jokes, trying to unnerve her young brother, only relenting when she was shushed by her mother. While driving Fin occasionally flicked his eyes up to glance at them in the rear vision mirror and was doing his best at keeping a straight face, only to break concentration and join Yvonne in her teasing.

Lillian, worried by the gossip and rumours about some of the unsavoury characters who often found work in boarding houses, had discussed the possibility of encountering paedophiles with Noel. Fin had offered his son advice on what he should do if Noel found himself being preyed upon, but Lillian soon squashed any of that sort of talk. Instead, he listened to her advice on things to look out for, and how to avoid such people.

Lessons at Wangaratta Tech were different to those he had experienced before. In the small Kiewa school, everyone was in the same room and the grades interacted. Here it was structured, everyone at the same level studying the same things and for the first time he was in a class of a similar age group. Without the distraction of overhearing an older child being instructed by Mr. McKenzie, he found himself excelling, always pushing toward his dream.

Making friends during those first weeks in the hostel, started building lifelong associations and only one boarder ever gave him any trouble. Noel remembers him as not being much of a bully, only ever being on the receiving end of him once. After leaving Wangaratta Technical, he lost touch with him, hearing years later that he had been murdered.

Being able to look after yourself, is what you learn when you have a brother five years older than you. Roughhousing, always on the edge of going too far had taught him how to deal with those who were bigger. Sometimes they were stronger, even faster, but never smarter, and backing away from a fight is not in the Howard DNA. If the cause was right, the battle was on, a strength Noel would call upon, either when pressing his point, or defending his staff in later years.

For most of his free time, Noel could be found in the study hall where House Master Ray Clarke would be supervising students who were swatting for exams or doing homework. Ray ran a good home, allocating tasks to those in his care and making sure of alternating everyone's responsibilities, so that no one got into a rut by doing the same thing every day.

Having your homework up to date and achieving good results meant there was an opportunity to go home for the weekend. Wanting to get back to Kiewa was one thing, being able to another. Constant planning and scheduling found Noel catching a train to Wodonga, then changing onto the Huon to Tallangatta line where his family would meet him. Sometimes when money was short, he would begin the trip by hitchhiking to Wodonga. Getting back to school on a Sunday was easier as he could often catch a ride with the road tanker from the Model Dairy of Kew.

Model Dairies discouraged their drivers from picking up hitchhikers, but giving a ride to friends, or the child of a client was another matter and at the discretion of their driver. Noel saw this as a lesson not learned in school, but more of a case of, *who you know is as important, as what you know.*

Living in the Kiewa Valley with its abundance of farmers who bought their bread and other produce from Fin and Lillian, meant that most people knew Noel. However, whether he deserved it or not, he had acquired the reputation of a thrill seeker and was suspected as being the pilot who flew George Reid's Tiger Moth under the Bethanga Bridge.

NOEL   An Authorised Memoir

Being part of a popular family had its benefits, but there were downsides too and, because gossip is gold in a small country town, Lillian and Fin did their best to show their children how to separate their public and private lives.

When a question from the tanker driver came, asking if he was the kid crazy enough to fly with George under the bridge, Noel was skilled enough to divert the enquiry without lying. Instead, as they passed what is now the Rocla concrete site, he pointed it out, telling the driver how he, with his grandfather tailing behind in the horse and gig with their supplies, had driven a mob of cattle from Kiewa to Wodonga. Trying not to boast, he was just recalling the small events that happened as he and the dogs pushed the mob along. His strategy worked and supposing the driver didn't need his ear ringing with the stories of a high-school kid, was pleased that the question of George, the bridge and the Tiger, never came up again.

Back at the boarding house he flopped onto his bed and folding his hands behind his head, he was completely unable to stop smiling. So, someone had seen George's Tiger fly between the pylons of the bridge. It was probably crazy at best, or downright dangerous and stupid at worst, but the rush of adrenalin and the feeling that came as he pulled the stick back wheeling skyward was worth it. Hearing about feats of daring by Biggles and his chums on ABC Radio, may or may not have inspired the stunt. However, with George handing over the controls and encouraging him, Noel asked himself, who at three thousand feet in the mid-afternoon wouldn't want to have a bridge, a body of water and a Tiger Moth at their disposal. Besides, what competent flyer wouldn't want to give it a crack?

He and George had flown over, alongside and across the bridge often and, on every flight, Noel had taken notice of the way the wind moved over the water. Having an interest in engineering he was always interested in the bridge and its construction. Wondering how the designers worked out the span width and the desired height above the water. Now being faced with the opportunity he didn't have to think if the Tiger would fit. It never even entered his mind to think about his airspeed, or how low could he fly above the water without risking the aircraft? The weir was low so he knew he had room and dropping to a few feet above the water he lined up. Now there was no time to change his mind and in the blink of an eye they were through and climbing to his right, gaining altitude.

His biggest challenge now, was what would happen when his parents found out?

Lillian was not amused; she was worried that an antic so stupid in her eyes would bring shame on all of the Howard family not just Noel. Her remarks stern enough for him never to mention it to his family or anyone, the Bethanga Bridge becoming his to relive in his mind and a personal memory for him to enjoy for years.

The weir and the bridge had been part of his life, either swimming, fishing or driving across the bridge, he was at home near water. Fin would often take him fishing and with the lines in the lake, he would show Noel how the wind would push small ripples and waves across the surface. From the size of the ripples and their direction he could see how strong the breeze was. Noel learned that if it was important for a fisherman then it was useful for a flyer too.

Closing his eyes, he relived the moment. George swinging on the propeller when Noel yelled contact. The engine firing, bouncing from one cylinder to another at low speed, every splutter shaking the fuselage and wash from the propeller swirling dust, stinging his face.

George and his brother Gordon lived with their mother at their beef cattle farm on the Kiewa Valley Road. George maintained he saved time using the Tiger to check the cattle at Balldale, but being an avid aviator, it also offered him a chance to enjoy his hobby.

Every flight with George sitting behind him and offering instructions via the radio in his helmet, had been improving Noel's skills. He soon reached the stage where George was confident in his young friend's abilities, handing him the controls from take-off to landing, the whole flight. For some time, they had been letting the Howards know they were home by doing a loop-the loop manoeuvre over his Kiewa home, then a long lazy circuit over the town and back to the Reid farm. Fin would be arriving just in time to help push the plane back into the hanger, have a quick yarn with George and head back to Kiewa in time for dinner. At the table everyone would drop their eyes as Lillian would caution Noel about the dangers of his aerobatic flying, but she never forbade him from going with George.

He sat up and thumbed through his science text book hoping there might be something inside to take his mind away from flying under the bridge. It was proving a difficult task.

He had, as always on the return journey, taxied to the end of the rustic airstrip, turned into the wind, gunned the throttle and eased into the air. Flying to their preferred altitude of three thousand five hundred feet always gave George time to relax and look at the countryside.

On this flight like all the others, Noel knew George would soon be reaching into the pocket on his right, fishing around for a bottle opener and flipping the top off a long neck. They had worked hard and George always said he had earned it. Noel didn't need to worry, from now on, George was a passenger.

After twenty minutes, the weir with Bethanga Bridge reflecting in its surface was almost below them.

'What do you reckon?' George's voice drawled through the speakers in Noel's flying helmet.

'Sorry?' Noel said.

'The bridge, reckon you can make it between the pylons and fly beneath the bridge.'

'The weir's down, so yeah.'

'Better go for it then.'

Now on his hostel bed with his head on the pillow and eyes closed, Noel was reliving every minute of their adventure.

NOEL   An Authorised Memoir

Pushing the stick forward and to the right the little plane started descending. He levelled out, squared the Tiger up and knowing he had enough distance between the pylons, all he had to concentrate on was staying close enough to the water without dipping his wheels into it. Having flown the machine so often it was now an extension of him, there was nothing more to think about. The shadow from the bridge flashed over them and they were through.

Pulling back on the stick and adjusting the engine for the climb, he held the aircraft on a long lazy arc until they were back on course. Only then, did the gravity of what they had done begin to dwell in his mind. The thrill and excitement mixed with a foreboding, at some time he would need to explain it to his mother, in fact the sooner he told her, the better.

Lillian's response was not exactly as he had expected. She found the act to be laced with stupidity. Warning that under no circumstance should he mention it to anyone and that included family. Always conscious of maintain the family's good name she stressed how important it would be to his future and becoming a doctor. Laying there, he thought she shouldn't have been concerned, from an early age he had learned the value of discretion.

The next three years flashed by in a rhythm of lessons and journeys home playing to the beat of the seasons. Each offering an opportunity to pick up seasonal work and always adding to his savings account.

CHAPTER FIVE

Holidays on the Mornington Peninsula in January was always a time for Noel to relax, reconnect with old friends, make new friends and become lost in adventures for a few days. Over the years his holidays with the Grants were to provide a turning point that would establish an enduring partnership to cornerstone his career.

The Grants were the Howard's neighbours in Kiewa and as orchardists and farmers the Howard children often helped during the fruit harvest, or hay season. The orchard had been sold and the Grant family had moved to Rosebud and every year from the time he was eight they had invited Noel to stay with them while they were there.

Travelling from the high country of Kiewa to Rosebud, was quite an undertaking for a lad not yet ten and travelling on his own. For Noel, he was excited by the prospect of adventure and thoughts of getting lost or running into trouble never crossed his mind. It was like he was one of his storybook heroes heading for the big smoke, seeking the fun of the seaside and bound for adventure. What could go wrong? All he had to do, was follow the notes his mother provided and he would arrive at the Grant's Rosebud home, safe and on time.

The night before his first trip south and laying on his bed staring at the ceiling, Noel could feel his stomach churning. He tried to ignore the thoughts that began swirling in whispering whirlpools of excitement and filling his mind. He squeezed at his eyelids, willing them to stay shut to help him sleep. When it did come, it seemed like he had only been asleep for a few moments before the alarm called for him to get up.

From that first morning until he was able to drive, each journey to Rosebud would begin with Fin and Lillian driving Noel to Albury, waiting on the platform with him until he boarded. Through the windows he would watch his mother waving to him as he made his way along the carriage until either the train moved away from the station, or he found his place.

Travelling on the Spirit of Progress was another experience of wonderment for a nine-year-old, as at the time this was the pride of the Victorian Railways and the most modern and fastest train in Australia.

The hiss from the steam and smell of burning coal filling his nostrils he marvelled at the shape and contours of this behemoth, thinking about how much his friend Bob Brown would have envied all that power. It was much more powerful than the steam engines used to drive the electricity generators and other machinery at the Kiewa Cheese and Dairy Factory. Riding the Spirit of Progress from Albury to Melbourne, a fast train with airconditioned carriages and filled with the prospect of adventure, this was the stuff boyhood dreams were made of.

Armed with a list of instructions of where to go and what to do, the first he knew that they were under way was watching his mother moving with the train. Dodging past other people on the platform. Walking at first, then pacing until she was running alongside the carriage window blowing kisses and waving him farewell.

Now moving at speed, the carriage smelled of warm cigar smoke, brewing coffee and perfume from the lady in the seat behind him. Noel felt for the ticket in his jacket pocket, it was still there. He looked at the list of notes his mother provided, she had given him a schedule of station stops, which ones were important to remember and those that weren't. Once the conductor had clipped his ticket, he knew he could fold his jacket up and use it for a pillow if he wanted to sleep, the ticket could stay there until reaching Melbourne's Spencer Street Station.

Watching the changing panorama from his window as the train moved and swayed with elevation changes, he took in everything. The mountains to his left changing colour with the dawn and yellowing fields of wheat waiting for harvest to start. As they raced into the lowlands past Wangaratta and onto the wide plains of the Goulburn Valley, how could he know that agriculture would play such a big part in his working life.

For now, all he had to concern him, was which platform to change to when the train stopped in Melbourne. There he would catch the train to Frankston and from there, it would be a ride on a Peninsula Bus Service coach to Ninth Avenue Rosebud and walking to the house where the Grants would be waiting for him.

NOEL   An Authorised Memoir

Having sold their Kiewa Valley Orchard, Frank and Cecil Grant had moved south, Frank to Rosebud and Cecil had taken up an apple orchard at Pakenham. Frank's home was large and right alongside that of his sisters May and Beryl. Most of Noel's holidays were spent with Frank and his family, however there were times when he would be accommodated with Frank's sisters, Beryl and May. A caring woman, May cut a serious figure as the district nurse, her are covered the peninsula and all the small towns and farms as far away as Flinders.

Noel's desire of becoming a doctor was known to May and she often invited him to accompany her as she made her rounds. After a short time, he became adept at spotting an ailment, taking blood pressure and applying bandages. May had been giving him a crash course in first aid and those early years of helping and observing became something that he found helpful when joining the ambulance years later.

The friendship with the Grant's grew past that of friends giving a neighbour's child a holiday at the beach, to one where Noel was accepted as an equal. These end of year excursions punctuated his growth from childhood through adolescence and on to becoming his own man.

When Fin and Lillian could manage it, they would stuff Fin's big, black, 1939 Packard with provisions and gifts, bundle the family in and head off for Enfield in Sydney's inner west to stay with her brother Geoff and his family for a couple of weeks.

This year the Packard, still powerful and spacious was pressed into service again because of what Fin believed to be rationing and religious discrimination. He had placed an order with the local Ford dealer and while waiting for his car to arrive, saw the Catholic Priest parking a new Ford, exactly the same as the one on his order. A few subtle enquiries led him to believe this to be the car he had ordered and, with the dealer being one of the Priest's flock, the dodger had made Fin's Ford available to the church. During these times, religious discrimination was rife in Australia and Fin's family were not of Catholic faith. Supposing it was the dealer's connection to the priest had made Fin angry enough to cancel his order and so the Howards were heading east in the Packard again.

The trip to Enfield was always broken with a stopover at the Murray Valley Travel Lodge in Canberra. These holidays were a time for Lillian to challenge her children to absorb as much of Australian life as possible. Believing by immersing them in the place of government she could show them how politics worked, she would take them to what is now Old Parliament House. Whether she did it through prearranged contact, or by luck, Lillian would find an usher or someone in security and coerce them into showing the family around the building. all the while explaining its function to an eager Noel and his siblings. If Parliament were sitting the family would make its way into the gallery to sit silently and watch bills of the day being debated.

NOEL   An Authorised Memoir

Across from Old Parliament House, The Australian War Museum held special interest for Lillian and she made it her mission to show her children how the many generations of her family and other Australians were remembered.

The Quonoey family are well represented in both World Wars and, as a custodian of that history, she made sure everyone knew it. Often, she would find a volunteer and ask them to explain the exhibits, even urging the unsuspecting individual to allow the family into closed off areas.

From Canberra the Packard would cruise east to Enfield where the Howards would holiday with Lillian's brother and his family. Geoff Quonoey was not without influence and without Noel realizing it at the time, his uncle was another beacon setting an example for him to follow. Geoff being the manager for ANA Australian National Airways and knowing of Noel's interest in flying, suggested that his nephew might enjoy an overnight trip to Brisbane in one of the company's freighters. With his parents agreeing, soon Noel was standing on the tarmac looking at this huge four engine aircraft and embarking on another adventure.

Sitting on the jump seat between the captain and co-pilot he could feel the thrust of the engines as the two men eased the throttle levers forward and sensed the rumble of the undercarriage as the moved toward the end of the runway. A lasting shove coming from the back of his seat, Noel tried not to smile as he waited watching for the moment the controls would be hauled back and the plane would climb into the air and soon be flying over the northern outskirts of Sydney. He listened to the conversation between the three men flying the plane while keeping a vigilant eye on the numerous switches, levers, wheels and gauges that adorned the flight deck.

Every movement of the fuselage responded to the captain's hands pulling, pushing and turning the open topped steering wheel, brought with it visions of an orchestral conductor at work. Everything moving, working in unison as the captain commanded.

The Douglas Skymaster was a big change to the open flying environment of the Tiger Moth and although there was no air-wash from an open cockpit, the noise and vibrations coming from four Pratt and Whitney 14-cylinder, 1,450 horsepower radial engines excited him. This was the future of flight in Australia and tonight he was part of it.

Geoff Quonoey reasoned that the Skymaster would be in the air for quite some time and, concerned Noel may become bored, had planned another surprise for him. In Brisbane an animal carry crate was loaded, addressed to his uncle's attention and emblazoned with the word MONTY. It carried a cattle dog pup. The captain's smile wide, he showed Noel, letting him know Monty was for him. However, the pup had to remain in its crate until they were well out of Brisbane and had reached their cruising altitude of 9,000 feet.

Eager to meet his new friend, Noel spent his time on the cargo floor and let Monty lick his fingers through the wire mesh in the crate door. Heading south from Archerfield and climbing Noel saw the captain moving back into the cargo area. He passed by, telling Noel it had been a long day and as his number two had control of the aircraft now, he was taking an opportunity to stretch out on the freight floor and catch up on some sleep.

The captain's snoring was the cue for Noel to slip the latch and meet his new friend properly. Out of his cage Monty bounded around the confines of the freighter, sniffing, searching, following his nose back to Noel, only to bound off again and return tail wagging, ready to play.

Being in the transport box may have got the better of the dog, and for one quiet moment. Noel couldn't believe what he was seeing. Monty stopped and began sniffing at the bed roll where the captain was laying. Then turning, half squatting, half cocking his leg like pups do when they pee, settled and started to empty his bladder, drenching the poor man's face.

Dragged from those dying seconds of sleep, the captain sprang to his feet, hands searching his pockets for a handkerchief to wipe himself dry. Continued making spluttering noises until he gained composure. Noel could feel the man's words as much hear them and picking the pup up by the scruff of the neck consigned Monty to the cage of shame. Captain, crew and passenger, remained quiet for the rest of the trip home.

Lillian, upset that the jaunt to Brisbane had ended that way, made Noel aware that he should have used some restraint and kept Monty in his crate. However, she was still able to laugh about it and because the captain had left his post, no more would be discussed. Geoff and Finn were left to share a beer and it always became a source of humour whenever the families holidayed together.

Sydney visits often included a train ride into the city and the family would then catch a ferry to Taronga Park Zoo, where they would picnic before returning the same way. On one of their city adventures Fin took Noel to meet a friend who owned Pratten Bros, a business that made calendars, distributing them Australia wide. Telling his eager young brood that there were good chances a Pratten Bros calendar would be hanging in every mother's kitchen during the 1950s. Fin's friend Basil Cameron offered to take Noel and his siblings to watch the printers setting type and they could hear pride in his voice as he described the action of the presses moving. To a young Noel Howard, it was as if their motion had been choreographed, the clattering and the swoosh of paper on rolls rising above the solid thumping rhythm, to him it sounded orchestral, industrial music at work.

Another treat was their visits to the Sydney Law Courts where a relative of Finn's practiced as a magistrate. This may have even been the genesis for Noel's younger brother, Barrie's interest in the law, practicing as a solicitor in later years and becoming well regarded in both his legal and local circles.

In Enfield and with their Sydney holidays behind them; the family would pack everything into the Packard, say their goodbyes to the Quonoey's and start returning home via the Canberra Travelodge.

Other than road trips to Sydney and holidays with the Grants, there were times when Noel holidayed away from the family. One of these was when in 1949 he attended The Lord Mayor of Melbourne's Holiday Camp at Portsea on the Mornington Peninsula.

The camp first emerged during 1943, as the idea of Sir Thomas Nettlefold, Lord Mayor of Melbourne. His original thought was to offer rural children from the Mallee District of Victoria a seaside holiday. The Lord Mayor of Melbourne's new incumbent Councillor, F R Connelly took ownership of the idea and with his predecessor's help moved the camp from its original canvas billet at Balcombe Army Camp Mount Martha, to their new permanent venue at the Port Franklin Barracks, Portsea. When Noel arrived in 1949 and because of corporate donations along with an army of volunteers working weekends to make it suitable, the camp was ready. Over the next few years volunteers continued converting and improving the accommodation and grounds providing a wonderful holiday experience. Something children would remember for life.

Embracing this time away from home and on his own, Noel enjoyed every opportunity offered. Rural children from across Victoria would be offered medical and dental checks that were otherwise inaccessible, either because of distance, or financial difficulty. Fortunately for Noel his background was a little more prosperous and therefore if any conditions were found during the medical checks no treatments were necessary.

After the train trip from Albury to Melbourne and then by bus to Portsea one of the first things he saw when they arrived, was a Lancaster Bomber sitting in the middle of the camp. This was an aircraft he knew from the letters his mother's brother Des Quonoey sent home to the family. His uncle, a Squadron Leader stationed in England during the war, was flying similar aeroplanes on bombing raids into Germany. To find such a machine available for him and his newfound friends to explore, had him imagining hearing orders to scramble. The aircraft's shakes and vibrations running up through the seat and into his hands, as the squadron taxied in formation and flying alongside Uncle Des on another sortie into enemy territory.

Starting off from Wodonga the only person he knew on the train was a distant relative, Ted Barton. Ted being older than Noel may have looked out for him, however new friendships soon developed and grew. Apart from the rounds of medical and dental checks, the camp divided into teams and all kind of activities from athletics, ball sports and swimming including Naval excursions onto Port Philip Bay.

Too soon the busses would arrive and the return journey would begin.

CHAPTER SIX

Sitting, watching his father Finlay and first employer Charlie Milthorpe signing indenture papers, Noel waited until it was his turn to put his signature to the document. Having secured employment before his fifteenth birthday, a new era of life was unfolding.

Working for Charlie became a meeting of minds. Quitting his role as the sales and service manager at the Albury Ford dealership, Charlie, like a lot of young men in the fifties, began hearing the entrepreneurial siren's call to action. Whispering to him, telling him this was the time not to be working for an employer, but to become independent, building his own business. The only thing he needed now was somewhere to work, he had leased the land and now all that was left to do, was for him and Noel to build it.

Building the workshop and serving his apprenticeship as a motor mechanic specialising in farm machinery became new skills Noel added to those taught to him by Bob Brown, the maintenance manager at the Kiewa Butter Factory. Bob had played mentor to this skinny baker's son who was full of questions. Always asking how this worked, what that did and how to do that, practical skills like learning to solder, to understanding the workings of the large steam engine that powered the whole place. Noel found Bob patient, teaching him how to do everything, and waiting until he was proficient before showing how to do something new. These might have been every day skills to some at the time, but Noel was adding them to the ever-increasing bank of knowledge he would draw down on in later life.

Mister Bob Brown was a patient mentor too, and for years after Noel left Kiewa, gave Noel his time. Freely listening and offering sound advice to any quandary his former pupil was ruminating over.

Starting with Milthorpe's in Albury meant Noel would need accommodation, a place where commuting to and from the workshop would be easy to manage. With help from his parents, he found a room at a local boarding house. He knew letting him set out on his own was hard for his parents and they knew the fare for his room was taking most of his meagre wage. Although cash was limited and it was teaching him how to budget, Noel was always grateful if Fin or Lillian left him a small allowance to help meet his obligations.

More than one boarding house had proved less than satisfactory for Lillian and within hours she had found a room at Mrs Smith's in Albury. This was the last time he needed to move and it proved to be a sanctuary away from work. However, with other boarders coming and going at all hours, it was necessary to find a quiet place to study. Knowing from their very first meeting that Noel had set his hopes on becoming a doctor, Charlie came up with an idea.

Between nailing sheets of corrugated iron onto the new workshop roof, repairing leaky radiators, soldering patches onto weeping fuel tanks and rebuilding the occasional Chev six-cylinder engines, Noel and Charlie built a successful business.

Working with Charlie this way, Noel was learning how an entrepreneur went about assessing risk and maximizing opportunity. Like other mentors, Charlie was happiest teaching his young charge what he knew while they worked and in Noel, he found a willing recruit.

Every two weeks, as obliged by the laws of his apprenticeship, Charlie released Noel for tuition and training at the Albury TAFE. The lessons were structured to ensure all students became confident technicians and went away with an understanding of every facet of automotive repair. The hands-on training made logical sense to Noel and completing his homework assignments became easy for him.

As the business grew and more staff were needed, Noel's responsibilities increased. Breakdown call outs to farms covered everything from all manner of repair to tillage equipment to replacing a vacuum pump on a milking machine. There were long days in the workshop performing everything from the regular service and maintenance of hay and harvest machinery, to tearing down a tractor transmission. It was all part of his trade. By the early fifties, pouring and scraping the big end bearings on a pump engine had become a disappearing skill, but it was a skill Charlie insisted Noel become proficient at. Shutting down any objection by reminding his young charge that although he may never need to do it again, it would be part of his *"Skills Encyclopedia"*. Whether it was replacing windmill gears, or rebuilding worn milking shed equipment all these tasks provided challenges that once overcome, added experience to Noel's abilities.

Charlie had learned from his myriad of contacts, that his daughter's school and the Australian Army had designed a new course for helping a percentage of their recruits complete their Intermediate, Leaving and Matriculation Certificates. Believing one more student wouldn't make much difference; he pressured the Dean to make the course available for Noel to study alongside them.

Noel's classmates would be soldiers, who through location, or financial restrictions had not had the chance of a better education. The same schooling that would have seen them complete high school years, Ten, Eleven and Twelve. Now, with this new course in place, they had an opportunity to undertake these studies out of hours as a further part of their training. It was of mutual benefit as it offered them a far better future with the army, and the military was also advantaged by increasing its ratio of better educated officers.

Noel took this as another of Charlie's lessons and by being an expert salesman, Charlie had managed to convince the school that a place should be found to include his apprentice. Arguing that it was only fair that he should be given the same opportunity as the recruits. And that was how, in 1955 Noel Howard began night studies at Albury High School.

Working with Charlie had opened the way for Noel to enjoy many benefits of his employer's wide circle of contacts. One of those came in the form of Charlie's brother-in-law, Ron Wild.

Employed by the New South Wales Ambulance Service Ron was Deputy Superintendent of Albury-Wodonga Corowa and Berrigan. Noel had told him he had an interest in medicine and it prompted Ron to suggest that he join the service as a cadet.

'You can't go wrong by learning the ambulance trade,' he told Noel.

Determined to improve his opportunities, Noel now had his apprenticeship theory, and the demands of fitting in three years of academic study on top of his day job. However, Ron had a plan to ease the burden and suggested Noel use his office to complete the increased homework load. This became an excellent arrangement as Noel would retire to Ron's office of an evening, which relieved him of the routine of living with strangers in a boarding house and, it meant he was able to respond to an ambulance call if one came through.

Being stationed on the Hume Highway, which during the fifties was probably Australia's most dangerous road, provided many a young ambulance officer reasons a plenty to resign from the service. However, Noel's interest in medicine was making him a keen student. Watching and assisting the more senior staff sum up the situation at a road accident, he started understanding the efficiencies they employed. Actions he would replicate later when running on his own and facing a triage situation.

On one occasion, a young Albury man joyriding with friends had rolled his convertible car not far from the Ambulance station in Dean Street. The accident left him with multiple injuries, the most serious of these being a ruptured femoral artery. Noel, having assessed the situation and relying on his training triaged the scene. Not wanting to watch his most serious patient bleed out, placed a surgical clamp on the artery to stem the blood flow.

While working to ensure the others injured in the accident were being cared for, he was relieved to see one of the more senior of Albury's doctors arrive to take charge of the situation. Understanding his place Noel acquiesced to his superior, even though he could smell the man reeking of alcohol and appearing worse for it. What had been an event under control, now began to dip into chaos as the senior medical man remonstrated and belittled Noel for fixing the clamp.

Removing it, the doctor now had nothing to stop the blood flow and a life that Noel may have saved was now draining onto Dean Street. The young man's death made it a bad day for him and the service, and because he had acquiesced to the senior medical officer who was intoxicated and incapable, staying quiet was something he vowed he would never let occur again.

The next four years for Noel, disappeared in a flash of mechanical work, study and running the Hume Highway in an ambulance. Having completed his apprenticeship with an A pass, Noel followed Ron and Charlie's advice and applied to Victorian Civil Ambulance in the City. Working alongside Ron Wild as a volunteer ambulance officer and using his office as a sanctuary away from the distractions of the boarding house, had allowed the habits of self-sufficiency creep into his routine. Habits which would follow him into the next phase of his career.

CHAPTER SEVEN

Each year, the pull of the waves and the fun of sand, surf and sun carried him to Rosebud where he would holiday, catch up with old friends and enjoy the hospitality offered by the Grants. However, having completed his New South Wales qualifications as an ambulance officer it was now time to move away from his roots. With an offer of employment from the Victorian Civil Ambulance, he packed a few sticks of furniture and the rest of his belongings into a friend's ute (*sedan based pick-up truck*) and headed for Melbourne.

With Charlie's encouragement reinforcing the views of Fin and Lillian, Noel understood that tertiary education was important. It offered a pathway for those men and women wanting to make a difference. As a doctor Noel was sure he could make that change and by doing so would find a way to overcome his concerns about speaking in public. All he had in front of him now, was to reach the city and settle in to the Box Hill boarding house that would become his home, while he completed his studies.

Ron Wild's time in the ambulance service had brought a wide range of influential contacts and his reputation allowed him access to people who could help. Speaking as a reference for Noel with a former colleague, Station Officer Kelly, Ron told him about the reason behind Noel's move to Melbourne and his desire to study. Kelly suggested that if Noel was to bring him his lecture schedule at the beginning of each month, then his shift roster would be arranged to fit around them. For Noel this was a perfect arrangement, he would be able to study between working his normal roster. His wage of twenty-five pounds per week was enough to cover his outgoings, leaving a little left to bank and giving him a set amount, he could use for leisure.

Becoming self-reliant had always been important. Looking back, he had been living away from home for extended periods since the age of ten and loved it. However, his desire to put some security into his life, saw him selling the BSA Road Rocket and buying a much smaller, second hand BSA Bantam motorbike to replace it. The Bantam, although sound in the frame its 150cc two stroke engine required an overhaul, which for a qualified mechanic like Noel was a simple exercise. Taking little more than an afternoon to replace the piston and rings, the Bantam was pressed into service. Buzzing around Melbourne trailing a wisp of blue smoke became normal after a while, and although the call of the country was always with him, it didn't take long for a routine to develop. Study was a great distraction from his work and more spare time was becoming available. Committing to a diary, he found keeping track of his commitments ensured his study and work life was structured.

Keeping a diary and detailed note taking was becoming a habit and one that helped him squeeze a little free time into his week. Because of the trade he was in, he soon developed a friendship with Cliff, at Mannings' Government Undertaker. Most undertakers have a well-developed sense of humour and Cliff was no different. Being offered an opportunity of assisting Cliff meant a little extra income boost and not something Noel was about to pass up.

Laughs with Cliff while earning Five Pounds a day for a funeral, was like being paid for a day off.

On one occasion Noel was driving Manning's' Hearse to the Fawkner Cemetery with the mourners' cars following. Cliff pointed and instructed him to turn into a side street. It turned out to be a circular court and before they could avoid it, twelve cars, a hearse, and mourner's coach were grid locked. There was nothing to he could do but watch as Cliff directed everyone out of the court and back onto the road.

At the graveside and with great reverence, Noel took one end of the coffin with Cliff at the other.

Noel had no idea of Cliff's dietary habits, but whatever he had eaten that morning was causing a gas build up. He watched his friend tense to lift the coffin, but by this time Cliff's bowels had reached their pressure point. Looking across the silent crowd all he could hear was Cliff's never-ending passing of wind. To him it sounded like a minute had passed before Cliff was done and the silence broken. One by one the mourners began breaking into laughter and it was taking Noel all the strength he could muster not to join them.

Only when they had settled the casket, could Noel again look at Cliff.

His acceptance into Melbourne University brought with it an opportunity to re think his goals and as he had already gained an A Grade Motor Mechanics Certificate, Noel began weighing up different options. Mechanical Engineering captured his interest. It became simple decision in the end; having already experienced the basics in his earlier studies and his natural empathy for machinery made his choice easy.

Australia was beginning to boom after the end of World War 2. The rebuilding of Japan's economy and a war in Korea, had driven a demand for wool and the increasing price for it, was giving Australian farmers an opportunity to modernise. The country's machinery industry would need many qualified, capable people to lead the sector and the idea appealed to him.

Once the decision was made, engineering study would begin and at the end of four years, he would be ready to take his place among all the sales reps who had called at Charlie's. Men and women who designed and built equipment that helped farmers find economies through mechanisation.

Having watched his lecturers deliver their lessons and Cliff managing funerals, Noel realised these people were confident in front of an assembly. Listening to them deliver their dissertations in a clear and positive manner. He began how they assembled their words and delivered their intonation; it was enlightening. Noel understood if he were to deliver on his ambition, he had to overcome what he saw as a personal weakness.

Determined to become stronger when giving a speech, led him to enrolling in a public speaking course being offered by Dale Carnegie Training. It was a revelation. After a few night courses and practicing speaking exercises in his mind, and in front of the bathroom mirror. Within a couple of months he became confident and proficient, dissipating all earlier fears of speaking in front of a crowd.

Being with the Ambulance Service in Melbourne had good days, bad days and days when you wondered just how far humanity had slipped. Occasions when Noel attended road accidents, or a broken leg became routine and attending a patient suffering a bout of asthma, could all be written off as being all part of his day's work. Then there were the simple patient transfer runs, these were often within the suburbs and managing traffic was the hardest issue to deal with.

One patient transfer involved a much longer trip. TA grazier having been discharged from the Freemasons Hospital in East Melbourne; it was Noel's task to deliver him back to his property in the Casterton district at the far western corner of Southern Victoria's sheep country. After six hours of driving, they arrived at a sheep station bustling with the business of shearing. Having delivered the patient into the care of his family, the owner's daughter asked Noel if he would like to watch a bit of shearing before heading back to Melbourne. Glad to be able to stretch his legs and take in the sights and smells of a busy shed, he accepted.

He had witnessed plenty of shearing in his time, but often in smaller sheds. Australian shearing sheds had by now, become one of the most efficient workplaces in the world. A workplace, where the toil of everyone involved was driven by the desire to harvest the fibre in the least amount of time. The buzz of the handpieces, the bleating of sheep and the clatter of hooves mixed with shouts and cursing of the shed-hands surrounded him. Absorbed in the activity he jumped out of the way when the wool classer shouted at him to move and as he did, one of the hands tossed a fleece onto the skirting table.

Leaning with his back to a door frame and through shafts of light angling through the south facing roof lights, he could see the presser working in a darker corner of the shed. Moats floated and danced in the sunlight, as the presser unaware of the beauty Noel saw in the scene, toiled away from the others at one side of the wool floor. Dressed in the traditional blue singlet, sweat shone on the man's shoulders as he forced fleece after fleece into the wool-press. Always moving and working its pins and levers until he had reached the owner's desire for a full bale.

Turning his attention to the activity outside, the bustle of the yards began filling Noel's ears. It was a scene of dogs barking and running along the backs of sheep as their handlers encouraged them to push the flock up from the outer pens, into those in the shed.

This was rural Australia at work. The familiar smell of sweat and lanolin reminded him of his rural roots. Going over the day in his mind as he drove home, he found himself laughing in amazement as he envisioned the shearing shed's toilets. These were open at the front, with the pan and seat set at a sixty-degree angle, meaning the occupant would need to back in to complete their business. Looking for a reason anyone would build a convenience this way, he asked the farmer's daughter and she explained, it was designed like that because her family wanted to discourage loafing. Noel didn't see this as the way to treat your workers, but imagining the sight of someone using it, had him smiling all the way back to Melbourne.

Not every day could be handled as easily. Being called to a murder scene or suicide took incredible strength and he could only deal with the situation in front of him. It was difficult at times, but he could not let questions about who, what, or why enter his mind. There had been many road accident call outs that he had attended on the Hume Highway before coming to Melbourne, often the vehicle occupants were treated for cuts, bruises and the occasional broken limb. However, he had also attended many accidents where drivers or their passengers had been maimed, losing their lives in a grizzly way. Ron's training in Albury had prepared him, teaching him to analyse everything he saw and understand that he had no control over what happened. He just had to do his best for those patients he could save. Everything outside his control was in the hands of the gods and he could go home knowing he had done as much as he could.

The four years as a Melbourne paramedic had him witness enough of the worst of human nature. Too many murders, drug overdoses, suicides, and murder-suicides, each one of them leaving a scar on his memory. However, it had been being called to a drowning at the Fitzroy Baths that caused him to draw in a few deeper breaths before rushing in. A baby had drowned, its tiny body still in floating the pool as he arrived. No-one had tried to even rescue the tot. Making it worse the baby had been dead for some time before the alarm had been raised.

That had been a hard day and so had been another when called out to a gruesome scene in the Exhibition Gardens. A young husband had used a shot gun to murder his wife before turning the weapon on himself, there was nothing anyone could do, but retrieve their bloodied body parts.

These were the bad days and yet there were the good days too. During his time working with the service in Melbourne, Noel had delivered eleven babies, all desperate to make it into the world before reaching hospital. Those were the days, moments to savour and days for creating memories almost good enough to rub out any of the bad ones he had witnessed.

CHAPTER EIGHT

The end of year break in 1959 was to take Noel's life in an unexpected direction. Instead of going to Rosebud on his BSA Bantam, for this trip he was returning Frank Grant's big black Nash. In Albury he had rebuilt the engine with one of his Milthorpe's workmates Colin Koschel. They had completed the work in a borrowed shed behind a mate's house and using a few tools from Charlie's, the Nash was running like new. He had been looking forward to seeing his friends at their home on the Mornington Peninsula, filling his days at the beach and socialising at night.

Leaving Albury early on Saturday and knowing the importance of running-in the reconditioned engine and transmission, he listened for noises that may be coming from an ill-fitting part. With a keen eye on the gauges, he sniffed for any smells that might be coming from an oil or water leak. Satisfied all was well with the work, he loped the car along at an easy pace until he reached Euroa and after filling with petrol, checked the water and oil.

Despite the Sydney to Melbourne traffic and torrid road conditions the Nash was quiet, comfortable and for overtaking, fast. Keen to see the Grants and catch up with acquaintances who also spent their holidays at the beach, he watched for the mile posts. Counting down the distance to Melbourne's centre,

Noel was keeping his mind occupied by calculating and recalculating the time he would arrive at Rosebud.

Christmas holidays continued to draw more people to the outer beachside towns and in the days when the hotels closed their bars at six o'clock, one of the entertainments offered were a circuit of Saturday night dances held in local halls across the city and suburbs. These evenings were often headed up by a master of ceremonies who would keep the action moving, ensuring that the dancers were treated to a mix of music to please their young and the not so young audiences. The music, often supplied by amateur bands from Melbourne's outer suburbs played a mix of old-time waltz and ballroom dance music, offset by the emerging Rock and Roll tunes coming from Britain and the United States.

Before leaving for the dance, Mrs Grant beckoned Noel over to her and flicking a speck of lint from his jacket smiled. She had always welcomed him and it warmed him to know he could receive so much kindness from them. Frank dangled the keys to the Nash from his fingers, they jangled as he shook them. Smiling he suggested Noel use the big black car to go to the dance this time.

Approaching the Rosebud Town Hall, his eyes were drawn to a young woman in white and being distracted by her beauty, searched for a place to park. In his desire to meet her, he charged the Nash into the only park he could find.

Watching the Nash from the town hall steps, Joan Brown told him later, that she wondered who this poseur audacious enough to park front and centre, was?

Noel, turning the engine off, hadn't seen this was a place reserved for official cars was just happy to find a vacant spot. He checked his image in the rear-vision mirror, ran a comb through his hair and getting out, locked the car and followed this vision inside.

It took a while for him to convince Miss Joan Brown to dance. However, after the awkward start to their evening, the music and enjoying each other's company, a fire sparked and it kept them together until the dance was over. Ensuring they were together for the medleys, Noel'. Her hand in his, they took in the scene at the top of the steps before gliding down to where his car was parked. Joan waited for Noel to open the Nash's passenger door while she smoothed the skirt of her dress before sliding into the large car's front seat.

The Nash had a radio and Noel asked her to find a station of her choosing, while she gave him directions to take her home. The radio centred in the dash was just out of reach and Joan slid closer to Noel changing its frequency to 3DB. Noel, who having rested his left arm along the top of the bench seat gently dropped his hand onto her shoulder and when she didn't object, rested it there. Here was a woman warm to his touch, soft, gentle and yet blessed with a sharp wit. He found himself beginning to like this person who made him laugh and liking her a lot.

After what may have been the long way home, they stopped in front of a house with a caravan parked in the front yard. Joan explained she had moved out of home after her parents split. Now she was living in the caravan and the house belonged to her friends. Almost hidden behind the van he noticed a motorcycle and curious about who it belonged to, asked her about it. Relief swept through his mind when she said it was hers, not only was Joan all she seemed from that first moment he had seen her, she rode a motorcycle too.

They sat in the Nash and talked for hours.

With dawn beginning to break as he parked at his lodgings, plans for the future charged his mind. How would he sleep and complete his waiting assignments with Joan Brown's laughter and the memory of his arm around her shoulder, fighting their way to the front of his mind? Her scent seemed to fill every crevasse and surface of the Nash, it was as if he was surrounded by her, even though she had left the car fifteen minutes ago. She had agreed to see him again and with her address scribbled on the back of the ticket stub in his pocket he wanted to waltz to the front door, but that would be too audacious for Rhoden Noel Howard, so fighting against desire he walked inside and still dressed, flopped into bed.

Electing to spend time together at every opportunity, confirmed their relationship was strong and had Noel thinking about his next career move, a move that would include Joan. Lillian and Fin, pleased for their son finding someone he wanted to share his life with, made sure Joan was treated as one of the family and trips back to Kiewa were a joy for the pair. Noel wondered if his city girl would ever be comfortable with a life in the country. However, he need not have worried, Joan had a connection with the bush as her forebears had come from farming stock. Normally quiet around people she didn't know, Noel soon found Joan at ease among his family and their friends.

While he still had lectures to attend and shifts to fit around his assignments. Whenever he had been rostered on and not out on a call, he took advantage of the down time to bring his study into the break room. The room was large and a long table divided it from the ambulance officer's lockers that lined the south wall. Having a kitchen at one end complete with a refrigerator where employees could leave their perishables, made for easy mealtime arrangements. A sink over cupboards for crockery with a stand of drawers at one end locked the décor in the late forties.

Spare space in the 'fridge meant the midwifery nurses made a practice of keeping milk expressed from new mothers in the ambos' there. This opened a prank opportunity for experienced members of Noel's team to offer a recruit a glass of milk, only to tell them after they had consumed it, its origin. It may have created a lot of laughs, but theirs was a difficult job and something like this was harmless and relieved the tension at the end of a demanding shift.

Mulling over options which would include Joan, Noel had a greater incentive to complete his studies and look for a move into a role that would employ his new skills and soon to be confirmed, engineering qualifications.

While the study was demanding, mechanical engineering was logical. Helped by Noel's understanding of mathematics and with his lecturer Bill Brown being open to enquiry and offering assistance whenever asked, he flourished. Bill's generosity becoming another example of leadership for Noel to tuck away in his memory bank. Everything experience that he thought would be something he could use in the future, especially if it was extremely poignant, or a life lesson would find him making a short note in his diary.

During the last weeks of 1959 Noel received a phone call from Ray Ruch of Gippsland and Northern Co-Operative telling him about a position being advertised for a Department Manager at Wangaratta and suggesting that should he apply, then the job would be as good as his. Talking things through with Joan and deciding to finish his degree by correspondence if successful, he forwarded his application and waited.

CHAPTER NINE

Honolulu Time: -- 02:21:00
Passenger voices break over the roar coming from the hole in the fuselage, 'city lights, city lights,' and hope rises.

Joan's father Russel loved the male line of his family and treating his daughters as secondary to that of his sons, made for a prickly relationship with Noel who worshipped his bright and bubbly girlfriend. Strong willed and confident, Joan found her father to be a bully and no longer willing to witness the way he treated her mother, had moved out of home at the first opportunity.

Starting as a clerk with Shell Oil in the city had offered her a means of escape, then finding a friend willing to let out their caravan was all she needed to break away from the tension at home. Having saved a little extra money from her salary, she was able to purchase a sewing machine fulfilling her desire to keep up with the latest fashion styles by drafting and making her own clothes.

Whether it was her father's association with the white goods trade, or a love of design, Joan had a natural affinity with fabric and her eye for fashion, meant she would be stepping out in a new dress whenever the occasion called for it.

Having sewing skills also provided an opportunity to make a little money by sewing for her friends and associates. For a spirited young woman, a means of transport was another priority and rather than take up the new fad of owning a DKW scooter, she preferred a more masculine machine in the form of a BSA Bantam. While a DKW may have been more suitable and easier for zipping around Melbourne in a full skirt, the BSA was British and a machine her brothers approved of.

Russel may have considered his second child wilful, but for Joan, she always felt he undervalued and underestimated her. It seemed to her, that he considered his wife and her daughters as secondary to himself and his sons. His attitude, a constant source of tension within the family, made Joan's decision to make a life on her own an easy one and once free of the angst, began to flourish in her independence.

Most of her spare time was now being spent with Noel visiting his family and friends, or stealing away to somewhere quiet, designing their future. In the Howards, she found a different family structure to the one she had grown up in. Lillian and Fin were open and welcoming, there was nothing to show Lillian being at the disposal of her husband. To her, Fin was laid back and a bit of a larrikin.

Noel had described Fin to her as a poacher, someone who enjoyed a laugh. A bloke content in his own skin and a chain smoker, who on Saturdays could be found in the pub working as a penciller for his mate Bob Northey, the local SP Bookie. To Joan however, even if he had been all these things, she found him respectful, a gentleman. What she admired most about the relationship he shared with Lillian was that each saw the other partner as equal. The Howards were a family where each sibling was valued, encouraged and inspired to be the best they could be. Most of all they had warmed to her and were welcoming too.

If she and Noel desired it, Lillian promised Joan she would be easily accepted as a new member of Kiewa's Howard clan.

The next year blitzed past in a flurry of industry. For the couple were spending more time together, growing closer, planning, discussing and redrafting their future. They both believed their destiny was to see old age together and for Noel, becoming restless with his duties as a paramedic began thinking about his next opportunity for advancement. His curiosity and the call of the country had him perusing job advertisements in his mother's local paper whenever they were in Kiewa.

His dilemma was answered when a contact from Gippsland & Northern Co-Operative said they confirmed they were opening a new branch at Wangaratta and asked Noel if he would be interested to managing the machinery side of the business. The Victorian Civil Ambulance had been good for Noel and as much as he loved it, he saw the offer from Gippsland & Northern Co-Operative as an opportunity for advancement, Joan agreed and committed to following him as soon as she could secure a position too.

Gippsland & Northern Co-Operative were a huge agribusiness with farm machinery dealerships, land sales and stock and station departments to their business. This type of commerce fitted his aspirations and had been part of his life for a very long time. He knew the ropes; his grandfather owned a stock and station business in Kiewa and his first employer Charlie Milthorpe had been their agent in Albury. His engineering studies added more strength to his confidence and beginning with them as a manager in Wangaratta would be something of a homecoming.

With their plans to move settled, it was time to announce their other more important intentions to the world. They decided an evening out for dinner in a Collins Street Hotel with friends, Jack and Iris Skinner as the right moment.

The Old Edward Hotel's dining room of hushed voices punctuated by cutlery clattering onto plates stopped the moment Noel slipped down onto one knee and took her hand. For Joan, it seemed everyone was staring at her. Every eye seemed to be watching them and all she could do was look at him. She knew this moment was coming, they had discussed it. However, tonight with Noel on one knee and Jack and Iris cheering him on, pressure mounted.

Noel knew her answer, but waiting for her to say yes after asking if she would be his wife crushed. She seemed to be taking her a long time with her answer and the wait was excruciating. The room, loud a moment ago had now become silent. The weight of stares falling on them made his heart race and he willed her answer to come.

Joan lifted his hand and kissing it whispered she would love to. She stood pulling him into an embrace. It didn't matter that the Skinners and others near them were standing and applauding, they had announced their engagement. Champagne was ordered and the couple's health toasted, making this a time for them to savour.

Brimming with excitement and ready to share their plans with their family and friends they decided to follow a traditional route. Noel would ask her father's permission. Russel Brown, his attitude prickly from the outset had never warmed to Noel. However, and knowing this would not be an easy task, Noel approached Russel and asked him for his blessing. Russel with typical misogyny agreed.

Taking the cheque for twenty-five pounds from him, she looked at it somewhat bemused. He was talking, but not saying anything that made any sense to her. He was at pains to tell her in his most pompous manner, to enjoy the money. Explaining that it would be the last money she would see and as it was likely that her mother was going to be there, not to bother sending an invitation, adamant that he and his de-facto wife had no intention of attending the wedding.

Joan angered by his attitude tore the cheque into pieces, threw them at his face. Telling him she didn't want his money and, as far as she was concerned, he could stick it up his bum.

Leaving Russell's house, the lovebirds tried to calm down and drive to her mother's home. It was time to let Joan's mother know their plans and, in the afternoon, they drove to the other side of the city to tell her. Her attitude was more welcoming than Russel's, but again the buoyancy of the couple's news foundered, as Doris displayed a cold and unfeeling aura.

If that was to be the Brown's attitude, Noel and Joan saw it as an opportunity, deciding it would be them who paid for everything. They were free to have the wedding they wanted. If the patriarch of the Brown family, or his estranged wife didn't care, they did. However, there was much to do before wedding bells would chime in the September of 1960.

Relocating to Wangaratta had meant living in separate boarding houses. However, this arrangement did allow the couple time to concentrate on finding somewhere to live after the wedding and between them they found a cottage at 58 Murdoch Road, Wangaratta. It was a quiet street and the property backed onto the disused narrow-gauge railway line that once ran between Wangaratta and Whitfield. Although the tracks were gone the embankment shielded the fence and allowed access to the back yard, something Noel would need as the house was heated by a fireplace in the sitting room. He knew that trying to keep the place warm and dry during winter would require the couple to burn a mountain of wood and in his mind, he saw things going full circle from Fin supplying wood for the bakery ovens to himself heating a home.

The terms of rent settled, their landlord happily agreed to them painting some of the rooms and restoring the gardens.

Joan knew, when they left the cottage, it would be in much better condition than when they found it. Between work commitments and making her wedding dress, she made curtains. Now with new paint on the walls Joan hung the new window coverings and placed new and borrowed furniture. Making the house on Murdoch Street feel like a home.

Noel through his contacts with suppliers, had been able to purchase new white goods for the kitchen and laundry. Their pantry filled with tinned fruit and a fridge-freezer stacked with meat, they had everything ready for the moment they returned from their honeymoon.

CHAPTER TEN

The management role at Gippsland & Northern Co-Operative came with a new FC Holden utility and it was available for his personal use. However, with those perks came responsibility and being on a salary, Noel found his time now belonged to the company. Not that working long hours bothered him, but he had arranged to finish his engineering studies by correspondence and having invested the best part of the last three years working toward a degree, he wasn't about to give up on it.

Staying on top of his workload meant his diary habit became even more important and completing company reports when due became a priority. It was excellent training, teaching him the value of direct and to-the-point reporting. A talent that would later become a feature of his commercial and community life.

Joan, having moved to Wangaratta took a position working as a clerk in the accounts department of Wangaratta Woollen Mills and finding separate accommodation to Noel, quickly settled in.

Noel now used any spare periods he had to study and finish his assignments. He would have liked to spend more time with Joan as he knew she had much readjusting to do and they both worked hard to capture as many special moments as they could. As well as creating a social life for herself in her new home, Joan faced the much bigger task of designing and making the dresses for herself and the whole wedding party. It took much of her time and the months flew by.

Gippsland & Northern Co-Operative's Wangaratta branch covered a large part of North Eastern Victoria, and as Machinery Manager, Noel was expected to follow leads provided by the company's dairy representatives as well as generating sales of his own. Travelling through a changing dairy landscape brought opportunity and growing his market share of David Brown tractors produced good returns for the shareholders. Particularly satisfying, he was outselling the leading hay machinery company New Holland with a similar line of Bamford equipment. His strong work ethic began to pay off and the new branch's profits soared.

Selling more product meant the spare parts business began generating enough profit to support a person to manage the department, a service technician and a company secretary. A dairy fitter soon completed their small, but effective team.

The products sold were almost as diverse as the territory's farmers, the nationalities, or crops they produced. Just as his line of David Brown tractors proved popular with tobacco farmers and the dairy farms in the Ovens Valley and around Myrtleford. The graziers who moved their cattle to the high country during the spring months also bought a good number of Bamford hay machines.

NOEL   An Authorised Memoir

In Australia, tobacco farming had always been subjected to intense scrutiny from both Federal and State governments determined to eliminate any tobacco, (chop-chop) reaching the black market. Unannounced visitors often made for unwelcome questions and an endless quota of forms for growers to complete. Therefore, a salesman cold canvassing the area dropping in to say hello were often sent packing without ever getting out of the car. Noel made a habit of phoning ahead, introducing himself and making an appointment. During these calls he often asked about the neighbours, or if there was any one the man knew of who was looking for a new piece of equipment. The Co-Operative rewarded this information with a commission, sometimes called a spotter's fee, which was credited to the spotter's account immediately a sale was made. This practice had been used to great effect and Noel's detailed note keeping continued bringing him leads.

If at times an appointment could not be made and if he was in the area, Noel would call cold. Until he became known, this could often lead to a prickly encounter. However, if a farmer became wary at his arrival, an introduction and a story about his boyhood experience with his Uncle Des in the high country became something he would use to begin a conversation, putting the man at ease.

From Myrtleford his sales territory stretched through to Moyhu in the King Valley, across to the mountains to Mansfield near Lake Eildon, down onto the plains of Benalla and back home to Wangaratta. Selling a similar product range to those in other of the Gippsland and Northern Co-Op's stores allowed the company a greater buying power when compared to those of competitive independent retailers. Smaller machinery dealers also had the disadvantage of not having an ability to offer credit in the same way the Co-Op could.

For a farmer, the Co-Op was often a buyer, a seller and a bank rolled into one. For manufacturers, the ability to get your product onto the Co-Op's books eliminated risks associated with high inventories. Benefits included having their debtors' ledger sheltered from ever-increasing aging of accounts receivables and the number of staff required to administer the paperwork. For agricultural machinery suppliers the Co-operative would benefit from a farmer's access to a one stop shop too.

What it meant for the Gippsland & Northern Co-Operative's sales teams was an expanded catalogue of products and the finance mechanism in place with which to make a sale. Noel had all the various models of tractors from David Brown, Minneapolis Moline, and Deutz plus a range of hay machinery from Bamford and Eclipse Milking Equipment to choose from. Along with hardware, nutrition products and stock feed. it was a full catalogue.

However, life was changing for dairy farmers and with the adoption of herringbone milking-shed design, an opportunity for Noel arose when he was charged with designing and managing the building of the first of these in the Kiewa district.

The herringbone concept was an adaption of a milking parlour, or shed, designed by New Zealand dairyman Ronald Sharp. Sharp experiencing pain in his legs, back and shoulders caused by stooping to clean udders, apply cups and then remove them, thought there had to be a better way. Using an image of cars angle-parked in the New Zealand city of Hamilton's Victoria Street, he imagined if he could train his cows to stand in tubular steel bails at an angle on either side of the shed, milking could be done in batches. If the dairyman worked in a pit with the cows' udders at chest height much of the back breaking bending down and reaching would be eliminated. With friends at his dairy in Waikato he built the first shed of its kind, and soon his creation was being copied worldwide.

A timely chance for Noel to put his engineering skills to use soon had him clearing a space near his sales office and setting up a drawing board. Working from the client's sketches and research material from his friends in the department of Agriculture, it was only a few weeks before the first drawings, complete with material lists were sent to Geelong Company, Striproll Industries a steel fabrication company. Noel had contracted them to build breech rails, holding yards, everything from roof trusses, to gates and stalls for fitting into the milking parlours.

Council approval and the preparation of ground works needed completing before the concrete contractors could be briefed with notes and several drawings for the complex floor. It was a time of pressure as each step required decisions whether it was the position of the plumbing, or where electrical cabling was required, working with the builders demanded his constant attention. Even the positioning of the tanks and the fit-out of the plant room meant checking and double checking his plans. The Dairy Department of Gippsland & Northern could be relied on to provide the specification and assembly notes for the milking equipment. However, it became Noel's responsibility to provide a fitter and see the job through, while attending to the sales enquiries for other equipment too. While a huge outlay for the farmer, this type of milking shed reduced the time taken to milk the herd and did so in a more hygienic and labour-saving way.

The success of Noel's design had the Co-Op's dairy reps talking to their clients and driving them across to the Kiewa farm, to see the concept for themselves. Therefore, it was with much joy that Noel learned his next shed to be constructed, was for his old friend from Kiewa, George and Gordon Reid.

George was keen to hear of his young friend's achievements and for Noel it was a time to renew an old acquaintance as they had much to catch up on from after he had left the valley. When the deal was done, Noel started thinking of all the ways he could thank his friend when they would celebrate the end of milking on the shed's first day. Driving home from work one night, he came up with just the plan.

Remembering George's preferred refreshment and on the weekend before the shed was finished, Noel cut a one-gallon oil tin in half and after turning the sharp edge over, fashioned a handle.

Sitting with their backs against the tree over-looking the dairy at the end of milking and watching George's cattle string their way back to the paddocks Noel seized the moment. Lifting a loose, hessian cover off his bucket, revealing two longnecks of beer, each wrapped in its own half of the Weekly Times. Waiting and careful not to drip water, he offered one to George and took the other for himself. Tapping the necks together, the mentor and his young charge sat content in each other's company enjoying the evening view.

CHAPTER ELEVEN

Joan's easy-going manner attracted friends and fitting in to life in Wangaratta took little time. She liked the casual and inclusive nature of the country, yes there were busybodies, but she soon became aware of the people to avoid. Noel's family calling in on her when they were in town was welcome and Lillian taking an interest in her life was a refreshing change to her family home. Her future mother-in-law asking what she could do to help, while keeping enough distance as not to be interfering, felt lovely.

Their wedding plans were taking shape, Noel wanted to use the church at his old school and asked Archdeacon Dicker to preside over their nuptials The Archdeacon had overseen Wangaratta Cathedral during Noel's time at school there and living as a boarder. His Best Man would be his brother Graeme, while Joan had asked her ten-year-old sister Julie to be flower-girl. Graeme being the accountant at Albury's Holden dealer, had offered two shiny new FB Holdens for the couple to use as bridal cars.

The pair had prepaid for a reception for seventy-two guests at the Wangaratta Hotel and Noel's friend and past president from their Jaycees days, David Bool agreed to perform the duties of Master of Ceremonies. With everything locked into place, all they had to do was complete furnishing their future home and confirm their honeymoon accommodation.

Fin being a Ford man and having recently taken delivery of a new Ford Zephyr Mk2, surprised Noel with an offer to swap his car for the Co-Op's ute, telling the couple that the driving aspect of their honeymoon would be more comfortable in a sedan.

Leaving the reception, Joan could hear cans rattling behind the Zephyr, their sound seeming to wash any stress or anxiety she may have had about moving to Wangaratta and planning their wedding, away. Their life as singles had died and as a newly married couple, she knew they made a formidable team. With the car driving out of town and away from their guests she shifted across and to Noel and snuggled into his shoulder. The Howard's had a world to conquer.

With his arm around her, Noel too relaxed, they had accomplished much together. He recalled watching his bride, his hand on hers as they cut their cake had sent a tiny shiver through him and holding her now, he had the same feeling. She was making him smile. He couldn't help it, he was happy.

His mind shifted, cataloguing everything that had happened since meeting her. Nothing would ever be the same, they had been married in the company of friends, hosted a reception they had paid for and cut a wedding cake Joan had designed. They had a nice home to come back to and their new refrigerator was stocked with all they would need when they returned.

CHAPTER TWELVE

Arriving home was not the welcome they had envisaged; the new refrigerator had broken down while they were away and everything including a freezer full of meat had spoiled. Joan began laughing as she opened windows to air the house and opted for unpacking their bags, suggesting Noel's career with the ambulance made him better qualified for handling decaying material.

It was an easy time; Joan was enjoying her work at the Woollen Mills and Noel continued building to his sales tally. They had a widening circle of friends and it was easy to make time for trips back to Kiewa to visit family.

At the end of 1960, a letter arrived from the University of Melbourne informing Noel that his studies were complete and he had achieved his engineering degree with Honours. Being baler season and planning for the arrival of their first child, little time could be spared. However, this was a reason to mark the occasion and the couple enjoyed a quiet celebration with Noel's parents.

While relishing the responsibility and success of his role, he had always been keen to improve the way the business presented itself. To him Massey Ferguson had burst onto the scene in 1958 and began setting a presentation display and demonstration benchmark for others to follow. He was intent on learning more.

However, you needed a similar budget if you were going to match their show. What you could do was to compete in the field with a skilled demonstration and it was here that Noel perfected his ability to match a tractor to the machine being trialled.

Gippsland and Northern Co-Operative fielded a team of sales people and dairy representatives at the National Ploughing Championships, which were held at Wangaratta in 1962. One of the many demonstrations Noel completed over the three days, was operating a David Brown 850 tractor with a Massey Ferguson grader blade coupled to the three-point-linkage.

Noel's objective was to demonstrate the David Brown's hydraulic capability while constructing a "**V**" drain. The local area receiving high rainfall in winter months often meant for waterlogged paddocks, making them slow to produce the fodder needed for stock and dairy production. By creating drains and cleaning them in either spring or autumn, was a means of shedding water off boggy ground more quickly. In the past contractors had carried out this work using heavy machinery which would compact the heavy soils and that too would lead to slow spring growth. Having an inexpensive alternative meant rather than wait to fit in with the contractor's schedule, it could now be done when the farmer decided.

Cliff Freer, the Victorian Branch Manager for Massey Ferguson, had been observing the demonstrations and at the end of Noel's run approached, asking if he would be open to talking about taking up a position with Massey Ferguson as a Sales and Service Area Manager.

Noel recognising Cliff's approach as another opportunity to advance his career, agreed to attend a meeting at Massey Ferguson's Sunshine Head Office. However, he made a point of telling Cliff that his role with the Co-Op was extremely lucrative and would need to consider any offer carefully to weigh up if a move would be worthwhile.

After hearing Cliff out, Noel began to quantify those things he admired most about Massey Ferguson. He looked down at the grader-blade he had just finished demonstrating and cast his eyes toward the flags and bunting that bordered the Massey Ferguson display. In the carpark, company cars and service vehicles were lined along the fence as if they were tanks on some imaginary wartime front line.

Leaving Gippsland & Northern's stand he wandered across to the demonstration plots, a Massey Ferguson demonstration was underway and the number of company people supporting the operator, exhibited the power of their organisation. He alone, was the whole team for Gippsland & Northern, completely different to the local Massey Ferguson dealer, who had a raft of people from Sunshine on hand to assist with their display.

After talking the proposal over with Joan, the couple decided it was in their interest to at least hear what Massey Ferguson would offer. The next morning Noel contacted Cliff Freer arranging an appointment.

Winding his way around the roads bordering the massive, Massey Ferguson manufacturing complex at Sunshine, Noel parked the car and checked his watch. He was early and it gave him time to look around. Across the road, sounds of industry floated above the trains whisking their way through the marshalling yards of the Sunshine station. He secured the top button on his shirt, tightening and centring his tie. A quick check in the rear vision mirror and once out of the car he shrugged on his jacket, pulled on the cuffs of his shirt making sure to show just enough colour.

Striding up the few marble steps to the huge revolving door that marked the entrance to an oak panelled reception area, he took a quick look around. Pushing through the door he passed his business card to the receptionist, telling her he had an appointment with Cliff Freer. The woman indicated to a chair and asked him to wait.

The building had been built in the art deco era and was a monument to H.V. McKay's industrial vision. It was one man's testament to tenacity. Gazing around the walls his eyes fixed on the narrow oak and glass door of the lift. He had heard that when H.V. McKay arrived at work, his chauffeur would open the rear door of the car and wait for his employer to alight.

Then after walking behind his boss would open the lift door for the waiting McKay and when the industrialist was safe inside, he would run up the oak staircase ready to open the first-floor door as the lift arrived. He smiled to himself, vowing should he ever reach such heady levels, he would never treat another human like that.

Cliff arrived asking him to follow upstairs and ushered Noel into the boardroom. Here too, the oak panelling was expansive. This seat of new century, industrial power was only broken by seemingly out of character photos of Massey Ferguson's new product lines. Only a trace of the company's history and a photo of a young Queen Elizabeth hinted to the building's past.

He introduced Noel to a man waiting for them. He was the newly appointed Victorian Manager, Dereck Keeble-Johnson. His handshake firm, his hands soft, Dereck explained he too was new to the office having only recently moved across from Western Australia to take up his new role. Cliff pointed Noel toward a chair and suggested they all sit down. Over the next twenty minutes Dereck, who came across as a stiff and unwelcoming personality outlaid the company, its policy, the company ethos, where they were headed and, what was expected of an employee.

Cliff described the role, saying this would be something of an experiment. Up until now they had maintained separate sales and service representatives, but with Noel's experience they saw an opportunity to generate a new position by combining the two. Noel immediately thought that as they wanted to employ one person to do the job of two, he should ask why. However, he decided to let it rest until it came to negotiating the terms of his salary.

All through Keeble-Johnson's spiel he noticed the way he wove the borrowed history of McKay's vision and drive into the new Irish American owned conglomerate. Massey Ferguson was truly a multinational company. Although Noel would be low in the pecking order, this was a company with a structure in which, if you applied yourself, you could reach the top.

Asking to know more about the size of the sales area he would cover, Cliff and Dereck explained that his territory would take in the northern end of the Wimmera. Going as far west as Warracknabeal, then turning north to Hopetoun, calling on the dealer at Sea Lake then continuing to Mildura, taking in Walpeup and Ouyen. From there the territory crossed the river into New South Wales, following east along the Murray River until crossing back again into Victoria at Robinvale and on to Manangatang. Back into New South Wales and it was wheat and grazing country to Balranald, where the boundary turned south toward Swan Hill onto Kerang and back toward Wycheproof.

While they were talking Noel listened, summing this up as an area that had a huge number of dealers to assist and a large disparity of farms and practices. This position would take an enormous effort to stay on top of every concern that may come up, not only from a sales perspective, but service problems too.

Realising they were expecting him to perform the duties of two men, he couldn't stop thinking that theirs was a big ask for anyone.

When it came to deciding a salary, it was expected he would eventually move to Melbourne and provide his own vehicle. It was to be a new vehicle for which he would be compensated. The terms of his employment and salary agreed, and after a walk around the factory he accepted their offer, shook their hands, and drove home. It was a big day and one he was keen to tell Joan about.

CHAPTER THIRTEEN

Taking up his new role with Massey Ferguson in 1963, meant buying a vehicle to suit the purpose and following his father's preference of vehicle manufactures, he chose one of Ford's brand-new XL Falcon Vans. Knowing that he would be covering long distances, he reasoned the extra power offered by the Falcon's optional 170 cubic inch engine outweighed that of Holden's EJ model. The holden was still using an updated version of their original 138 cubic inch grey engine and was outdated. Knowing the weight of the tools, workshop manuals, service bulletins, brochures and any other marketing material would be significant, the carrying capacity was significant. Noel reasoned the vehicle would always be loaded, so decided opting for the extra power made a sensible choice.

Still living in their rented home in Wangaratta and reporting to Head Office in Sunshine, as well as having part of his new territory in the greater Melbourne region, it made no sense to live in the country. It didn't take long for the couple to decide that with baby Geoffrey to think of, it was time for them to make their move back to the city.

Happy with his role within the sales and service team, and Noel confident the sales side of Massey Ferguson would offer him a pathway to management, he and Joan placed a deposit on a new display home in the East of Melbourne suburb, Vermont.

A typical week for Noel would begin with leaving home at 5.00 am and arriving at his first dealership as they opened their doors around 8.00 am. His visit would often involve inspecting parts that had failed and replaced under warranty, checking that the identification tags matched the warranty claim and either approving, or denying the paperwork.

If there was an issue with a customer who was unhappy with the performance of a tractor, or machine they had purchased, then a visit with the dealer would be necessary and a solution found. Customer relations were a big part of his role and as he had learned with Charlie in Albury and Gippsland & Northern, expectation often had little relation to performance and it was his role as the company representative to *put things right*.

He soon identified that the difference between putting things right and a customer's expectation was likely to be an issue of poor sales training and incomplete installation procedures. Sales people often worked on a low salary topped up with sales commissions and they were likely to have had it drilled into them to achieve as many calls in a day as they could. This bred a shortcut mindset with most dealer sales people and the focus on making a quick sale often resulted in poor profits. The desire to get the order and get out, having been bored into them during training, they kept repeating the process with the next client leading to even more shortcuts.

Often the installation process followed a similar vein. Delivery to the client happened either at the dealership or on farm and on many occasions, time was of the essence. This meant a lot of little things that the owner or operator should have been comfortable with, were being skipped over and leading to poor performance. The owners were at times angry, as the machines didn't operate as they anticipated, especially when they compared it to the expectation they had at the time of its purchase.

Noel found himself bemused by this; Massey Ferguson had some of the most advanced publications on farm machinery operation in the world. They had a complete library of training material and books on Australian farming and as part of his indoctrination, he had been exposed to it.

If he wanted to simplify his dealer visits and make more sales calls, he needed to begin educating the sales staff and holding farmer information nights. All these initiatives would be discussed with the dealer and at times conducted without seeking approval from head office.

A typical example of the problems he faced happened after the introduction of Massey Ferguson's new MF-585 SP Combine Harvester range. Having been on sale in early September 1965 the machines were arriving at the dealerships without workshop or operator's manuals. The new MF-585 models had been received with great fanfare at their release, the demand being so strong that farmer's sons were being hired on a casual basis to build machines in time for the cereal harvest. For experienced operators a lack of manuals was less of a problem, as they could tune their way around the settings and make the machine work. However, the harvester was new to many and Noel, as part of Massey's service team came under pressure to solve service issues on the run. Had proper manuals and training been available many of the problems may not have appeared.

However, he found an ally in Ian O'Rourke, who while being an area service manager for a large territory that included Western Victoria, had also spent time with Massey Ferguson's test team. Noel being allocated the Mallee area to manage, had been tasked with breaking the news to Ian. He arranged to meet him out on the road and found Ian was relieved that part of his territory had been re-assigned. Their meeting went well and a friendship developed. Ian had been one of the team who proved the MF-585 and was happy to discuss service issues with his new colleague.

Intent on rectifying product concerns at the source, Noel always kept notes on his, or any dealer, or customer fixes. Referring to those notes, he used his time at night to write reports that he hoped would filter back to engineering, trusting that manufacturing would appreciate the number of photos, sketches and drawings he supplied.

While ensuring the service side of his role was running smoothly, Noel knew that sales was the major reason for Massey Ferguson engaging him and in this he was determined to prove their faith in him. As well as carrying a required range of special tools in their own carry cases, he had his own tool box equipped with a wide range of spanners in both SAE and Whitworth sizes. Also taking up quite a lot of room in the rear of his van, a range of brochures and sales material in their own dust proof boxes set in easy access. The same style of cases held workshop manuals and binders for sales and service bulletins.

Arriving at a dealership, if it were a service call, he would meet with the dealer principal before finding the service manager and begin working through the numerous service bulletins that had accumulated since his last visit. Then he would meet with whoever oversaw the workshop to review any failed parts before analysing the warranty claims and approving them.

Always having an ear out for sales prospects, he would canvas the service personnel asking about the machinery they had recently worked, or were working on. He wanted to know if they thought any of those clients were likely to be in the market for a new machine. Making a note of their names and contact details, Noel would then review those prospects with the dealer, or the sales team. His attention to detail soon saw an increase in sales across his territory and proving Cliff Freer's early faith in him.

CHAPTER FOURTEEN

Purchasing a new display home in the Melbourne suburb of Vermont, the Howards continued to live in Wangaratta for another three months while they waited for their plumbing and septic system to be completed. Noel being on the road from Monday to Friday made for a busy time for Joan. As well as managing to care for their new son Geoffrey, packed all they didn't need into boxes in readiness for moving day. It might have been a busy time for her, but as a family they were happy.

Noel may have worn combination overalls to protect his clothes when attending a service call, but the dirt still crept in. The dust, the grime and his need to be always well presented, made for a full laundry basket. Often there would be four pair of overalls and five, or six shirts to iron every week and yet, together they managed.

Noel had always thought Joan was the more romantic of the two and he loved arriving home after a week on the road. He would be welcomed with little touches like a posy of flowers on the table. Her ability to make any space their home always made him smile

As he pulled the door on the Murdoch Street house for the last time, he saw Geoffrey in his seat and his wife waiting by the car. For them, much had happened since the rush of Mid-March the year before when their main concern was making sure the nursery was ready in time for the birth of their first born.

Deciding it would be better to give birth at the Women's and Children's Hospital in Melbourne, before the end of March Joan had packed a case for her and their baby. She would stay with her mother until the baby was born and only move back to Wangaratta when confident all was well with mother and child. Geoffrey Noel Howard joined the family on Wednesday the 11th of April of 1962, and the continual flashing of Noel's camera proved his proud parents were smitten.

For Noel it was over too soon and he found tearing himself away was harder than the task of driving through the night to be back in Wangaratta in time for work next morning.

Now he started walking toward Joan, who already had Geoffrey settled in his seat, he held the door and waited while she sat down before closing it. Pulling his own door closed he started the car and with Geoffrey in his seat between them, thought about how things had changed. He would have liked to have felt Joan slide across to the centre of the bench seat and lean against his shoulder again. However, sitting here now with Geoffrey between them their family had grown, there were three Howards now. These were days to savour.

That first night in their new home is one they will always remember. Welcomed by the lingering smell of fresh paint, their furniture that they had brought with them from Wangaratta looked lonely in a so much bigger space. Although the house sat on land yet to have a proper landscaped lot, Noel knew it wouldn't take Joan long to design the gardens she wanted.

He could do the grunt work, moving soil and making borders. Once the grounds had been completed, he knew Joan would soon be filling the empty spaces with shrubs, bordering the lawns with annuals and perennials.

For Joan those first few weeks flashed by in a flurry of nesting and preparing the home, ensuring she had everything they would need to nurture their new family. Her weekends were an opportunity to spend time with Noel. Even if it meant his laundry taxed the washing machine and a growing number of white shirts often needed their collars and cuffs to be scrubbed by hand, they were both working to secure their future. With the washing dry on Saturday, she would iron his shirts and overalls Sunday morning before packing his bag for the new week. This division of duties worked well, Noel laboured in the front and back gardens landscaping, while Joan was managing the household chores.

Weekends were over too soon. However, Noel knew he had been gaining the confidence of the Massey Ferguson dealers in his territory and as that trust grew, the number of service problems began reducing and even more sales came.

Due to the expansion of irrigation in both Victoria and New South Wales, vignerons and fruit growers had begun expanding their farms. This expansion coincided with the release of Massey Ferguson's agile, light and powerful MF35. As a replacement for the much loved *"Little Grey Fergie,"* the

Thirty-Five became a sales sensation and easily outsold rivals from David Brown, International, J.I. Case and Ford Motor.

Being home for the weeks leading up to and including that of the Royal Melbourne Show that opened in September brought Noel a little of the home life he had been missing since leaving Gippsland & Northern at Wangaratta. Although the training days were long, planning and setting up the show stand in the same area that HV McKay had occupied for five decades, meant only a few miles to a warm home in Vermont and was wonderful. Helping too was Joan's mother, who wanting to be near her daughter, had moved to Vermont allowing her to be on hand if needed.

Fifteen months after being approached by Cliff at Wangaratta, Noel had won the confidence of his dealers and the respect of his colleagues in both sales and service departments. Now he was at the Royal Melbourne Show, working alongside every man and woman on the Massey Ferguson sales and marketing team. Everyone was busy fielding enquiries from customers across every spectrum of Australia's primary industry. It was a total team effort.

The hangover from Prime Minister Menzies' Credit Squeeze had receded and Grain Growers from across the Mallee formed huddles around the machinery on display. Farmers would group around a member of the sales team firing questions at them, with most of the enquiry being centred on the New MF 585 SP combine harvesters. On more than one occasion Noel took a customer to their temporary office and completed an order on behalf of his dealer. It was the same with men and women from across Gippsland and South West Victoria, all seeking answers to questions about tractors and hay machinery.

Sharing spruiking duties with his colleagues during the Grand Parade, would find him sitting behind the microphone in the public address truck and describing the Massey Ferguson machines that were taking their turn. Led by the Grand Champion, the parade was a snaking stream of cattle, sheep, goats, other livestock, cars, trucks and all kinds of farm machinery that followed in a long and colourful line. The line kept circling ever inward, only stopping when the Grand Champion reached the centre and the parade had turned the arena in a slow-moving carpet of colour.

Creeping around the arena could be a stop-go affair for vehicles that were high geared and they sometimes succumbed to overheated engines and clutches. Not a problem for the Massey Ferguson Team.

For someone who years earlier had worried about his confidence when speaking in public, he had come a long way. In 1962 almost ninety thousand people per day attended the show and after lunch, most would wind their way through the many exhibits to watch the Grand Parade, it was a big stage.

When the show finished, Noel and the members of the sales team would visit their dealers, to discuss the high number of sales leads the show had generated. Together they would triage the urgent from the casual and separate out the tyre kickers. Not everyone was as diligent as Noel and his sales results proved it.

Sunraysia Motors at Mildura were experiencing one of their best years. Australia's credit policy had changed and money to buy equipment was becoming more readily available. Between them, Noel and the dealer principal Syd Mills, designed an easy payment plan tailored to meet the region's grape growers' needs. It was simple to understand and very successful. After schooling the Sunraysia Motors sales team, it didn't take long for the dealer to secure enough sales of MF-35 and MF-65 tractors to place a bulk buy order with Noel. The knock-on effect was that on receiving an order so substantial, the outward goods department of Massey Ferguson had to secure a complete train to supply the dealer.

At the time Syd Mills held the position of mayor for Mildura and he and Noel became kindred spirits. They enjoyed each other's company and this conviviality rubbed of onto the Sunraysia sales and service staff too. By working closely with everyone at the dealership, Noel encouraged Syd's people to try that little bit harder and soon he found the whole dealership buzzing with enthusiasm. Sales came and service issues were dealt with. It was a good time and for a period, Mildura became a jewel in Massey Ferguson's crown.

The weekend before the tractors were to arrive in Mildura, Noel travelling through Melbourne in his Falcon van. He had Joan and Geoffrey in the car with him and a major intersection they were hit by a red-light runner, destroying most of his vehicle. Although they escaped with minor scratches and light bruising, it had ruined his plans for traveling meet the train load of tractors in Mildura.

Intent on meeting his obligation to Syd Mills, Noel arranged to borrow one of the company's Fargo one-ton utilities. The Fargo's ride was hard and passenger comfort sparse, but with tools and sales boxes loaded, he would manage his commitments. It would be a busy week. Meeting Syd and his sales team as the tractors arrived was his priority, however he had other calls to make along the way.

To add to his already crammed schedule, Victorian Manager, Dereck Keeble-Johnson advised Noel that he too would be there for the train's arrival. Representing the might of Massey Ferguson in an official capacity he would like it if Noel requested Syd organise arrangements for a special handover ceremony.

From the moment he met him, Noel had thought Johnson pompous and making this delivery an opportunity to be at the very centre of importance, just another example of the man's arrogance.

Keeble-Johnson had instructed Noel to drive the 140 miles or 223 kilometres across from Mildura, collect him from the dealership at Swan Hill and take him on to Mildura. Insisting he have everything to be in place for his arrival, adding that Noel could brief him on the way back to Sunraysia Motors.

Seeing Noel's borrowed Fargo utility truck at Swan Hill, Keeble-Johnson stood looking at it, a picture of disdain on his face and telling Noel he expected better from one of his staff.

Noel thought even less of the man than he had, when after giving an explanation of why he had to use another car. And, without enquiring to the health and well-being of either Noel or his family, Dereck asserted that for the good image of Massey Ferguson, it was imperative that Noel buy a new car on his return. It was easy for Noel to think this man with a hyphenated name was an ass.

Hoping it was the end of Dereck's pompous rant, Noel answered questions about the dealership, the region and Syd Mills. He asked questions as to why no dealer had been appointed at different little spots they passed through along the highway, suggesting Noel appoint them Massey Ferguson dealers. Noel responded to this endless barrage by giving reasons why the existing number of dealers represented the territory well. Insisting those already in place understood and serviced their market areas, explain that by increasing the number of dealers may in fact reduce the total of Massey's sold. The mood within the Fargo became frosty.

Things didn't get any better at the dealership when, to big-note himself, Dereck began belittling Noel to Syd by using the Fargo as a point of attack. Syd Mills, Massey Ferguson dealer and Mayor of Mildura reiterated to Dereck that a train load of tractors arriving in his regional city was a big event and that the work Noel had put in to help it happen was more than welcome. Every one of Dereck's assertions, jibes or put downs about Noel being rejected by Syd. He backed Noel's reasons for using the Fargo and praised the way he went about his work.

The big day done Noel couldn't wait to put his boss onto his flight home. The man was an idiot and not someone you could trust. He thought back to the story about HV McKay, his driver and the lift. Dereck Keeble-Johnson belonged to that time, not to the present. Noel knew his time representing Massey Ferguson under Dereck's leadership was moving into difficult times. However, he also knew men like Dereck move on, He only had to wait him out.

CHAPTER FIFTEEN

Now settled into their Home in Vermont, Joan was looking forward to re-joining the work force. After finding a place in a child-minding centre for Geoffrey, she had started working as a ledger machine operator for Turner Industries, a tool manufacturer based in Mitcham. After dropping Geoffrey off and satisfied he was settling in, each morning she made Turner's in time for her day to start. It was a good time for Joan and she knew the money she earned would build their savings, increasing their readiness to invest in something better, should it come their way.

The Falcon back on the road, Noel's life consisted of dealer visits, customer enquiries and putting right whatever service issues arose. He sensed an uncomfortable mood of growing disappointment among his colleagues. Massey Ferguson might have had the right product, at the right time and even though the sales and service staff loved the company, the leadership lacked initiative and courtesy.

On Wednesday 11th December 1963 Ford Australia's classified advertisement in the Herald newspaper announced they were ready to embark on something big in the agricultural machinery industry. and after their embarrassment with the Ford 6000 model, they intended to do this right. In Friday night's Herald, another advertisement took up three columns. Ford were employing, and by the size and number of classifieds, they were extremely serious.

For Noel, Dereck Keeble-Johnson's attitude and his total lack of transparency, made picking up the phone to Ford and seeking more information an easy move. Applying in writing on Saturday morning was an easy choice for Noel. He may have liked Massey Ferguson, even seen himself leading the organisation in the future, but Ford was a much bigger company and if they were going places, it made sense to go with them.

Although his colleagues were gossiping about Ford's push for people and Massey Ferguson's management were aware of it, his bosses didn't seem to care. Why should they, Ford only had two aging models that over the past few years, only enjoyed mediocre success. The Super Dexter, a competitor for the MF35 and the Super Major, which sold alongside the MF65.

The recently released Ford 6000 with its dubious Select-o-Speed transmission was a low powered flop. Massey Ferguson knew they had a new range of machines coming, so management's attitude was one of, why would anyone with a sound mind wish to be part of Ford Tractors?

However, Ford did have something and it seemed exciting. Arriving for his interview, Noel was shown into a waiting room where he found one of Massey Ferguson's trainers waiting. Dave Cunningham who had also applied for a position. That evening Noel received a phone call from Ken Fitzmorris his sales manager, who informed him his services had been terminated. It took no imagination for him to suspect who had told Fitzmorris about Noel's appointment with Ford.

Rhoden Noel Howard was no longer, a Massey Ferguson employee.

Honolulu Time: -- 02:29:00

Captain Cronin and his flight crew, knowing they only have one chance to bring their wounded craft in safely, have been dropping fuel all the way back to Hawaii. The plane is still overweight and flying on its two port engines.

Completing a wide circle on approach, the crew are fighting to control the Boeing 747. Cronin knows they need to make a high-speed landing, hoping to bring the plane down without the landing gear failing.

The only thing you can give a man without hurting him is opportunity"

        Henry Ford I: Interview: New Orleans Times

CHAPTER SIXTEEN

Honolulu Time: -- 02:34:00

The captain calls for evacuation and within 45 second the 747 is empty. Passengers and crew are being herded toward waiting busses and away from the wounded aircraft.

Around him disembarking people looked dazed, bedraggled. His mind in a similar place, Noel thought back to his medical training and assessed the situation. One of those herded like dairy cows toward waiting busses he turned back to look across the tarmac at the stricken 747. A row of ambulances had already lined up, loading those too injured to walk and tending to others locked in distress.

    Boarding the bus, he took the first vacant seat and tried to understand what had happened, could they have been players in an unsuccessful terrorist plot, or was it something even more sinister? He expected at some time, he too would be required to give an account of what happened, or one of the medicos would find him to enquire if he were injured. Thinking it would be *company protocol for the airline to do so and thereby minimize United Airway's exposure to litigation.*

A few of the busses had already arrived in front of a large hangar on the far side of the terminal building. Knowing there would be interviews to attend, he took a slow look at those around him. Some of his fellow passengers appeared agitated, others were zombie-like and some were crying, the more spiritual among them praying, thanking their various gods. They had become an all-shuffling line of survival headed for the plastic tables and chairs, all dwarfed by the gaping chasm of this industrial hall, that a few hours earlier they never expected to see. Everyone weary, and out of place.

Hushed tones raised and fell as officials went about asking people to come forward and give a statement. Noel, still holding his flight-bag gave his summation of the event to an official and agreed to appear at an inquest if called.

After what seemed hours, he was boarding another of the busses and taken to another hotel where the airline had arranged for people to rest. United Airways had organised for another 747 to be flown in, ensuring those passengers who could, would be able to complete the journey everyone had started five hours before.

When he reached the lobby, hotel staff were offering tea or coffee. His mind was ahead of that and he making his way to the desk, arranged for a taxi. While he waited, he noticed the sombre mood among his fellow passengers. When he did overhear a conversation, the voices sounded quick. To him, it seemed as if each of the speakers were thankful they had lived, and yet the same voices remained hushed. The feeling in the room seemed spiritual, almost in reverence for the poor souls who had been sucked out of the plane.

Some seemed to draw away from the crowds and he recognised their desire to be alone, particularly the woman who had crawled under the grand piano. He stared at her for a second and walked past. Doubting she would have welcomed his intrusion even if had he enquired after her.

While waiting for his taxi, he found a seat near the desk. Opening the briefcase, which he had kept in a strong grip from the moment he stood at the top of the exit slides, Noel took out a note pad and began making a summary of the report he had given to the authorities. To keep his mind on the present, he started on a list of things to do while he waited for the flight home.

He had come so very close to death and cheated it.

Looking at his hands, he found they were still steady and went back to his to do list. All those hours riding with Ron Wild in the ambulance and his experience in Melbourne, now helping him deal with this tragedy, but it didn't stop him wishing he had woken Joan when he thought of it earlier. He so wanted to hear her voice telling him she loved him, to tell her how much she meant to him... and to say that he loved her.

The taxi stopped at the doors of the Royal Hawaiian Hotel where he had secured a suite. They knew him here and seeing the doorman holding the cab-door open for him, today it almost felt like a home-coming. Checking in at the desk and moving into his room he sat for a moment gathering his thoughts. His mind steadily sorting his priorities, he called Qantas and booked a first-class seat on their first flight from Honolulu to Sydney with a connection to Albury.

Calling down to reception, he asked for a wake-up call, booked a taxi to the airport and made a note to call Norwood's in New Zealand to cancel his calendar for the next few days.

Stripping out of his clothes, he brushed his trousers to remove the dust and marks as best as he could and put them on a hanger, his shirt too, went on a hanger which he took into the bathroom. With luck they might find his luggage and deliver to the hotel, if not he could buy clothes from one or more of the stores within the hotel.

Standing under the shower in that Honolulu hotel, no matter how hard he soaped and scrubbed, the stale smell of aircraft dust would not disappear. The adrenalin had left him now and sliding into clean sheets, he thought about what had transpired. Overall, this had been a shitty end, to a shitty week.

CHAPTER SEVENTEEN

Arriving at Ford about twenty years after Henry Ford II, in his role as President had begun the greatest reorganisation of the company since the Model A, Noel's timing could not have been better. The company was energized, working toward cementing its image as vibrant and youth driven.

Beginning as early as the end of World War II, Henry II had embarked on changing the relationship between management, the workers on the factory floor and that of thousands of Ford dealers across the globe. Although his focus began in the United States of America, he knew change had to happen across the Ford Empire.

Ford's greatest threat had always come from General Motors and the war being finished, cashed up servicemen were returning with money to spend. Company Management knew an energized Chrysler were also poised to take huge bites out of Ford's market share. To reduce the impact of competitors picking off Ford's biggest and best dealers, Henry II toured the USA, meeting with dealer principals from both large and small dealerships, assuring them the company was positioned for change. However, while it would take time, he needed them to remain loyal and patient. Explaining it wouldn't be easy, his assurances were well received. For the next ten years the management of Ford Motor oversaw the modernisation of every department.

Even before his appointment as President, Henry II's strength was knowing the things he didn't know. However, he was determined to rid the company of the corruption and back-handers that had worked its way into every corner of the company under his grandfather's watch. In this he was decisive and his actions swift. The subsequent purge of executives and factory floor foremen, who had become bullying stand-over men, made it a time of great change.

First to go was Harry Bennett, the confidant of his grandfather and one of the men the Ford family knew had made Edsel Ford's life hell. Bennett had supporters within the company and the board. He might have even challenged Henry for the top job, but as much as Bennett could have found support with outside directors, the family represented forty two percent of the voting stock. The family were united and Henry II assumed the role he was born to.

Understanding that no one man could hold it all in his head as his grandfather had tried to, and knowing the company needed talent as well as leadership, Henry approached General Motors executive Ernest Breech.

Observed by the industry as an excellent man with a head for organization and numbers, Breech had great people skills which resulted in his uncanny ability to get the best from his managers. Ernest Kanzler, Henry's uncle and a powerful Detroit personality in his own right, provided the introductions.

Telling young Henry how Breech had turned Bendix, one of GM's poor performing companies around, he thought Breech was the man that Ford Motor needed. Breech had climbed through the ranks of General Motors to a point where he was touted as a possible future President, only to be overlooked by the GM board.

Convinced Earnest Breech was his man, Henry acted, recruiting Breech and charging him with the reorganization of the company. Ernie as he liked to be addressed, was not the only new recruit to Henry's team; at the end of the war he had received a telegram from Charles (Tex) Thornton introducing himself as spokesperson for the Whiz Kids.

Thornton, a former aide to Assistant Secretary of War Air, Robert Lovett, offered the services of his cadre of Army Air Force management experts, who because of their experience in tracing millions of Air Force details, believed if they stayed as a team, they could do for business what they had for the Air Force. Impressed with their presentation and having John Bugas, a former FBI man interview them, it's rumoured Henry hired them at their asking price.

Things were changing at Ford and it showed.

Different to the old regime of the original Henry Ford and Harry Bennett, members of the press were now welcomed into the once secretive world of Ford Motor. Reporting this openness, slowly began to change Ford's image. The company appeared modern and forward thinking. Where the management had once been antagonistic toward its workforce, attitudes were open and inclusive. Now university graduates from across the country began flocking to Ford Motor, each one searching for a career and wanting a part in a forward thinking, fast paced organization that valued its employees.

For the next fifteen years, the younger Henry would personally drive the company to improve their competitiveness by offering a modern model line-up. Having asked his dealers for their patience while the company re-tooled after the war effort, he made good on his promise with the first all new Ford in 1949.

And once, where his grandfather would have demanded, he asked for their loyalty.

Henry II knew it would take time. However, he had started introducing his *Human Engineering* philosophy into the way the company engaged with its employees. Gradually improving relationships with everyone from the canteen staff to senior management, the company began attracting the brightest and best, all wanting to be part of the Ford Motor Story.

By the beginning of 1960, Henry Ford II and Ernie Breech had become convinced the company had to grow. Market share had always driven the sales teams, however gaining a few percentage points in the United States, while important, would not bring the same opportunities that Henry was sure lay in Europe.

Ford's overseas business was built up of small to large manufacturing and marketing operations, that were co-owned by Ford Motor and a mismatch of other interests. At times, the structure which had grown almost organically under his grandfather's watch was now working against the parent company. To Henry and Ernie Beech, it was a mess.

Charged with a desire to untangle the company from a series of cross ownerships and unprofitable operations, Ernie Breech completed a review of each company. From there they built a new strategy, it would mean the parent company buying out the remaining stockholders. Not an easy task, but if the company were to thrive, it had to be done.

Henry started by convincing the board to buy back all of the second-party interests and consolidate Ford Motor's World Wide Operations into one truly multinational company and offering stock options to managers. This would put it onto a similar footing to General Motors. It would mean bringing all the shareholders along with his vision, and selling off some of the smaller unprofitable companies. Ford France, was one that went, being offered to Simca which was later absorbed into the bigger Rootes Group.

By the time Noel walked into his role at Ford Australia in the early sixties, the Ford Motor Board had been witnessing over seventeen years of Henry II's drive for change and bringing with it, increased bottom line results. The once sloppy accounting and data systems were now a distant memory and the purge of aging ideas had made Ford an enticing place to forge a career.

In Australia, Ford Motor Company operated under the auspices of Ford Canada and the vehicles sold were a mix of models from Britain and Canadian plants. The Australian Government, in a desire to encourage an expanding manufacturing industry, had created a raft of regulations, which after the second World War, made Ford's ability to compete against General Motors Holden in the Australian car market financially difficult.

Enter Charlie Smith, an enterprising English born Canadian executive, who had climbed his way through the ranks of Ford Canada, to where in 1946, he landed the role of Managing Director at Ford South Africa.

In 1950, Charlie moved again, taking up the position of Managing Director of Ford Australia and so began a period of great change. If Ford were to gain Market Leadership as Charlie desired, investment and government co-operation was required.

Although by 1958 Charlie's vision of a modern manufacturing operation had been realised and he had the new manufacturing facility at Broadmeadows pumping out Fords by the train load, the early days of the new Falcon were marked with suspension failures and minor manufacturing quality issues. That coupled with the number of different languages made communicating with new and unskilled immigrants working on the assembly line difficult and added to Ford Australia's woes. However, there was hope in the number of young Australians gaining experience in the numerous Ford plants around the World

One of the men Noel would work closely with during his years at Ford was Brian Inglis. A young Australian engineer, who after resolving the reliability issues with the first Falcons, rose through the company to become Ford Australia's Chairman. Later knighted for his services to manufacturing and in particular the car industry, Sir Brian Inglis was one of many young rising stars who shaped the face and style of Ford Motor Company Australia.

CHAPTER EIGHTEEN

When Noel first arrived, Ford's tractor people were working out of the Coburg offices at the end of the tram terminus on Sydney Road. They would continue to work from there until new and larger premises were sourced further north in Campbellfield. Noel's introduction to Ford was in complete contrast to that of Massey Ferguson; here he found a company transitioning from the staid British way of doing things to a company operating with a multinational organization's focus.

Like those who were coming to Ford Tractors from other machinery companies, along with men and women from alternate industries, everyone went through a selection process. From the initial interviews, some were placed immediately into roles within the company, while others were singled out to complete an aptitude evaluation at Chandler McLeod, a human relations specialist established in 1959.

Ford Motor Company Australia, like its Dearborn parent, wanted to get their next big move into the agricultural machinery sector right. It required careful selection to find the right people who could be trusted to take the company into the future.

Many of his Massey Ferguson colleagues had chosen to take up Ford's offer and after the first few weeks of training, headed overseas to Ford farms in North America and Europe.

For those employees, it would become an exhausting if illuminating few weeks, they would undergo intensive training, making them expert in every aspect of the new Ford 6X tractor range.

For all, this meant knowing more than how to operate the tractors and understanding every single feature of each model's differences. They began learning to argue a Ford feature against that of a competitor. Aspects of a short-stroke diesel as fitted to the Ford, were compared to that of the Perkins engines in the different Massey Ferguson tractors and, the even longer stroked American engines of the International Harvester tractor range. Piston speed was compared to wear rates at different RPM, these and other facts, were all tools which would be relayed to the dealer network when they arrived home. It was an intensive time.

For sales teams it meant learning how to sell something different. Until the 6X range arrived, the most popular tractors had always been built with long-stroke diesel engines, driving through a variety of manual transmissions. At best, they had live, dual clutch, power-take-offs. Some smaller horsepower models were even more basic. Ford's new offering had been designed with a clean-sheet of paper. All the popular features of the Fordson would be a baseline. However, the complete range would be designed to eliminate parts duplication. New modular, direct injection engines, had to be as easy to start as their predecessors and, although a pre-heater was fitted to the intake manifold, it was rarely needed to make a Ford tractor start under Australian conditions. An independent hydraulic system to operate the three-point-linkage and remote couplings for trailed equipment, was offered either as standard on the higher horsepower models, or as an option on others. These features coupled with the independent-power-take-off, made the model range desirable; Ford knew it, and their planning department had customer surveys to prove it. However, to get over the disappointment of the Ford 6000 released in Australia during 1964, the tractors had to prove their place in the market. And now Australian sales and service teams were overseas, not only learning how to make them winners, but to make certain any teething problems could be quickly remedied.

In Australia Noel found it was a time to learn new things too. After his Chandler McLeod IQ survey, the Ford men, Norm Logie and Gordon Russell who interviewed him were impressed. As well as having an engineering degree, he had ambition. Aware of his thirst for study and self-improvement, Norm suggested that if he was prepared to grow with Ford, studying for a management role could be the start to fulfilling an even bigger dream.

Part of his induction into his role, with that of his colleagues, involved participating in at least one Ford Marketing Institute (FMI) module. Even from their early years, educating their executives had been a priority for management at Ford. As the company matured it developed many courses to help new team members and dealer representatives understand their roles.

The Ford Marketing Offices being situated on St Kilda Road Melbourne, meant Noel had an easy run to work each day. His day began at 8.00 AM and he would finish at 4.30 PM. Not far from the office was a café come coffee shop that Ford would use as and when catering was required. It didn't take long for the staff to recognise Noel and his new colleagues, calling them by name whenever lining up to order lunch. Often, if things had been hectic on his leaving home, Noel would stop in there for breakfast before heading into the office. Doing so allowed him to clear his mind from the study of the night before, ensuring he was able to start his day fresh.

On his first day he assembled with his colleagues and FMI lecturers, Gilson Moore and David Lockwood to wait for an address from the managing director, Brian Inglis. Their mood was sombre and although a few of his contemporaries tried to crack a joke or two, the pressure of the boss's visit could be felt in the room. More than once Noel straightened his tie and checked his shoes, looking across to his colleagues he found they were doing the same.

Brian was almost Ford Australia royalty; his father and uncle both worked for the company, not only that, but Brian had served during the Second World War flying Spitfires in a protection detail for the Normandy Landings.

After his time at Massey Ferguson and remembering the advice given to him by his grandfather, Noel knew you needed to know as much as you could about the man employing you. However, as important was knowing who were the key people in your line of command. Those who could help you and the others determined to hinder you. So before starting at St Kilda Road, Noel researched the people he would report to; everyone, up to and including Brian Inglis, the man at the very top of the Ford management tree.

Noel took a lot from that first address, the way Brian was presented, the easy manner with which he spoke, never once making anyone feel his position was more important than theirs.

The address focused more on the way the company functioned, explaining how management's role was to ensure that if the right people were engaged in achieving their personal goals, then the company would prosper too,

However, he also stated that many candidates had applied for this course and it was Noel and his classmates, who were the chosen ones. He was sure each had been selected because of their aptitude, their willingness to work and that the selection panel believed each brought with them an individual set of skills, which would see them benefit Ford Australia in different ways. Because of that, they owed it to the many who were overlooked to not take this opportunity for granted. It may be one thing to be selected, but you had better work hard to prove to your peers that your selection was the right one.

After the address a morning break was scheduled offering an opportunity for the boss to mix with his newest intake of

FMI trainees. When it came time for Noel to meet Brian, it was a meeting of similar minds and Noel was left with the assurance the boss's door would always be open to him, if he earned it.

Of the courses Noel undertook, one focused on Sales Management, another on Dealership Management, and the other was Finance. At times either Norm Logie, or Gordon Russel from tractors would call in to check on Noel's progress and it was during one of these visits that Noel realised they had marked him out for bigger things. Norm had taken Noel away from the others, explaining to him that if he were to apply himself, identify and eliminate any deficiencies in his education, then Norm could see Noel's career skyrocketing him into senior management roles. He also pointed out that for an executive to push his way into senior management, Ford insisted the candidate have an accounting or finance background. He knew credits from Noel's engineering degree would apply to a Business Management course and recommended he investigate acquiring that training with haste. Saying he knew it would require sacrifice, but if Noel wanted to find out how far he could go in Ford, he needed this.

Noel wasn't so sure; Massey Ferguson had made similar promises. However, Norm assured him if he were to do it, then his effort would be well rewarded within the company, but only if he applied himself.

Noel remembered his first interview and how Norm and Gordon were impressed that, during his apprenticeship he had taken Charlie's advice and completed his high school studies after hours, receiving marks high enough to secure a position at the University of Melbourne.

They had complimented him on his drive and the organisational skills he had applied, allowing him to earn a living with the Victorian Civil Ambulance while completing an engineering degree. They were just as impressed that Noel had undertaken a Dale Carnegie training course to eliminate his nervousness when it came to public speaking. Gordon was smiling when he told Noel that Ford always expected perfection from a manager when addressing a group, no matter its size. These two men would have a great influence on Noel's life at Ford, always finding their door open whenever he needed advice.

Although participating in the three FMI courses showed him how to read a spreadsheet and transfer that data into forward planning. It revealed another tool and although complex, Ford's famous DA-60, designed to help dealers manage their finances and predict trends. To him it demonstrated how important it was for a manager of any business, no matter its size, to understand and evaluate a cost/benefit analysis.

Those first few weeks of study and the memory of Brian Inglis' first address gave him cause to analyse his own strengths and weaknesses. Taking Norm Logie's advice and knowing his accounting skills could be further improved, Noel believed if he pursued advancing his career at Ford, he needed those extra qualifications. Discussing his situation with the Melbourne University admissions officer, he had it confirmed, that having completed his engineering degree, meant credits would be applied to his Business Finance studies.

After talking it through with Joan and with encouragement from his FMI trainers, found him enrolling with the University of Melbourne. Along with time allowed him by the Ford Marketing Institute and studying at night, he added a qualification in Business Finance to his Engineering degree.

The Ford Marketing Institute was a great time for Noel, and being the only representative from the tractor division among the nine students from different departments of Ford Australia, had marked him as one of the people who could lead the tractor company one day.

Having a love for routine, meant he was at home every night and this with a much-improved salary, over the offering from Massey Ferguson and not having to maintain a car meant the Howards had more money coming in and securing their future.

His mind focussed on finishing his FMI training Business Finance degree, Noel had little knowledge that behind the scene, Norm Logie had been working on his next challenge. Norm wanted to test him selecting him to step into the Riverina Zone Manager role.

Modes of transport have always been important to commercial travellers and company reps, no matter which organization they worked for. Their vehicle separated them and gave them status. Of high importance, your vehicle had to be different from the cars driven by the thousands of government employees, who were forced into driving vehicles with the personality of a window envelope.

On joining Ford, Noel had chosen a Mk 1 Cortina GT from the allowed vehicle list. Different roles within the company meant your choice of car was tied to your salary group. As a manager moved up the different levels, their salary scale changed offering a more extensive range of vehicle choices. Starting in a junior management role decided your choice between a range of base model Falcons, Zephyrs and higher specified Cortinas.

The Cortina GT offered him a great car to zip around the city with. A willing 1500cc engine, close-ratio 4speed gearbox and high-speed differential, made it a comfortable cruiser for country trips too. Chrome strips that arrowed into a point near the headlight flashed along its flanks, telling onlookers his was a car from a family of winning competition vehicles. It didn't hurt the young executive's prestige to be seen driving one either.

Being introduced to the way Ford was forward planning, identifying and selecting its executives for an upcoming role was interesting to Noel and many of the key people he met at Ford only had encouraging words for him. These were people whom he reported to. They expected nothing short of truth and to him, that was a safe place to be. Truth and trust, had always been his stock in trade.

Unlike at Massey where it was impossible to get an audience with top management, at Ford he felt secure and able to criticise a policy or product, providing he had an alternative to back up his fears. Access to people above and below him in the management chain was encouraged, far different to that of his last employer.

Although he had access to people in the Engineering Department and others in Supply at Massey Ferguson, there was always a *Red Ceiling*. Senior managers believed they were a class above others and not prepared to mix with those a rank or two below them. At Ford, change was in the air and the atmosphere was light. Everyone, readying themselves, improving their knowledge, learning new skills in readiness for the prospective new model release and their coming battle with Massey Ferguson, International Harvester and Chamberlain.

CHAPTER NINETEEN

Still living at Vermont Noel's time with FMI had finished and by the latter half of 1964 he found himself working out of the Coburg offices. When Norm offered him the chance to step into the role of Riverina Zone Manager, he grabbed it. It would mean a lot of time away from home again, but he was confident that in doing so the rewards would come.

He would still be leaving home early on Monday mornings to arrive in time for the 8.00 AM sales meeting. After that meeting had concluded, there would be time set aside to go over things with other department managers. These meetings would follow up on the questions he'd asked during his dealer visits the previous week, how they were being actioned and if it was necessary for him to report back to the source of the enquiry. Those rounds completed, he would be free to set off for the zone and often he would return home until late Friday.

The territory was a large one, going into New South Wales as far north as Ardlethan. East to Gundagai taking in the rural wheat and wool country around Urana and Lockhart. West to the river towns of Narrandera, Leeton and Griffith. South as far as Wangaratta. This was a territory comprising of smaller dealers selling cars, trucks and tractors, then there were the major dealerships focused on either vehicles, or tractors. To make the mix even more confusing, sometimes these large dealerships were very organised and marketed the whole Ford product offering. Although this would be enough to keep one zone manager busy Noel's Victorian Sales Manager Gordon Russel added Tasmania to the portfolio. Now he had Hobart, Launceston, Devonport and Burnie introduced into the mix.

Malcolm Moore were the Ford tractor dealers charged with providing an excellent distribution through its centres in several large Tasmanian towns. A Melbourne based manufacturer that had a history making roadmaking and other industrial machines beginning with horse drawn equipment. Fordson tractors had offered motive power for an early range of road graders and front-end-loaders. Like Ford they had developed into a very structured and vertically integrated organization, promoting the bright and keeping the skilled. As distributor for Ford's Industrial tractor range, they were also well placed to manage the parts distribution to dealers.

This would be a big area to cover and knowing Ford wanted nothing less than market leadership, Noel understood the importance of the opportunity. He knew Ford Australia having split the management of their tractor and implement business away from the vehicle side, it would be the tractor company offering him a chance to shine. His Grandfather had always told him to look for opportunity in change, committing himself to meeting the goals of his immediate boss and those his boss answered to would secure that goal. If he could do that, his career with Ford Tractors could lead to anything, and for now, time was of the essence.

Each Zone Manager knew that while market share was their key to being considered for a loftier position, a dealer's outstanding whole-goods account and, unresolved service issues would slow that ambition. Being a man given to organisation and tidiness, Noel made sure he had a co-ordinated and planned routine for his dealer visits. He organised his monthly itinerary around those dealers who made the most sales, the dealers who, with his help would make even more sales and the smaller ones who he knew would respond with a visit. Then there were those dealers with outstanding credit issues, they found themselves being his top priority.

Once he had a plan, it could be changed, without it he knew his time as a Zone Manager would end in chaos. His itinerary saved time and at the completion of each visit, having the dealer sign his report made sense. For the dealer, he had a carbon copy record of the one Noel would be presenting to his superiors and there was trust in that.

Having a good grounding in the ways of a dealership, through working with Charlie Milthorpe and the Gippsland and Northern Co-Operative was helping him understand the dealer's position. However, his Business Finance degree and FMI training gave him the tools to assess the potential for each dealer and their business. All he had to do was find the dealers who would agree to sharing the information required to complete the DA-60. Once an agreement was reached, he would show the dealer's bookkeeping staff how to find the information and then collate it on the official Ford spreadsheet.

Completed information would be returned to Ford's computational staff who entered the data, completing a report comparing the dealer's performance against State and National averages. Receiving the reports, Noel, the dealer principal and his team could see the strengths and weakness of each department. Noel believed it important they agree on a plan of action to strengthen departments that needed attention and on his next visit they could review the performance again. Eventually dealers began making more sales and retaining more margin. During the sixties, the DA60 became an important management tool to many Ford dealers.

Knowing he had the trust of Gordon Russel and direct access to Norm Logie, gave Noel the confidence to build a case for keeping or letting a dealer go. His role had a much larger scope than dealer development. If the dealer operated selling both Ford vehicles and tractors, then the zone manager would work with the dealer principal to identify one person within the business who would oversee everything to do with their tractor operation.

For Ford Tractors to maximise sales, they needed a single contact within the dealership before going to the dealer principal. If it were a large business, then zone managers were expected to help them to design a working structure complete with responsibilities and actions. Smaller dealerships had to prove a willingness to move with the company and its desire for both growing market share and profitability.

Long before he had been assigned a zone, Gordon Russel not wanting to send young charge out cold calling on dealers without at first learning how one of his more experienced Zone Managers worked, had asked Keith Eagle zone manager for Western Victoria, to *show Noel the ropes*. This became the first introduction to the Ford way and it started Noel thinking about building his foundation on his FMI studies.

Noel remembers the day well, it was Tuesday the 25th of February 1964, he and Keith were visiting dealers in Western Victoria and travelling from Nhill to Warracknabeal when news of the Australian Aircraft Carrier, Melbourne cutting the Destroyer class Voyager in two during a manoeuvre, A misunderstanding had gone horribly wrong and eighty-two people died. The mood in their car immediately filled with a sombre and crushing silence.

Gordon sending Noel with the experienced Keith proved to be an inspired move, it showed his young charge how Ford expected its officials to conduct themselves. During this week away they covered quite a lot of territory stopping at Nhill Warracknabeal, Horsham, Bendigo, Hamilton, Ballarat and Geelong.

At every dealership Keith introduced Noel as the recruit, someone who was well versed in both dealership and zone manager duties. Keith explained to each dealer Principal, that

Noel had completed his apprenticeship as an A Grade motor mechanic, and put himself through school and university by working as a paramedic with the Civil Ambulance. When asked what Noel had studied Keith's face would light up as if he had unearthed hidden treasure explain it was an engineering degree. It made Noel feel as if Keith had genuine pride in his new recruit's achievements.

By the end of the week Noel knew how one of Gordon Russell's well-regarded zone managers operated and promised himself to follow suit. Keith was a legend in the western district and his induction would stay with Noel until he too had to manage a zone.

Having responsibility of his own zone, Noel wasted no time and within the first few weeks he had built trust with those he reported to and the dealers he served. Taking a lead from Keith Eagle, he structured his itinerary in a logical way, calling ahead and making appointments with each dealer. Allowing enough time to complete a dealer visit and travel to his next appointment, was little different to the habit he developed at Massey Ferguson. However, without the added burden of service, it did feel more relaxed. Making accommodation arrangements at the same time eliminated the need to worry about where he would spend the night.

Through his months at FMI and his university studies he learned how the DA-60 could be an illuminating document and an essential management tool. Within a few weeks he began a campaign to ensure all his dealers were using it.

The document was an accounting and planning tool. Data from each dealer across Australia would be collated by the company's finance division and would have information from the Bureau of Statistics added. It gave Ford a seasonal picture of their current position, comparing each dealer's monthly and year to date forecasting against that of neighbouring dealers and the Australian market average.

For the dealer, it showed them where their business was positioned compared to others of a similar size and their closest Ford Tractor dealers. The key to getting dealers to co-operate was confidentiality and confidentiality was not just a strength for Rhoden Noel Howard, it was the main tenet of his modus operandi.

Originally, he experienced some resistance and after careful explanation most could see the benefits on offer. Others believed their business had the right checks and balances in place. Then there were the small dealers who sold only a few units a year, dealers who couldn't justify the bookkeeping expense. These were few and their numbers may have either been thrown out because of their volatile nature, have jaundiced the report through laziness of the reporter, or the dealer's bookkeeping may have been too limited to have their data required.

Noel continued to work with each dealer, interrogating suspect lists and determining if they could be hastened into prospects. Identifying the people who were likely to buy, those who if the *Man from Ford* visited would help result in a sale and those who were time wasters. Sometimes these clients would be single tractor farmers, but as the wine growing and tobacco farms were becoming industrialised, multiple tractor sales were made.

Meeting and adding influential people outside the company found Noel's circle of influence growing with acquaintances. These were people who he could help, and those he could be helped by.

Hughie Condon was one such man, Hughie's father had started his business Great Southern Motors in Wagga Wagga New South Wales in 1928. Over time Hughie had helped many an aspiring young farmer to achieve their goals. Well known throughout his own area he knew the great and the good as well as those who were lesser known. Noel found in Hughie; a friend able to open doors of people who were unlikely to greet a tractor salesman.

Larger dealerships like Great Southern Motors were receptive to using their premises for training purposes and as others before him had done, Noel took the opportunity to conduct sales and product training there. As Ford progressed into hay harvesting machinery, Noel having sold and serviced mowers and balers since his Milthorpe Motors days found himself and leading the training on occasion. Although his work had become more administrative, he would jump at the chance to use his mechanical skills.

If his diary was his Bible, then his briefcase and the files in it, were his battle plan. The diary dictated his itinerary, the itinerary determined which dealer was important and his briefcase held the secrets of engagement.

Having a logical and easy to manage system made his managerial life easy. Like everything he had learned up until now, it was order that created easiness. However, something being easy didn't mean hard decisions went away, but order helped determine how to treat a problem and how to identify factors that could turn the tables into finding an opportunity in a bleak situation.

He remembered as a boy, telling his grandfather when the old man was searching for something he had on his desk and couldn't find, how one of his teachers had told him, memory was like any other muscle, it needed exercising. His grandfather only smiled, lifted a hard-worn clipboard from under a pile of papers in his in tray. On it a list of names some in pencil others in ink and others crossed out with a new name scrawled below them.

'Yes,' and Noel saw the glinting joy of victory in his grandfather's eyes as he replied, 'however, if you employ a lever, the exercised muscle is even more powerful.'

It took a while for Noel to understand the full weight of his grandfather's remarks, but he knew his diary, all the dealer files and his lists of names, all of these tools were a lever for his memory. Weapons of routine, for it was always routine and order that kept his memory sharp.

Understanding how important it was to keep the confidence of competing or neighboring dealerships, Noel ensured only the relevant files would be in his briefcase. It would only ever contain the results and a copy of the dealer's last DA-60, the action plan they had agreed to on the last visit, any recent sales bulletins or new marketing programs, a copy of the dealer's last finance statement, and a list of any company stock the dealer may be holding. Other paperwork of importance the dealer should have had on file and if a letter or bulletin had been missed, Noel always kept his copy in a separate file-case locked in trunk of the car. It was a system that worked for him and brought results.

Working with key dealer sales people followed a similar pattern. Reviewing the latest price list and sales bulletins. Ensuring the prospect list was up to date and comparing notes with those from the last visit. However, targeting and going to visit the prospective buyers they had determined important, offered the most joy. Shaking a farmer's hand as the *Man from Ford* was always rewarding. If they had a new machine to introduce, helping to explain the product and offering insight into the way it would benefit a farmer, often became the beginning of long-standing friendships. On many occasions the deal the salesman had offered didn't change, but having Noel present as the company's representative to help write the order, did.

It was pleasing that by employing the same system he had used to great effect during his time at Massey Ferguson was now getting similar results at Ford Tractors. Working with Sid Mills had been a wonderful experience and he was grateful for Sid's friendship. Often, he would find a smile crossing his face as he remembered the two of them calling on a customer and almost becoming a double act while they explained the purpose of their visit. Theirs became such an effective routine that it helped Sid and his business to achieve an unprecedented call to sales ratio, and now Noel was repeating that experience with a new and different group of people.

CHAPTER TWENTY

After serving his time as Riverina Zone Manager, in late 1966 Noel had applied for and secured the position of State Manager, South Australia. It required a move interstate where he would base himself in the Franklin Street office in Adelaide. His first management appointment meant he had several staff reporting to him, and for the family, it meant shifting interstate and therefore finding another home had priority.

The house Noel had found for them was less than it appeared. Before the furniture had arrived, they were staying in an Adelaide Hotel and had only one fleeting visit to the modern brick home, situated on a pleasant street in the suburb of Paradise, before moving in.

While unpacking boxes, Joan was intrigued by one of her neighbours and the number visitors they received. On the first day, cars came and went at regular intervals and it started to unsettle her. How could the growing number of male callers have been relatives? Phoning Noel, she told him she was convinced they had pitched camp alongside an illegal brothel, and he had better drop everything and find them another place to live... She told him she didn't care if they continued staying in a hotel, but the family would be moving out that day.

The estate agents showed them several places before they decided on a nice home in Tranmere, one of Adelaide's better suburbs that provided good transport options for travel into the city.

South Australia was a sound appointment for Noel, he was replacing Don Mason who had moved to take over a similar role in Sydney. there were a couple of less than competent staff members to deal with. The state was emerging from the ravages from the drought the year before and although great tracts of the state were taken up by cereal and sheep farming, other than the tropics, it presented him with a microcosm of Australia's primary industry.

Higher rainfall areas like the Adelaide Hills, Mount Lofty Ranges, the Coorong and Mount Gambier regions, were often small acre dairy farms who bought smaller horsepower tractors. The Barossa Valley and to a lesser extent Clare Valley were populated with vineyards surrounded by grain growing areas. Dealers in these areas were likely to sell the whole range of Ford tractors. So too the dealers selling to buyers in the vineyard and citrus orchards that peppered the "Soldier Settler" irrigation blocks dotted along the Murray River from Renmark to Morgan. Although some wheat farming was popular it was the grapes citrus and numerous stone fruit growers dominated the region's primary industry.

When Surveyor General, Goyder presented his report to the South Australian Parliament in November 1856, the agricultural land within state's borders had been divided into arable and nonarable land. He described an imaginary line, which would later to take his name, it almost duplicated the ten-inch rainfall line and he declared the land to the south suitable for farming. To the north, he believed a living could be made only on large acreage sheep or cattle operations.

Outside of his line, it was only the irrigation schemes that flourished along the Murray River that had made the dryland along its fringe profitable. The farmland to the south of Goyder's line was good productive wheat country and by the time Noel arrived, he found wheat farmers intent on increasing their productivity were looking for higher horsepower tractors and wider tillage equipment, who were buying competitive products. A lot of this plant and equipment Ford Tractors were unable to supply. It was not an easy role he found himself in, but Noel squared his shoulders and prepared to meet the challenges ahead of him.

During the meeting confirming his appointment with the General Manager of Ford Asia Pacific, Ewan Scott-McKenzie and General Manager Australia, Jim Parker, both told him South Australia would not be a territory he could take for granted. The dealer base had remained almost unchanged for a long number of years, often different dealerships were owned by members of the one extended family with every father, brother, or cousin determined to out-manoeuvre the other. At times they would try to draw the company into territorial disputes. However, should the company try to push one of them in a direction they were uncomfortable with, then it wasn't the one dealer, it was the whole family who pushed back. Jim clapped him on the shoulder saying he knew Noel would have it all sorted soon after arriving. Noel thought it all sounded hopeful and something he could handle. That was until Scott-McKenzie, walking him out of the office, shook his hand, smiled and said Noel should know that if he mastered the Murdocks, then the often-rowdy South Australian Dealer Council meetings should be a doddle.

The South Australian tractor market was divided between several importers including Ford and the West Australian company Chamberlain. Chamberlain were on the march and had been since pleading with the Australian Government when subsidies were slashed from forty percent to twenty-five, in 1952.

Chamberlain had been selling direct to the customer until then. However, with the collapse of wheat prices and the arrival of the new Fordson models their business model was under threat. Fordson Majors being offered in both kerosene and diesel models, offered a strong reason to buy. They also saw that by buying local, their dealer carried spare parts and if a part wasn't on the shelf, they were able to take advantage of a national parts supply. Combined with the dealer's technicians having factory-based training and being further supported by company personnel, was an ongoing objection Chamberlain's sales staff had to overcome.

Without a distribution network Chamberlain would never be ready to compete on even terms. Although with the current subsidy arrangement, even terms would never be to their advantage. However, an increasing inventory of unsold tractors called for a different plan. By the mid-fifties, Chamberlain needed to change strategy and change fast.

Change fast they did, within a couple of years, Chamberlain had recruited hard and charged their sales people with incentives to secure a dealer network. At the same time, they knew that tractors alone would not sustain a dealer network and began designing and building tillage equipment to complement the Chamberlain tractor line.

With the dealers in place, it was important to take advantage where they could and so began an aggressive plan for Chamberlain to become sales leaders across all categories. By the time Noel Howard reached South Australia, Chamberlain, taking advantage of the tractor bounty system was gaining the upper hand.

Although he had met some of the dealers at conferences and product release meetings, this time it was he who had charge of Ford Tractor's representation and he who had to reinforce the benefits his leadership would bring.

After his first tour of the primary dealers in each zone Noel leaned on his FMI Training and began preparing an action plan. As he had done with his dealers as a zone manager, he determined the team should triage their territories as a priority. It wasn't easy getting the confidence of his team and having them divide their dealers into those who would grow the Ford Tractor business without intervention. Those who could grow with assistance. Then there were the others, the more difficult, small dealers who would remain stagnant no matter how much company support they offered. These were dealers who through personality or circumstance, were people who would continue to trundle along at their own speed. He suggested adding a fourth group to the plan, dealers who should go.

Some dealers in the last category, were likely to be the ones who could become a more difficult group to action. Often, they were rusted on family businesses, which were by now managed by sons, or grandsons of original dealers who had an association with Ford going back to the first days of Henry Ford I

CHAPTER TWENTY-ONE

Having a love of order, Noel had taken pride in being well prepared before he left the office to travel into the zone. Now in charge of a sales team, he wanted to be sure they all possessed similar attributes, making it his priority to interview and review the performance of everybody. He identified those who needed help, those who didn't and organised specific training to bring everyone's skills into line.

For him dealer visits should be simple, almost going back to his days as an ambulance officer and establishing a triage of an accident scene. Back then, he needed to know who was too sick to be helped, those who could wait for help, and those who were unlikely to make it with or without intervention. To be effective he needed to access the situation and act accordingly.

A visit to the dealer should be managed the same way, he knew his Zone Managers had great resources to support them. However, he recognised how they went about sorting the urgent from the important, may require direction from him and he started a program to develop the skills of everyone under his command. Contacting the personnel department, he requested everyone's position description and pay-scale, including his own. The information before him, Noel began the task of reviewing each staff member's file and making notes for a later review.

His actions were warranted as rumours of misconduct had reached him before taking the position and he wanted to stop it before they disrupted the efficiency of the whole team. One instance, revolved around a service manager who had developed a gambling addiction, a problem which had been interfering with the ability to function in his role. Suspecting the individual concerned would attend the Balaklava Races, another staff member was dispatched to confirm his attendance. Advising him to seek help and get his life back in order Noel under advisement from personnel, had no alternative than to fire him.

Only one other staff member was found to be working outside of the rules. Ford Motor Company had specific policies around the sale of company vehicles and it proved a valuable lesson for Noel. One of the Adelaide team was responsible for disposing of company cars that had reached, or were near their end of service date. The employee, seeing an opportunity, pocketed the funds he'd received for a vehicle, intending to pay it back later. Unfortunately for him the fraud was found and luckily, he was saved by loyal friends and colleagues from the same office who agreed to pooling enough funds to cover the debt.

After the matter had been brought to Noel's attention, he called the staff together and spelled out the company's policy on such a matter. Including going on further to explain the incident was a criminal offence and had it not been for the fact that the money had been paid, the company would have been expected to report it. Insisting that everyone know by covering up the fraud by paying the money, they too had been complicit in the action. From what could have been a stain on his record and that of the staff, became a lesson well learned.

When Noel first started interviewing everyone for their annual appraisal, he had a paper for every staff member, their history and what they thought their future with Ford might be. The interview complete, the employee would have their position description agreed and signed and the performance review sheet marked with any role they wished to pursue noted. Noel would point out the markers each employee would need to achieve to enable them to move ahead in the company and suggesting any training they would need, which would help them in achieving their goals

Zone managers were expected to follow Ford's strict practices for working their zones, which many complained about it being tedious was not as difficult as it seemed. However, Noel further supported them by insisting wherever and whenever possible, each manager should follow Noel's own habit when visiting dealers. It wasn't mandatory that they followed his method to the letter, but it was mandatory that all Ford's policies were completed and by employing his routine they eliminated ambiguity.

He showed them his methods of keeping track of everything from how to determine if machines were being sold

"Out of Trust", Ford's term for fraudulent conversion. Warranty transfer was another area for concern and although often hard to detect, Noel showed his managers how keeping a graph to track sales of spare parts and comparing it to delivery and end of warranty dates for equipment sold could often point it out. However, he assured them by judging every claim on its merits would be the best way of ensuring its integrity.

Nothing should have been new to them; everything was out of the FMI training handbook and every one in his sales and service teams had completed the training. However, reminding them what was expected and marking it on their performance review helped him set everyone a base line.

Having a picture of who was who, their capabilities and bolstering their training built a solid sales and service team. That done, achieving an increase in sales became the team's priority. Although there was no sharing of sales statistics between companies, an overall picture could be gleaned by averaging the reports of lost sales and recording each by competitor. It wasn't offering the accuracy vehicle registrations provided, because in South Australia farmers were not required to register their tractors. However, he expected the Zone Managers to be diligent and after compiling their dealer reports should be able to collate competitive sales numbers as close to accurate as possible.

Getting a proper picture of the company's sales performance within the state and then itemised by dealer territory and horsepower was a priority. At best, it was a guess, even worse was if the Zone Manager had not followed their market and the work hadn't been done. Hoping several DA-60s would offer a better picture he found South Australian dealers were not easily given to any paperwork they thought unnecessary.

Knowing the state was emerging from a drought and working with dealers complaining that the now one-year old 6X range of tractors were lacking in horsepower to a few of their competitors, his Zone Managers were reporting an undercurrent of unrest among the dealer body. The 6X Ford 5000 tractor had been released with an eight forward speed transmission, unfortunately two ratios were the same effectively making it a seven speed. The optional Select-o-Speed transmission offered ten *"Change-on-the-Go"* forward speeds and this meant the tractor was a perfect match for operating equipment like balers and trailed power take off driven combine harvesters. However, the machinery manufacturers were now making bigger machines and with farmers growing heavier crops, the 5000 needed more power and the newly released 8000 was priced higher than its competitors.

The tractor that should have been a saviour for dealers in broadacre areas, the 8000, with 110 horsepower had everything. From its flat floored operator's platform, power steering, multiple remote hydraulic couplers. Its styling was of the moment, making it a head turner when it first appeared at the Adelaide show in 1968. However, without a delete option for the three-point-linkage, the tractor was expensive when compared to machines offered by the competition.

Noel knew there were better options coming with the release of the 6Y range, but he was unable to reveal it to his sales team, or to the dealers. Reports from the field confirmed what he expected, the wheat areas needed more horsepower and a more comprehensive range of tillage and seeding equipment. Tractors needed better cabins, a delete three-point-linkage option for the 5000, complete with a sizeable reduction in the recommended retail price. While front end loaders were available from many separate suppliers, most dealers would prefer a one stop shop with everything being supplied on the one Ford Motor Company invoice.

At the time, machines were invoiced to the dealer's Floor Plan Company the moment they left the factory gates. The preferred financier was Australian Guarantee Company (AGC), who charged interest on the unpaid balance on each machine. To prevent tractors or machinery being retailed and the wholesale invoice remaining unpaid after the prescribed forty-eight hours grace period, stock checks were performed at irregular intervals. To offset the floor plan interest dealers were encouraged to sell retail paper (finance/consumer mortgage plans) in return, their outstanding balance would be credited with a commission for offsetting the floor plan charges. For a small dealer in the wheat area, selling retail paper for AGC was often more trouble than it was worth and, with the banks actively discouraging their clients to do so, often resulted in a sale lost to the competitor's more flexible and easier terms.

Noel understood the advantage he had experienced when retailing farm machinery with Gippsland and Northern. That ease of invoice and payment had helped him make sales. At Massey Ferguson, their Early Bird and tailored finance programmes drew the customer to the dealership, helping the retail salesman to close the deal. When representing the Riverina Zone, he had experienced the same push back from dealers in the cereal growing areas, but there, Ford had other dealers servicing a higher volume of farmers who were receiving monthly payments for products. For these dealers, retail paper was the lifeblood of their sales effort. In South Australia, most of the farm income came on an annual cycle and his staff were reporting this was often the primary reason behind the high number of missed sales.

His South Australian appointment coinciding the fiftieth anniversary of Henry Ford I's first mass produced Fordson tractor, Noel knew it was time to call everyone together for his first dealer meeting.

CHAPTER TWENTY-TWO

For his people in administration, a dealer meeting had become just another routine and one they had done many times before. His personal secretary Anne Mullins had secured the Arkaba Hotel on Glen Osmond Road in the inner Adelaide suburb of Fullerton as the venue. Capable and organised, she met with the caterers and made suggestions for the menu. However, this was his first dealer meeting as a state manager and he wanted everything to go smoothly. Being prepared and waiting for dealers to arrive, Noel knew meetings like this could cause a few nervous moments for him, but only if he let himself be sidetracked.

When Noel arrived, he looked around the room and saw jugs of water surrounded by glasses on each table. In front of each chair, a corporate pen and notebook. The scene confirmed Anne's capability. He had nothing to concern himself with here.

From his days at Massey Ferguson and the number of times he had spent preparing material for similar meetings as a zone manager, Noel knew just how much work went into a day like the one before him. Working through each slide for the overhead projector, selecting a promotional film to relax the meeting. His South Australian team had rehearsed it all. Remembering all the other meetings he had attended, either in his dealer capacity with Gippsland and Northern, or as territory manager for Massey, and later as Ford Tractor's Zone Manager, only two areas ever caused friction, product availability and service issues.

Product was his domain and he had prepared well. From information relayed to him by the zone managers of his own state and the data shown in the national figures, Noel was as ready as ever to talk to any questions relating to product supply, forward sales and outside sales pressures.

Ford's immediate future was always front of mind and at this meeting representatives from Head Office in Melbourne would reveal the coming marketing and advertising programmes. Also, for this meeting his General Manager, Jim Parker had flown in from Melbourne and would address the dealers on this front. Accompanied by the National Marketing Manager, John Blyth who would be addressing the company's national sales position. He would also field questions on the impact of Chamberlain and International tractors on the Australian market and explain how the bounty system subsidised their position in the sales arena.

Overseeing his service managers' presentations and ensuring the figures surrounding warranty claims and actions being taken to remedy Service and Parts concerns were accurate, Noel felt confident this side of their business would be handled with little controversy.

Glad-handing began thirty minutes before the scheduled start time and as he watched Jim Parker move around the group with the swashbuckling style of Errol Flynn, Noel took particular notice to see Jim had started with Gavin Sanford-Morgan. Gavin was the Managing Director of Malcolm Moore in Adelaide, Ford Tractors' industrial products dealer and official spare parts distributor. Not lost on Noel was Jim's intention to begin his get-to-know-you session with their largest and most powerful dealer.

In South Australia, Malcom Moore's industrial machinery business possibly commanded a bigger turnover than Ford Tractor's did, so it was appropriate Gavin received Jim's attention first. Gavin had many capable lieutenants running Malcolm Moore's different departments, his tractor sales manager and spare parts manager were expected to attend these meetings, and today they accompanied him.

The meeting followed the plan without much dissention, service and product issues tended to merge. Dealers asking what progress had been made to rectify the 6X 5000 having duplicate gears in the eight-speed transmission. Those selling Select-o-Speed tractors were interested to know if anything had been planned to smooth out the lurch as the transmission changed through its fourth and fifth gears. John explained that, Laurie Curtis of Tumby Bay and his mechanics, working with those from A S Harris & Sons of Bute, had developed a remedy. He called on Laurie to offer an outline and added that Ford's Roger Ferkins, a New Zealand based engineer was coming to Australia to investigate their modification and write a report for Engineering's approval.

One of the sticking points for all dealers from cereal growing areas were concerned with was the lack of a competitive Ford tractor for Chamberlain's Countryman, Massey Ferguson's Super 90, the larger models being offered by International Harvester, Allis Chalmers and Case. Noel understood their side of the discussion because he had experience with the Massey Ferguson Super 90 and knew it well.

Before Ford and being the Area Manager for Massey Ferguson at the time, Sid Mills the Mildura dealer had asked him to show one of his tractor customers from Nangiloc over the Super 90. It meant meeting them at Massey Ferguson's Sunshine factory. So impressed was the vegetable grower that he placed an order with Sid that week. This sale becoming the first of a MF Super 90 in Australia.

The dealers in the room pointed out that it wasn't just the tractors where they were being beaten. Their closest competitors, Massey Ferguson and International Harvester, had tillage and seeding lines to support their tractor offerings. The Ford Motor Company's single franchise policy set them at a disadvantage and this made for a difficult sticking point.

Jim Parker had heard these arguments before and brushed them aside with a well-worn candour. He knew too, that others before him had tried to pressure Ford's Head Office to allow Australian Ford Tractor dealers to find alternate farm machinery to support their businesses. He also knew, management above him had approached Australian manufacturers to single out those who may be prepared to make specific products available for branding under Ford's Blueline range. While it would have made better sense to be branding as Ford, that would have presented another problem because everything carrying a Ford brand had to be officially signed off by the company's team of engineers. An exercise which would add many layers of interference and making the item far too expensive for the Australian market.

While Ford had a relationship with a few of these companies and they did supply a limited number of trailed and three-point-linkage models, what was offered did not meet the needs of a Ford dealer in broadacre wheat areas.

Of the broadacre machinery manufacturers, several were based in South Australia. John Shearer with their factory in the Adelaide suburb of Kilkenny enjoyed a superior reputation, David Shearer of Mannum and possibly the oldest of all manufacturers, Horward Bagshaw, whose origins went back to the first waves of white settlement in Adelaide.

Other and much smaller Australian manufacturers were maintaining a toe hold in their respective states. However, John Shearer was well established and by following strict engineering protocols had forged a solid reputation. Conner-Shea, the company founded by one of Noel's idols and ex Massey engineer Tom Connor, were now well established in Victoria, and distributing nationwide through Dalgety's Stock and Station Agency network, they too had begun taking big chunks out of the market share in S.A.

John Shearer had been manufacturing agricultural machinery products for over seventy years and by 1967 had developed one of the strongest dealer networks to market them. Offering inspired, Buy Now-Pay Later, programmes had been making them ultra-competitive for almost a decade. However, with Dalgety's hundreds of outlets across the South Australian wheat belt, farmers could tailor many different finance options when it came to buying a Conner Shea machine. Every main competitor offered finance options, from deferred floor plan interest to targeted retail plans, all of which were unavailable to every Ford Tractor Dealer in the room.

Dealers from along the Riverland and wine growing regions had similar concerns. International Harvester had a great little competitor to the Ford 3000 in their 414 and 454 tractors. Benefitting from local manufacture and the bounty system, International Harvester dealers too, not only had less expensive tractors to retail, but they had better finance options to support their sales push.

For Ford Tractors, the Riverland dealers were often established in several co-operative stores. Saved by the finance options that these Farmers' Co-operatives offered their members, helped to put a lot of Ford's little blue tractors onto fruit blocks from Morgan to Renmark.

When it was time to close the meeting, Noel looked at his notes. He had underlined, more horsepower several times. Quality tractor cabins, affordable front-end-loaders, better tillage range and finance options also highlighted his list.

However, on the opposite page, he had written TRACTOR BOUNTY in capitals. Below it two columns one for International Harvester and one for Chamberlain. While the meeting had progressed, he had been listing the number of impediments dealers raised regarding the bounty system and the Federal Government's policy of propping up two competing tractor companies. One column for make and model, the other with points illustrating their attempts to gain market share.

West Coast dealers were backed by those in marginal areas and explained that when it came to sales, every attempt Ford was making to increase their share of the market was falling short. The Ford 5000 was too expensive in its current configuration and being unavailable as a drawbar model they had little ability to argue it was a feature with a benefit. Farmers saying as linkage was an option on the Chamberlain, paying for it on a Ford was an unwarranted expense.

Chamberlain were strong performers in wheat areas and he understood the mood of those dealers along Goyder's Line. To lessen Chamberlain's advantage, the 5000 had been offered with a bench seat as an option, but the seat interfered with the linkage controls and not making them any easier to sell. To him it was unfair that their market share was being compared to their contemporaries from more mixed farming and wine regions, in an open forum. International Harvester was able to offer large horsepower tractors sourced from the United States, it added to their armoury, leaving Ford dealers exposed without a competitor to meet the market.

On the West Coast of South Australia, if Chamberlain were winning on price, Massey Ferguson were winning with range. They had everything covered, small to high horsepower tractors complemented by a full range of tillage, seed-drills, mowers, rakes, balers and combine harvesters. It was a competitor Noel knew well. For a farmer, their local Massey dealer was a one stop shop.

Different dealers explained that they had been regularly approached by Jeff Howell from Allis Chalmers and others with offers to supply another tractor or machinery line with which to boost their product offering. Although some were tempted, they were unwilling to give up any ground they had made with earlier sales of Fordsons. The thought of dual franchising was tempting however, with Ford demanding loyalty and their dealers to be single franchise, eliminated that option. A clause in the dealer agreement that was likely to remain there.

Closing the folder and taking a quick look around the room, although a lot of issues had been raised, he felt satisfied that his first meeting had been successful. However, he knew if Ford were to succeed in their quest for leadership in Australia's tractor market there were many more obstacles to overcome. Almost from nowhere he remembered a quote someone once told him, "Credit is the oil of Industry," they had said. He wasn't sure who, where, or why he had even remembered it, but driving home with the sound of his tyres and the beat of traffic, the phrase fixed in his mind. Sound and light flashing like a neon sign at the front of his thoughts.

If he were to make a difference to Ford Tractor's fortunes, better focused sales and credit plans would be central to their dealers making more sales.

CHAPTER TWENTY-THREE

Noel was not expecting the call asking him to fly to Melbourne for a meeting at Head Office. He was not to discuss it with his staff, or those they reported to. To him, it all sounded like the script of a cloak and dagger movie, but when the head of the company calls, he believed you had better answer.

For a fleeting moment his thoughts returned Massey Ferguson. Sure, he had started casting his eye over the list of upcoming staff changes, hoping there would be a vacancy in pay scale above his present position. It was not that he didn't like the dealers, or the people under his command, he just believed he was ready and in need of another test. A bigger challenge, one where he could employ everything he had learned to date and something that would stretch his abilities even further.

While this demand sounded like the one, he experienced when sacked by Massey, because of his interview with Ford. This call had a different feel, more congenial.

Noel knew if opportunity came, it would be because his superiors had been impressed with his results. The company had a policy of always casting their net for suitable candidates among their current roster of managers and selecting them for different roles. In his last performance review, he had indicated his willingness to take on new challenges and in reinforcing his commitment, asked that it be noted on his personal Development Plan. On the flight to Melbourne a dozen or more scenarios play through his mind.

When he reached Ford Tractor's Sydney Road offices, he made his way to Jim Parker's office where Ewan Scott-McKenzie and another man stood to greet him. Introduced by Parker as the Ford of Europe, General Sales Manager, Harry Watson. Harry had landed in Melbourne as part of his visit to Australia. Noel had met Harry before, but it had only been fleeting.

Harry explained, that as part of Henry II's plan to bring all the territories into one multinational company, like Australia before it, South Africa remained a priority in need of restructure. It was not a challenge for those who had to home every night, or for those without heart to invest in it. However, World Head Office was looking for someone prepared to travel to the remotest of African areas and study the market. That same executive needed to be capable of writing and delivering a comprehensive restructuring plan for setting Ford Motor Company South Africa on a path to profitable dominance.

Having learned the skill of selling from dealing with customers in the Kiewa Bakery, to negotiating multiple tractor deals with Sid Mills and sitting at the table with the descendants of German grape growers in the Barossa Valley, he remained quiet. He could hear his mother telling him he had two ears that listen and only one mouth. It had been a great lesson and a quality he wished a few of the sales people he knew possessed the same. Only responding when asked a direct question. He could see the challenges ahead, but believing it was within his skill set and ability, he kept his excitement in check. learning Harry would accompany him knew while he was away made saying yes to the proposal would be easy. However, being away for a full quarter of the year had its downside too.

When it came to his turn to speak, he asked if others had been invited to apply and was met by poker faced colleagues saying only, that Ford World Head Office would always seek the person most suitable for managing the task. Knowing he would be one of a small number invited to apply, he asked what academic criteria would be expected of the applicant. During the answer Noel knew the chance of having another person in the company with his exact suite of skills would be a longshot, he pressed to know more.

In little over an hour, he had agreed to an increase in salary and the terms of compensating him for being away from home for a period that he was told could take as long as three months. Ahead of him now would be to discuss the pros and cons of the appointment with Joan. Although, he understood the difficulties she would face at home on her own with young children to manage, he knew she would be able to see the appointment's rewards too. When they had talked about their future before, she supported every stepping stone in his career and he knew she would encourage him to take it. Having their family financially secure had always been a goal they shared.

Accepting the post, Noel told Harry it was on the proviso Joan agreed and it would depend on the time he would be away from home. However, the challenge ahead thrilled him and as difficult as it was to remain calm, he could hear excitement in his voice. To travel to foreign places, hear the languages of different people and see beasts he had only seen in picture books or on zoos visits, had his heart hoping Joan would see it for the same opportunity he thought it was.

While Murphy's Law was a hard-worn set of rules he learned when growing up, often hearing his father say, 'before you start anything, you must do the other thing first.'

Charlie Milthorpe, not unlike Noel's grandfather had another set of laws too and it always brought a smile when he remembered Charlie's, Law of Opportunity.

'Opportunity is a shy mistress... and if she knocks at your door, you'd better be ready to let her in,' he would say.

South Africa would be no different, this was an opportunity where Noel believed he could extend himself and prove to his superiors' that their confidence in choosing him for the task, was the right decision.

###

Offered this chance at a more influential assignment, Noel started to believing that he had been marked as an up-and-coming talent within Ford Tractors. Somewhere, someone within the organization had spotted something in him and now they were willing to put their faith in his ability. Under the guidance of Harry Watson, he would develop a plan to restructure Ford of South Africa.

Harry had given him a management chart with roles and responsibilities of the people working for Ford of South Africa along with a set of notes briefing him on the task ahead. The region was vast and while they would be based in Port Elizabeth, most of their time would be spent either driving or flying from one dealer point to another.

They would be reporting to Paul Gillis from World Head Office in Michigan and drawing on the resources of the various sales and service teams from South Africa to assist them. At the beginning of the assignment Ron Scott the then Managing Director of Ford South Africa said his door was always open to them. Bill Bayford the General Manager, had offered the services of his personal secretary Lyn Strong, who would type and assemble their findings and bind everything into his report.

Ever since Harry Bennett had been ousted from Ford Motor, the company's executive had employed what later became to be described by the author, Dr Laurence Peter, as the Peter Principle. In his book of the same name, he describes how companies chose to blood, up and coming managers by continuing to promote them until they were out of their depth.

Whether, The Peter Principle, is a thoroughly researched document or not is not important, what it illustrated was Ford's willingness to educate and test managers to find the limits of their capability. The fact that Noel had by his own initiative found a way to educate himself, fund his way through university and eliminate his fear of public speaking, checked many of the boxes he would need if climbing Ford's management ladder.

CHAPTER TWENTY-FOUR

Joan on hearing from Noel that by the time he returned from South Africa, Ford Tractors Australia would have redrawn their regions. He would not be returning to Adelaide to manage the South Australian territory and a return to Melbourne would be imminent. Relocating to Melbourne excited her, even though the timing could have been better. By then she would be well into the last trimester of her pregnancy when Noel arrived back in Australia, but it did give her time to make arrangements for welcoming the arrival of their fourth child at The Women's and Children's hospital in Melbourne. It was, for her, a happy confidence building coincidence, as this was where she had given birth to her first three babies.

Having their Vermont home still let out to Dr Glasson and his family, priority one for Noel and Joan would be finding a new place for the family to grow. Their preference would be for a house in a quiet suburb, somewhere with tree lined streets and to the east of Melbourne. This time Joan insisted she would have the final say on any home they found. There would not be a repeat of the debacle she found on her first introduction to living in Adelaide.

Having lived on the east side of the city most of her life, Joan thought they should search for a house somewhere around Mitcham. It would mean an easy run into the city for her if she ever needed access to the hospitals and other specialist services. Shopping also headed her list of needs, if she could walk to the supermarket, it would offer her small family exercise and fresh air. Next, a good school for Geoffrey, who as a boy with energy to burn and a natural inquisitiveness, needed the type of school which would challenge him intellectually and socially.

Now six Geoffrey had been in his second year of a primary school in the Adelaide suburb of Hectorville and while Noel readied himself for South Africa, Joan searched for a school so their move back to Melbourne would lessen the disruption to the early years of Geoffrey's education. She also hoped that by finding one close to home, walking him home after school would offer everyone exercise. If Geoffrey was too tired to rouse up his siblings all the better, it would leave her more time to concentrate on preparing their evening meal before settling all three of them for the night.

Being close to family would be a welcome change from living in Adelaide, where although the Howards had made friends, it never felt permanent. This time she wanted to be sure she had the welcome mat out to all her family and friends. Having had her mother visit for four weeks while they were in South Australia was a blessing. Joan treasured her company and help with the children. Now everyone hoped to see more of her once they were back in Melbourne.

With a list of things to do and being in the last months of a pregnancy, Joan had her hands full and any help that might come her way would be most welcome.

Watching the doors close on the removal van, Joan settled Tania into the child seat in between the boys. Their car was a Fairlane that Ford had assigned to Noel for personal and business use. Leaning over the back of the front seats, she locked the rear passenger doors and happy that the children were as safe as she could make them, slid into her own seat and looked across the yard to find Noel.

Taking a final walk through the house and after pulling the front door shut, Noel walked to the meter box turning the power off at the main switch, next were the gas and water mains. All that was left was to drop the key into the estate agent and begin the drive east past the Toll Gate and into the Adelaide hills.

The road, although being four lanes, snaked its way along the face of the Adelaide Hills, always twisting, always climbing. At Eagle on the Hill, Joan caught a glimpse of the city behind her. From Glenelg, north past West Beach and futher north to Outer Harbour, St Vincent's Gulf seemed to be waving them goodbye. She raised her left hand to use the vanity mirror in the sun-visor and watch it a little longer. When it next came into view, she called to the children to look out of the window and wave Adelaide farewell. She too, raised her hand and twinkle waved goodbye to the City of Churches.

Small towns dotted the highway to Murray Bridge and as usual semi-trailers packed the narrow road, shuffling east and west with freight. Approaching Littlehampton, they caught the furniture truck and it was another chance to distract the children as Noel used the power of the V8 engine to overtake it.

In Littlehampton, Noel concentrated on the different machinery and lumber yards at the eastern end of town, he slowed to look at the presentation of the second-hand tractors in the front of "Garwood Machinery". No matter how personable the original dealer, Les Garwood may have been, Noel's first impression was that this dealer appointment had never been the best representation of Ford tractors.

One of his early tasks when arriving in South Australia was to replace Les as the dealer and installing Peter Murray as the Dealer Principal. It was a sensible outcome, Les kept his mechanical workshop while Peter owned the sales side of the arrangement. For Ford, Peter was a perfect fit. He had a keen sales focus and coming from a large Massey Ferguson dealership in Warrnambool, knew what was required to make a profit from both the new and used sides of the business. Service and warranty work would be carried out by Les and his experienced team of technicians.

They were almost through the Mount Lofty Ranges as they left Littlehampton and heading toward Melbourne. Still winding down to the plains country and on toward Murray Bridge. Here the Murray River flowed under the first railway bridge to cross it in March 1879. There they could stop, give the children a chance to play while he refuelled the car before they continued their trip home to Victoria.

Knowing Ford of South Africa was yet to separate the tractor and vehicle business, his eyes were pulled to the Ford Dealership on his left. A showroom bristling with Falcons, Fairlanes and Cortinas, this was a picture of corporate strength. Alongside the showroom a yard full of second-hand cars, alongside that the tractor yard where the range of new 6X tractors lined the street. Beyond that Ford trucks and Falcon utilities stood in strength. He thought about the attitudes of both dealers and how different their approach to servicing the customer's needs was. Garwood's, a tractor and machinery only dealership, had evolved from an opportunity to serve the locals around Littlehampton and their need for mechanical repairs.

He had visited Esmond Park Motors before. In them, Noel saw a dealership used to maintaining a high sales volume across all profit centres. Cars, trucks, tractors, parts and service all had to perform. Salesmen here were well versed in selling retail paper for new and used vehicles of any description and it was a great illustration of what easing the lines of credit could do. However, he also knew with the vagaries of drought and pressure on interest rates, everything in a business of this size could be volatile. In a bad situation, who was the dealer working for, himself or the credit company?

Passing over the river the boys laughed at the noise the tyres were making as they rumbled over the bridge. Soon Noel was winding up the rise of the far bank and stopping in the parking lot of the first roadhouse on his left. Roadhouses dotted each side of the highway on both sides of the town. Murray Bridge had become a huge transport Hub by 1968 and an endless line of interstate trucks hauled into town by the minute. Roadhouse cafeterias offered a restaurant service of three course meals to ordinary travellers and the truckies often had a separate dining room. Over the counter, a menu of pies, pasties, milkshakes and other take away food were always available.

Stopping in Murray Bridge allowed the Howard family to stretch their legs and after letting the boys burn off a bit of energy, Joan and Noel were soon back on the road. The noise of road lulled the rear seat passengers to sleep and as the car became quiet, Noel began to think about the difference in the two dealerships they had recently passed. He understood their different approaches to financing their retail sales and how it either secured their business or placed it at financial risk. Picturing both dealerships in his mind, caused him to ruminate on the financing question until they reached the South Australian and Victorian border. No matter which scenario he ran through his mind, an easy answer under Ford's current retail finance policy wasn't easy to find.

On either side of the road through the West Wimmera, farmers had been cutting swathes around fence lines in readiness for baling. The wheat was still green and random paddocks of oat crops had been mowed and raked ready for baling. The Wimmera was primarily wheat country. Country that demanded bigger tractors, wider machinery and faster harvesting equipment. Here, as on the West Coast of South Australia, farmers demands were consistent. Australia's success in farming wheat, had caused the same industry to introduce quotas for grain being delivered to the Australian Wheat Board in 1969.

Dealers were reporting to their Zone Managers that most of the wheat areas were having a better than average season and that, although their customers would be cashed up, they would struggle to make tractor sales in the coming year. Because of Ford's one franchise policy they would miss sales because their clients were intending to put the bulk of their capital expenditure toward on farm grain storage and Ford was not in that business.

There had been rumours of a vine pull due to a coming grape glut and this along with the wheat quotas, added weight to the proposal before him. Now, was exactly the right time to make his move into Special Assignment. Looking across to Joan, he caught a reflection of his face in her sunglasses, he was smiling. Embarking on another adventure lifted his spirits, he reached across and rubbed his wife's arm and she smiled back, it was a happy time.

All the way east in towns large and small, dealerships from a myriad of different manufacturers lined either side of the

Dukes Highway. These were all businesses of different sizes, in one town there may be a General Motors Holden dealer with a Chamberlain franchise, in the next town a Chrysler dealer may have a combination of BMC cars and a Chamberlain, or Massey Ferguson franchises. All of this had him thinking about Ford's one product policy and how if strengthened could make them a formidable force.

Having passed through Bordertown, Joan leaned across and with a hint of sarcasm in her voice asked why Noel hadn't stopped to call in on his former Massey Ferguson colleague Dave Cunningham. Dave was now the Massey dealer in the town. She knew just how betrayed Noel had felt by Dave's actions after seeing him at the Ford Motor Company interviews.

Noel just smiled back at her; Ford was now offering him opportunities he thought would have been out of reach at Massey Ferguson. In his own way, and if he had let his employer know about Noel and the Ford interview, then Dave had done Noel a favour.

From Murray Bridge to Bordertown the topography had been reasonably flat and since discovering South Australia's south-east country required trace elements, farmers had begun to see better returns from grain and wool. Now the country changed again, still sandy but with more loam, they had reached the Wimmera where he started his career with Massey and later, been shown around the Ford dealers with Keith Eagle. He was home in Victoria and familiar sights now seen through experienced eyes had him noticing things in a different way. Near a farmhouse before Pimpinio, all the farmer's machinery was from International Harvester, the truck, the tractors, the tillage and seeding, even the mowing and baling equipment had been built by the same manufacturer. Only in the driveway alongside the house, a Valiant car from Chrysler appeared to be a concession.

The mood in the car remained quiet as they motored through Horsham and as he looked at the different car dealers, truck sellers, and machinery yards. He thought about how striking it would be if Ford Tractors could have a consistent image and, at the same time offer enough product to prevent other companies pressuring dealers to take on different tillage seeding and harvesting franchises. A one stop shop made sense to him, and although it seemed logical, Noel knew that it was not a decision made at his pay-scale, but it didn't mean he should forget about it either.

Arriving in Melbourne before dark, it would be a few days before their furniture and Joan's car arrived. For the next few days, they would stay with one of Noel's colleagues from his Massey Ferguson days, Jack Thake. Jack had been living alone after going through a divorce years earlier, his was a lovely large home in Doncaster East and not far from their Mitcham rental.

Jack being a close friend made sure everyone was welcome. Apart from her normal chores, during the next day Joan found a school for Geoffrey and enrolled Peter in kindergarten. Once their furniture arrived at the house on Orient Avenue in Mitcham, she set about making it into their home. Directing the removal men to put the furniture into the right rooms and after making doubly sure they had carried each box to its proper destination, she waved them goodbye.

It took a while to settle in, but with a few homely touches and some afterhours handiwork from Noel, they soon began to see the place as home.

CHAPTER TWENTY-FIVE

After their first meeting, Harry Watson had intimated to Noel that because of the wheat quotas and general decline in the agricultural sector, World Headquarters would be looking to use this as an opportunity to restructure Australia's staffing requirements. He also pointed out that one of the benefits of being engaged on an assignment to World Head Quarters, Noel would be moving out of the Australian management firing line.

Although there was some time to prepare for this assignment and he had relocated from Ford Tractors to a private office in, *"The Glasshouse,"* an affectionate name Ford dealers and staff had given to the new Head Office building in Broadmeadows. Knowing exactly what was to be required of him would be a different matter. He remembered the FMI course and Brian Inglis's offer of an open-door policy. Brian had served some time in Ford's Port Elizabeth operation and thinking he may need help with the preparation, Noel sought his counsel.

Brian happy to pass on his knowledge of who was who in Ford of South Africa and how the territories had developed, gave Noel an overall view of the operation as he knew it. He also agreed to provide letters of introduction and brief biographies of people Noel would be likely to encounter.

A few days later, Brian Inglis' secretary brought him a file. On opening it, he noticed a list of personnel and alongside each name had been placed a tick or a cross, which Noel took to be an indication of how Brian had found the same men when assigned there years earlier.

Working in separate regions meant communicating by sending telex messages back and forth. After numerous phone calls, it took a little over two weeks for Harry and Noel to assemble their tool kit of planning materials. Noel was as ready as he could be.

CHAPTER TWENTY-SIX

For his first international flight, Noel would fly from Melbourne to Sydney where he would meet with Harry. They would stay in the Wentworth Hotel overnight, before flying on to Perth the next morning. From there they boarded a Qantas Boeing 707, which would take them on to Ivato Airport in Madagascar and after refuelling, would continue to Johannesburg. On landing they were to be met by Colin Cullingworth who would drive them to the Kyalami Ranch where they would stay and recover from the flight.

Stepping into a Port Elizabeth night, Noel watched a porter take their luggage to where their driver was waiting. Colin was serving in the role of Regional Manager for the Transvaal Region at the time and as he lived close by in Johannesburg, had agreed to show Harry and Noel around a few dealers in his region. While driving toward each dealer point, it was agreed that Colin would give them an overview of their performance before arriving at each one.

For now, it would be rest and recover before their scheduled meeting which would include a comprehensive review of his separate zones.

Standing in front of this glass airport entrance and taking it all in, Noel thought he could have been in any city anywhere in Australia. However, while some of the lights, sights and sounds were familiar to him, they now mingled with those of the exotic, South Africa was no longer a place on a map for him. For the next twelve weeks this country and its people would become even more familiar.

Arriving at the Kyalami Ranch just outside the city limits, this was the International Hotel to the rich and famous. It sat alongside the Formula 1 Grand Prix track of the same name and hosted Grand Prix teams who tested there all year round. Although there were teams practicing for a race scheduled the following weekend, motor racing wasn't something either Harry, or Noel had time for; they had people to visit and places to see.

Different faces greeted, them and some were even bowing. This was apartheid South Africa and here, everyday life could never be the same as in Australia. Tipping their porter, and making their way inside, they found reception, registered before each made his way to their room.

Prior to Noel arriving, the South African Government had been in debate, planning and drafting new laws that would exclude black settlement in white areas. The Black Homeland Citizenship Act would see all Black people denaturalised, making them aliens in designated white areas. There would be some minor exceptions, but the laws were being designed to exclude anyone of colour owning land in South Africa.

Even when their properties had been bequeathed to them, Blacks would need to seek special permission from the Minister for South African Information and Interior to have their names registered on the title. To Noel, these rules were difficult to understand. However, only two and a half years earlier, he and Joan had both voted in Australia's referendum to ensure indigenous people would be recognised as Australian Citizens. Never had he thought about skin colour, or race should be a barrier in life. Yet here in South Africa the stark difference between Australian and South African law saw to it that those of colour and race were being officially segregated from white communities.

The coastal fringe of east Africa has always been a place where one race, tribe, or culture has displaced another. Even going back to times when the Iron Age, Bantu people either invaded, displaced, or absorbed the Khoikhoi and San peoples, there had been change. The Bantu continued to push south conquering and absorbing other groups who had yet to discover the properties of iron, eventually the melding of cultures included bringing words of the conquered into the Bantu language.

History is sketchy at best, however it seems Europeans first discovered this side of the continent during May 1488, when Portuguese explorer Diogo Cao after sailing out to sea to avoid a storm missed the cape and landed somewhere around Groot River, He called his anchorage, *Rio do Infante,* it was on his return that he first saw the cape which called *Cabo das Tormentas* only to have his King, John II, re-name it Cape of Good Hope.

It would be at least another one hundred and fifty years before Europeans began to settle on this side of the continent. Two Dutch East India Company seamen who, after surviving a shipwreck, had lived there for several months bartering for meat and water from the local tribes. They had managed to sow crops in the fertile and abundant soil. On return to Holland, they reported the cape would be a good site to create a warehouse garden to service passing ships that were ready to take on fresh supplies. By 1652, a victualling station had been established at the Cape of Good Hope by Jan Riebeek on behalf of the Dutch East India Company.

Company employees seeing out their time also settled in the cape and Dutch traders imported several thousand slaves from Madagascar, Indonesia and East Africa. Unions between the peoples led to the creation of a new ethnic group called Cape Coloureds, who were mainly Dutch speaking Christians.

These frontier farmers were referred to as Boers and to protect their lands from raids by the Xhosa people formed into loose militias which they called commandos. European settled South Africa was always a place where violence and turmoil bubbled just below the surface.

The British arrived in Cape Town in 1795, determined to conquer it, protect their spice routes and secure it from falling into French hands. They stayed until 1803 when it was ceded back to Dutch rule under the Batavian Republic. This everchanging rule, and ongoing wars fought against different ethnic groups caused many of the Dutch to relocate into other areas.

After the Zulu Wars where up to two million people were either killed or displaced, the British had another war on their hands. The Boers in the Transvaal rejected British rule, fought for and won their independence. Known as the First Boer War, they had used guerrilla tactics to resist the British. However, the British returned with a different strategy and a more experienced military. At the end of the campaign over twenty-seven thousand Boer women and children had perished in British Concentration Camps and for the Boers, their independence was lost.

A little over fifty years later, Noel faced the reality that this was still a troubled community where old prejudices run deep. Dutch speaking Boers were wary of the English, and that sentiment was mutual, even to the point of Noel being told by a senior manager that even his Dutch colleagues were not to be trusted.

One of the first images he found burning deep into his memory was the quantity of prominently placed brass plaques, printed in English and Afrikaans, declaring these were public premises and for the use of White Persons only. He knew of the White Australia Policy, but here in South Africa, Apartheid shouted an even uglier way of managing the population.

However, working for an organization the size of Ford had taught him that commerce doesn't differentiate by colour or creed. If there is money to be made, business is blind to discrimination. During his career, he had never witnessed a need to see differences in people based on their colour or race, people were people and if he treated them well, then he expected to be treated the same way. He remembered his grandfather reminding him, when in Rome, do as Romans do. This was not a place for judgement they were here on business.

Settled into his room, all he and Harry had to do was to criss cross this part of the continent, meet with hundreds of already suspicious people from a hundred different backgrounds; draft a plan to pull all these groups together, and unify them to work together as a reinvigorated, if not new, Ford workforce. They had also been charged with identifying new sales opportunities, recommending the right products to be manufactured by Ford of South Africa and bring it into line with Henry Ford II's vision of a Global company reporting back to Dearborn.

Should be easy enough.

CHAPTER TWENTY-SEVEN

Honolulu Time: -- 10.30am
Now secure in his hotel room, Noel has tried for hours to empty his mind. Dispatching in minute detail his every memory of Flight 811 to a distant cavern of his cognisance. Intending to leave it there totally undisturbed, until he was sure he would need it when called to testify at a gruelling inquest, sometime in the future.

Waking to a knock on his door, Noel found the hotel porter outside with the new clothes he had ordered. Relieved he wouldn't be spending the next thirty odd hours in yesterday's attire; he took the parcels and offered the waiting man a generous tip.

QANTAS flight QF-4 to Sydney was due to depart at 12.40 AM. He had time to phone Norwood in Palmerston North and cancel his appointment. After showering, he dressed and went down to breakfast, he would have been happy to eat in his suite, but after what had happened last night, he needed to push himself to be among people. It may have been seen as a distraction technique to take his mind away from what had happened, but he knew he needed to draw his focus back to *concentrating on the future. The rest of his morning would be spent trying to laze by the pool. Before lunch he would swim a few laps to maintain his fitness and try to concentrate on which career choice would be best for his family. After that he would write his reports in comfort and try to decide on which employment offer would determine his future.*

His reports completed by early afternoon; Noel decided to find a cabana lounge by the pool intending to relax enough to doze part of the afternoon away.

Ever since arriving back in Honolulu he had been trying to forget about Gor Cowl and the way the man had completely discarded the wealth of talent Noel had built with Ford Tractors Australia. It angered and frustrated him, that he himself, had been cast aside. Instead placing a team of overspending, underperforming and incompetent managers in charge of the company he had nurtured to market leadership during the last two decades. Lounging in front of the pool and hunting for sleep all he could see in his mind was a picture of Gor Cowl, Nels Craig and Peter Morris gloating over a headline and a picture showing the disabled Boeing 747. He knew it was all fiction and totally wrong, but every time he closed his eyes, it was as real as any other thing in his mind. Treachery had been at play and Noel knew it would be impossible to trust his Ford colleague again.

Eventually he did manage to find a more pleasant thought and imagined Joan moving through the cattle on their farm. She was talking to each one of her young heifers, handing out hay, rubbing her hand across their hides, whispering and laughing as she did. It was enough to send him into a doze.

When he woke, the colours of a Honolulu sunset were painting everything around him in rich reds, splashes of orange and purples that lit up storm clouds of a tropical Pacific sky. Even after the events of last night, Hawaii was still a place of paradise.

Hoping Joan would be near the house and knowing she would soon be stopping for afternoon tea. He picked up his towel and slinging it over his shoulder, pushed his sunglasses back into place and headed for his room. The thought of a quick a call home excited him and he hurried through the lobby toward the lifts. The easy lilt of the piano caught his attention and he turned to see where the music was coming from, the pianist caught his eye and nodded to acknowledge Noel was listening. He was playing the Glen Campbell song, Gentle on my Mind, and that was exactly how he thought of Joan right now and couldn't stop thinking about her. Right then a cloud lifted and as he waited for the tune to finish, his mood lightened. Just hearing Joan's voice was better than any drug that money could buy. Walking over to the pianist, he pushed a tip into a jar already overflowing with greenbacks. His steps seemed to glide him across to the elevator. He punched the floor number and skipped a little before stepping into the lift. Another flight and tomorrow he'd be home.

CHAPTER TWENTY-EIGHT

Knowing Colin would arrive at the Kyalami Lodge at the agreed time, Noel had woken early determined to watch the dawn and soak up everything that his first morning in Africa could offer him. It was all he hoped and more.

After WW II, former bomber pilot, Captain Bill Forssman founded the Kyalami Ranch. Bill, a South African native later became a captain with Dutch Airline KLM and it was during this time he recognised there would continue to be a need for a high-class hotel catering for the needs of Airline staff on their layover.

Building it alongside a word class Racing Circuit had made it a perfect place for Formula One racing teams escaping a European winter. Plenty of clean accommodation and being able to test in the warmth of the southern hemisphere, made it attractive to F1 teams. World famous racing drivers relaxing in the expansive lawned gardens mixing with sun tanned, bikini clad air hostesses around the swimming pool, made the Kyalami Ranch a place for the affluent to see and be seen.

Even though this was where the jet-set came to play, for Noel it was for only a day and now he had to find Harry. They had work to do.

After a day touring the Transvaal with Colin and meeting some of the larger Ford Tractor dealers, they returned to Johannesburg. The following day they spent most of the day meeting Colin's staff in the local office Ford of South Africa, where they ascertained the strengths and weaknesses of the region with various department chiefs. Without fixing on anything in particular, Harry and Noel had been taking photos and making notes, already forming ideas of how to present the current structure of this region for their report.

Next morning was a flight to Durban for a meeting with regional manager, Brian Dearden and his staff. Brian, like Colin, showed his colleagues around the areas closet to Durban. Foremost in Harry and Noel's minds was the need to stick to the guidelines offered to them by World Headquarters. The goal being to bring everything into line with the vision set out by Henry II and Ernie Breech over twenty years before. Not wanting to declare their hands until a report could be formalised, this meeting was for the time being, very much a glad-handing operation and getting to know the way the regions operated. They needed an outline of who the main dealers were, meeting the dealer principals and then sketching out a rough background of the major players within the industry. The real work would come later when they reached Port Elizabeth.

After a day with Brian, both Harry and Noel were impressed by his eagerness and ability, he appeared ambitious and willing to become a bigger player within the Ford Motor

Company. Something Noel made a note of in the back of his diary.

During the last leg of their flight from Durban to Port Elizabeth, they discussed the organizational charts that he and Harry had brought with them. Then with energy and opportunity pulling him in every direction, for the final time, they went over Harry's briefing notes. Noel felt satisfied and prepared, having sweated over similar notes he had made before leaving Australia.

Although it wouldn't matter to the job that faced him, he knew whatever he had heard, or any perceptions he may have made, experience told him nothing would be as he may have been led to believe, or had once thought.

When the flight landed in Port Elizabeth, they were met by Ford South Africa's General Manager, Bill Bayford, who would take them to the hotel that was to become home for the next three months. The following morning, a car called and took them to Ford House, at the time the largest building in Port Elizabeth. It was an impressive statement of Ford's investment and the top floor would be where Harry and Noel would work while in South Africa

Being introduced to the Ford directors and outlining their task, they continued with equally brief meetings with the team of executives they would work with. At the end of a seemingly endless round of names, handshakes and explanations, Noel and Harry began the process of settling into separate offices.

Ron Scott the Managing Director occupied a suite of offices with an outer office for his secretary on the same floor. Having been introduced, Ron suggested Bill Bayford make his personal assistant Lyn Strong available to tend to any of Noel's secretarial needs. By the end of their first day in Port Elizabeth, both Harry and Noel began the task of segregating Ford Tractors from the automotive business.

Their brief, although simple in its objective, required an extensive survey of the dealer body. Determining how they ran their businesses and what would be required to increase market share without sacrificing profit. South Africa being a vast country with an extreme range of diverse farms, crop types and agricultural methods, it was important for Harry and Noel to divide the tasks. Review their findings at the end of each week and remain in lockstep to ensure nothing was missed or contradictory. Where they differed in opinion, they had agreed from the outset to go over the data interrogating each line and asking each other why, until any issues were resolved.

To ensure they could trust each other's assessment, they worked from an agreed data collection framework. Gaining the trust of regional and zone managers and allaying any fears that he and Harry were there as a part of a Dearborn razor gang would be difficult. He knew too, approaching this in the same way as he had with his staff when taking up his management role in Adelaide, they would achieve the success they planned.

Never losing the vision of what they wanted to achieve at the end of their project would be how they approached their plans to build a stronger Ford of South Africa heading into the seventies. By the end of the first month, they had identified that one of the biggest risks to maintaining Ford's market share would come from large corporate dealerships. These companies not being reliant on the sale of Ford products as their majority income stream and, by having representation through multiple outlets across wide regions diluted their reliance on Ford Tractors. These were businesses, which through a monopoly of branches and centralised purchasing, could take up to ten percent of Ford's production each. Noel and Harry saw this as risky because losing one or two of these corporates to a competitor would have a disastrous consequence on Ford's financial position.

For Harry and Noel, the next six weeks travelling to every corner of South Africa, flew by. Meeting with each Regional Manager and spending time in their territory recording any opportunities and roadblocks each manager identified. Initial meetings finalised, they would find themselves travelling with the Zone Manager to meet every dealer in his territory, again using the time to drill into the obstacles and opportunities.

The gathering of information was important, however winning the trust of the dealers was even more important. The segregation of tractors from vehicles called for a diplomatic solution, one both Harry and Noel had set as one of their tenets of the operation.

If the dealership was of a corporate style, the regional manager would be requested to schedule a meeting with the dealership's manager and one or more of the directors. At these meetings Noel and Harry would explain the reason for their visit, the desire of the company to divide the responsibilities of their vehicle and tractor division. They would explain why this was important to Dearborn and how they believed the dealer body could be better served. These were two-way discussions where Noel or Harry would invite suggestions from the dealer, asking them to justify their recommendation as being advantageous to both parties. This pattern followed for several weeks, with each new week taking them in separate directions.

Noel kept noticing how there were several layers of affluence in South African society. Apartheid certainly divided the different groups. White South Africans for the main part, were at the top of the tree, with coloured or mixed-race people slightly better off than those of the indigenous population. However, as his father had often told him, "When in Rome, Do as Roman's do." It wasn't up to him to run a commentary on the politics of South Africa, he was here on behalf of the Ford Motor Company and unless the plight of the people would show an advantage, or put sales at risk there would be no need to include his personal feelings within the dossier.

Harry having been away from home and with his family now located in the Melbourne suburb of Toorak for an extended period, had decided to fly back to Australia catch up with them. While he was away, he suggested Noel that he take advantage of a free weekend and join a tour of the Kruger National Park.

Talking about their suggestion with Joan during one of his daily phone calls, had her encouraging him to make the most of it.

From his early days as a sales and service rep for Massey Ferguson, they had made it a habit of talking each day he was away. Noel always wanting to know that Joan and the family were okay. She wanting to know where he had been and how his day had treated him. It always lightened his mood hearing the excitement in Joan's voice as she told him about any milestones the children were making in their development. Joan too, was always keen to know where he was, who he had met and to listen to his news. Their calls were nothing more than small talk, and they loved the intimacy of hearing each other's voice on the end of telephones several thousand kilometres away. A three-minute investment in time that strengthened their bond.

To join a tour, Noel caught a South Africa Airways flight from Port Elizabeth to Johannesburg where he would join another twenty-eight tourists boarding a Comair DC3. This flight would deliver them into the park. For Noel it became a wonderful experience that would burn in his memory, a time to be relived. For years after he could be found telling his young family about the weekend when he saw all manner of exotic animals. He and Joan had taken the children to the Coburg Drive-In, just outside Melbourne where they watched the John Wayne movie Hatari when it had first been released. Now seeing those same animals so close he could almost reach out and touch them, made the hair on the back of his neck tingle.

Having time away from thinking about the task at hand re-energised him and the company of people outside those at Ford, was a welcome distraction. Even though the accommodation reflected the bush country, the hospitality was first class and too soon he found himself boarding the same DC3 to make the flight to Johannesburg where he would layover until boarding another Boeing 737 to Port Elizabeth.

Three days away had cleared his mind, he had spoken to Joan and the boys, telling them all about the wildlife he had seen. Describing the elephants and lions. Joan sounded upbeat even though Natalie was cutting teeth and Joan had spent a lot of the night before trying to ease her pain and make her comfortable. Everyone was missing him and hoped he would be home soon. That night he slept the dreams of a contented man, knowing tomorrow he would be reviewing their progress with Harry and setting new goals for the coming week.

CHAPTER TWENTY-NINE

Reviewing their progress Harry and Noel questioned everything looking for patterns and, although swamped by detail, began dividing the data collected to date into a SWAT Analysis.

Their initial desire to write a prospectus for every dealer was at this stage going to be a mammoth task and although they had some resources available to them, looking at the project by region would enable them to model the segregation project more easily.

The evidence of their SWAT analysis pointed to the difference Noel had witnessed travelling from Adelaide to Melbourne a few weeks earlier. Describing the differences between Garwood Machinery with its hurriedly painted hoarding and one metal Ford Tractors sign tied to the chain mesh fence. Comparing it to Esmond Park Motors in Murray Bridge, who displayed their Ford alliance with pride. Although theirs was a bit of a mish-mash of styles, it was easy to see it as the business where you came to buy a Ford product.

Asking Bill Bayford if they could acquire a blackboard, or similar for each of their offices he obliged. They began working on the strengths and weaknesses of each dealer by the region they thought would benefit most from change, listing them by priority.

Looking through the photographs of the dealerships, they agreed corporate identification had to be employed if any benefit was to come from common advertising. For Ford Tractors of South Africa and their dealers to thrive, then every business representing the product needed a similar look. At present the corporate appearance was haphazard and, in some dealerships, non-existent. Noel wrote Dealership identification at the head of the Weaknesses column.

While they both agreed it seemed unfair to be concentrating on the negatives, Harry was quick to reinforce it was not their job to sugar coat the report. They were charged with drawing up a long-term plan for Ford's success, therefore, only after they had prioritised the problems, could they concentrate on offering a menu of solutions and have any hope of predicting an outcome.

The week continued reviewing a nominated region by zone and then by dealer. Working out where the weaknesses were, what strengths each one had, where any threats would come from and importantly, where opportunity to grow individual market share might be found. By Friday they had a picture of one zone and a solid format to follow all they need do now was put it into the reporting format as laid down by Paul Gillis and make their recommendations.

The next three weeks flew by, Harry visiting dealers taking photos, talking with regional managers and their staff in one region and Noel in another. On occasion they would accompany the dealer to meet with farmers both large and small. At other times they would seek out a local department of agriculture official to attain a perspective of the opportunities and trends for future farming that might be different to those offered by the dealers or zone managers. Everything became useful fodder for their report and as the weeks passed the detail grew and a bright picture began to form. Segregating the business could return good profits for both the dealer body and Ford of South Africa.

It was easy to see how the vehicle side of the company could evolve into a more streamlined, fast paced and immediate business if the more sedate agricultural side was segregated from the dealerships. It would not be an easy sell to some of the directors of the larger companies who saw the tractor business as a profitable sideline to their car and truck operations. However, Harry and Noel had based their findings on the DA10T reporting system. The system had many benefits but flaws could easily creep into the report, affecting the results. It was just as Noel had found with untrained clerical staff using the DA 60 in Australia. What the dealers required was a simple method to demonstrate how beneficial it could be to operate the tractor business as a separate entity.

Noel had seen older style ledger machines being operated by Joan when they were courting and, later too, when she had gone back to work after Geoffrey was born. Back then there had been rumours about Burrows bringing out an electronic accounting machine. It would be a desk sized computer, designed around the needs of small business, which would cut accounting time and offer profit and loss trading summaries over multiple departments.

Having managed to build a sample of their plan by focusing on one region, it was now time to push for all the data and complete the comprehensive report they were charged to do. More travel both inside and outside South Africa added to the miles both men were covering each week. Hands shaken, faces to remember and put names to, dealerships visited and council area profiles built. It all added to an ever-growing mountain of data. Even though the data base grew, the basic information stayed true to the first sample.

Ford of South Africa needed to segregate its car and tractor business and do so in an equitable and profitable way. The weaknesses within the dealer body were like those experienced world-wide. A good number of dealerships were undercapitalised, others were small well financed who however only cherry picked their deals. Then there were several that were owned by wealthy companies for whom the Ford Tractor business was not that important to their end profit and their overall focus was elsewhere. These were corporate organizations, businesses where their multiple Ford Dealership Agreements gave them access to enormous sales areas. In several cases to retain the Ford Tractor business, these were dealers who needed to do more.

Around the tenth week into Noel's time in South Africa, he and Harry had everything they required to assemble a comprehensive report. Bill Bayford again offered the services of his secretary who was willing to take dictation and manage their typing. She had experience to ensure their layout was in line with company reporting regulations and bind the document for presentation.

As they worked through their notes it became evident that Pareto's Principal was in play within the tractor dealer groups. Eighty percent of the sales were made by twenty percent of the dealers and the sales of spare parts was similar. Warranty issues could be divided in a similar way, eighty percent of the service concerns came from twenty percent of the dealers.

Drilling down into the data was providing hard reasons for an action if based on a purely profit, but Noel and Harry knew there was more to being a Ford Tractor dealer than just a financial aspect. Noel remembered his South Australian experience where small, welded-on dealers always flew the Ford flag. They may not have been super-efficient, or sold many tractors, but for him, they had heart and, heart was something much more difficult to measure.

To manage this class of dealers, their material would make up an appendix divided by region, by zone and by dealer. An appraisal of the company staff member responsible for each, would be included.

Between them, Harry and Noel looked at every instance where they could make recommendations of what was needed to either improve the dealer's market share, or at times discontinue their dealer agreement. Dealers were ranked by a set criterion based around the effort needed to improve their efficiency in gaining their nominated market share. Zones were treated in the same manner as were the regions.

Paul Gillis, Director of Affiliates and Direct Markets, flew in to Port Elizabeth from Dearborn as they were putting the final touches to the report. Noel felt happy to see him and even more pleased when he commended both he and Harry for their efforts. Paul offered a couple of suggestions to include in the final document. By the end of Paul's stay, Harry and Noel had finished the Executive Summary and Recommendations. Now they had to wait to find out how World Headquarters would receive it.

Having read its contents, Gillis tapped the report as it lay on the desk in front of him and looking Noel in the eye said, "How about bringing the family over and staying on here to implement your recommendations." His face expressionless, he paused for what Noel felt to be an eternity before adding, "I don't need your answer now, talk it over with Joan and let me know in a few days."

Noel understood the difficulties ahead if he accepted. However, he tingled with anticipation, knowing Joan would agree to joining him on a new adventure. The only sticking point being her anxiety about flying.

CHAPTER THIRTY

Now confirmed in the role of Manager on Assignment, and being charged with implementing the plan he and Harry Watson had devised, Noel and his family had much to do before leaving Australia for Port Elizabeth in late February of 1970.

His South African experience had been exciting and had taken numerous twelve-hour days to bring it all together. It wasn't until he had arrived back in Australia that Noel noticed how draining the assignment had been. Although an extended holiday had been planned there would be no time to rest, their fourth child was due around his birthday on January fourth. Any celebration would have to be reserved and among all the excitement of the pending birth, their trip to South Africa bubbled away in the background.

January fourth 1970, a sharp little cry and Natalie Anne Howard let the world know she had arrived the day after Noel turned thirty-one, a perfect present. However, he still had a lot to do,

A few days later and after Joan and Natalie had been discharged from the Women's and Children's Hospital in Carlton, Noel and Joan continued with their preparations to leave. It would only be another four and a half weeks before the family were scheduled to depart on the first leg of their South African adventure.

Their timing was perfect, Australia had become a victim of a world wheat glut and Australian farmers were being locked into a quota system as allowed by the Wheat Industry Stabilization Act of 1968.

Either through planning or coincidence, Ford Australia Tractors had embarked on the reformation of the number and capacity of management and white-collar staffing roles. However, being on special assignment, Noel's position was not one of those to be considered for review.

For Noel and Joan, there were cases to pack with everything they would need to set up house in Port Elizabeth. Although the house the company would be providing was to be furnished, clothes for a growing family, baby's needs and a million other details lay outside Noel's remit and yet Joan managed to make sure everything was being ticked off this ever growing to do list.

They had planned a sea voyage to give the family time to enjoy each other's company and treat the trip as an adventure. A time to read to the children, play board games and enjoy all the treats a cruise ship had to offer.

At the beginning of February, with all their luggage stowed the Howards and their guests were being shown to their first-class cabins by the purser. P&O treated their guests to an onboard, farewell party prior to leaving South Melbourne for Adelaide and then onto Perth. Watching from the railing they saw their friends joining everyone going down the gangplank to wait and wave them goodbye. Joan had made sure the boys were loaded with streamers to throw and she held Natalie on her hip while Noel managed an excited and wriggling Tania on his. A long blow of the ship's horn and in moments they felt the streamers they had thrown tighten and as Auld Lang Syne played; P&O's Arcadia began moving away from the dock.

Sailing across Port Phillip Bay and toward the heads between Point Lonsdale and Port Nepean Noel took the children to the nursery while Joan fed and settled the baby before dinner. He let the boys explore the racing car amusements while he took Tania to a toy box where she busied herself dragging everything out until she was sitting in the centre of toy town chaos. With Natalie down for a nap, Joan found the family, explaining that they were being called for dinner soon and she wanted everyone tidy before they went in.

With dinner done, it was time to settle the children and enjoy a leisurely evening cruise across Bass Straight. The twenty-nine-thousand-ton vessel would motor on a south-west by southerly course before reaching the Southern Ocean, changing to a west by northwest bearing until docking into Port Adelaide in South Australia the next day.

Disembarking in Adelaide the family was met by Ford Tractor colleagues John Macpherson who had agreed to be their taxi for the day. While in Adelaide it was good to catch up with old friends and allow the children some time to run around in the parklands. This time they were seeing Adelaide through a tourist's eyes. However, as much as Noel tried to rest his mind from tractor business, it always bubbled near the surface of his consciousness and knowing the scale of what needed doing in South Africa, he studied every different car, truck or tractor dealership they drove past going to and from the ship. In his mind, he and Harry had been right to make corporate identity a high priority.

John picked up and delivered the Howards back to the Arcadia in time for sailing and a routine, the same as in Melbourne followed before the family went to dinner. Leaving Port Adelaide, they sailed south into Investigator Straight before bearing west and toward the Great Australian Bight. Sheltered from the Southern Ocean by Kangaroo Island, the sea of the straight was smooth and a gentle swell lulled them to sleep after turning in for the night.

P&O's SS Arcadia's had her keel laid by John Brown & Company of Clydebank in Scotland in 1952 and launched on 14th of May 1953, took her maiden voyage from Tilbury in the United Kingdom to Freemantle in Western Australia on 22nd of February 1954. This passage was to be an almost return copy of that first sailing.

By dawn the ship had been rolling in a heavy southerly swell for over three hours. By breakfast the westerly chop had the bow of the ship lifting and lolling a circular motion, there would be only a few takers for breakfast. Lunch came and went with half of the passengers registering with motion sickness. For the Howards, only Noel had escaped being nauseous, he called the ship's doctor who with the ship's company were doing their best to assist everyone in need. Restricted to their cabins for the remainder of the trip only Noel ventured out to find supplies or grab a bit of fresh air while the children rested. Natalie had by now gone off breast feeding and Joan's condition had deteriorated that much that on reaching Freemantle, the doctor ordered her off the ship.

Ford Motor Company's Western Australian operations were not far from the port and on leaving Noel arranged with the people in the tractor office to borrow a car while he found a hotel and rearranged their trip, booking flights to Johannesburg.

Within a few days Joan and the children had recovered enough to travel and as anxious as she may have been about flying Joan was in no mood to entertain the thought of another sea adventure for a while. Retracing the route Noel and Harry had flown, the Howards were weary from travel when they arrived at Kyalami Ranch. Here they would take a couple of days before flying to Durban and down to Port Elizabeth.

After a night and breakfast on South African soil, Noel took the family down to the pool area to allow the children to play and give Joan some time alone with Natalie. When she joined them later, she surveyed the grounds and feeling better, allowed a wicked little tease ruminate in her mind. Seeing the airline hostesses and other crew lazing around the pool area, Joan strolled across sat down onto the cabana alongside Noel, reached over and stroked the back of her hand along his forearm.

"Easy to see why you Ford guys like staying here," she said drawling the words in her best pouty tone and having gained his attention, nodded to where the air hostesses were sunbaking, "it's like a candy store for randy commercial travellers."

Noel did his best to explain why staying here was their best option, but the more he talked, the louder she laughed. In the end he knew he had been had and was more than happy that after the start to their travels his wife was back to her old self. They both knew their marriage was very sound, but occasionally taking the mickey out of him, was sport for Joan.

The family refreshed from their overseas journey still had the leg from Johannesburg to Port Elizabeth to complete and all were excited to see their new home. Noel would have a few days to help Joan settle in and they would be expected to employ staff to help Joan with the chores around the house. This type of intrusion to family life had not been something she had expected and it took her a while to understand the way this country worked. Noel too although used to working with staff during the week found his weekends were empty of chores, the benefits came in the way their family had more time to enjoy each other's company.

The house, located in Walmer comprised mainly of neighbours like themselves, transient staff on assignment to multinational companies working in South Africa. They knew the home would be sparsely furnished when they arrived, therefore the Howards based themselves at the same hotel where Noel had been staying only a couple of months earlier.

Having arrived a few days earlier than if they had still been on board the Arcadia gave the family time to adjust to their surroundings before Noel was due to start work. While he looked after the children at home, Joan embarked on a mission to furnish the house, searching new and enticing stores for furniture and other comforts they would need to make this place their home. Then came the task of interviewing staff to help her in the home, not that she either wanted, or expected to employ help. However, this was South Africa and here certain customs were to be respected and observed. Their first week became one of frenetic activity, but in time, the boys were both enrolled at school and the company house was ready for the Howard family to move in.

CHAPTER THIRTY-ONE

The first week at Ford of South Africa disappeared in a whirlwind of meetings and interviews. For Noel, priority one had been finding a team of people who would stay with the task until completed.

Again, he read the letter from Paul Gillis advising him of the establishment of the new Market Representation Department and confirming his authority. It boosted Noel's confidence and again as with every other time he read it, he found himself rushing to read through the finer parts of his now approved Master Dealer Plan for South Africa. He and Harry had taken a lot of time carefully working through each detail before finalizing their recommendations. Even more exciting to him, was to see that World Headquarters in Dearborn had changed little in the original document they had presented. That, made the task ahead easier. Between them, they had questioned and interrogated every thought or discussion so many times before putting a recommendation into the report, that Noel could almost recite it line by line.

From his office on the top floor of Ford House and taking occasional advice from Bill Bayford, Noel began breaking the plan into manageable tasks. Nominating who in his team would lead each part of the change and those who would then be tasked with ensuring the completion of each phase, for every zone. Knowing the individual attributes of each manager among the Ford of South Africa personnel became a high priority. He then set about recruiting the most appropriate zone, or regional manager, who would be charged with ensuring that their dealers understood the benefits of the proposed change and help in them to transition.

He was sure that if everyone approached their tasks with an easy, but determined manner, then most of the plan's major elements would be easier to complete. Succeeding to convert one dealer at a time and creating a knock-on effect. He knew a logical approach to problem solving should see the project to arrive at a speedy conclusion. Getting dealers with combined tractor and vehicle businesses to agree to separating them would require a different approach. Convinced that by showing them how Ford's new and improved electronic accounting system would provide them with immediate data collection. That more accurate and immediate data, could then be used as a planning tool, making their businesses more profitable. He was sure if the team explained the system properly, dealers would jump at the opportunity being presented.

The proposed machine could produce data from which they could analyse the performance of any number of revenue streams. By employing the new accounting system dealers would now have an ability to cut the time in dividing their business into independent profit centres. Central to this new accounting policy, the recently released Burrows L-2000 Desktop Accounting Machine would mean a reduction in accounting costs. Rather than using several bookkeepers, the speed and versatility of the computer meant highlighting waste, or finding another profit opportunity would be speedier than before. Even for dealers of size, seeing their business as many small departments would often be beyond them because of the labour involved. Noel knew if he could convince his team on the benefits, then when they approached a dealer showing them an opportunity to save money, it would eliminate cost as an argument.

A probable contentious issue would be getting them to agree to appointing new dealer principals, installing separate accounting, banking and financing procedures. However, he and Harry had devised a strategy to minimise these objections too.

They applied a lot of thought into listing such objections, providing affirmative answers, which showed dealers where more money could be made and agreed it was also important to highlight, where a dealer's expenses would increase. The working document they produced for Ford Tractor's sales and service teams was comprehensive and if someone did receive a question they couldn't answer, Noel and his team would be on hand to provide it.

The Watson/Howard proposal could be viewed as ten pillars of management, each one designed to lift the sales and profits of the current dealer network. Where a new dealer had to be appointed this document would offer a management blueprint to assist them to become both effective and profitable.

For Ford of South Africa, these would be businesses operating at the highest-level and capable of bringing longstanding market leadership to their primary market areas.

Noel selected Brian Dearden, George Strydem, and Peter Reynolds to be the nucleus of a team effort in delivering the project. They agreed from the outset their attitude toward the dealers should proceed with a spirit of co-operation, using more carrot than stick wherever possible. While he and Harry had been drafting the report, Noel had often thought about that last drive from Adelaide to Melbourne and envisioned a time when all Ford Tractor dealerships would have integrated identifying hoardings and signage. Other parts of the Ford world had been encouraged to adopt a common branding philosophy and Ford was growing its logo to become the second most recognised brand in the world at that time. Between them, they agreed to make this initiative a cornerstone of their marketing plan. Reasoning, that it would sow a seed for South African Ford tractor dealers to grow more opportunities that would come from having a better corporate image.

With approval from Dearborn and his role now a reality, it was time to call the tractor division staff together and outlay the company's vision for a new Ford of South Africa. Expecting there would be some resistance to change, he began preparing question and answer sheets to head off any confrontation before it could gain momentum.

Reports of rumours circulating the dealer network were coming back to head office from their field staff. His and

Harry's visits to dealers while fact finding had left some of the dealer body uncertain of their future and what the company might be planning. Noel knew it was important to stop any inuendo, or incorrect assertions from competing manufacturers whipping through the market and causing their sales teams lose business because of it. For that reason alone, everyone in the company from general manager to janitor needed enlightening fast and then the dealers needed to know.

The Master Dealer Plan for South Africa had been built around sound business principals. While nothing in it was entirely new, they had found that checks and balances of most Ford of South Africa dealers had slipped below their contemporaries in other parts of the Ford world. World Head Office decided if it was going to show profit from their investment in South Africa, it was time to take stock and install policies and procedures which would improve the benefits for everyone in the supply chain, from manufacturing to the retail customer.

The ten main points of the plan were to:
1. Create the most effective and profitable tractor dealers, by creating a planned step by step management structure for all departments. Improving every dealer's accounting and management practices through increased focus on delivering market leadership.
2. South African Ford Tractor dealers had to agree to raise the sales and income capacity through

creating separate profit centres and establish goal setting initiatives for each.
3. Improve the Service Department space, employ only trained service technicians and have access to all available specialised equipment.
    a. At least 60% of available service labour revenue should be directed toward retail sales, the remaining 40% reserved administration and warranty.
4. Each dealer would be expected to provide a neat and clean "Customer Lounge", alongside and apart from their service area where customers could wait.
    a. In certain cases where service or parts issues had an opportunity to damage the company's reputation, "Loan units" would be made available on request.
5. Dealers would be strongly encouraged to participate in the Dealership Identification and Appearance scheme.
    a. Employ an annual competition supporting this initiative.
    b. Open to dealer principals, zone managers and regional managers. The objective being to quickly wash an image of corporate strength across the whole tractor dealer network.

6. Along with the outside image of each dealership being updated, service vehicles, delivery trucks, parts departments, all internal and external signage would carry the same corporate theme.
7. Dealer Finance: An assessment of each dealer's position would be made by company staff and a prospectus written to form the basis of an operations manual for each dealership.
    a. Ford Credit being available to Ford of South Africa dealers, a high level of importance will be focused on training. Ensuring all sales staff understand how each element of retail finance can be provided.
    b. Regional finance training schools will be scheduled to ensure new staff are capable, confident and accurate when completing paperwork.
    c. Dealer Principals to be instructed on the profits available through capturing commission from the sale of retail paper.
8. Utilization of modern technology to track and analyse dealership performance across all profit centres.
    a. Ford of South Africa will arrange training of accounting and bookkeeping staff in the use of the Burrows L2000 Desktop Accounting Machine

9. Regular After-Hours Reviews are to be conducted at the dealership level with either the Zone Manager or Service Manager in attendance.
10. Regional Managers will oversee a continued review and assessment of implemented changes. With the dealer principal or his nominated executive, they will interrogate any previously nominated additional opportunities to increase sales, profit, and competitive activity.

The question-and-answer sheets in rough draft, Noel started working on a set of dot points for the slides he wanted to use in his staff presentation, relying on Brian to select suitable South African backgrounds. From these headings and dot points, he sketched a draft of the text, prompting the next slide. As this was to be a team effort, each member was assigned a part of the presentation which would become their area of expertise. While Noel gave them the main points to be covered, he also left enough room in the speech for each one to make it their own.

Later, scheduled press releases would begin to appear alongside dealer advertising, heralding in a new era of sales and service for anyone who bought a Ford tractor or piece of equipment. The initiative would extent to industrial equipment markets too, no dealer would be left behind.

Practice makes for a perfect presentation, and had been a Ford Motor Company catch cry for years. For this management meeting, it was even more so. Noel relied on his Dale Carnegie Institute training to coach and encourage his team where needed. Even insisting that they should call out anything that seemed out of place.

Artists' renderings of two or three of the more important dealerships based on samples supplied by Dearborn, required corporate approval. Then came the arduous task of preparing a comprehensive handout for the tractor company's field staff. Early in the project, high importance had been given to ensuring this would be an easy document for field staff to understand and use. Although a confidential document, nothing it contained needed to be hidden from the dealers. It was to be a working tool, something that could be opened and referred to while at the dealership.

Everything they covered had been directed at making this a smooth transition to the ideals of the new, Tractor Dealer Plan for South Africa. Another important point was getting agreement that this was the best future for the dealer body and ensuring everyone could be confident delivering the same message.

Regional Managers were invited to Ford House and attend a presentation a day prior to the Zone, Service and Parts managers. Splitting the numbers into two groups meant more intimate discussions could be held with those responsible for their regions and if an unseen difficulty was highlighted, Noel's team had a few hours to find an answer or solution.

While they were practicing and retuning the presentation, senior personnel within Ford House were encouraged to sit in with the team a few days before the scheduled staff meetings. Noel asked for, and welcomed input from Bill Bayford whenever he dropped in on one of these sessions. A buzz of excitement was building within the walls of Ford House and the curiosity of employees grew.

CHAPTER THIRTY-TWO

His mind at ease and confident in the team's preparations, Noel led the tractor company's regional managers through a vision of Ford's "Master Dealer Plan for South Africa". Outlining step by step goals and expanding on the revision processes, he explained each of them would be expected to report on their region's progress. It was part of a belt and braces approach in monitoring each region and territory. Looking around the room he tried to spot any dissenters. As he expected people were taking notes and he could feel excitement coming from those seated in front of him. Knowing it was time to put them at ease, he explained how their answers to the questions he or Harry had asked during their visits last year, had given them the data that was now driving the project.

Explaining that when everything had been collated, it built a picture of the whole South African market. Now they were certain of places where weaknesses lay and where opportunities could be found. Discussion would be invited with regard to targeting those dealers where the biggest improvements might come from. However, during the morning session, they would concentrate on overall elements of the plan.

He handed the meeting over to Brian Dearden who worked his way through a few slides. These showed the ad hoc way in which Ford Tractors were being represented by dealers of different sizes, across their various regions. When the murmurs died down, he spoke about the urgent need to achieve a consistent image through branding and produced slides showing images of a range of Ford dealerships from different countries.

Early in the history of the Ford Motor Company, it had created a separate department to ensure that their dealers worldwide presented a consistent identity. Brian introduced several slides that showed an artist's impression of the same dealerships as before, only now, they were cloaked with the corporate look.

The next slide flashed a copy of Ford's DA-10 accounting report and Noel smiled as he heard an expected moan and muttering circulate the room. Brian went on to describe why an accounting document like this was important not only to the dealers, but the company too.

However, as Noel watched him take them through the basics, he noticed those present who were taking notes and as they had discussed before the meeting today, there were a few managers, who's attention Brian had lost. It was time to change slides.

Brian took them through the list of dealers they had identified as those where immediate gains could be achieved. They wanted these discussions to begin a process of welding the whole Ford Tractors team together and as they had planned, asked for the regional managers to offer suggestions.

After a robust discussion, Noel outlined the afternoon program and called a break for lunch.

Paul Gillis had warned him he could face some resistance to the changes being proposed. However, Noel believed his small team being well prepared and inclusive would head any push back off, and this morning had proved him right. Listening to the discussions floating around the tables at lunch Noel sensed the optimism in the room, managers were discussing different opportunities for individual dealerships if they signed up to the program. Although his mother may have told him eavesdropping was a form of rudeness, his grandfather pointed out the value of using it for research and in this case, it was one way of gauging support.

The afternoon meeting started with the same slide of the DA-10 they had used earlier and Noel began the session by introducing George Strydem. It was now up to George to explain how the computing program the company had investigated for its dealers to use, would make the dreaded accounting document obsolete.

George's next slide showed a photo of the Burrows L 2000 and then using more slides, he proceeded to explain how the machine could analyse data streams. He described how the inbuilt program would save staff having to complete the many longhand functions that were required to complete the DA-10. The computer would mean a large capital investment for dealers and the company had put in place payment plans to assist with the purchase. However, once a dealer's database started to grow, they would soon see how targeted sales initiatives could benefit from more accurate data. Simple tasks like prospecting for new sales had the opportunity for a better success rate. Breaking down their business into numerous revenue streams would help the dealer to understand departments that were profitable, departments which needed more attention, those that were frivolous, or others that needed eliminating. A round of applause marked the end of George's presentation and at that moment, Noel knew they would succeed. That evening the discussion around the dinner table was lively and laced with encouraging suggestions to get reluctant dealers to agree to any additional investment.

Next day, the Zone Managers meeting followed the same format and with more people in the room, took a little longer to cover each topic of the presentation. Discussions over signage, warranty, service issues and retail finance required Noel to divert the distraction and bring the focus back to the benefits of the program.

To ensure everyone had understood the message, Noel took them through the handout, answering every question he could and calling on George, Brian or Peter, if it was in their area of expertise. More time had been scheduled for the lunch break, as it allowed the group to mingle. This way they could get to know George and himself as they were both new to Ford of South Africa.

After lunch, Noel introduced another handout and took the room through the marketing competition they had designed to entice the dealers to participate in the program. The lack of groans told him it was being well received.

Deliberately and until now, his team had held back releasing the details of a competition and a range of other incentives that would only be open to company personnel. The last session of the afternoon was spent going over the points of the competition, explaining that it had been tailored to give every manager in the room an opportunity to win expensive prizes. Each of them had been set a personal qualifying target and their schedule of rewards. This document being confidential, would be sent to their homes later that week.

At dinner that evening, Noel gave a speech thanking everyone for making him welcome and reiterating the points covered in the earlier meeting. He promised he was committed to making Ford Tractors the market leaders in South Africa and asked very one in the room for the same commitment. His request was met with rowdy applause. All he had to do now was capture the same enthusiasm from the dealer body.

With the wheels set in motion, focus turned to the Regional Dealer Meetings they had scheduled. This would entail flying the whole team and their equipment to different venues across the country. Their mission that week would be rigorous. However, Noel knew his team would be buoyed by the excitement of doing something special in Ford of South Africa's history.

Their dealer meetings over, the next phase of the plan began. After the regional managers' suggestions had been received and before the dealer meetings, Noel, Brian, George and Peter had revised their dealer priority list. They had already targeted those dealers where historical data implied early sales gains could be made and cross checked them with the expiry date of their dealership agreements.

Recognizing not every dealer would be either in a position or have the need of such a complex computerised accounting system, they moved them into a separate listing. Then they looked at those dealers who would benefit through the Dealer Identification and Appearance program and separated them by priority. Giving each dealer a rating between one and five they built a picture of where the immediate results might come from. By the end of the week a comprehensive schedule, complete with an action plan including their individual responsibilities had been prepared.

In conjunction with the action plan, a draft dealer manual had been produced. This could now be tailored to each individual dealership and offered any representative from Ford Tractors a comprehensive document they could show to their dealer. Photos of the current buildings and areas would be set against the artists rendition of how Ford Tractors imagined the premises would appear when tailored with the corporate look.

The strongest part of the document held the section describing the steps required to determine each profit centre and then how to calculate the profit or loss from each. While most dealers knew their accountants could extract these numbers from their own bookkeepers, it was cumbersome and expensive, which is why they had ignored the company's pleas to complete the DA-10 over the years. Now their excuses could be eliminated.

Until the 1970s, computers had been bulky and expensive, often requiring a large air-conditioned room, or complete floor of a business dedicated to them and, it took a team of trained professionals to manage them. By using the Burrows L 2000 and incorporating the accounting program the tractor company had designed, dealer principals now had an easier means for advancing the profitability of their business. They also had an ability to track performance of, and for the rewarding of employees. A dealership using this system could now log customers purchasing patterns and it would also show when a customer's tractor was due for service. And at the press of a few keys, it would generate a letter telling them so.

George, a university graduate with expertise in computing, had been selected to train staff on every aspect of the new machine. Noel wanted him to be an expert capable of training not only dealer staff, but the company personnel too. To ensure a wide range of zone and service managers had required the right skill-set for their departments, one member from each region would be chosen for the training. These people could then go on to take the dealership's accounting staff through the many tasks required in building basic databases, like customer records and then transferring any existing data to the various income and expense streams. Noel could see a time when every Ford of South Africa Tractors manager and administration employee had become skilled in both the benefits and operation of the Burrows machine.

However, dealer principals had to agree to the capital outlay required. Having agreed their top ten and believing these were the dealers from where the biggest benefits would come, he and his team began working on dedicated strategies with the regional managers. Everyone agreeing that both the zone and regional managers would accompany Brian and Noel for their first meetings.

It was important to Noel that more than one of his team were adept in the function of the computer and its systems. Brian who had both a good knowledge of accounting and business management agreed to be involved with the training. George had been selected because his university studies had exposed him to the intricacies of coding software and, because he had an understanding of the emerging market for small computers and their systems, would as well as conducting training, be the go-to person for a dealer's technical issues. To further support the team's drive to ensure every dealer complied, George had written a comprehensive user's manual. Working with the DA-10 as his guide, it covered every aspect he thought a dealership might need or encounter, no matter its size.

Although the systems could be provided through a lease arrangement with Burrows, Noel saw no point in carrying unassigned inventory. Supplying the equipment was not an immediate priority, because only when a dealer signed a new agreement including the installation of the computing system, could the order be placed, a view also shared by his superiors.

One of the early dealerships to agree to the plan was Perry Ford based in Stranger in the Natal Province. To the Accountant Don Christie, the introduction of this new Tractor Computerised System made sense and he was quick to see the number of clerical hours it could save when producing a similar report to one that the DA-10 supplied.

At first, the technicians from Burrows found a few teething problems while loading the operating systems and in reading the data it supported. However, in time they became adept at identifying the source of the problem and wrote new coding to bypass the problems overcoming further issues.

CHAPTER THIRTY-THREE

Before visiting any one of the larger dealers and reviewing the dealer agreement with the dealer principal. The first goal set by Noel's team, had always been to have the dealer agree to the company's requirements for them to continue as Ford Tractor dealers. Knowing every dealer would be different and have demands of their own, they prepared an answer for every objection they thought would be raised. To support their endeavours, the team would commission an artist's rendering of the dealer's premises including showroom, parts and service departments, used machinery yards and company vehicles.

Examples of advertising bromides were included in the dealer manual, these could be used as a base to create tailored classified and display advertising for not only new Ford products, but used tractors and equipment along with service and spare parts too.

Little was left to chance and for the next eighteen months, Noel and his small team crisscrossed the country. Flying or driving to meet with regional and zone managers who would take them to another dealer where they would outline the marketing plan. Most satisfying was being on site to witness many dealers accepting their own objectives and where they related to in the Master Dealer Plan for South Africa. Even more pleasing for Noel was counter signing their new dealer agreements.

There were times when Noel or one of his team members would sit in on an out of hours staff meeting where the dealer principal would take his staff through the business's progress department by department. With the aid of the new accounting system, some of the earlier dealers to install their computers, began exploring their data. Often taking the opportunity of showing their staff where gains had been made. By highlighting their gains, as well as showing which of the dealership's departments were making losses, mostly, discussions were frank and open. Dealers were now inviting their staff to suggest areas where they thought management should focus its attention, then asking them to explain further, backing up their ideas with how they thought it would help that department to improve.

Encouraging employee involvement before it became a term Ford Australia would later discover. Now every dealer principal was encouraged to give everyone in their own businesses an opportunity to forward suggestions on how to increase the performance of each department.

On occasions, bonuses would be given and at times dealer employees were invited to Ford House. There they would be booked into one of Port Elizabeth's better hotels before being given a chaperoned tour of the factory. Staff who came from lower socio-economic backgrounds grabbed at the opportunity. If they didn't have a suit to go to dinner, Noel's department had a budget to fund a properly fitted suit. The system was working and with it came results.

Sales were beginning to improve across all market segments and as he had experienced in his first dealer meeting in Adelaide, South African dealers were competing with the same products as their counterparts were in Australia. However, no longer were the triple red triangles of the Massey Ferguson logo dominating the tractor dealer landscape. A sea of blue on white buildings complete with modern signage adorned Ford tractor dealerships. Now the Ford logo with its history going back as far as 1903, started calling buyers to tractor dealers across the country.

The Ford Dealer Identification and Advertising scheme was working. Dealers through their agreements were expected to present a consistent clean and tidy image. In sales competitions, dealer principals would have points added or deducted for their business's appearance. Use of the Ford logo when advertising, added to the new dealer look too. Properly serviced and presented used machinery had improved not only the image and reputation of Ford Tractor dealers, but these sales were now adding profit too their bottom line.

Although they were only twelve months into the program, independent customer surveys were proving that the initiatives laid down in the Master Plan for South Africa were working. Larger dealerships that had adopted the computerised accounting system were reporting a lift in individual profit streams. In those departments that were not performing, dealers were actioning plans to either ease their way out of them, or more importantly had identified the areas where they could improve the performance.

When it came to product availability, their concerns were like those raised by Australian dealers. Not enough equipment to sell, and the tractor range required a wider selection of options to suit the extensive variation of crops grown across their primary market areas. As much as things for Ford dealers in South Africa were changing for the better, some of the old issues remained.

One area where South African dealers had better opportunities than their Australian counterparts, was the accessibility to Ford's own finance arm, Ford Credit. When it came to designing consumer mortgage or leasing plans, Ford Credit was chartered to serve the dealer body in securing sales of Ford branded products. Therefore, along with retail finance plans, corporate leasing could be written to include a variation in terms to serve customers with irregular income patterns. Of greater assistance to a dealer was the ability to secure finance for their customer at the point of sale. In larger dealerships, the salesman would introduce their customer to the business manager, where those staff would guide the customer through the financing process.

However, Ford Credit offered training to smaller dealers and their sales staff too. Many salesmen began adding to their commissions by offering something they would have asked of a broker or bank, knowing there was ever a risk of losing the sale by doing so. Ford Credit had always maintained a reputation for providing finance at a competitive interest rate and this was beginning to bring enquiry through the dealers' doors in increasing numbers.

For Noel and his team another by-product for Ford Tractors, was having the product paid out from the dealer's wholesale floor plan finance, at point of sale. Without there ever be a chance of it being sold out of trust if the dealer's own financial position was in jeopardy, headed off a potential problem.

Although a separate entity, Ford Credit existed to make money for the Ford Motor Company's shareholders and maintained strict controls on the way finance was made available. Noel's South African experience, now showed him this could be an opportunity Ford Australia had been missing.

New dealers had been recruited to fill areas that had proved lacking in sales. These dealers may have existed with competing franchises and had come over to Ford because of the energy being generated by the marketing team. In others, regional and zone manages may have spotted a sales or service person from a competitor, or from within their own Ford dealer base who had dreams of being a tractor dealer. If their credentials and enthusiasm fitted with a need, the marketing team working with Ford Credit would partially fund the prospective dealer into their own business. Although nothing new to Ford, it became a winning formula for Ford of South Africa.

In the last quarter of 1970, Noel was called to Detroit to present his report on the progress he and his team were making. To be delivered in front of the directors from various departments at the Ford Motor Company's World Head Office, this would be a moment to shine or fail.

Failure was not an option and as he had done on so many earlier occasions, Noel steadily worked through the objectives in the original brief, marking progress and status of every point. While there remained a pressure to visit and revisit as many dealerships as he could, Noel sorted through photographs he could use as background for the slides he would use in his presentation. Dealer visits now gave him an opportunity to collect testimony from various dealer management team members across a variety of departments. It was early in the program and his notes to accompany each slide would declare he was confident their success would continue to grow.

Packing his documents into the pilot's case that always travelled with him, he heard a knock on his office door and saw Bill Bayford, Ford of South Africa's General Manager waiting to be invited in. Bill perching on Noel's desk, one leg swinging in lazy circles waited until the case had closed, then among the small chit chat asked Noel that while away in Detroit, would he consider assuming the role of General Manager – Ford Tractors, when the incumbent retired. Flattered by the offer, Noel declared he had more work to do as he wanted to see the current project completed, adding that even though he could be tempted, he needed time to work through the opportunity it presented. Bill said he respected Noel's position, telling him they would examine the question further when he returned and shaking his hand, wished him well in Detroit.

At home, Joan as always was supportive, reiterating that she would go along with whatever Noel thought was best for his career and in turn, what was best for the whole family. They had started to settle in to life in Port Elizabeth and although they didn't see much of the neighbours, she and the children enjoyed the company of friends they had made through school and the families of Noel's Ford colleagues.

Registering his details with reception of the Holiday Inn in Dearborn, Noel stretched and without thinking rolled his head trying to relax the stiffness in his shoulders. It had been a long trip from Port Elizabeth and more than once on the flight he had gone through his presentation in his mind. Following the house porter to his room Noel played the greenback he had ready to tip the man through his fingers. He knew tipping was a custom in the United States of America and he was grateful for Bill Bayford's guidance on what was appropriate to offer. Following the Porter into the room and throwing his jacket onto the bed, Noel took note of his surroundings as the man took him through all the suite's features. Taking the tip and thanking Noel the porter left the room. Noel watched the closing door and wondered about the number of times the man would have made a similar presentation to other guests.

The more he thought about it the more he could relate the porter's routine to Ford's tried and tested "Six Position Sell" logic. Everything worked, one feature prompting the salesperson, in this case the porter leading onto the next. He smiled to himself, all he had to do was appear confident when called on, his report was only another six-position-sell and he was the salesman describing its features.

The Holiday Inn was one of the preferred hotels Ford would accommodate their overseas executives when visiting and, on this occasion, Bernard Sarfras had been billeted there too. Noel knew other Ford Tractor employees from across the globe would be staying there for their annual review of international operations. He phoned reception and asked if there were any Ford employees already registered, hoping to find someone he may know among the guests. Being told Mister Sarfras was expected later Noel asked if they would give Bernard a message to call his room.

That day Noel and Bernard began an association that saw their paths cross often during his Ford years. Even though the next day, Hollywood actor Tony Curtis had invited them to join him and his family while they relaxed in the pool area, nothing could eclipse finding a lifelong friend within Ford Tractors.

The round of meetings Noel attended that week had been exhilarating and tiring at the same time. Having been applauded for his work in South Africa, had proved the company's faith in him and boosted his confidence. Before leaving he met with Paul Gillis to say goodbye and tell him that he had been offered the role of General Manager Tractors in South Africa. Paul, being surprised by Bill Bayford's offer to his protégé, suggested to Noel that it would be a poor move, as he had bigger plans for him home in Australia.

Paul offered a timeline of events that World Headquarters had in mind for their Australian operation. It involved the coming retirement of Clarence E Ragan, who had stepped into the role of Interim General Manager after the retirement of Ewan Scott-McKenzie. Noel would be expected to tidy any loose ends he had in Port Elizabeth and move back to Melbourne. Once there, he would step into the role of Marketing Manager left vacant by the resignation of Jim Parker.

Paul told Noel, that Clarence would stay in his role as a mentor until they were confident, he had acquired the skills to replace him as General Manager. Noel imagined what Joan's reaction would be when she heard the news. Knowing a smile would cross his face if he stayed too long, he thanked Paul for placing his confidence in him and promised to do his best for Ford Tractors. With the words he had learned during his Ford Marketing Institute training," Get the Order and Get Out," burning in his ears, he thanked Paul again and left.

His first adventure to a World Headquarters review having been met with success, it was now time to relax, catch his flight and return to Port Elizabeth. He and Joan had much planning to do.

Tidying up loose ends in Port Elizabeth took time and yet Noel was still being asked to accompany regional or zone managers to assist them when introducing prospective dealers to the Ford Management and Marketing system. Bill Bayford while understanding Noel's decision to return to Australia, had sought his counsel on candidates for the appointment of a General Manager Ford Tractors South Africa. They had three strong candidates who had the ability and qualifications to manage the role. As difficult as it was to separate them, Noel thought the experience Brian Dearden had gained during the segregation of the vehicle and tractor businesses, made him the best candidate. Bill agreed and later Brian assumed the role.

For Noel and the family, they had much to do. There was a home to find back in Melbourne, packing to do and a holiday to plan. Every couple of years since they had set up house in Wangaratta, Joan had moved house. She had started getting ready the moment she knew they were on the move again. This time she would have help from the African lady who had been engaged to help within the house.

    Believing this would be their chance to build a forever home, they started scouting for a block of land on which to build. Noel had engaged Melbourne architect John Taylor and provided a description of their requirements. He and Joan had agreed on a desired floorplan and along with a few sketches Joan had produced, were ready to begin building. John, was the son of one of Noel's heroes, Headlie Taylor of harvesting machinery fame and by the time their contracts had been signed, the couple had settled on a block of land in Templestowe's Athena Court.

Working with the property manager from their Melbourne real estate agent, they were able to find a home on a ten-acre property on the outskirts of Vermont in Melbourne's North East. Everything in place to leave South Africa, Noel and his family could relax and enjoy their last few weeks there.

Planning the schedule for the holiday they wanted to combine with the return journey became an exciting distraction. By the end of 1971 Noel had left Port Elizabeth. He and Joan along with their young family started home, taking in the sights and sounds of Nairobi, Greece, India and Thailand before arriving at Customs in Perth. From there flying on to Melbourne and to their newly rented home in Vermont. The Howards were now back in Australia and ready for their next adventure.

CHAPTER THIRTY-FOUR

"It's a funny thing coming home. Nothing changes. Everything looks the same, feels the same, even smells the same. You realise what's changed is you."

F. Scott Fitzgerald

Honolulu Time: -- 7.30 pm
Approximately four and a half hours until a QANTAS jumbo fills with passengers and departs Hawaii for Australia. One of them, Noel Howard, is a survivor of last night's ill-fated United Airlines Flight 811.

The dining room was a murmur of consistent chatter, rising and falling over the clatter of cutlery, ice clinking and rattling, slowly melting in warming glasses of spirits. Occasionally a wine bottle cork would pop and a cheer would pierce the clamour. From the bar, you could hear the beer taps snap back when the barman finished filling each glass. Sitting alone at a table for two, Noel's mind was elsewhere. He had survived a flight that should have plunged into the sea. Now, in a matter of hours he would again take his place in the front of a Boeing 747 and flying out over the expanse of Pacific Ocean. However, this time it would be a QANTAS aircraft and he would be heading for Sydney.

Dinner was a lonely affair and although surrounded by people he was not in the mood for company, he wanted to feel the warmth of home, just to sit on the veranda, look out over the farm and have a coffee with Joan. Sanctuary would give him a chance to sort through what had happened and what to do next. Time to think had always been important to him, usually, he closed the office door telling his secretary, he didn't wish to be disturbed. Sitting alone on long flights also offered a similar opportunity. However, for this flight, Noel knew himself well enough to know his senses would be alive to every little creak and rattle the aircraft had during the flight home.

He finished his beer, folded his napkin and pushed up from the table, thanking the waiter who came to clear up. He had time to kill and all he wanted to do was to fly home, but the flight was hours away. Lost in thought, he wandered through the shops in the foyer, stopping to look at a few ornaments tempting him to take them home, but he and Joan both decided years ago, they had enough souvenir shop trinkets. He smiled at the attendant and put the item back in its place.

While he only had a few pieces of clothing, it had now come back from the laundry, he purchased an airline carry bag from the souvenir shop and as he had done a thousand times before, packed for the journey home. He thought about packing the complimentary toiletries kit, but elected to put it aside thinking that if there was a flight delay, it would be better for him to have it in his briefcase. After all, nothing he had done on this trip until now, had worked out the way he had hoped.

Honolulu Time: -- 8.55 pm

After settling his account at the desk, he handed the receptionist an envelope addressed to the hotel's manager. Noel appreciated they were able to accommodate him at such short notice and was grateful for the help that the staff had offered him. It hadn't taken him long to write the message, but he knew it would be appreciated by the manager and everyone who had helped him.

He moved to the far side of the foyer and waited for the taxi he had booked for 9.00 pm. As much as he tried to remain calm, the events of last night and the treachery of Gor Cowl continued to dog his memory. Looking his hands and searching their lines for answers, he found none. However, he was relieved to find they were steady.

The ride to the airport retraced the journey he had taken twenty-four hours before, the only difference being the cab was now stopping closer to the QANTAS International terminal. After checking in he settled into a back corner of the QANTAS Club and pulled out his list of reasons why he should take the job Cowl had offered. While the proposition was appealing, this opportunity was likely to evaporate when the company found a buyer for the Ford New Holland business. He closed the notebook; he still had his position at Ford Asia Pacific in

Australia and an offer from Case. Until now he had always responded to requests from his superiors in Ford with a positive answer, Gor Cowl's pay scale was at a much higher level and his position in the company carried more weight than Noel's. However, what he couldn't know was how well Noel was connected to people within the inner sanctum of Ford World Headquarters and in this, he underestimated Noel's influence. Believing Cowl's blind obsession for keeping his Australian New Holland stooges in place was wrong, Noel could have called on those contacts, but as it had been Cowl who made this test personal. Besides, running tittle tattle to World Headquarters, it just wasn't Noel's way.

Honolulu Time: -- 10.55 pm

The whole journey to the airport, checking in and walking through the international terminal made for a déjà vu moment. Noel knew it was highly unlikely another aircraft would suffer the same fate as Flight 811 and although his personal flying experience told him flying was still safe, images of chaos and noise from kept flashing through his mind. There were moments when he could divert his attention back to Joan and the family, but the throng of people in the airport brought sharp memories back as if running on an out-of-control looped recording.

Thanking the waiter for his beer, he looked at the glass as he lifted it from the table, his hand didn't shake as he thought it might. Putting his mind toward the choice he would make *regarding his career, pushed flying into the background of his thinking. While the familiar began dissolving any irrational thoughts, a confident easiness started to wash back over him. He only had one choice to make, stay within Ford, or negotiate the best offer he could with Case.*

CHAPTER THIRTY-FIVE

After saying goodbye to South Africa, the Howards now full of stories from their holiday knew arriving home in Melbourne would be the easy part. First, they had a rented house to make into a home and boys to enrol into a new school. It was time to pick up their lives from where they left them before their move to Port Elizabeth. As usual Joan stepped straight into organising their homelife, leaving Noel to take up his position as National Marketing Manager, of Ford Tractors Australia.

The couple did have other matters to attend to and finalising the plans for their new home with architects John Taylor & Son took time. They expected the home would take eighteen months to complete as had with their first home. Noel would landscape the grounds, building paths and terraces where Joan decided. After building the garden beds, she would complete the planting.

For Noel the choice of architect had been an easy one, John Taylor the son of his agricultural machinery heroes. Buoyed about their choice after he and Joan had spoken to him, they felt confident that some of Headlie's instinct for innovation and problem solving was evident in his son.

Headlie Taylor, a farmer's son from Emerald Hill near Henty in New South Wales, had become a self-taught engineer, a man who had a keen eye for problem solving. Having experience with flattened crops and the inability of early strippers such as those produced by HV McKay to harvest rain, or wind damaged wheat or barley, Taylor set about designing a better machine.

Beginning with his own design for a header front in 1910, his first machine proved unsuccessful. He persevered, always improving by constructing new generations of the original. After successful field trials he finalised the design, applying for and securing a patent in 1913.

The new header incorporated an even longer fingered comb than the earlier stripper style. This feature now allowed the header front to get under tangled and beaten down crops. By employing a reciprocating knife like those used to mow hay, it scythed through the incoming crop while a twin spiral auger conveyed it to the elevator and into the threshing system, completing the separating process.

In 1916, a manufacturing deal was struck with HV McKay who's Sunshine works would manufacture and sell it as part of their new Harvester. As part of that deal, Headlie moved his family to the city and took up a role in McKay's Sunshine company as a production supervisor in the header works.

Headlie however, was only one of Noel's idols who had served time in the employ of HV McKay.

Coming home to Australia, Noel could feel the mood of those working in the agricultural machinery business had changed.

Newly introduced wheat quotas that had been squeezing the finances of cereal growing farmers when he had left for South Africa, had lifted. In the lifting, he felt the attitude of his colleagues had lightened and optimism buoyed a now smaller team at Ford Tractors.

Settling into his new office at Ford Australia's tractor and equipment operations on Sydney Road, in Campbellfield, Noel had much to do. Apart from meeting a few new staff members and rekindling his relationships with others. In his role as marketing manager, came expanded responsibilities. As in South Africa, it was important to meet with those managers and dealers who could quickly respond to any new initiatives that would increase Ford's tractor and equipment sales.

Knowing he had support from Paul Gillis and other senior managers in the upper reaches of Ford had given him confidence in his own authority. For now, he would have to prove himself capable of climbing into the General Manager's role when the incumbent, Clarence E Ragan retired.

Clarence proved to be a great sounding board and mentor for an aspiring Noel Howard. Now charged with building a more profitable division within Ford Australia and competing with companies like Massey Ferguson, Chamberlain and International Harvester, Noel knew he was hamstrung with such a small product list.

However, as Clarence pointed out in an early meeting, Noel's marketing manager position now offered him an opportunity to start solving some of those problems his South

Australian dealers and others had highlighted before his leaving for South Africa. Therefore, understanding the gaps in the product offering took priority, importance being given to areas that could produce immediate results.

Using the price book as his guide, Noel made a list of each model in the tractor range and in another column, wrote the options dealers had asked for in the past. Checking those options against manufacturing's price list, if they were available, he placed an "F" alongside them. If unavailable internally, he wrote an "O" to remind him to research all opportunities for supply from manufacturers within Australia. In further columns he graded them by their importance, accessibility and the opportunity to create extra profit streams.

To further understand the sentiment among dealers and what they required as standard build options within the tractor range, he requested copies of minutes from the last few Ford Tractor Dealer Council meetings. These meetings offered Ford tractor dealers a chance to meet with senior management and raise any issues or concerns in an open format, knowing their requests would reach the people who had the authority to approve or deny them.

The minutes confirmed what he had learnt from dealers he had met in South Africa and those few Australian dealers he had visited since returning home. If he took the emotion out of everything he had heard and relied only on the data he would have a list of what was required. However, people's buying decisions often included emotion and therefore it was important to include a scale which would rate an offering based on how it appealed.

The tractor price list had fewer holes in the range and he thought by starting there, he believed there had to be areas where extra profits could be found. He knew that without listing every build option and whether it be included as a standard specification, or offered as a factory fitted option, it would be impossible to keep track of it all. Even more difficult would-be providing features that were unavailable from Ford's internal build list.

Clarence believed that if it was impossible to have Ford's own engineers sign off on new products that they needed to bring in to bolster the model line-up, then there had to be another way. He drew Noel's attention to their supply arrangement with Napier. Noel however, wanted something the tractor company could promote more vigorously, a product finished in the same blue and white livery as the tractors and carrying their own brand. After several energetic discussions they decided to register a new company and sell other than Ford manufactured products under the BLUELINE name. This initiative meant Ford Tractors Australia could continue to market a third-party product through their dealer network as before. Without the need to incorporate the Ford name or trademark, more machines could be designed and built under Ford Tractors, Blueline banner. It proved to be an elegant solution to what might have become a paperwork night mare.

Over the last few years and as farmers became more successful, they were using more than one tractor on their farms, often buying a higher horsepower unit and keeping their original machine. Australian farmers had long established a reputation for innovation and now as the handling and transport of cereal crops had changed from bags to bulk, a growing market emerged. A second more utility tractor that took over the lighter and daily tasks around the farm. An increasing demand for bulk deliveries of super phosphate proved the need for the second tractor to be fitted with a front-end loader.

Noel saw this development as an even bigger opportunity. With increased sales of new tractors, dealers traded in tractors and these units were not only clogging up their used inventory. This put pressure on a dealer's cashflow, limiting their ability to make more new sales. He reasoned that some of these traded machines could be repurposed and by fitting a loader, they could be marketed as a farmer's necessity. A multi-use second tractor.

At the end of the week and after many meetings with his colleagues, Noel had a picture in his mind of what needed to be done. His report for Clarence outlined each opportunity. Concentrating on the Blueline machinery range as Noel now began referring to it, tillage and seeding machines for broadacre farming were his highest priority. Because of the farmer's initial capital outlay at time of sale and the prospect of ongoing sales of spare parts, this was where their biggest opportunity for increasing profit lay. However, finding a partner willing to supply would be their biggest challenge and that could take time.

Tractor cabins were another area where broadacre farmers were demanding that their new tractors should offer more comfort. In Australia's marginal wheat growing areas, farmers were beginning to work around the clock and driving an open tractor through the chill of a winter's night held little joy for them. Chamberlain had offered a rugged canvass covered option on earlier models and since the early sixties, dealers had been reporting lost sales because of it. A. F. Gason in Ararat had improved their cabin offering over the years and many Ford dealers had bought from them, but more needed to be done to improve the comfort of what was considered at the time to be very basic. More design and development work were needed on the local unit, or the alternative was to import tractors already fitted with cabins from Europe or the USA. Supplying factory fitted cabins to compete with those rumoured as being introduced by John Deere in the next twelve months was something to be tackled at a higher level.

Front end loaders were a different matter, these could be fitted by the dealer and if designed properly, would offer a consistent quality and good warranty support. Australia had many manufacturers offering products of variable quality and backing, even a few of the Ford dealers were producing loaders under their own brands. However, Noel believed the company needed a single manufacturer, an independent, focused only on loader production would be a better fit with his vision.

Even before Noel's return to Australia, Clarence had begun talks with Domenic Barbaro of Robot Equipment located in Salisbury North, a suburb to the north of Adelaide in South Australia. After being briefed by Clarence about those discussions with Robot Industries, Noel made it a priority to meet with Domenic and his engineer Horst Boetger to explore all opportunities for creating a profitable future together. He had heard reports from dealers that the company was in trouble financially.

Learning that if it had not been for a large front end loader order placed by Ford tractor dealer, Laurie Curtis of Tumby Bay, Robots may have folded. Unlike their competitor Superlift, Robots were still clinging on. Noel sensed both urgency and opportunity. If a deal was to be done, it should be of mutual benefit and one where Robot Industries would secure their position financially. It made no sense to him to build a product's profile only to see it fail because of the manufacturer's failure to make profit.

Superlift had been another company on Noel's original list. He had seen their products on display at the Adelaide Show and at the time, had been impressed by their quality. Based in Adelaide, Superlift had won a military contract to supply fifty-six loaders and backhoes mounted to Chamberlain Mk III tractors to the Australian Army. Fulfilling the army order caused delays in supplying the needs of their loyal dealers.

Eventually the army deal and the loss of wholesale orders to dealers, caused the company to fold.

At the end of a day of strong negotiations from both sides it was agreed Ford Tractors would supply the full range of Robot loaders under the Blueline brand at the current pricing. This would give Ford's tractor dealers a full range of loader products they could fit to a competitor's machine along with an offering for the full range of Ford tractors.

Domenic welcomed Noel's offer to install Frank Pitney, a Ford UK trained engineer, within Robot. Frank's brief was to find economies in all processes and write a quality policy for all aspects of manufacturing. Frank's work brought a more modern look to the whole loader range and continual efficiency improvements were being made.

To further strengthen Robot's sales efforts, Domenic had employed Graham Ainsworth, a draughtman who had experience with Superlift. When Superlift collapsed, he found work where he could, even taking work at Simpson Pope. However, it was a role at Fairey Australia in the Weapons Research establishment that had taken him north of the city. Fairey were close to where Robot Industries were based, and believing he had experience to offer, Graham arranged a meeting with Domenic. His easy going nature and permanent smile made him a perfect fit as someone who could take Domenic's Robot brand further and be the ideal point of contact for dealers.

For Domenic and Noel, a common respect for business and the dependency of working together meant their relationship has grown into a long-lasting friendship. Often, they travelled on joint overseas missions to visit international suppliers and trade shows. On one of these overseas visits to Europe, Noel had invited Domenic to join him and meet with executives from Ford Tractors from other parts of the globe.

They were in Rome enjoying a free weekend and Domenic being of strong Catholic stock, explained to Noel how he had always wanted to see the Vatican. While they were wandering through Vatican Square, Domenic joked that he wondered what it would take to arrange a meeting with the Pope. However, at this time of the year the Pontiff as always, spent some of the season at his Summer Residence.

Without his friend noticing, Noel slipped away, leaving Domenic exploring the square. Noel sought out one of the Swiss Guards. Asking him if there was an official protocol he could follow, so that he could arrange for his friend Domenic to meet with the Pope. The guard smiled and nodded, saying in his broken English that he knew of a way. Palms were crossed and information exchanged. A quick relocation and another payment to a policeman. Domenic and Noel found themselves being invited into the Vatican and having audience with the Pope. It was a fine end to a great trip to Europe.

The loader business in place, it was now time to review other opportunities for the Blueline range and cull those machines from the price list that were not selling. Although an arrangement with A F Gason Pty Ltd had been in place for some time, Noel felt a more formal arrangement like the one he had struck with Domenic was needed. He wanted to be able to provide dealers with an opportunity to access Gason's cabin catalogue under the Blueline brand.

The company had been selling products built by Napier for over a decade or more and many scarifiers and wideline cultivators finished in traditional Ford Blue, dotted the Australian countryside. However, with an increase in tractor horsepower, traditional sowing systems were changing.

With the introduction of Connor Shea's mass produced Airseeder, Australian farmers were now able to take advantage of using wider machinery for seeding. This intensive development had resulted in farmers being able to improve their opportunity for sowing their crops at the most efficient time of the season.

Manufacturers like Massey Harris, John Shearer, Chamberlain and International Harvester had dominated this market for decades and now change was coming.

Albert Fuss had been experimenting with different methods of distributing cereal seed across a wider area than available using conventional seed drills. Albert's experiments led him to consider conveying the seed through synthetic tubes across the width of his twenty-foot Graham Holme chisel plough. To overcome the distance from the metering unit to the seeding boot, he employed a sealed seed box. The grain fell from a metering unit into a low-pressure air stream, taking the grain along the flexible tubing to where it entered the seeding boot and dropped into the furrow. Having patented the machine in 1956, Albert's original design is now found in seeders manufactured on every continent.

Wider tillage machines were becoming more sophisticated. Farmers who were chasing a more regular seedbed preparation had begun demanding a level lift feature on their wideline cultivators. This created an increased demand for tractors with multiple remote hydraulic systems. Ford's 8000 and 9000 tractors featured dual remote hydraulics and Noel could see a time coming when the smaller horsepower buyers would want similar hydraulic systems, to those developed for higher horsepower machines

In his favour, the Blueline 520 Wideline cultivator incorporated an ingenious design, it employed only one eight-inch hydraulic ram to operate both the depth control of the tines and the wing fold mechanism. While this feature helped sell the wideline in great numbers, he knew its time was limited. He knew from development in other markets, that a single remote hydraulic outlet would soon be superseded, meaning multiple outlets would be offered as standard equipment on the Ford 5000 and 7000 tractors. With such sophistication came a demand for wideline machinery to perform more than one task and Noel needed to understand what Napier had planned for the future.

Finding the right product partner with capacity and desire to help Ford bolster their tillage line became a matter of urgency. Therefore, during his frequent trips to Adelaide to visit Robot Industries, he secured appointments with management of John Shearer, David Shearer and Horward Bagshaw.

Horward Bagshaw had supplied equipment to Ford before and buoyed by his experience, Noel was hopeful of a good hearing.

CHAPTER THIRTY-SIX

Meeting with Napier Brothers in Dalby, Noel was shown sketches of a new airseeder they had been working on. They wanted to add a seeder option to their already extremely popular wideline cultivator range and take advantage of the growing pneumatic seeder market. The rapidity of growth in this sector, was now enticing them to employ their own ideas into those principals Albert Fuss had patented in 1956.

Noel was aware of the inroads into the wide seeder market Connor Shea had achieved against the conventional forty and sixty row wide seeding drills being offered by John Shearer. These drills followed time proven Shearer designs and gave exceptional results in conventional fine tilth soils. However, with the entry of better pre-emergent herbicides, farming was changing and a trend toward less intensive tillage principals began to emerge.

Weeds and stubble trash had always been a problem when *working back*, a term used by Australian farmers to describe secondary cultivation. Before herbicides this cultivation may have taken up to ten passes to destroy emerging weeds after the ground had been opened with heavier scarifiers or chisel ploughs. Dependent on rainfall and persistent weed growth, until the early seventies fallow ground often returned to a very fine tilth, the loss of organic matter leaving paddocks easily subject to erosion.

Finding a way to lower the amount of draft horsepower required to haul any tillage machine was the objective of every designer. Napier Brothers reasoned that by employing a slimmer tine, a lower amount of earth would be pushed ahead of it. The forged slimline part they designed, set a trend for others to follow. All Napier Widelines and therefore the Blueline machines employed a slimline tine as standard equipment in their design. John Shearer would soon follow Napier's lead by offering a similar narrow tine as an option on their cultivators too.

Farmers looking for a better way to quickly cultivate scarified ground, began buying the lighter tined folding sectioned "widelines" to replace their seed drills for secondary cultivation. It was these machines to which the airseeding manufacturers started attaching their own design for a seeding unit.

In Adelaide, John Shearer Limited were aware of the benefits of Albert Fuss' patent, had designed an air-seeding system employing a grain and fertilizer hopper. The hopper designed by engineer John Thomas and his team was an elegant system. It trailed behind the tractor and from there, a forward hinged swinging drawbar, reached back to engage the drawbar tongue of Shearer's own wideline cultivator. Seed and fertilizer dropped via a metering system into the airstream, which then carried it along the cultivator frame and onto the secondary distribution heads. These were mounted at the top of vertical tubes bolted to the cultivator mainframe and its wings. Every effort had been made to make the system simple, a low mounted power-take-off driven fan close to the drawbar pivot, reduced the angle of universal joints in the PTO shaft during cornering.

Noel had hoped his appointment with John Shearer's Managing Director, Bob Shearer and Sales Manager, Keith Porter might extend to a visit to their experimental farm at Wanbi in South Australia. However, he thought that would be akin to a Ford executive being invited to a General Motors proving ground. Luckily for Noel, on one of his visits to South Australia he was met at the airport by John Shearer's chief engineer, John Thomas before driving to their farm. Happy in each other's company, they drove out through the Adelaide Hills and toward Murray Bridge and once over the river turned north into wheat country. While they travelled, they discussed the difficulties facing Australia's broadacre farmers, exchanging thoughts on chemical weed control and how it affected current tillage manufacturing companies. Although this new development had a long way to go, John had mused that tillage companies wanting to stay relevant may need to find other opportunities to combat an emerging weed-sprayer market. Noel thinking that in solving some of these challenges, over the next two decades the industry would witness traditional farming undergo radical changes.

Impressed by what he had seen, now came the homework required to draft a proposal that explored the possibility of getting John Shearer's management to agree to supply machines that Ford dealers could retail as Blueline.

However, if an agreement with Shearer could not be reached, his trip to Wanbi had given him a better understanding of air seeding principals. Having John Thomas explain the layout of the Shearer Airseeder, had raised his confidence in the design that Napier's had shown him. Although Shearer's were employing a three-tonne hopper and the Napier designed machine held only two and a half tons, it was comparable to machines by Connor Shea and other competitors.

Napier had no plans to produce three-point-linkage vineyard seeders, or trailed seeding machines suitable for smaller farm holdings, which did limit their ability to extend the Blueline offering. However, for Noel, having an airseeder available soon presented a much bigger opportunity. If they could make it available by the mid-seventies, he thought it should satisfy the needs of Ford's broadacre dealers.

The farming model employed in these dryer areas and from where most of the cereal production came, had been rapidly changing. Different methods began evolving and so did the equipment.

Three-point-linkage machines that had been previously supplied from Napier's range had been reviewed and only those that had a continuing sales history, remained in the new

Blueline catalogue. Market share might have been judged on overall unit sales; however, every salesman and accountant knew it was from higher value products, that their larger profits came from.

In Napier, Ford Tractors Australia had a partner with whom they had a great and long-lasting relationship, a good range of products, an airseeder in planning and an agreement to supply. It made sense to Noel that the company continue with that arrangement at least for the time being.

Having delivered his latest report to his General Manager, Noel left the office unsure what to make of the chill he had sensed in the old man's demeanour. Hadn't he proved his leadership skills and an ability to delegate in Port Elizabeth. Taken on extra responsibility when asked, and spent hours of his own away from home while putting a more modern implement strategy together? Hearing the ghost of Charlie Milthorpe telling him not to concern himself with things over which he had no control, he walked back to his office via the factory, before starting to work through the number of items on his to-do list.

Noel, having tried to see Clarence a couple of times later that morning, only to be told his mentor would be gone for most of the day, resigned himself to trying again tomorrow. What he couldn't know was that his future was the subject of discussions taking place between Clarence and Ford Australia's Managing Director, Brian Inglis.

The following morning, Noel found himself seated opposite his mentor as summoned. Not knowing why he was there, one of his grandfather's old sayings kept rolling on a loop in his mind, *"Noel, if you ever find yourself in a hole, it's always best not to keep digging,"* so, he decided to sit quiet and listen to whatever Clarence had to say.

The old man, his thinning grey hair revealing a reddening scalp had nodded when Noel walked in, and with his pen, pointed to the chair for Noel to sit, while not taking his eyes from the papers in front of him. The office was different to every other time he had been there, today all the visitors' chairs had been pushed back against the wall and whatever Clarence wanted, or was reading made Noel uncomfortable. These were not the ways of a man he thought of as a friend and mentor.

Clarence began stressing how important it was for his protégé to design a long-term strategy that their sales teams could build a future from. Noel understood what he was saying, but this was a sales role and he already had his hands full of negotiations with suppliers for the Blueline range. Clarence pushed the paper he had been studying across the desk. It was a copy of Noel's last appraisal. Again, using his pen, Clarence pointed to a comment where Noel had stated his desire for more responsibility. Tapping the pen on the same sentence as if to add value to his words, he said he wanted Noel to take on the National Sales Manager role while continuing in his same position. Elaborating that he knew Noel was more than capable and, if he was willing to accept the extra workload, then the job was his.

Noel believing this offer to be another test for him to prove to Paul Gillis and those senior to him, that he could handle the role of General Manager. While they may have discussed it in South Africa, nothing is ever certain until formalised with a written notice of appointment. He knew that accepting the position of National Sales Manager, while continuing to perform his marketing duties would mean more pressure on his time. However, he also knew that in Ford's world, that with the extra responsibility came rewards. It was time to negotiate.

In June 1972 Clarence only had about eighteen months left on his assignment to Ford Tractors Australia and in planning for his departure at the end of 1973, needed to be sure there was a candidate in place to take over his role. By then and being a Ford veteran at sixty-nine years of age, he considered it time to retire.

For Noel, Clarence had become much more than a friend or work colleague, always encouraging his young charge, his guidance pushed Noel to seek further horizons. Therefore, he saw nothing unusual in Clarence's request to see him in his office before work next morning.

It was near the end of June and although Ford Australia reported on a different financial year cycle, there was the constant murmur of paper turning interspersed with the crash of filing cabinet drawers being opened and closed. This morning, Ford Tractor's somewhat compact office in front of the tractor assembly plant on Sydney Road, Campbellfield, had been invaded by auditors from head office when Noel arrived.

He knocked on the General Manager's door, Clarence murmuring something about the intruders, waved him in and indicated he close the door. Passing a pamphlet across his desk one morning, he suggested Noel think about applying to study International Business at Harvard University in Boston. Noel looked at the fee structure that accompanied the brochure and shaking his head, told him it was too rich for his salary. Clarence held out his hand, took the papers back and spread them out on top of his desk pad. He leaned back in his chair, folded his hands behind his head and saying nothing, just smiled.

Noel thought the old man was up to something. He had often heard rumours of other executives within Ford gaining a Harvard education at the company's expense. Was this what Clarence had in mind? He stood expressionless and waited. Clarence leaned forward and producing another sheet of paper placed it on the desk in front of him. Noel knew the document, as National Sales manager he had approved tertiary education expenses for several staff. Clarence using his thumb to anchor the page on the clear expanse of desk slowly turned it toward Noel with his fingers, the fountain pen in his other hand pointing to where Noel had to sign the application.

Clarence would not be in Australia for much longer. His brief when accepting the role had been to leave the tractor company in a better way than when he arrived, putting in place the right people to carry on building a future for Ford Tractor Australia.

Tapping his finger impatiently he told Noel to sign the document. Ford would pay all fees including travel and accommodation. Noel seeing Clarence had completed the form held his composure and suppressing every desire to grin, signed the places where Clarence's sausage like finger pointed. For a man given to self-improvement, Ford had once again offered him an opportunity to study and even though he knew it was in the company's interest, he felt honoured by the gesture.

On more than one occasion in the past, both Paul Gillis and Clarence had impressed on Noel that he should formally apply for the General Manager's position the moment it became vacant. After setting the Harvard papers aside, Clarence pushed a hand written message across the desk. It was a short simple draft suggesting how Noel should frame his application for the General Manager's position. Noel asked when he should make his application formal.

Standing, Clarence walked around the desk and putting his hand on Noel's shoulder told him there was time and he would give him plenty of notice before any official notice of his impending retirement would be made.

Joan on hearing about this extension to Noel's duties, could see the family staying in Australia for the next few years and again fully supported her husband's decision to accept it.

She understood he would be spending even more time away from home, committing to both interstate and overseas obligations. She would continue managing the family's day to day needs, happily knowing they were securing a more comfortable future for their growing brood.

On the 26th of June 1972 Noel officially made his pitch for the General Tractor Operations Manager Ford Australia, handing it in person to Clarence. While sure his request would be accepted, as his superior had suggested Noel had indicated a second option by requesting an overseas position should another candidate for the General Manager role be appointed.

The Howards had by now, moved into their new home in Athena Court at Lower Templestowe. As before, Noel's weekends were given to making paths, garden beds and preparing the earthworks for a swimming pool. This was their fourth move in three years and while building and dressing a new home was exciting, for Joan, it meant there was little time for anything outside the needs of their young family.

For Ford Australia's tractor dealers, until now cabins to suit the tractors they sold, all came from aftermarket manufacturers were extremely rudimentary and those offered by Gason, were a retrofit proposition too. Noel knew that when compared to new designs coming from John Deere, they had to find another way for Ford Tractors Australia to meet this new challenge.

Ford's higher horsepower 8000 and 9000 tractors were now being offered in the USA with a square sided cabin as a factory fitted option. These spacious, two door, flat floor cabins offered the operator a panoramic view and their rubber insulated mountings lowered noise levels to 83db, an unprecedented level for a cabin equipped Ford tractor. With these machines already in the marketplace and the 1972 release of John Deere's 82.5db cabin equipped tractors, farmers started putting sales pressure on Ford Dealers to provide a better option for their 7000 and 5000 models.

At Dealer Council Meetings, reports from regional staff and the numerous calls he was receiving from dealer principals, Noel kept hearing how they were losing sales of their high horsepower models to John Deere. More worrying were the rumours surrounding Chamberlain releasing a high specification, low noise cabin on their new model range due for release in 1975. The thrust of these concerns was that unless a solution could be found to provide a better cabin option, then sales of 5000 and 7000 models would face a serious disadvantage. Something had to be done.

Noel knew better optioned "Q Cabins" were coming with the release of the range of 8600 and 9600 tractors. These cabins had been designed to meet increasing pressure from regulators in Europe and the USA to provide operator environments with more protection at lower decibel levels. However, if the European "Q" cabs were imported as a factory option to Ford's three and four-cylinder models, setting a competitive selling price became impossible.

Establishing the Blueline brand had changed Ford's relationship with AF Gason in Ararat away from that of an Approved Supplier. Where once Ford dealers had placed their cabin orders directly with Gason and paying on Gason's invoice. Now with the Blueline Agreement in place, Ford Tractor dealers could buy Gason built Blueline cabins on Ford paperwork. Gason still shipped cabins and ROPS direct to dealers, but now they invoiced Ford for these units at the end of each month.

From Noel's perspective, the Blueline business had brought extra profit opportunities and the tractor company had started building toward a complete product list. However, taking tractors to market with an Australian version of the "Q" cabin soon became the main consideration for Noel's focus. Finding a way to develop Gason, or another cabin manufacturer into an OEM supplier came with its own set of challenges. Undeterred, Noel believed he had the people around him who could find a way through the extensive maze of internal company regulations.

The imposition of wheat quotas had damaged Australian tractor sales from the end of 1969 until June 1972 and as the end to quotas became imminent, confidence grew among Australia's wheat belt farmers.

During that wheat quota period, Ford Tractor's sales teams had worked hard to keep their dealers in horticultural, wine growing and dairying areas focused on competing for market leadership. Now as National Sales Manager, Noel began focusing his attention on Ford's higher valued, broadacre tractors and Blueline equipment. Now, as Australia's wheat areas headed toward a harvest free of quotas, farmers in these regions began enquiring about bigger tractors and showed interest in buying more efficient equipment.

An emerging threat that Ford still had to face came from their old wheat area enemy, Chamberlain. In the late 1960s Chamberlain had secured the distribution rights for John Deere tractors and their company was changing. Now already market leaders across the wheat belt, Chamberlain dealers had a range of green, high horsepower, silent cab tractors appearing alongside their traditional, orange liveried models.

Without a "Q Cab," Noel understood his dealers would be at a disadvantage and heading off a charge from John Deere in the high horsepower had its challenges. John Deere was not alone, other North American companies had developed high powered tractors for their domestic market and the similarities between wheat farms in Australia to those in North America, meant Massey Ferguson, Allis Chalmers, Case, White and others had product to dump in Australia, should they build surplus stock.

It could be a time to worry, or one to search for opportunity, Noel chose to be aware of the challenges and do the latter.

At the end of December, a letter he received from Emery Dearborn, Vice President and General Manager of Ford Tractor Operations based in Troy, Michigan, raised his spirits. Noel was overjoyed that the efforts of his sales team in Australia had been recognised. Emery stated that for 1972 the company had delivered more tractors than in their best year, 1966. More importantly the company had secured profits higher than ever before.

Sure, Emery had asked for a similar effort during the next calendar year, but that was always expected and Noel felt confident 1973 would be even better.

With a Blueline Agreement in place with Gason and knowing that Les Gason had also secured an order to build cabins for Chamberlain, made Noel wary. Concerned about the increasing share John Deere were taking from the large tractor market Ford needed to change its approach to the way they sourced their products.

Noel firmly believed that an Australian Government paying a bounty to a multinational company to produce tractors grated against the very ideals of an independent Australian Manufacturing Industry. If there were opportunities in the bounty scheme for Ford, he didn't pursue them, instead believing there had to be a better more sustainable way to tackle the challenge and began planning his next move.

However, the thought of a bounty still nagged at him and he decided to study every sentence of the act of parliament that carried it into law, desperate to find any loopholes in the documentation. After skimming through the act, he decided it would be a long-drawn-out battle and one for another time. However, because the bounty scheme offered an unfair advantage to his competitors, it irritated him and every day he kept writing a note to lobby against it on his to do list.

CHAPTER THIRTY-SEVEN

Having support from Paul Gillis and now with Emery's letter offering recognition of his team's efforts was a comfort to Noel. However, it had also explained that the company expected the year ahead to be more of a challenge and asked for an even bigger effort.

To top the results of the last year would be difficult. Noel's suspicion told him Chamberlain would be gearing for their new John Deere powered models to come on line within the next three to four years. He knew if Chamberlain John Deere (CJD) as the company had now been named, could still take advantage of the Bounty Scheme, then every tractor and machinery dealer who wasn't selling Chamberlain would be disadvantaged.

One of the management clichés Noel had always held onto, became the driver of his next initiative. *"If you can measure it, you can manage it."* The cliché rang as true to him then, as it had when he first heard it. Sales figures for the Australian tractor and machinery market were at their best, an ad-hoc set of numbers. Dangerous for those taking them too seriously as a baseline for product planning. Knowing it was probably a fight for another day, Noel asked his secretary Wendy Orchard, to build a list of names and contact details for all the General Managers of each manufacturer in their industry. It would take time and much coercion, but he was certain that if he could get everyone into one room, he may be able to talk them into strengthening their Tractor and Machinery Association.

He believed the association should set out to be a body powerful enough to lobby government, but to achieve that, it would need to produce accurate production and sales numbers. However, for now it remained just an idea, but the spark that came from it, grew into a passion that burned bright and he never failed to mention it when talking to other colleagues within the industry.

While Noel's list of goals continued to grow, Australia's lifting of wheat quotas heralded new signs of optimism among cereal growers and the orders for broadacre tractors and equipment started gaining momentum. Although confidence began creeping into the order books, sales of higher horsepower John Deere tractors began having a devastating effect on the traditional tractor market. International and Massey Ferguson were no longer the biggest villains, as the spectre of Chamberlain John Deere loomed large in the profitable, high horsepower market segment.

In 1970, a targeted and strategic move by John Deere signalled their intent, when the American machinery giant purchased 49% of the now publicly listed Chamberlain Holdings. Ford's old foe Chamberlain, armed with a modern range of high power, drawbar only tractors equipped with industry leading quiet-cabins, had positioned itself to take advantage of the growth in that market segment.

Believing that their bottom line was bolstered by the rewards of the Bounty Scheme, Noel reasoned CJD's American parent could now afford to wage a price war on their rivals. He believed too that holding onto such a scheme would become a sales advantage that CJD would fight tooth and nail to keep. However, there had to be ways that Ford Could combat the coming challenge from Welshpool and he set his mind to finding them.

A F Gason Pty Ltd had in conjunction with W F (Bill) Baillie of Melbourne University's engineering department, and who was now the Officer-In-Charge of the Tractor Testing Station at Werribee, had been experimenting with a test rig for roll over protection systems.

As early as 1971, Frank Gason had asked Ford to provide a suitable tractor for the testing of a cabin and two four post ROPS frames. This Ford 4000 would serve as the mule to which the ROPS and cabin would be fitted and verified by the testing station's officer, Bill Brown. By June, Ford had secured safety certificates for their Blueline cabins and ROPS frames not long after the 4000 tractor had arrived at Gason's.

Just as much as it had been a triumph for Frank Gason, so it was for Noel. Now, farm safety had become another driver in their marketing plan. Even more interesting to Noel, was learning from Frank that Gason had tested its rubber mounted

ROPS under the same conditions, and these tests too, had met Australian Standard Requirements.

About the same time as they were testing cabins for operator protection, Gason invested a considerable sum into purchasing a Bruel and Kjoer noise meter. This was an expensive investment and it now gave Gason the capacity to accurately measure decibel levels in targeted positions within the tractor cabins. Frank also understood his market and that a tractor buyers' demands were becoming focused on operator comfort. Where once air-conditioning had been the big driver of the last decade, both he and Noel believed noise reduction would drive cabin research and development for the next ten years.

Being able to present their four post ROPS as a safety feature demonstrated Ford Tractors Australia's commitment to farm safety. This new development, created yet another idea in Noel's growing list of possibilities. Possibilities that could support his desire to produce an equivalent locally built unit, to those of the new "Q" cabin equipped 700 series tractors when released in Australia.

Australia's Bounty Scheme became a continual niggle. Every time a dealer called to talk about a lost deal to a Chamberlain, or an Australian manufactured International, Noel often found himself agreeing with dealers. Men like Norm Feigert from Cleve in South Australia's wheat belt, but he too was stymied by regulation. Dealers in broadacre areas were concerned about the ground rules being biased against them and it rankled. Ford offered nothing to fight with. Norm would often tell him it was like playing football uphill, while kicking into the wind. He was not alone, reports like his were coming from everywhere. Chamberlain of old, had by now, fashioned a successful dealer network and the trickledown effect was having a devastating consequence on the morale of Ford's broadacre dealers.

Noel needed to understand the scheme better. Which of Ford's competitors were receiving a bounty, how was it applied and, did any of their Blueline suppliers qualify? If Ford were to compete on equal terms, then he had to know every aspect, every policy decision. To plan an attack on a government decision would be difficult, so he needed to know the policy inside and out. He needed to understand how the dollar amounts were applied and who decided them. While he had met a lot of people in his time within the industry, until now, he had little or no experience when it came to lobbying government. His role as Sales Manager offered him some prestige and quite a lot of freedom, but tackling government was another matter. Noel was about nine when his father had given him some advice. They were discussing a problem Finlay had had with his flour supplier. The old man advised Noel, saying, *"if you need to fight, be sure of the facts and always, be sure you are on the side of right."*

He remembered their holidays and when as a child exploring Canberra with the family, his mother would often find someone in power who could get them access to different parts of the War Memorial. It was all about finding someone who knew the right person and he was determined to follow her example. A person who could get him an introduction to the right people in power. First, he needed to find an executive within the company who could provide an introduction to the right person for him to talk to.

Clarence E Ragan had now been winding down his role and although he had been passing more responsibility onto Noel, he couldn't help, other than suggesting that Noel speak to someone in Ford House. Noel understood this was another of Clarence's lessons, if his mentor couldn't help, then it was time to test Brian Inglis' offer of always having his door open to him.

Brian as good as his word welcomed Noel and while free with his advice, recommended Noel start with the Tractor and Machinery Association. Ford Tractors already had John Blyth registered to represent them at meetings, but in Brian's opinion the association had become, or always was, a toothless tiger. His comment left Noel of the opinion that if Ford were not taking their role in this association seriously, then it would only be those with a personal stake in keeping the status quo, whose voice would be used to lobby government.

While Brian had been talking, another of Noel's mother's little homilies popped into his thinking, *"if it's going to be, it's up to me."*

Believing their conversation had ended, Noel stood to leave the office when Brian asked him to stay. He said, he and the board had been deliberating over the need for Ford Motor

Company Australia to enter the world of wholesale and retail finance. Brian saw an opportunity to fight for a bigger share of the Australian vehicle market, by meeting General Motors Holden head on with Ford's own credit company. Brian knew Ford Credit in South Africa had often helped to finance prospective dealer principals into new dealerships and he understood how effective that direct link had been. However, he wanted to know how Noel had found the South African experience and get his opinion on how a similar finance arm could help Ford sell more tractors.

Brian asked him to formulate a proposal that would demonstrate to his directors, how Ford Credit Australia would be a benefit to increasing the sale of tractors. Impressing on Noel their need to keep their plans confidential, Noel agreed, saying he understood and was very much aware of the risks the company would face should, while researching such a project, lose a finance partner. The loss of a financier would force them and their dealers into a position of uncertainty, something they didn't need. Noel assured him of his willingness to work diligently, stating that although he knew that as highly confidential the proposal was, to get a good picture of how important this initiative might be, he would need to take a few trusted dealers into his confidence. After all, his proposal required balance and solid numbers, rather than be based on hearsay

Buoyed by his discussions with Brian, Noel now sought out John Blyth to better understand the workings of the

Tractor and Machinery Association. John too, had realised the TMA was a group of people going through the motions and without proper leadership would continue to do so. They agreed that Noel would accompany him to the next meeting where John would introduce him to the group.

The to do list in Noel's diary had grown, he entered the date of the next TMA meeting, and opening a new page in his notes section, headed it with Ford Credit, underlining the words in red.

Noel's time in South Africa had reinforced his belief that Ford's current arrangement to provide floor plan finance to dealers had become cumbersome and bogged down by third party regulation. He felt it was holding back their ability to meet their competitors head on. While he had raised the issue with Clarence on many occasions, the older man would always point out there were other obstacles to overcome, before tackling a proposition with the enormity of creating a credit arm. Clarence agreed that finance had been the reason behind many of their lost sales. Always, it had been their competitor's easy finance terms. While that had been the major factor and until now, Ford Australia had no appetite for entering the finance market. However, after his meeting with Brian, offering dealer finance through Ford Credit Australia, had just become that little bit closer.

CHAPTER THIRTY-EIGHT

On October 26th 1973, Clarence E Ragan, General Tractor and Equipment Manager circulated a memo advising all members of management, that pending his return to the United States, he was pleased to announce the appointment of Mr. R N Howard as his replacement effective from the 1st of November.

As predicted by Paul Gillis and under the guiding hand of Clarence, Noel had proved himself worthy of their trust.

Before leaving to go home to America, Clarence had asked Noel to take him for a drive into the countryside north of Melbourne, he wanted a *"no bullshit tour,"* just a drive with a friend. Somewhere they could go where he could ask about Australia's colonial history and just enjoy the drive. Saying, that by just being in the car there would be no phone calls, or people knocking at the door to interrupt them. It was to be a day when Noel could ask him anything and for him, the same of Noel.

For Noel, it offered an opportunity to ask questions within a confidential environment. Nothing was to be excluded and besides, they enjoyed each other's company. It would be a day where the master passed on knowledge to his pupil and a time for Noel take advantage of that experience. Leaving the office in Campbellfield they travelled north out along the Hume

Highway, drove around a few farm roads at Wallan and then followed the road to Bendigo.

Driving into Kilmore, Clarence pointed to a large empty plot of land that backed onto the Catholic Church. It was in the main street, and wondering out aloud, said if he wasn't returning to the United States, it would make a great investment. Noel remembering those comments made a mental note that once they were back in Melbourne, he would set about finding the owner and make an offer. Within a short time, he and Joan, had become the proud owners of four acres in the main street of Kilmore. It was a great asset and one that rewarded them more than ten-fold when they sold it only a few years later.

On their drive, Clarence advised Noel to ensure he had the ear of the king-makers in the company, both in Australia and at World Headquarters in Michigan. He would offer some letters of introduction, but felt it would be better for Noel to assert his position as a General Manager in the Ford Motor Company hierarchy. Noel understood what he meant, as he always looked for who the decision makers were in any company he dealt with. He didn't decline the offer, saying only that he would call Clarence if he ever found himself needing his help.

Clarence talked about the threat of Chamberlain John Deere and his belief that they were using the bounty system to prop up their marketing push for selling high horsepower, broadacre tractors in Australia. After all, the bigger the power rating, the higher the retail price and the bigger the profit. His words were no different to those of the dealers in wheat areas across Australia. Even during his time at Massey Ferguson, Noel knew how difficult it had been, coming up against a Chamberlain, or International deal, in a dollar for dollar fight.

It was Clarence's view, that the only way the company could fight them was selling feature for feature. Only then, did their Ford tractors hold the better cards. However, he also had come to understand by the way Australian farmers had been ordering their tractors, that unused ancillaries, held little value to them. Power steering proved a good example, it had been an option in the pricelist years before he arrived and yet, while farmers took it happily as a standard fitment on International Harvester's 554 tractor, the price as an option on the Ford 5000 often caused the loss of a sale. It demonstrated a case of, if they could get all the comforts on the tractor at the same cost, they would take everything. However, justifying the value of three-point linkage to a man who wouldn't use it, became a hard task.

Noel knew this too and he had often tried to find ways to reduce the Australian company's buying price with his superiors, but the parent company never wavered in its cost structure. He applied his mind to working through alternative suppliers for ancillary equipment that might be added under the Blueline brand to improve their competitiveness.

With Clarence no longer in charge, Noel held the reins. Being charged with steering Ford Tractors & Equipment Australia into a bright and prosperous future, would need fortitude, empathy and initiative. Noel's performance would only be as good as the people under his command and while he supposed there would be difficult staff decisions to be made, it was critical to ensure he only placed people with the ability in strategic roles.

Wanting to find those who had asked for a challenge, he asked Wendy to gather the last performance reviews of all staff. He needed to know who could meet their aspirations and those who needed training to bolster their knowledge. For now, a comprehensive dealer review could wait.

Along with his current to do list, he had now inherited all the glad-handing and constant roster of head office meetings to attend. Meetings that until now, he had only attended as a guest of his superior. Moving into the General Manager's role, now meant he carried all the glory into such meetings, but it also meant all the company's problems stopped at his door too. He relished the challenge, for not only did he assume the responsibility for everything, he also had the power to instigate change.

It was important for Noel as the General Manager of the tractor division, to visit Brian Inglis to take instruction and seek advice. Brian had to be made aware of who Noel had in mind for the position of sales manager, and discuss the suitability of candidates who could be appointed to the marketing manager's role. A position made vacant due to his promotion. Taking his boss's advice would be both a political and strategic move, he had people in mind for promotion, but knew allowing Brian to consult on their selection made sense.

After cutting his discussion topics down to a short list, Noel made sure he thought through the questions Brian might ask. Working on the assumption that if their roles were reversed, he interrogated each question and answer fully. "*Be prepared,*" might be a Boy Scout motto, but for Noel, it had always mattered in business too.

Brian's secretary Mrs. White, aware he was coming, moved around from her desk and congratulated him on his appointment. She knocked on her Managing Director's door and pushing it open, stood to one side, ushering Noel in. Brian stepped out from behind his desk, his wide smile and outstretched hand greeted him. To Noel, it felt as if Brian was as excited about his appointment as he was.

The mood in the office was welcoming and informal, Any nerves he might have been feeling before he saw Brian, melted into the ether. Apart from discussing staff appointments, Noel wanted his boss's advice on lobbying government.

Brian as good as his word, welcomed Noel to sit and while free with his advice, recommended he start with the Tractor and Machinery Association. Ford Tractors already had John Blyth registered to represent them at meetings, but Brian again articulated his opinion that the association had become, or always was a toothless tiger

While Brian had been talking, another little homily from his childhood popped into his thinking and he could hear his mother saying, *"if it's going to be, it's up to me."*

Coinciding with the scheduled three-week factory shutdown over the Christmas period, most of the salaried staff were encouraged to take their annual leave too. Leaving a skeleton staff in the office to manage any incoming enquiry, which traditionally reduced as the hay harvest finished. Noel took this time to leave a monumental year behind him.

He may have left the building, but as he and the family knew, it would be unlikely that his mind would stop running hundreds of different future scenarios in a never-ending loop.

Now was time for him and the family to have a holiday.

CHAPTER THIRTY-NINE

Over the holiday period, Noel had built a list of names to make up his desired management team. For some who were comfortable in their current positions, moving sideways or down a couple of spots might be a tough adjustment. Ford had a policy of looking after their executives and as such, one should never say never. For those who were being promoted, they might find their new roles equally as tough. Noel knew his promotion had given him the authority to drive the company forward and their expectation would always be to see an increase in market share and profitability.

In 1974, sitting in what was once Clarence's office and at Clarence's desk, Noel began his first task of getting plans in place to move the best people into those roles where he thought they would be best skilled. He would be asking them to work hard to help him lead the company into a brighter future.

When Bill Bourke arrived at Ford Australia as Sales Director in 1964, one of his first strategies had been to conduct a staff performance review. Hitting upon the idea of driving a youth culture through the management of the vehicle business, Bourke educated and promoted those people who he knew would go on to run the company long after he left. Noel thought this idea even though not new, had merit. A lot of the old-guard had left before he arrived home from South Africa, but he too, had identified a few positions where a younger, more skilled manager could be more effective.

Moving into the General Manager's position had created a vacancy in the Marketing Manager's role and while his desire was for a young executive. He understood youthful exuberance sometimes needed harnessing, and mentoring their creativeness in pursuit of better outcomes. To Noel, Ford Tractor's stalwart, John Blyth had served the company in many executive positions and he knew by putting John back into that role, he would have a man capable of offering sage advice without breaking the spirit of young men given to rash ideals.

John, an impeccably dressed Englishman, had come to Australia to escape the harsh rationing and find somewhere more bearable. Having some experience with Ransomes, Sims and Jefferies, machinery manufacturers, a young John Blyth left them as most would be young war heroes did in 1944. He joined the Royal Air Force and stayed with the RAF until 1949. Returning to Ransomes and training as a draughtsman.

Searching for a better opportunity, he convinced his new wife they should emigrate to Australia. After six weeks at sea they arrived in Victoria, where in no time John found draughting work with a small agricultural manufacturer in Melbourne. He stayed until securing a job with Ford Motor Company Australia as an Implement Specialist. Initially his position entailed finding manufacturers who could supply Ford with products that their dealers could sell. Machinery that would help them to bolster their profits from sales of wholegoods into the growing machinery market, particularly those products which were suitable for use with Fordson tractors.

John's biggest test came about six years later when he was offered an opportunity to transfer to Adelaide as the branch manager. It would be a promotion, but not sure he was cut out to be a tractor man, he asked his superior Ewan Scott-McKenzie about alternative positions. Being told there were none, John knew a job in South Australia would be better than returning to scraping by, as he had done in England.

Although terrified from the stories he had heard about South Australian dealers being difficult and hardnosed Crow-Eaters, John accepted the position. His worries had proved unwarranted as in time he established an office and warehouse independent of his city-based vehicle colleagues. This move and by targeting new areas to install Fordson Tractor representation had helped the South Australian region increase tractor sales and for the company a profit.

Noel understood how difficult his time would have been, having taken a posting to Adelaide himself only a few years after John. However, a sage individual with a good nose for marketing would help take the company forward and having John agree to the role, set the first cornerstone in his personnel plan.

Another young man Clarence had recommended to Noel and who he thought could add even more value to his sales team was David Hosking. A person of bright intellect and ambition. Noel took time to look over David's history. In it he found repeated, the great passion he displayed when talking about Ford, its tractor division and his part in the future of both. David had a varied career until joining Ford, first working as a clerk for Ted Barry in the finance department before Clarence had asked him to transfer to Queensland as the Northern Region Service Manager.

Noel too saw something special in David, but he also knew appointing a young man without tempering him in battle first, would be unwise. The best thing he could do would be to move him into a tough region as a zone sales manager. There he could watch him negotiating with dealers who he knew would introduce him to hard-nosed and single-minded farmers. If he thought it necessary later, Noel could provide guidance for David to improve his tertiary skills through a company funded education plan.

He set the folder aside attaching a note to remind him to place David in one of the zones he had worked as soon as a place became available. Remembering his own experience with Hugh Condon mentoring and encouraging him, made him think that this might be exactly the region for David too. He added Riverina onto the note, underlining it twice.

Steve Officer's file caught his eye and while he had observed Steve's conscientious application to his role in the company's Supply and Distribution office, there was something about him that stood out. Noel thought there was more to his character, it may have been charisma, but it was more than that. Steve seemed to possess a similar inner confidence to his own. Noel thought he may have even seen a little of himself in Steve and maybe it was that, that made him stand out from his colleagues.

Like he had with David's file, Noel made a couple of notes inside the cover. He thought targeting Steve for training with Ford in the United States would benefit the company on his return. Steve had worked as part of Jack Nasser's team as financial analyst before moving across to Tractor Operations. Noel thought about a North American posting and how it would provide Steve with the opportunity to grow within the company, make friends and useful contacts within the global reach of the Ford Motor Company would be of benefit too.

Replacing Roy Pinney as Tractor Manager, New Zealand Steve returned to manage Ford Australia's truck division where he came to the attention of Graeme Bignell, the principal owner of the Stillwell Ford Group in Adelaide. Graeme offered Steve the Dealer Principal's position and an opportunity to invest in one of their biggest dealerships. Stillwell Ford were centrally located in the inner Adelaide suburb of Medindie, with their Stillwell Trucks division situated in a new facility in Regency Park.

It was a great move as the Stillwell Ford Group expanded rapidly and became the publicly listed company, Adtrans in 1987. Steve ultimately moved into the role of Managing Director and CEO of the Adtrans Group. The company flourished under his stewardship until being bought by A.P. Eagers in late 2010.

While disappointed Steve would be leaving tractors behind, Noel found it hard to hide his satisfaction when he learned of Steve's appointment as Graeme Bignell's replacement. For Noel, Steve reaching such a highly regarded position vindicated that original faith he had placed in him.

Returning the folders to personnel, Noel believed that until he could find another executive who was battle ready, it would mean leaving Ian Blow in place as National Sales manager. He had time to wait for someone to stand out and when they did, Ian could be posted into a Regional Manager position.

Ian Blow had transferred to Australia from New Zealand and as Noel learned later, had also applied for the General Manager position only to be overlooked in favour of Noel. Leaving Ian installed as Sales Manager until Noel thought David would be skilled enough to handle the role made sense. Wholesale changes to a company's management structure always had an opportunity to cause unwanted confusion or worse, conflict.

Noel had until the next round of performance reviews to decide those individuals who could be moved into better suited positions. For the time being he decided it would be logical to leave most of the key people in their current roles.

Three of the things on his "To-Do List" were:
- Removal of Australia's Bounty System
- Establish Ford Credit
- Develop an Australian Cabin to combat Chamberlain.

Noel had long believed that people often confused the urgent with the important, and knowing how to differentiate them led to better decision making. While having a credit company had high importance and he believed it to be urgent, he knew it would require time. Creating a credit company would require years of planning, initiated by Ford Australia and then gaining approval from the board in World Headquarters.

Working to defeat the bounty system that was propping up two of Ford's competitors would mean months or even years of lobbying the Australian government to effect change. Noel had read and re-read the act that passed the bounty system into law and couldn't see an easy way to have it repealed. His ability to think laterally and act logically, showed he only had one choice to market more four-cylinder tractors. Ford had to provide a comfortable, ergonomic cabin with a decibel reading far below the Australian standard of 88 dBa.

Noel had been made aware of the style of the cabins fitted to the 6700 and 7700 tractors were like those fitted to the six-cylinder 8700/9700 models. However, it was unlikely these would be available to Australia until Europe's needs were met. This would leave a huge hole in Ford's armour if he wanted to meet Chamberlain head on. Therefore, it became a matter of urgency that the company explore the possibility of building something similar locally.

To him it made sense to bring a cabin builder like Gason into the process as early as possible. The company were still selling 5000 and 7000 model tractors with Blueline cabins as dealer fitments. Even though Noel knew this had to be done properly, an original equipment manufacturer would ensure consistent quality control. Knowing that only a small number of the company's dealers would be capable of undertaking such a task, he could not guarantee consistent quality and as a result, reputation damaging problems could set in.

However, sourcing a cabin built by a third party as an OEM product would require much more than shipping tractors to their factory. However, once the cabin had been installed it could be shipped from there and onto the dealers. If it were possible to overcome the engineering hurdles and if they agreed to jointly research the idea, selecting Gason as a partner made sense.

An OEM specification cabin would require an immense amount of planning, sourcing of components and ensuring that the build quality surpassed Ford's strict engineering standards. Unsure he could leave something as important to the vendor, again he called on the expertise if Frank Pitney for advice.

Frank agreed with Noel's assessment that the only company in Australia with the expertise and capacity to mass produce a cabin of OEM quality would be Gason in Ararat. However, he saw numerous obstacles to clear before they even started to talk with Les Gason. Noel wanted this Australian product to rival or better the cabins due on the 700 series tractors in 1976. A planning group would be selected from those managers whose input would be required to keep the project on track. Noel decided they should take the imported cabins as a starting point and asked Frank to set about collating all they needed to know before he called a team meeting.

Obtaining pre-release information is always difficult, so working from sketches they had seen, the forward ordering parts price list and their accumulated knowledge gained from the 8700 /9700 model release material, they agreed on a baseline. Knowing the optional cost to buy a cabin already fitted to the tractor, indicated the price an Australian OEM cabin would need to come in at. Noel believed on insisting that if Gason were to agree to manufacturing them, then they should aim for, "Best in Class," when setting parameters.

Noel believed a starting point to imported cabins should be to meet their competition head on, but in the past and selling against Chamberlain in particular, had proved to him that selling like for like had not always been successful. In Australia, price was often the driver. For this exercise to prove if it were possible to manufacture a local cabin, had to be more than another futile function. Now would be an opportunity for Noel to prove that he too could forge his name among others who had held the title of General Manager Tractors.

At best, the exercise would take a wholistic approach to product development and pricing. At worst, he and the team involved, would have framed a template on which they could base future initiatives. There was little to lose and a lot to gain.

CHAPTER FORTY

Sitting in Les Gason's office in Ararat, Les recounted how he had been unsure what to expect the first time he agreed to meet Noel. Noel had only recently been appointed Marketing Manager Ford Tractor Operations, Australia and had arrived on a getting to know you tour. This time he had phoned Les, asking to meet and discuss something he believed it would be of benefit to both companies.

A.F. Gason had been doing business with Ford and their dealers for almost twenty years and for all that time every man in the general manager's role had been almost twice Noel's age. Noel striding toward the office complex, straight backed and moving with purpose had been unaware of Les watching him. The changing of the Ford Motor Company's old guard ushered in a period of optimism that had started with the arrival of Bill Bourke and his desire for Ford Australia to chase the youth market.

The same thing had happened to Les. He was not much more than an apprentice when he had been forced into the manager's role. His father had taken ill and being advised by his doctor to give up work or at the very least, took an extended holiday. Frank had arranged for a manager to take his place. However, that gentleman only stayed in the role for a short period and without anyone to lead the company, not yet twenty,

Les had assumed the responsibility of their company's management.

On their first morning Noel believed if they could engineer a good working relationship, then it would be a great move to further strengthen an old relationship. This morning however, he would be visiting not only a business associate, but a friend.

Before Noel reached the main door into Gason's offices, his mind went over his reasons for being there. His priority as always, was selling more tractors and to do that, the urgency of delivering a "Q" cabin on Ford's four cylinder 6700 and 7700 was paramount. He knew there were other businesses Ford could turn to who would be able to supply such a cabin, but they were untried. He had made numerous visits and had meetings with some of these other parties over the years, however, partnering with Gason remained an obvious choice.

Holding the lion's share of the tractor cabin market, along with their financial stability and Les Gason's willingness to invest in new technologies, made Noel's choice an easy one. However, there remained a lot to discuss before they would see a draughtsman's pencil scratching its first line.

Today's negotiations would be the first step on a long journey to building a cabin capable of delivering a quiet, clean, safe, and comfortable environment for the operator. The "Q" cabin's design brief, was to engineer and build a high volume of Australian cabins with a level of comfort that would eclipse the market.

The cabin would need to be similar in design to the two door units coming from Europe or the USA. They should be quiet, set a new Australian benchmark, reaching to or below 83db. Have an opening rear window and side glass, but most important would be an air-conditioning system that led the industry.

Even though Noel knew these were lofty goals, he had to get Les to agree to have someone from inside Ford Tractors who could oversee their interest in the project. The incumbent would ensure quality targets were being met during each phase of the design and manufacture. Noel knew there was a lot for Les to consider too. Gason would need to invest heavily, providing additional resources into developing an Australian made equivalent to Ford's imported "Q" cab. Noel argued, there would be much prestige attached to being considered an OEM component supplier to Ford Australia, but he also knew that a business could not survive on prestige alone.

Noel and his team in Campbellfield had created an easy-to-read spreadsheet showing strengths and weaknesses of Fords imported cabin, and those of their competitors. It illustrated the cabin features that would give them a, "Best in Class," target for which to strive. Achieving such lofty goals would be a huge accomplishment for an Australian manufacturer and both he and Les Gason understood how much would be at stake.

Securing an agreement from the Ford Design Centre to provide assistance for the project, became a strong pillar to his plan. Ford Tractor Operations Australia too, had a strong engineering presence in Frank Pitney, Les Stack and others with quality surveying and draughting experience. Les had come to Ford from Malcolm Moore, a manufacturer of road making and other industrial equipment. However, it was the quality control skills that Les possessed that Noel knew would be necessary, on a project of this importance.

Armed with photos, drawings, brochures and specifications of the new cabin for the 700 series tractors. Noel spread them out on top of a table in Gason's drawing office. Some things could be duplicated and others would require a new design. Gason did not have a facility to provide stamped panels, so where the imported cab utilised a stamping, Les Stack working with Gason's engineers would draw a fabricated replacement.

Pre-release photos of Chamberlain's 80 Series tractors had arrived on Noel's desk from an unknown source almost twelve months before its proposed release date. Industry pundits had slated the eleventh of November 1975 for the big reveal. These new Chamberlain tractors would be the first to be designed from a clean sheet of paper under John Deere ownership and were aimed at the 80hp, to 120hp broadacre market. Powered by John Deere's six-cylinder engines driving through a Chamberlain dual range transmission and completed with an airconditioned and noise reducing cabin, these tractors followed the design principals employed in John Deere's imported offering.

From those early photos and specification rumours, Noel began to understand how dangerous these new Chamberlain models would be to Ford's market share. Those details would be the driver of his team's "Best in Class," sheet. Sourcing information from other competitors and wherever they could find it, the design team's spreadsheet started taking shape. Looking at all the tractors as complete units, proved there was one area where as Australians, they could match or better the competition and that was the cabin.

The team saw it as a challenge and looked for ways to better Chamberlain's offering. CJD would now be locked into producing enough cabins to achieve a break-even figure on their investment. Noel knew that would mean Ford and Gason had a huge opportunity to better the single door unit on the yellow tractors from Western Australia.

Operator comfort had now become a huge selling point when Australian farmers began weighing up what was important when they started looking around to buy a new tractor. Every tractor company in Australia soon began to understand that a generational change had begun taking control of buying decisions. Older farmers who had persevered with high noise levels, heavy steering and all kinds of weather, wanted better for their sons and daughters who would be taking over their cropping duties. In most cases, this new generation of farmers had been better educated than their parents, often attending agricultural colleges and becoming qualified in the latest farm production methods.

Higher farm wages had also influenced the number of farm workers employed in broadacre areas and as these people left farms to find better employment opportunities, farm owners began working even longer hours to cover the lost labour. A warm cabin, better lighting and more power, hauling larger machinery, equalled better on-farm efficiency. However, tractor comfort was also often driven by a parent's need to provide their offspring with better than they had lived and worked with.

All these notions had driven the ideas behind the ideal specification Noel and Les were contemplating during that first "Q" cabin meeting. At first, they looked for all the things in this new Chamberlain cabin that they determined were outstanding features. At the time, it was also tempting to compare International Harvester's new 86 Series with its "Control Centre Cabin," which had been planned for release a few months earlier than Chamberlain. However, as it had a higher cabin noise reading, for now they decided it would be better to focus on the tractor they thought posed the biggest threat to market leadership, the Chamberlain. Further into the planning process they would place the IH cabin and every other competitor's offering into account.

Deciding to keep the overall shape of the cabin as close to that of the import would mean stocking a lower glass inventory help the project's long-term costing. However, using the imported windscreen posed a problem. The glass on the "Q" cabin fitted down into a gap between the rear engine cowling and the instrument binnacle. The open 6700 /7700 tractors already had a flat floor, but the engine cowling and instrument binnacle had been manufactured as a one-piece part. This meant the Australian design for the cabin required raising the windscreen above the cowling, resulting in a taller overall cabin height. The team looked at the expense of importing or locally manufacturing the parts needed to keep the look the same, however, every way they looked at it, the subsequent cost made it uneconomical.

With drawings signed off by the automotive design centre, Les Stack from the Engineering Department took up his chance to lead the project. Having been charged with helping Gason streamline their processes and manage a high-quality control throughout the Ararat factory, the project began.

The Q-cabin's air-conditioner had been adapted from one of Gason's own designs and if anything, had a performance advantage. The engineers had managed to achieve a more than satisfactory decibel reading of 83 dBa and buyers in the 70-80 horsepower market found them most attractive.

At the end of production, this new two door, flat floor Australian cabin, offered six hundred Ford tractor buyers a unique and comfortable operator's platform. For Noel, Les Gason and the teams led by Les Stack it was quite an achievement.

Ford's influence had begun to change the way A.F.Gason approached manufacturing. The company could proudly claim to be an Original Equipment Manufacturer and many of the processes Ford Tractors through Les Stack's time with them, soon began filtering into their own design and manufacturing philosophy.

By the end of production, Australia had been entering a period of inflation and the country's monetary exchange rate eventually put an end to building a local "Q" cabin for Ford Tractors Australia. It felt disappointing, but something out of Noel's control. At the end of the project, it had proved to Noel that if he could make the numbers work, then it was possible to build world class equipment in Australia.

CHAPTER FORTY-ONE

QANTAS Club: Honolulu Time: -- 12.55 pm

Noel knew reliving the events of the past twenty-four hours would serve no purpose and yet there were so many things that out of necessity had to be repeated. The cab ride from the hotel, the check in and customs routine all felt too similar and for someone less pragmatic, may have even presented as some terrifying nightmare. Luckily, he had never been given to such thoughts, his engineering background and the bush practicality of his upbringing had strengthened his mind. The chance of a similar incident happening to the same model of aircraft from a fleet with a far better reputation for safety would be billions to one. He understood the rationale for his thinking, but then, the brain worked on emotion and now sitting in the lounge surrounded by opulence, he allowed his mind to run through his own pre-flight check list to keep those same emotions under control.

Again, tonight he sat with a beer in front of him while waiting for the call of his flight. Just hearing Joan's voice when he had called to let her know his flight details had changed, had warmed him and now he imagined her arms closing around him. She always made him feel loved, even halfway around the world her voice had a magic quality that put him at ease and gave him strength to meet the challenge of his work.

He flicked at a tuft of lint that had landed near the crease of his trousers and reached for his beer, the cool of its creamy amber liquid soothing his throat as he swallowed. During the day he had decided he wouldn't be staying on with Ford. He knew his friends in the upper echelon of World Headquarters couldn't prevent him from losing out if he did take the job on the North American West Coast. By then the company would have sold the tractor and machinery business, as had been their plan all along. An automotive position within Ford Motor Company, would probably be even more risky. He took out his pen and on a spare drink coaster started to doodle random shapes that soon resembled images of tractors.

Although the offer from Case IH had been generous, it had come from outside the Australian operation. Yes, the approach had come through a human resources company who knew him, but as the parent company Tenneco had requested it, it made him wonder what skeletons they may be trying to keep hidden. Even if there were skeletons to be unearthed, that would be part of his role and ridding Case of them would bother him. Only that the company had discounted their own staff for the position had piqued his curiosity and had him wondering why.

Opening the pilot's case that always travelled with him, he rummaged through the files until he found the Case IH prospectus. As with all documents of its kind, the glossy pages and glowing words hid more information than they provided. It was not what he needed now. However, he could find enough information to work with in the raft of tractor and machinery brochures they had included.

Again, it showed mainly a glossy story. However, the information contained when cross referenced it against his own knowledge of the Ford New Holland products, told him Case IH had a lot of weaponry he could use to compete against his old employer.

While studying the intricacies of Case's Axial-Flow combine harvester, he heard his flight being called. It broke into his train of thought and finishing his beer, he put the folder into his case, tossed his coat into the crook of his arm and strode toward the boarding gate. This had been a very long twenty-four hours and now it was time to go home.

CHAPTER FORTY-TWO

After Noel's appointment to General Manager, he soon started becoming accustomed to the roster of head office meetings and before he noticed, most of 1974 had flashed by. During their annual three-week break over Christmas, he and Joan turned their minds to what their future might look like and they daydreamed their time away, making plans for a life in the country. Noel had always told his children stories of his time in Kiewa and how on his own, he had ridden into the high country trailing a packhorse laden with supplies for the men rounding up cattle. Often one or more of the children would finish the story for him, it became a family moment as in unison they would chant his lines about bringing the cattle down to winter on the farm.

With the three youngest Howard children returning to Doncaster Primary School and Geoffrey going to Templestowe Technical College, Noel and Joan could feel pleased that the constant upheaval of changing schools as often as they had, had not shown any deficiency in their children's education. If anything, their children were more world wise than their classmates and this may have helped the more than respectable reports that had come home at the end of this year.

Turning thirteen this year, Geoffrey had taken to cycle racing and showing a keen aptitude for the music at the

Technical College. This interest in music had been continuing to grow outside of school hours often practicing music, instead of finishing his homework by the time it was due.

Geoffrey, bringing his Howard entrepreneurial skills to the fore, had found a few mates who could play enough notes to start a Saturday morning garage Rock band. The boys' playing continued improving and every new tune learned would be practiced over and over, every rendition performed with even more gusto than the last. It wasn't too long before the band had reached a point where they were coming to the notice of neighbours. Their teenage confidence bringing unwelcome attention from a local police patrol who arrived to investigate noise complaints from unhappy residents. Those same residents, the boys had deduced, were completely without an appreciation for teenage music, but to keep the peace, they agreed to turn their amplifiers down.

For Joan, the police arriving became enough motivation to convince Noel they should start looking around for their next project and they found it on the Kilmore to Broadford Road. Highfield Park was about halfway between the two towns and only forty-two miles, or sixty-seven kilometres north of the Melbourne's General Post Office. This road had originally been part of the old Hume Highway, until a new section of the road to Sydney now bypassed Highfield Park, leaving mostly local traffic flowing past the front gate. Joan agreed the new highway would also provide Noel with an easy drive to work at Campbellfield and probably continue saving him much travel time as Melbourne's northern suburbs expanded away from the city.

The small acreage may not have created a lot of interest for cashed up squatters around Kilmore, but for Joan and Noel it was *"just the ticket."* It came with an old weatherboard farmhouse and a few tumbled down out buildings, which they thought would do until they could rebuild. However, it was the acreage that had attracted them at first, not that big that Joan would need help to manage it while Noel was away, but large enough for the family to start living their farming dream. Happy knowing that owning their own farm would provide room enough for their children to grow, they placed an offer that the owners accepted.

After walking around the tired old farmhouse and judging its suitability, both Noel and Joan knew it would be difficult to live in its somewhat primitive conditions. It wasn't as if they couldn't live there, they could, but their aim had always been to provide a comfortable place for their children to grow and this house was far from that. While it might do for a few months over a summer, winter would be draughty and cold. Living on the farm in its present state was not practical and they decided to find another option close to Kilmore. They planned building a new home by demolishing part of the old house and extending its footprint, it would when completed, offer them a home equal to the one they had been living in.

A solution of where to live while renovating came from their friend George Benyon. George, learning of their plight offered his very comfortable and vacant Kilmore home to them for as long as it took to rebuild. Refusing to take Noel's offer of rent from them, was just another testament to the friendship he and Noel shared.

After moving to Kilmore, Natalie attended the State School in Broadford, Tania attended the Convent in Seymour, while the boys Geoffrey and Peter attended Assumption College also in Kilmore. It meant Joan had a challenge managing the needs of her children who were attending three different schools over thirty kilometres apart. For some, that would have been daunting enough. However, Joan had always been a homemaker and Highfield Park soon started showing signs of her touch.

Not long after taking possession, they began adding sheep, cattle and a menagerie of other animals, all purchased with an aim of providing an on-farm income. The Sheep and cattle yards were in fair order when the family arrived, but in true Howard fashion, over many weekends these were redesigned and rebuilt. Joan and Noel, working together to improve their ease of animal handling and overall farm efficiency.

Joan and Noel had both admired the look of Hereford cattle for many years and because the breed was hardy and well suited to the high rainfall conditions in the Kilmore region, Herefords became their breed of choice. A desire to be more than a woman who ran a few cows and sheep, Joan set herself a task of learning all she could about bloodlines and best breeding techniques. She wanted to understand the work it would take to hang out a shingle that said, Highfield Stud Kilmore.

Once they decided how to renovate, Joan stepped into her designer role again. Talking with Noel at night and making sketches of the floor plan whenever she had a few spare minutes. Working on designs for each elevation she soon had sketches of how she believed their home should look. Noel particularly liked her drawing that displayed their home when approaching it from the driveway.

This time they decided against using an architect and seeing the detail in her drawings Noel thought they should ask Les Stack who was living nearby, for an opinion. Les, complimentary about Joan's sketches, offered to take on the task of drawing detailed plans for their new home. He and Joan would work after hours and weekends to ensure their builder had detailed drawings to work from.

With the plans accepted by council, it was only a matter of time before they started looking for a builder to do the work. Contracting a builder, Jack, who had arrived in Australia from New Zealand a few years earlier and taking advantage of Melbourne's growing housing boom. One of Jack's stipulations was that he be provided breakfast each morning. Joan agreed to making breakfast for him, but the man's appetite had surprised her. He started his day with six eggs, a similar number of thick greasy bacon rashes, topped off with toast and a flask of tea. His breakfast never altering in all the time it took him to build their house.

By the end of the project the Howards had built a new three-bedroom two-bathroom bungalow with an expansive living and dining area. A sweeping veranda circled the house and felt as though it nestled into house yard as if it had always been there. A swimming pool came later, but when the family moved in it felt like their own little Shangri-La.

Living at Beynon's, every Sunday evening Noel would make a list of things he wanted done around the farm on the following weekend. Then he'd leave another list of jobs for Joan to manage while he was at work. This second list would lead Joan to joke, that his list of tasks would never be as big as the one he left for her.

However, Joan had been proved right about his travel time. Noel's drive from the farm in Kilmore, took no longer than his run to work from Templestowe had. However, whenever he had been required to attend planning meetings at Ford Australia's Head Office on Sydney Road, his morning commute often led to an earlier start than normal. It gave him time to go over his presentation in his mind and if necessary, practice his speech, working from the bullet points he had memorised.

Joining the Hume Highway, the traffic heading into the city flowed easier and the first set of traffic lights that could impede his journey stood at the junction of Cooper Street and Sydney Road. This new drive into work gave him an uninterrupted time to deliberate. Although Noel had learned to plan his next day before leaving work at night, driving along the freeway without the radio on each morning, helped him focus on the day ahead.

While the role of General Manager had a continuing list of demands, it did allow greater access to those people at the top of Ford Australia, Ford Asia Pacific and his superiors in Troy Michigan. There were times when visiting dignitaries would fly in from Dearborn and on these occasions, presentations would need to be made. Noel, far from the young man who had sought out a Dale Carnegie class to overcome his fear of public speaking had now become well versed and confident when addressing any number of people.

These meetings often took place in Ford Australia's boardroom although dependent on the dignitary's program and the number of guests, they would be held at one of Melbourne's five-star hotels. One occasion was a flying visit by Lee Iacocca, the new Ford Motor Company President who was on a whirlwind trip to meet with management of Ford Motor's outposts around the globe. Noel had not long been in the role of National Marketing Manager when Clarence E Ragan invited him to Ford House to meet the "Big Boss." Noel knew Iacocca was coming to Melbourne and although his role would normally have not resulted in an invitation, Clarence not only made sure his young prodigy went along, but insisted they sit near the front.

Lee Iacocca, charismatic, keen and forthright spoke about the company's plans, their current position in worldwide markets, their upcoming models and the expectation of an increase in sales. Noel listened as Lee described the importance of Ford Trucks to their business, the acquisition of Philco an electronics company that Ford had restructured and how it had now won a contract to supply NASA. He talked about new car lines planned for release in Europe and the USA. He touched on a relationship with Mazda and the board's plans for Asia, but not once did he mention Ford tractors.

This came as a surprise to Noel because, whenever he had attended planning meetings in Ford House, the tractor company had always been treated as a vital part of Ford Australia's business plan. He waited until Lee had asked for questions and stood, asking why the tractor division was of such little significance that Lee had not mentioned it. The reply Lee gave was in no way dismissive as Noel thought it might be, instead Lee gave a quick insight into what the tractor division was worth to the company annually, how rural life in America was changing with the development of larger horsepower tractors and the changes and challenges the board saw for the agricultural industry in the short term. He asked Noel his name and thanked him for the question.

Later Lee broke away from the group and sought Noel out and in the little time he had, he wanted to know more about Australia's agricultural market, Ford's part in it and plans for the challenges Noel could see coming. To soon, a minder called him away and Lee Iacocca left the building.

For Noel, meeting Iacocca and being able to speak with him, confirmed all his ideas of what a top executive should emulate. Lee had charm to burn and Noel had watched it working on his Ford Australia colleagues. However, the fact that the man could back it up with quickfire facts, showed he had a keen sense for every operation of the company and a wonderful memory for people, place and figures.

Lee Iacocca had joined Ford Motor straight out of Lehigh University in the Pennsylvanian city of Bethlehem where he graduated with two degrees, one of them being engineering. Ford had always been at the top of his wish list as a preferred employer, often joking with his friends that Ford needed him.

He would stay with Ford for thirty years. Beginning in the engineering office and seeing that as a slow path to the top, switched to sales where he was posted to relocated to Ford's office in Chester, Pennsylvania. Working in the fleet department, he honed skills he had learned by helping his entrepreneurial father hustle for extra sales among his many business interests. Fleet sales is neither a flashy nor exciting, but it does count for a large percentage of any company's business and that was not lost on Lee and his list of influential contacts grew. It was not long before his hard work started paying off and soon, he came to the notice of Ford's hierarchy.

Lee understood that to climb to the top of the sales teams, he had to cut through to another buying demographic and in 1956 he imagined a finance scheme where for US$56 per month, a customer could own a new Ford. Advertised as *"A 56 for $56,"* sales grew and the Iacocca Plan as it became known, rolled out across the United States with unrivalled success. At the time Chevrolet may have been more desirable, but Ford provided a great package at an easy to afford and understand scheme.

Lee now summoned to Dearborn, skyrocketed through Ford Management ranks. Famous for leading the team that designed and brought the Mustang into production, his reputation as a carman continued to grow. With his photo on the cover of TIME magazine and becoming known as the father of the Mustang, Lee Iacocca reach was about to reach the pinnacle of his career at Ford when he was named President of Ford Motor Company.

Understanding how Lee Iacocca had worked his way through the ranks had not been lost on Noel, for he could see similarities in both of their careers even if it was on other sides of the globe. If you put in the work, then you too could make your mark within the company ranks. If Lee had seen the youth market as having a big influence on car sales, then it was entirely possible that youth too would determine the path of the tractor market. An idea not lost on Noel.

The Iacocca Plan had resonated with Noel too. If he could create something similar that would encourage buyers to purchase a new Ford tractor, he figured it would take some sting out of their competitor's advantage and increase Ford's tractor market share.

CHAPTER FORTY-THREE

During one of these early meetings at Ford Australia's Head Office, Noel and his colleagues had found themselves debating Australia's economy. It had been having a turbulent effect on their ability to produce accurate planning and production targets across each of the company's factories. In 1972 Australia had seen its inflation rate rise to twenty percent and the government agreed to the Federal Treasury revaluing the Australian dollar in December of the same year. This revaluation had brought with it a squeeze in liquidity and banks having tightened their lending criteria, made predicting sales targets more challenging.

Noel being rather new to the process had promised himself to watch and learn for the first few meetings. However, it was not in his nature to take a back seat, so when asked to share his observations, he happily explained the tractor business was enjoying unprecedented success. Because of the government now abandoning wheat quotas and farmers making spring deliveries of held back grain, cereal growers were not feeling the effects of the current liquidity squeeze. In fact, for some, their bank accounts had swelled.

This additional income had also come from a great cereal harvest across growing areas. Devaluation of the dollar was also allowing the Australian Wheat Board to take advantage of a better currency exchange rate and this had increased the price they paid per delivered tonne. Enthusiasm emanating from Australia's wheat growers had now started building confidence and that was having a knock-on effect across wheat growing communities. Renewed confidence became the driver that had helped Ford Tractors to record good profits for the year and, season permitting, he expected it to continue into the next year. Tractor dealers who held vehicle dealerships were reporting to his field staff, that these factors had also resulted in a lift of sales across every facet of their businesses, including cars and trucks. However, he cautioned that their competitors were experiencing similar results and having a reliable supply of products would be necessary to capitalise on the opportunity ahead of them.

A couple of years after that early meeting, Noel had become a regular fixture arriving early and well prepared. From Noel's hand written notes, his secretary Wendy Orchard would prepare and bind a report for each attendee and attach type written dot points to the cover of his copy.

He had learnt from his early public speaking lessons at Dale Carnegie, to keep reports factual and to the point. If he needed to *err, or uhm,* he stayed silent until he had collected his thoughts and only then would he try to press on. This method had served him well and there were no uneasy silences. With dot points to keep his report pertinent and on track, there was often no need to open the folder.

During this whirlwind of meetings at Ford House and a raft of other duties that went along with managing a major tractor business, Noel also had his Harvard commitments to arrange. Travelling to the United States meant one long flight and a few changes of aircraft before he reached Boston. However, Harvard soon became an interruption in his diary that he looked forward too.

Finding that he had been enrolled with a group of other post graduates wishing to study International Business, meant they could share different life experiences. Excited by the possibility of learning from them too, Noel was keen to listen to everything offered, joining the conversations when and where appropriate. Before long he was making lifelong friendships with both lecturers and colleagues.

While the course was demanding with scheduling and assignment commitments, it wasn't to be without other interests, as these men shared their love of flying with him. The only difference in their flying to Noel's, was that they were Boeing executives, or long-haul airline pilots. Being the only one in their class from outside the airline industry, made his experiences unique and it often helped him offer a solution to a proposition from a different point of view.

Along with several other business cases they studied over the duration of their course, was the Boeing 707. This aircraft interested Noel as it had been the first one of its type and one he had flown overseas in. His curiosity for flying and aircraft, reached right back to George and that first flight in his Tiger

Moth. It had become a passion for flying machines that had never waned.

Over nineteen years between 1957 and 1978, Boeing built eight hundred and sixty-five 707s. It was to be the company's first commercial jet liner. When setting the goals for the project, Boeing Company President William Allen and his management team envisioned building a radically new aircraft that could disrupt conventional thinking toward air travel. Not wanting to alert competitors to their plans the prototype design was given a designation of Dash-80. In 1952 the Boeing Board committed sixteen Million dollars (US$16m) of the company's own funds to the project. A commitment which was later referred to, as betting the survival of the whole company on one aircraft's success.

Studying these processes with his fellows at Harvard, it was the secret development of Boeing's airliner that caught Noel's attention. It had been referred to, as being part of the company's *"Skunk Works"* protocol. The term, as well as the processes resonated within his thinking and having his own project in mind, he paid close attention to the details of Boeing's planning.

While he and his colleagues studied the secrecy that Boeing had employed to plan and develop the 707, there were a few times when a lull in their discussions would cause his mind to wander to tractors. There was a definite gap in the Ford tractor model range and it had a detrimental effect on sales. He knew if a solution could be found he could increase sales, market-share would soar and profitability increase.

Ford tractor dealers from Australia's cereal growing areas had continually told him, that the lack of a 100 horsepower, six-cylinder engine powered tractor, was one of the main objections they faced when selling against Chamberlain and International Harvester's bounty tractors. While the, *"what ifs,"* rolled around in his mind, he continued lobbying his superiors to provide a machine to meet the market, only to have his suggestion rebuffed every time by General Manager of Ford Tractor's European Operations.

Derating the power of an 8700 would offer an equivalent. However, the more he thought about it, he reasoned the cost of purchase would be far too high. Discounting a large frame tractor to meet the competition would mean selling each one at a loss and, that was not an option.

Noel knew County in Great Britain had used six-cylinder engines in front of the Ford 5000 manual transmission to produce their popular, equal sized four-wheel drive tractors. Ford Tractor Operations in Australia had sold them in good numbers and the office in Campbellfield had collated a comprehensive local service history.

Believing that being at Harvard among some of the smartest executives he was likely to engage would improve his management skills, he listened intently to the views of his lecturers and those of colleagues too. At times someone in a lecture or from the group would ask a question or say something and Noel began to see a way of cutting through his own obstacles. There would be an enormous amount of engineering to do, but the thought of building an Australian tractor with 110-horsepower that would sit between the 7700 and 8700 models began to gel in his mind.

In one of their more informal gatherings, the lecturer asked the group if they had a particular problem that they found they were unable to solve using the resources of their own teams. He developed the question further, asking them to think deeply about the problem, an ideal solution and the advantages that would come from solving the issue, asking them to quantify the risks. Explaining, that to give their proposals proper merit, they must put a value to both cash and indirect benefits.

Importantly if they were *"betting the company,"* what would be the costs of failure, again both in terms of the project's fixed expense and the financial damage to the company's image, assets and future. Their lecturer asked each of them to prepare a short discussion paper for the group to workshop the next day.

Noel seeing this as an opportunity to pose a question that might unearth something he hadn't thought of, framed a scenario far removed from the tractor business and waited for a chance to mine the intellect of his fellows.

Next day everyone in the room had thought of something different and after a vigorous discussion, solutions were found, proposals discounted and opinions offered. However, through it all Noel had his ideas confirmed. If the product planners in Dearborn were unlikely to offer the machines he needed, then he would *"kill two birds with the one stone."* As he had to write a paper as part of his course, he would show how developing a 110-horsepower tractor for the Australian market could benefit the fortunes of the Ford Motor Company.

Like Boeing developing the 707, he would use the Harvard course as a cover for his "Skunk Works" paper. The more he thought about it the bigger the opportunity looked. Believing that people in every Ford Tractor Operations department from finance to engineering would assist with any information he needed if they thought it was part of his Harvard studies, he framed his thesis around developing the missing 110-horsepower tractor.

Noel's time at Harvard introduced him to many colleagues from outside the vehicle or tractor business, who would become more than mere associates. Keeping in touch with them as their careers progressed and given time when overseas, making sure he called to see them.

Often after a lecture, he would find himself brimming with excitement at finding a solution to an engineering obstacle within the intricate detail contained within the County parts list, that he had asked be sent to his Boston hotel.

His Harvard studies always required students to stretch their knowledge of acquisition and takeovers. For this module, Noel used the case of Ford's buyout of Richier in 1972 as a case study to understand the way Ford went about it.

At the beginning of the 1970s Ford Tractors had seen a need to support an ever-increasing demand for industrial loaders and had purchased a controlling interest in the French company. This allowed them an entry into the articulated Load Handling market and by 1974 Ford's engineers had redesigned the product range. They improved reliability, operator ergonomics and style. Richier would remain with Ford until after a worldwide economic downturn had decimated industrial machinery sales, at which time, preparations were being made at Ford to divest their interest in their French machinery company.

Back in Australia, Noel returned to his wish list of products needed to offer a full line of farm machinery. He liked the idea of finding a supplier for harvesting equipment, but because of a tip off he had had by one of his senior colleagues high in Ford Headquarters, he discounted it. Learning that the company had been looking for an option, or options to purchase a constructor of harvesting equipment building everything from hay machines to combine harvesters, meant that wish would soon be covered off.

Understanding Ford's new tractor line-up would be launched in 1979, under the theme of, 80's Action, Noel continued looking at the models replacing the current 6 cylinder 8700 and 9700 tractors. The TW Series had a slight lift in horsepower, employing air to air intercooling and modified internals Ford's engineers had pushed the 401 cid, six-cylinder engine to 190hp. With these three tractors, Ford dealers finally had models they could not only sell feature for feature with John Deere and International, but could be shown as being best in class on many fronts.

After Ford's signing of an agreement with Shibaura of Japan to supply a full range of compact tractors, allowed Noel to look forward to fighting Kubota for a part of the compact market, which they had dominated since arriving in Australia. Around the same time, Dearborn had entered into a supply agreement with Steiger, but to Noel's disappointment, Australia would not be part of the distribution agreement. This meant Ford Tractor Operations Australia were without an articulated range to tackle John Deere, International, Case, Allis Chalmers and to a lesser extent Massey Ferguson.

In the heavy articulated tractor market, Steiger and Versatile were the biggest sellers, while smaller Australian start-up companies like Acremaster, Waltanna, and Phillips took a few bites out of the overall market share.

Looking over his wish list, Noel had seen a few holes in the new model range, but overall, it was a solid line up of tractors and far better than anything than Ford had offered to their dealers in the past. His idea to plug the gaping hole between the 7700 tractor and the new TW10 needed to be found. If Dearborn, Antwerp, or Basildon were unable to supply, then finding a solution to that conundrum might have to come from his desk.

They were just over four years out from a planned 80s Action release of the new tractors, and after losing a landmark legal case brought by several dealers against Ford that challenged clauses in the dealer agreement forbidding a Ford Dealer from entering a supply arrangement with a competitive company. For Ford Tractor Operations too, it had been quite a blow and as their dealers rushed to take advantage of new opportunities, Noel asked his sales teams to try to contain the damage.

However, Noel agreeing with the dealer's plight, felt the current single dealer agreement hadn't changed to reflect the advances made over the last few decades. He believed it was outdated and had become an impediment to Ford's ability to secure the best automotive and tractor dealers. He wanted Ford dealers to be viable, as the agreement stood, it not only restricted their options, but those of the Tractor Operations too and probably Automotive too. However, as a senior official with Ford Motor Company he understood the importance of adhering to company policy when questioned by a dealer, or whenever the press asked for comment.

Believing that with change came opportunity, Noel started planning to shore up their product offering with a newer and more dynamic Blueline range. It had to coincide with the 80s Action release. Implements and attachments had to be found for the compact tractor range, the broadacre tillage range had to be modernised to take full advantage of the rapid changes taking place in cropping practices. The menu of tasks looked huge when set out as a list, but once broken into timelines and action plans, he saw it all as possible.

CHAPTER FORTY-FOUR

At Ford House, as one of their monthly management meetings moved to general business, Brian Inglis held up a letter for those attending to see. Placing it on the table in front of him, he explained it had come from the Minister for Trade and Commerce's office. The minister's secretary John Stone had advised him how the new Fraser government would be creating an advisory body to Treasury and had asked the Ford Motor Company to nominate an executive to serve on the new committee. The letter said they were looking for someone with broad experience who would have both capacity and time to dedicate to the role

Noel had been attending these management meetings for about four years and while he knew he had the credentials to serve, felt sure someone among those gathered would jump at the opportunity. However, as he sat and listened to everyone's murmurings, he knew his assumption had been wrong.

As Brian worked his way around the table asking each of Noel's colleagues if they wanted to volunteer, all of them offered well thought out reasons why it would be difficult for them to make time available. When Brian's question came around to Noel, at first, he held back and then asked Brian what he thought it might involve. Brian who had been attending to Government committees for some years, explained what he thought it would take in the way of time to prepare and time to attend. Listening to Brian and only seeing positives and how the time spent on Treasury business might advantage the Ford Motor Company, Noel volunteered; telling Brian that although his calendar was more than busy, he would make the time.

Happy Noel had accepted, Brian offered his congratulations and with a round of relief driven applause from others in the room, people began congratulating him. Some even saying how pleased they were that the "tractor bloke' had stepped up and saying how it would be saving them hours of dealing with bureaucrats for no extra coin.

Brian called for the noise to die down and when he had control of the room again, he spoke about the importance of the invitation. Going on to say how difficult it could be for anyone in business to find a contact within government, someone who could introduce you to the people in charge and those who advised them. Wishing Noel well, he invited him to meet in his office later where he would provide letters of introduction to those in charge of the committees he served on. When Noel reached Brian's office, he had his secretary prepare a list of ministers and public servants with whom he had become friends and other people Noel could call to make things happen.

From that simple acceptance and having everyone in the room agree that with his background in agriculture, Noel had to be the best placed person among them to represent Ford Australia and the general manufacturing industry.

However, Noel sensed his colleagues were thinking they had dodged something time-consuming that would turn in to being a complete waste of effort. Although Noel did see it differently, being asked to serve the Federal Treasury would be more than just attending. He believed this would become the beginning of a relationship with the Federal Government and its public servants that would endure.

Arriving at University House in Canberra for his first Treasury Meeting, John Stone, a name Brian Inglis had underlined on his list of people Noel should get to know, greeted those attending and introduced then Treasurer John Howard. The Treasurer outlined the government's vision for monetary policy, designed as a platform for the country becoming a more fiscally assured and vibrant Australia. Although it had more of regurgitated election speech than substance, to Noel the Treasurer had a sense of honesty about him. However, whenever the politician paused or emphasised a point, he could imagine his mother saying to be careful. Reminding him that politicians like actors, rehearse their lines and lies until they believe every word of them too. He shook her memory aside, deciding he would reserve his judgement until he had a better understanding of this man who shared more than just his last name.

Being born in July 1939, John Howard not only shared the same surname, but at thirty-eight years, he had become the youngest Treasurer in Australia's history. Only six months younger than Noel and, in that first address his energy and intellect impressed. After everyone had made their introduction and outlined their reasons for joining the panel of experts, John Howard left Stone to continue with the first meeting of the Advisory to Treasury Board.

In that first and subsequent meeting, Noel presented on behalf of the Ford Motor Company Australia. His remit covered all aspects of the company's position and its current thinking, the state of the automotive market for which he presented monthly and year-to-date statistics along with current forecasting. Having delivered the automotive paper, he moved onto the sales of Ford tractors and machinery and his team's forecasting for the next period. However, he wasn't only representing the company he worked for, being one of the members who reformed the Tractor and Machinery Association, he offered the advisory board his report on the current the position and health of the industry.

Providing statistics by horsepower group for tractors sold, had always been a problem for TMA members' product planners. Relying on import numbers that he believed to be fudged, had messed with Ford Australia's forecasting for years and made model planning extremely difficult. One of his first intentions after joining the Tractor and Machinery Association had been to press for proper sales records. To achieve this, it took several attempts for members to get their companies to agree to provide accurate delivery figures.

While there had been some initial resistance, Noel believed there had to be another way and he worked on a plan to coerce them into seeing the many benefits. At the next TMA meeting he explained the work of the Advisory to Treasury, his role within the group and how their recommendations were data driven. If the Tractor and Machinery Association were to be effective as a lobby group, then they needed to provide accurate data.

Having reached an agreement to the report's criteria, their next objective had been to find a person from a tractor and machinery background. Someone who would be both competent and skilled in designing a data system to capture the industry's information.

Alan Murray of Massey Ferguson had been tasked with implementing the TMA's decision. At Massey Ferguson, Alan had become an early adopter of using modern computer technology to capture the company's own data. His employer now charged him with collating all the information coming from each associated manufacturer and importer. It took time and the although there were a few false starts eventually the data could now be relied on. Employed by Massey Ferguson to produce these results as part of his role, time taken on TMA business had become a bigger part of his day and eventually Alan made the move to form his own company. Agriview would now carry out the work of the Tractor and Machinery Association.

The information provided by Murray's company was now a foundation of figures that the Advisory to Treasury would use to inform future decisions on. In his time o the board Noel met many people who were to go on and have bigger roles in both government and industry. Various groups had representation and it was through this rotation of advisors or speakers Noel had been building his networks.

At one meeting he may have been sitting with someone like Ian McLaughlin who was representing the interests of Australian farmers, at another it might have been the president of the Cane Growers Association and so Noel's circle of influence grew. It would provide him with direct access to the people who made things happen both inside and outside of government.

During his time on the Advisory, he would see the Tractor Bounty system aborted, which he claimed was a great result for farmers and industry alike. Farmers would now have access to a wider range of competitive products and importers no longer had to withdraw from a market where their machines were uncompetitive against a product propped up by government subsidies. To quote his idol Lee Iacocca, he wanted a level playing field and, seeing the Tractor Bounty Scheme removed brought this a step closer.

Several times a Consumption Tax had been presented for discussion by members of Treasury and from both sides of government. Noel had come to know John Howard during his time as Treasurer and it seemed they shared more than just their name. Both men believed Australia's tax system required an overhaul and it had been a topic of discussion on more than one occasion.

John Stone's agendas had listed the idea of a consumption tax as points for discussion on many occasions, only to see it rejected by many in the room. By now Noel had been flying and visiting many parts of the globe for almost ten years. He believed that a consumption tax would be difficult to have accepted. Australian experience would need to mature and that could only come through travel for voters to gain an understanding of value added taxes in other countries, before it would be easily accepted at home. In Noel's opinion, it may take a generation or two before the time would be right for a similar tax to happen at home.

One of the more interesting items he had dealt with during his time on the board was wage reform, particularly Superannuation. When the Hawke / Keating government came to power, and because of his political leanings, Noel expected to be moved aside. During a break at one of their early meetings, he asked the Treasurer if he should resign, and remembers Keating being taken aback, telling him that he valued Noel's opinion and had no desire to see him leave the board. At the time, Keating's affirmation cemented him in his role, becoming a position where he remained right through until to the end of Howard / Costello years.

Keating had introduced the idea that a compulsory superannuation payment should be directly deducted from an employee's wage. Their employer would be legally bound to pay the deduction into an investment fund where it would grow untouched until retirement, offsetting inflation and minimizing the draw on the Aged Pension Scheme.

The elements of such a policy were debated at length by those on the advisory board and their recommendations were forcefully made. Many items had been discussed over the years, everything from product shortages, union troubles, and manufacturing and industry to trade agreements with foreign powers. Over the years Noel observed the different approaches by various lobby groups, studying the tactics and presentations that worked their way into legislation and those that didn't. Often the same issue came back for discussion. However, it had been reconfigured, presented in another more understandable way. Lobbying was selling, selling an idea and, in such a way that it showed the party being lobbied, how the policy would benefit them. Salesmanship was always running a fine head of steam in Canberra's corridors of power.

Serving on the Advisory had become an education and learning the difference in each lobby group's success or failure, became great fodder when it came to making his own point of view heard at Ford Head Quarters.

With the purchase of New Holland, Nels Craig insisted that Noel's position on the Advisory to Treasury be withdrawn and his replacement be Craig supporter, Peter Morris. Noel, could only see this as treachery and another tactic to reduce his influence within the Ford New Holland group. However, it would be far from the end, as Noel's circle of influence stretched further than any of his opponents could realise. Along with a friendship that reached into the very top of the Ford empire, he had fostered great relationships with people at the very top of power on both sides of Federal Parliament.

Gor Cowl and his cronies may have won this battle, but thinking Noel would quietly retire to his home and put his feet up at the farm, would be a great underestimation of his thirst to attack.

CHAPTER FORTY-FIVE

At the beginning of 1978, Noel knew he had an opportunity to fill the 110-horsepower gap in the model range and he began assembling the key people required to plan every facet of the project. Some of these men and women had helped him with fragments of the data he had needed for his Harvard thesis, a couple of years before. This time the project was no longer a work of probability, now Noel was asking them to turn their attention to working through the steps of running a *"Skunk Works Operation."* Everyone involved, knew their role had to be kept a secret until they had the design in place and could be sure their numbers would work.

This elite team would become the group who would develop Noel's vision for an Australian tractor that could fill the gap until the promised 7810 arrived. Their primary objective was to keep the needs of the Australian cereal farmer in mind, while building an Australian Ford tractor to compete with the bounty offerings. However, it would be product planners at Basildon and Dearborn who were huge obstacles and men who could thwart his plan in these early stages.

The US and European interests had always maintained the model range was complete, and having an upstart like Australia "going rogue" would not please them. Noel knew it, but if those above him in the company were true to their proclaimed desire for market leadership and increased profitability, then he knew he had to try.

Early planning stages saw the Skunkworks Team setting design targets for the project.

- This new Ford tractor had to fill the gap in the model offering and compete against the bounty models from Chamberlain John Deere and International Harvester.
- Because of the subsidies that gave CJD and IHC a price advantage, the tractor had to have a retail price that made it competitive against machines that practically had the market to themselves.
- Could be offered in both Drawbar and Three-Point-Linkage models.
- Be fitted with a comfortable, minimum noise cabin that incorporated a better filtration system than the imported "Q" cab.
- Modern and distinctive styling stating this is an Australian designed and built tractor.
- Have as standard equipment, similar ancillary hydraulics and other features in-line with those found on the soon to be released TW series.

Frank Pitney enjoyed the respect of his peers both in Ford Australia and engineers working in the overseas plants. Having Frank available for the engineering aspect of this new tractor made Noel's dream a much easier proposition. Asking Frank to start sorting through the myriad of options that Ford's manufacturing plants offered to other machine builders, would be the first engineering exercise. These components were described as "skids," an acronym for *"semi knocked down"*. Being a huge task, Noel understood Frank would need to spend many hours researching the build options to find just the right parts to specify a skid suitable for the basis for a110-horsepower two-wheel drive tractor. The team discounted offering an expensive Front Wheel Assist option, as Australian broadacre farmers had shown a resistance to that type of tractor. Knowing that the Australian bounty tractors only drove through the rear wheels, verified their decision.

Keeping the project secret until the last minute, meant communication with Basildon and Romeo had to be kept to a minimum. If questions were ever asked, Noel and his small team would only offer obscure answers until they knew the project was ready to be signed into production.

Working on the assumption that their project would work, it was imperative it be tried in the field. Two prototypes were produced in a back corner of the Campbellfield factory. An area had been cordoned off with huge screens and two 8700 tractors had been split at the front of the transaxle, as had two 7700 cabin models.

Ford had set out when designing the first 6X model range of tractors to keep the engineering simple. They had called it their modular design principal. A good number of critical engine components could be interchanged, pistons, connecting rods, pushrods, rocker arms, castings and a myriad of other parts along with the hardware were interchangeable. This meant that the factories could lower their cost by mass producing components that would serve many models. The Ford Parts departments around the globe would be able to carry suitable stocks, without unduly adding to the number of individual units in their parts inventory. It would be the same for dealers, they too could offer a better parts service while carrying a lower number of spare parts which would cover the whole tractor range.

This forward thinking at that early design stage was now working in Noel's favour, as with a few minor adjustments the front halves of the 8700 tractors were aligned and bolted to the 7700 transaxles. To some it may have been a mongrel, but in this early concept it became the answer Australian Ford tractor dealers were asking for. These would be no backyard hot-rods, or the bastardization of four brand new Ford tractors, for in that screened off corner and behind these heavy green canvas screens, stood the genesis of a new and exotic breed.

Trucked out after dark, these two test mules headed for field trials to a farm outside of Nagambie, one hundred and fifty kilometres to the north of Melbourne. Noel had found an isolated property belonging to a farmer he could trust. Ian Griffith had been a loyal, long time Ford owner and someone who understood the need for the complete secrecy.

Noel had learned from discussions he had had with his service manager Ian O'Rourke, that equipping these tractors with dual wheels would highlight any transaxle weaknesses. Without derating the engines and with the large wheel equipment fitted, these initial test hours showed the tractors were breaking the rear axle trumpet housings because of the reduction gears failing. If farmers ordered tractors equipped with dual wheels and water-filled the rear tyres, then these machines would require an upgraded epicyclic reduction gear set.

Australian 7700 tractors utilised the same three pinion planetary carriers that had been part of every Ford 7000 since the tractor had been introduced. As Ian predicted, the epicyclic gears although capable in the four-cylinder tractors, were now being verified as the weak link in the prototypes. Noel's risk was proving the concept and it was now time to push the start button for the next stage in the project.

To ensure this model stood apart from those produced overseas, Noel wanted the tractor to have an individual look, one that proudly stated that this tractor was an Australian endeavour. Taking a rough list of dimensions to the Design Centre at Broadmeadows, Noel commissioned a few draft sketches to workshop this new look tractor among a select group of his team at Campbellfield.

The automotive centre at Ford had been planning the release of their new angular look "Blackwood" range. This car was to be a departure from the rounded "Coke Bottle" shape that had been with the basic look of three Falcon variants that had made it Ford Australia's best-selling car.

Scheduled for an Australian dealer release at the Sydney Opera House in April 1979, this new Falcon with a model designation of XD, would appear square and smaller than its predecessor. Because the interior designers had been able to squeeze even more room into their new creation, Noel thought these principals should be a feature of their concept tractor too.

What he showed his Campbellfield team was an image of a long stretching engine cover reaching back to a huge flat front windshield that cast up into an overhanging roof. That roof extension was of deliberate design and it concealed far more than the hint of a visor, or a mere cosmetic touch. It covered the air intake and filter for the air-conditioner. The cabin had a broad expanse of tinted glass on all four sides. Completing the structure, a single door had been designed for ease of access by hinging it from the rear. The fenders were angular giving this beast the rugged masculine look Noel had been searching for.

Hearing about the roasting Henry Ford II had given Ford Australia's styling executives during a factory visit in February of 1971, Noel was adamant, the style of this tractor would not fall foul to the same mistake.

On that visit to the design centre, Henry II had been inspecting several clay models for the upcoming XC Falcon range when he became more than annoyed that not one of the representations carried a Ford name. Becoming a lesson learned, from the XC model forward every Australian built vehicle carried the distinctive Ford oval.

For this project, neither the commissioned stylist nor Noel would be repeating that early error. Therefore, in large white font, the Ford name was emblazoned over a bold blue stripe that swept back to the black cabin post. Another thinner blue stripe, broke the white base of the engine cover's flanks and its flat black top. The boldness of Ford's name hovering above the engine made no apology for its heritage, this tractor was different and bred for an Australian purpose.

On each side, three deep vertical slashes opened behind the flat proud nose. They allowed enough air to circulate past the fuel tank and be ingested through the overlarge air-conditioning and radiator cores. The design proving more than efficient at keeping the under-stressed 401 cubic inch engine's temperature controlled, even during Australia's hottest of summers.

Wanting to make an individual statement, this black and white livery departed from the upcoming TW series. All Australian delivered Ford tractors since the E27N had been blue. This new colour scheme suited the tractor, giving it a masculine and brutish look. Even the image appeared to growl. Although this was only one of a dozen different renderings, it was this picture that stood out, having an almost a cartoon quality about it. Everyone on his "Skunkworks Team" agreed it was an easy choice, and from that first sketch they started to believe their tractor would be, *"Of its Time"* and, for Australian Ford dealers, a game changer.

Having Les Gason agree to commit his company to investing in tooling for the building of this new cabin, meant another obstacle overcome. Noel knew that for Les, this would be another exciting project and he hoped it would take away some of the disappointment the Gason team felt at losing the supply of the "Q" cabin. However, Noel wanted to capitalise on the marketing of this new Australian styled tractor and at the same time, reward Les's loyalty. By offering the new square cabin on the 6700 and 7700 tractors A.F. Gason would be able to amortise this "E" cabin's investment over the three models. A move that would show both companies were reducing their overall cost of production.

Lessons learned with the "Q" cab made designing and building the new cabin more stress-free. Its boxy and on trend styling, made construction much easier. The flat glass panels could be sourced and cut to shape locally. Employing rectangular hollow section (RHS) tubing eliminated the need for many of the time-consuming, metal shaping processes they'd had to perform to achieve the lines of the earlier cabin. Their only difficulty would come when shaping the engine cowling. To finish it in steel would require a dozen or more folds and much welding to reach the desired shape.

Les Gason suggested an alternative could be achieved by creating it in fiberglass, the method was well tried and Ford Australia had been using it in their Louisville truck range for over a decade. Although a departure for the tractor industry, having a tilt forward engine hood would become another point of difference, making servicing the engine easier and a feature sales teams could sell.

While the Gason engineers were busy designing and building their proof-of-concept cabins and sculpting a buck from which they could take a mould for the engine cover, Frank Pitney had been spending hours combing through build options for prototype skid units.

It was imperative that the project maintained its secrecy and to keep visiting dealers and their inquisitive staff from spotting something unusual in the factory, Noel had arranged for Cliff Poole Industries in Sunshine, to assemble the initial run of field test units.

Time was of the essence; the project had started gaining momentum and Frank Pitney's engineering department were becoming swamped. The number of small details essential to keep moving the plan toward manufacture seemed to be growing bigger each day, so Frank asked Noel for help.

Company policy for justifying employment of extra staff had its complications and to get around the issue, all departments had budgets for out sourcing their extra labour requirement. Enter Tom Rae who on earlier occasions whenever they needed a drawbar designed, or ancillary brackets be drawn, he had been engaged. Tom had emigrated from Scotland some years before and had an understanding of agricultural machinery. Noel approved of Frank engaging him again as doing so would remove some of the extra workload from the engineering team.

Design and engineering work had started galloping in its momentum, only to have the added burden of solving any of the test mules' reliability issues. The team operating them were doing their best to find any weak links, recording and reporting their findings to Frank daily.

Gason engineers in Ararat had almost completed their first cabin and, along with input from Les Stack, been ticking off many of the elements required to bring this cabin closer to manufacture.

Frank Pitney had settled on the specification for his parts bin *Hot Rod.* Having designed it around a County 1164 SKID, he incorporated several delete options that removed any unnecessary parts, which would otherwise be scrapped when they arrived in Australia. Frank had contracted South Essex Motors whose business was located near the Basildon Tractor plant, to retrofit a crown wheel and pinion of higher specification which came from County Tractors.

South Essex Motors and Frank's team figured that they would be able to load ten units into a container, which could then be shipped directly into Perth, Melbourne and Brisbane for assembly. These were simple logistics, and a system that worked well. The plan contained savings for Ford Tractor Operations in Australia, particularly freight. By not carrying all the inventory in one factory and then shipping completed tractors across the continent became a key element of their plan. Not only was it a cost saving in freight, now tractors could be assembled and shipped to the dealer in a more expedient manner.

Needing a model number or a name for locking it into the mindset of everyone from its development team to its eventual owner had become another discussion point. Finding something suitable to call it had become a challenge.

The model designation for the Ford six-cylinder tractor engine range was an eight and its capacity was four hundred and one cubic inches. After many discussions and with several suggestions workshopped only to be discarded, the skunkworks team decided on 8401. Tractor manufacturers had been using numerals instead of names for decades and with this new Aussie Ford ringing in the Eighties, having an eight in the name seemed more than appropriate.

Tom Rae, concerned with rumours he had been hearing from the test teams, decided to send an unauthorised telex to the head of engineering in Troy, Michigan, USA, listing his concerns about reported rear axle failures.

Their secret was out and General Manager of Ford Tractor Operations Europe Gor Cowl, was less than amused. At every previous management meeting when Noel tabled the need for a tractor in this horsepower bracket, Cowl had vehemently shut down any discussion, thumping the table and stating his opposition to the thought of investing any time or money in it. Now with a copy of Tom Rae's telex, he had all he needed to put a stop to the project and sent his envoy, Gordon Guthrie to Australia to carry out his instruction.

Around the same time, Ford of Europe were contemplating the design of a similar tractor, but it would not be ready for release until the mid-eighties at the earliest and waiting was not an option for Noel. The tractor market in Australia would have moved on by then.

By coincidence, Australian Service Manager Ian O'Rourke and service managers from around the globe had been summoned to a world-wide service conference in England. As part of the activities, Senior Service Manager Bill Hepburn asked Chief engineer, John Foxwell to take the group on a tour through the Design and Engineering Department. Until now this had always been an off-limits area and, Ian for one was keen to see what improvements were in planning for newer models.

He questioned several items he could see which would remain an issue for Australian farmers. The new tractors were using the same "Q" cabin that first appeared on the 700 series, the position of the inlet for the air-conditioner filter hadn't changed and it troubled him. All his earlier correspondence highlighting this as the main source of air-conditioner failures had been ignored. Taking a couple of engineers to the rear of the prototype he explained how dust in dry conditions would be drawn to the low-air-pressure area on the outside of the rear glass of a forward moving tractor. Telling them that not only would dust cling to the glass, but because of the position of the intake for the air conditioner the filter could overload in as little as two hours. Not only did the rear position of the intake increase the repetition of cleaning the filter, the air-conditioner efficiency was easily compromised. If an operator could live with those things, then the difficulty of cleaning the filter without covering themselves and coating the cabin interior with dirt, would be the last straw. More than once it had resulted in a heated exchange with the selling dealer and their parts and service personnel.

He referred them to the number of times this issue had been raised at service meetings in the past, and Australia was still waiting for a retrofit program to rectify the problem on existing models.

There were other items he pointed to that he believed would need to be addressed for markets outside of Europe's more temperate, low dust conditions. However, as he had been walking around the design centre, he spotted several four planetary reduction carriers. Understanding the failings that had been coming from stresses being placed on the rear axles of the Australian test mules, he made it his business to check the part number of the item and the current stock-holding being held by the parts department. Learning just how many units the UK had in stock and knowing that this would see an end to any the reliability fears he harboured, Ian ordered fifty sets to be air freighted to Australia as soon as they could be shipped.

Waiting for Gordon Guthrie to arrive in Australia, Noel knew their 8401 project had passed most goals that Australia's engineering and sales teams had set as imperative. Gason had produced a prototype cabin and having fitted it to a 6700 had been dynamometer tested before and after the fitment of a turbocharger. This new E cab had achieved a noise level of 83DbA, some five decibels below the Australian requirement.

Once the four-pin planetaries had been fitted to the rear axle housings of the test mules, their early breakages had been overcome. That allowed the team to now focus on a specification for the tyre size they could offer as standard and others to be included on the pricelist as optional equipment.

In record time, Noel and those around him had managed to do what others in the Ford world had not. They had built several proof-of-concept tractors, that were specifically Australian and distinctively styled. The 8401 had been budgeted to retail at a price where it could take the fight to the bounty tractors from Chamberlain John Deere and International Harvester, without asking for a cent from the Australian Government.

When Gordon Guthrie arrived at his door, Noel knew it was with the intention of shutting down the manufacture of the 8401. By Gordon's tone, Noel could tell just how forceful Gor Cowl had been in delivering his instructions to his messenger. However, he didn't let it unease him, for he still had a few aces to play, choosing to let Gordon say his piece before taking him to the Australian Head Office to meet Sir Brian Inglis.

Whether it had been a mistake or bloody-minded interference, Noel didn't care, Tom Rae had overstepped his authority and within minutes of it coming to his attention, Tom's contract had been terminated. Gordon's arrival heralded a deepening of a rift between Noel and Cowl that would only become bitter over the next decade.

CHAPTER FORTY-SIX

Planning the 8401's distinctive livery had not been an accident, in 1977 Ford Tractor Operations Australia had an overstocking issue with the 6700-cabin tractor. Because of its retail pricing and perceived horsepower deficiency when compared to Chamberlain's new 4080 Sedan tractor, had not sold as well as predicted. In an inspired move, Noel asked his marketing team to brainstorm ideas. Their objective was to change the buyer's perception about the tractor's power and offer it as something deliberately Australian.

The marketing team were adamant their tractor should have a look of its own, something *sexier* than the run-of-the-mill Ford Blue with white wheels. They needed to create a new image and style those slow selling units into a "GT" version. Understanding that the engine could be easily upgraded to 85hp by fitting a turbocharger it would bring another edge to their sales strategy. They would have a tractor that sat nicely between their 67hp 6700 and the 95hp 7700.

Offering it as a limited edition complete with dual rear wheels would certainly take a lot of steam out of Chamberlain's marketing campaign, and build some much welcome publicity for Ford Tractor Operations. Costings completed, and a teaser campaign managed on a shoestring budget, field sales and service personnel had enough information to target the dealers who were likely to receive orders quickly.

To achieve the look required, Noel asked Andy Jacobsen from the Ford Australia Design Centre to complete a few three-quarter view sketches of a Q cabin tractor with dual rear wheels. He provided him with photos of a tractor of the same specification as a reference, saying he wanted to create a tractor as distinctive as the GT Falcons had been.

Within a few days Andy had delivered several plates for Noel to choose from. The one that the marketing team decided on was also Noel's favourite. Just as the design centre had with Ford's GT Falcon years earlier, Andy had only changed the cosmetics. The two-door cabin featured a flat black paint over blue, two-tone paint scheme. Starting just below the door handles and finishing at the white roof, the black paint harked back to the flat black bonnet paint-outs of the Falcon GT. Andy also employed a narrow red stripe to divide the two colours. The stripe continued running in one long horizontal line along the engine side panel, around the front of the radiator grille and along the other side. Below the flat black engine hood, in bold silhouette vermillion font was the model designation, 6700 FORD, with TURBO under it. Everyone agreed this was the look they were after, a complete departure from the graphics employed on imported tractors. Blue wheel centres with white rims, eight suitcase style front weights and a chrome plated side step, helped to make the tractor look larger than it had before Andy's makeover.

Naming it a 6700 Anniversary had been a masterstroke, with dealers snapping them up after minimal advertising expense. The fitment of a turbocharger had given it the GT status they had been hoping to achieve and their marketing ploy had been a success. All the 6700 Anniversary models sold in a matter of days, creating a demand and causing dealers to predict the tractor would become a halo model and very collectable in later years.

These efforts of Noel and his marketing team was a lesson not lost on the great grandson of the company's founder. Edsel Ford II had transferred to Australia after graduating from Babson College in 1973, from there he went on to study management at Harvard. Arriving in Australia after working at Ford as a product planning analyst and with experience in both the Lincoln and Mercury divisions, he was ready for an overseas deployment.

Ford Australia had found buyers had been losing interest in two door cars, General Motors Holden had stopped building the two door Monaro a couple of years before, while still had four-hundred of the coupe shape to sell before changing over to the new Blackwood range.

Addressing Ford executives at Geelong, Edsel explained another way that the company had found a solution. Using the 6700 Anniversary tractor as an example, he had suggested loading the coupe's specification with the 351 CID V8 engine, option it with an automatic transmission and raise the comfort level to that of the four door GXL sedan. He also recommended painting them white with two broad blue stripes running over the length of the car, applying Ford's snake logo behind each front wheel and harking back to past history, call them Cobras.

Marketing them as an individually numbered limited edition had seen dealers across Australia clamouring to buy them. The model had sold out in a very short space of time. The XC Cobra and the Anniversary 6700 tractor had once again proved that in the Ford world, turning a bad loss into a good profit would never go unrewarded.

When Gordon Guthrie arrived, Noel knew he had his job on the line with the proposed 8401, it may have been the right tractor at the right time, all Ford, and hugely profitable on paper. However, he still had to convince Gordon before he could move to production. In the week prior he had called the team together to update every facet of their business case. Finance included the latest pricing of components from each supplier, Australian assembly costings were figured in, along with the shipping charges from the United Kingdom. The report effectively demonstrated the savings made by landing into three separate ports for assembly and distribution.

Another report from the field detailing every failure and its reason, along with the cost to find a solution and rectifying the machines had to be included. Yes, as Tom Rae had reported there were failures, but they had been predicted and the remedy costed. Australian sourced components along with the exchange rate, had been extrapolated to show increases in materials costs through inflation.

The financial side of the report that Noel would present showing every expense for the life of production. On the other side of those numbers, the raw data showed the profit the 8401 would earn for Ford Tractor Operations in Australia. He didn't think it necessary to include a guess for the extra profit the parent company might be earning via the supply of every skid unit. However, Noel had estimated the figure and would be willing to point it out, should Gordon become pedantic.

Noel and Gordon exchanged pleasantries while they drove to Ford Headquarters for their meeting with Sir Brian Ingles. Apart from volunteering to serve with the Royal Australian Air Force during World War II, where he flew Spitfires during the D-Day landings, Brian had spent all his working life at Ford. His family were almost royalty when it comes to the history of Ford Australia, his father Scott had been General Manager decades before. His brother Malcolm too, had risen through the sales and marketing ranks to become a director before his retirement. Aged seventeen and straight out of Geelong Grammar, Brian started working for Ford's finance office in Geelong as a casual pay clerk, while waiting for the RAAF to call him to service.

Knighted Sir Brian, in the 1977 Queens Birthday honours list for his contribution to industry, he had become a strong mentor and supporter of Noel. Also unknown to Gordon, Sir Brian had been aware and an avid supporter of the 8401 project from its genesis. However, even though Noel was confident of his product and its viability, he had reservations about the outcome of this intrusion and its impact on his career. There was more than profit and loss riding on the 8401, if things were to go Gor Cowl's way, Noel believed he could be out of a job by day's end.

Sir Brian was waiting for them as they pulled into the "Glasshouse" car parking area, his broad smile and outstretched hand welcoming his old friend. Gordon had been a regular visitor to Ford Australia, as he too had been with the company for many years. In his role of General Marketing Manager Tractors, he had been instrumental in the introduction of Ford's 6X range and he knew the importance of having tractor models tailored to suit the demands of the current market.

Shaking Gordon's hand, Sir Brian said he was eager to show him Noel's new baby and ushered them both toward the Design Centre. Brimming with enthusiasm, Brian explained to Gordon their reasons behind the need for secrecy and how hard Noel's team had worked to take his original idea from concept, to test mules and then onto producing a production ready prototype. When he opened the door to the Design Centre's reveal floor, a production ready 8401 simply sparkled in the glare of the flood lights. Neither Noel or Sir Brian spoke, they just waited for Gordon to soak the moment in.

When Noel believed Gordon had taken in the presence of the tractor before them, he started explaining how their new

8401 would fill a gap in the current market and why it was so important to Ford Tractors right now. He gave him an overview of Chamberlain and International's ability to access funding for their aggressive sales tactics by way of Australia's Bounty System. Ford were selling into a market against competitors with an unfair advantage. This tractor would defeat that, and he hoped in turn would kill off the bounty system. Gordon listened and nodding his agreement stroked his hand along the fiberglass engine cover.

From habit, Noel guided his colleagues to the rear of the tractor and without realizing, in tried and tested Ford fashion, Noel began a six-position sell. Pointing to the drawbar, he told Gordon how Australia's wheat farmers had a resistance to buying things they didn't need; therefore three-point-linkage would be later offered as an optional extra at their time of purchase. Pointing to the top of the rear window he mentioned how the cab had been designed to include a large roof area with modest overhang, for shielding the windows and making it easier on the air-conditioning components.

Sir Brian made his excuses to leave, asking them to come to his office when they had finished at the Design Centre. Watching him walk away, Noel saw Sir Brian was still wearing the same smile he had when he greeted them and as he reached the door, he waved before he stepped through it. Noel saw this as a positive sign and continued showing Gordon around the tractor.

Moving to the right profile of the 8401, Gordon asked about the axle failures and why the tractor was wearing 23.1 X 34 Tyres, saying that in his memory that these trumpet housings and epicyclic reduction gears were not designed for such loads. Noel explained they had predicted the failures, the steps taken to remedy them with four pin reduction gears and how extended field testing had proved their capability. Axle failure was no longer an issue.

Gordon told Noel he liked the look of the tractor, yes it was a departure from the current design, but then he could understand why Noel's team had decided to move away from something that had been generated in the northern hemisphere.

Moving to stand beside the engine cover, Noel pointed out the front intake for the air conditioner above the windscreen glass, describing how the operator could easily service it by swinging it forward. Removing the filter was now a simple exercise and servicing it meant leaving the dust outside the cabin. Again, making it easier on the air-conditioning components. Explaining how the engine cover had been moulded in one piece to make servicing the engine and cooler assemblies easier, he watched Gordon look underneath it, testing the thickness with his thumb and forefinger.

Noel would have liked to tell Gordon a long and intriguing tale about deciding a name, but reckoned against it. 8401 had seemed to stick from the first time it had been suggested and the tractor appeared to wear it well. For the team, it had an affectionate ring to it. However, he did go into detail about the decision to work in fiberglass and pointing to pictures of upcoming Falcons on the Design Centre walls, showed him the reason behind the 8401's rectangular shape.

At the front, Gordon was pleased to see the Ford name emblazoned in bold white on blue capitals and they shared a laugh as Noel relayed the story of Ford emblems and Henry II's visit a few years earlier. Ensuring everyone knew who built a Ford vehicle, would be something no Australian designer was likely to forget.

Once in the cabin, Noel took Gordon through the process of working with A.F. Gason to complete the design, the reason for a single door and flat glass. For current Australian tractors this cabin was seen as ahead of its time, the bonus was being able to offer it on platform 6700 and 7700 tractors too.

From the driver's seat Gordon could see down the expansive flat black engine hood. Noel continued to explain the savings using flat black paint and the benefit its antiglare properties gave the operator. With following dust at seeding time, antiglare was very well received by those who fitted extra forward facing flood-lights above the operator's eye level. The finished hood was as good, or to a better standard, than that being offered by competitors at the time. Noel added he was extremely proud of his team for achieving, *"best in class,"* scores on over ninety percent of their targets.

In Sir Brian's office, Noel took both men through his business case for the project, answering every question and covering off the costs of development, showing predicted sales and a projected return to the company resulting in a healthy profit. They had calculated parts sales returns over the life of the tractors delivered, but it was such an ambiguous exercise, had not brought that figure into the profit projections. Everyone of Gordon's questions had been answered with a positive response, Noel could describe the tractor in intimate detail and offer figures from the cost benefit analysis, without referring to the business plan. Noel could never be sure of Gordon, but Sir Brian his old confidant was different, he was simply brimming with confidence.

Lunch with Sir Brian was in the Directors Dining Room where Edsel Ford II and Bill Dix joined them. Max Grandsen, who shared the role of Sales Manager with Edsel, had sent his apologies. For Noel it was interesting to listen to the conversation happening around him, everyone enthusiastic in their hopes of the 8401 and expressing how important it was to the company's profit over the coming years. They all agreed the 8401 may be a bit of a hot rod, but due diligence had been done and finance had approved it. If Gordon needed an example of success they could go back as far as the concept for Ford's first utility, that was how much of Ford Australia's success had happened.

By the end of lunch Gordon claimed to be a convert, he could see the project had merit and wished Noel well with it and if the paperwork stacked up, as they'd claimed, all they had to do now was release it to the dealers and watch the Ford 8401 sell.

CHAPTER FORTY-SEVEN

Gordon had asked to see one of the test mules working in the field and as they drove north along the Hume Highway, Noel explained why they had chosen Ian Griffith's farm at Nagambie, describing its proximity to Campbellfield and how the Goulburn Valley was becoming a more important wine area. Gordon showed interest and asking questions as Noel explained the Goulburn catchment area.

The region had been continuing its evolution as a food bowl and the variety of different crops expanded every year. Everything from wine growing and horticulture to cereal, stud livestock and racehorse breeders. It could all be found in the area. However, as markets were continually changing farmers would adapt their businesses to the prevailing conditions. Noel described the different crops Ian would be sowing that year.

Passing through the small road-stop town of Kalkallo, they discussed the marketing team's plans for the 80s Action release and who would be leading it. Noel said that although their plans were still a year or two from completion, he had a couple of young executives in mind. Junior executives who he thought might be tested with some responsibility. John Henchy was one, Garry Shepherd the other, but for now he would leave the detail of the release in the capable hands of the marketing manager Roy Pinney.

Gordon interested in Noel's succession planning asked more about John Henchy's history. Noel explained that John had come to his attention when he met him working as a sales representative for the Horsham dealer. There he had shown both initiative and a cheerful willingness to work. Then, having accepted a role with JI Case in Queensland John had worked his way through sales representative roles to managing the company store at Moree. Not long after Noel's return from South Africa their paths crossed again and when a role at Ford was offered, John joined Ford Tractors as Zone Manager for the Wimmera region and working from his base in Horsham.

It didn't take too long before John moved into Ford Tractor Operations at Campbellfield. As Noel believed him to be someone possessing great potential, John worked across many aspects of the marketing department. When an opportunity came to manage the West Australian office in 1974, John accepted Noel's offer for the role and moved his family across the Nullabour.

Gordon nodded his approval and began recalling the lead up to the 1964 release of the 6X range. The importance World Headquarters had placed on the tractors to not only be successful, but the company's desire to create a modern image, had placed huge pressure on the tractor team. After the disappointment of the early 6000 tractor, getting these new tractors right had been front of mind for him. Luckily the 6X range had been well received and Ford quickly started taking chunks out of their competitors' market share.

While they drove, Gordon told Noel that he had reservations and that maybe Australia's tractor market, had not been properly considered in the early development of the TW range. He scratched at his temple as if trying to pry the next thought loose, before asking for Noel's opinion on the coming TW models. Before Noel could offer his view, Gordon continued, saying he knew there would be changes coming, driven by farming procedures adapting to new herbicides across the regions. He pointed to a crop on the left and told Noel that new reports coming from America's thinktanks had Ford's management believing that with the use of newer, more advanced herbicides the number of tractor hours would decrease across cereal growing areas. Farms would drop in number as small farmers sold to their neighbours and bigger machinery would make small farming plant redundant.

Noel agreed, saying he had already noticed the trend beginning in Australia too and that he had people monitoring how the sale of chemicals and associated machinery from companies like Croplands, Hardie and others. These new agri-businesses had been taking a bigger share of the overall farm machinery market each year. He believed that if the trend continued, ever-increasing powered tractors would farm Australia's acreage. Meaning tractors would be working less hours. Particularly in cereal growing areas where the same acreage would soon be worked by heavier, high-horsepower, articulated tractors.

White, John Deere, Case, Allis-Chalmers and International Harvester had all been scrabbling over the small percentage of the current market, but with John Shearer rumoured to be taking dealers and customers to visit the Steiger factory in North America, change was all around them. Versatile too had been making approaches to Ford's dealer body, offering them tractor models ranging from 210 to 350 horsepower and access to several wideline tillage machines from Morris. He had received a few copies of different competitor's pricing and it was easy to see that these tractors were priced to sell.

Noel predicted Australia's tractor market would be very different by the end of the decade and although he needed to plan for that, he remained frustrated that his requests to Dearborn and Basildon were continually being ignored.

Gordon appearing interested but unable to commit the company to anything, continued to listen, while Noel expanded with his views on how the agricultural media and to a lesser extent, Ford's automotive arm saw the situation in Australia. Noel believed Australia to be a microcosm of world agricultural mechanization and, turning to Gordon used the analogy, *"if the world sneezed, Australia caught the cold."*

Conversely, when it came to experimentation in dryland farming regions and because of the vagaries of the climate, Australian farmers were willing to try new technologies. These advances proving successful for Aussie farmers, then those very systems would be adapted by others around world, thereby securing global food production. Ford Tractor Operations had to take the lead, or get lost in the drifting dust of new farming methodology.

After listening to Noel, Gordon talked about the case for even higher horsepower tractors and quizzed him about the specification Australian farmers were likely to need in new TW tractor series. Asking if they would they be working around the clock in wheat areas, as North American farmers were beginning to do. It had been an increasing trend in the short growing seasons of places like North Dakota and across into the prairielands of Canada. Explaining to Noel how the use of wider machinery was beginning to expose lighting difficulties. Some tractors were now working at the very edge of standard illumination and because he wasn't aware of any lighting options being planned, it concerned him.

Noel agreed it was an issue with Australian farmers too and different Ford dealers were experimenting with different solutions, none of it pretty or built to look as it should. Alternators were being replaced or upgraded to produce more amperage, but nothing was the same and the company was missing out on sales because of it.

Gordon moved from the new models and again started talking about the 6X launch and the obstacles they faced providing Ford tractor dealers with a new range of attachments to partner with the tractor range. He had often shown an interest in the Blueline range and asked how far Noel's marketing team had reached trying to source a variety of suitable equipment they could sell to cover this new tractor offering.

Noel outlined how the company would be sourcing of equipment to suit the Shibaura built compacts when the small tractor supply came on line. The marketing team had secured a front-end-loader and backhoe supply from Cliff Poole Industries in Sunshine, adding that this was the same company who had been building similar items for Kubota for several years. Cliff had been keen to partner with Ford Tractors Australia and could see how developing and producing a line of machinery to be branded Blueline would be extending his reach into dealerships, he otherwise would not be able to.

As much as he saw the compact tractors as being necessary, Gordon claimed it would always be the bigger machines that brought the company its biggest opportunity for profit and wondered what, other than the introduction of the 8401, would they be relying on. Noel him they had been negotiating with several suppliers to expand the Blueline machinery range. These companies could better cater for the higher horsepower tractors. However, for these bigger machines to be successful, the dealers had to be profitable and Ford focused. As the size of machinery increased, dealers fighting continual pressure from the many *"Buy Now-Pay Later"* schemes being offered by their competitors had become a much bigger thorn in their side. Noel thought Ford Credit could be their answer. If together they could create a better range of finance products, they might then minimise that effect. It would mean teaching the dealer sales teams how to sell finance, and with the Ford Tractor zone managers being supported by Ford Credit staff, it shouldn't be a huge task. The marketing team was working on the problem, but it was still in its infancy.

Gordon agreed and enquired what he could do to help in the quest for market leadership. Noel asked him to explain the more successful of the many different floor plan schemes being offered by different Ford Tractor Operations around the world. He wanted to know more than just what had worked and the things that didn't. He asked Gordon who to talk to, someone who could offer a more accurate vision of how their plan had been presented to Ford Credit.

Gordon talked about his experience as a young executive and how it sometime seemed his superiors thought he was too close to the dealers. Their alarm often came after he raised concerns or suggested programs that were new or unusual. Most of it driven by his suggestions around unexplored credit arrangements and how he thought easing the criteria would make it simpler for Ford dealers to retail more tractors. Finance had always been an issue and it was never easy to find a solution that worked for both the company and the dealer. Anything he suggested hardly ever worked well for both parties. However, looking at the problem another way, he had found another opportunity. By gradually easing the recommended selling price higher, he built the margin enough to cover financing his proposed scheme. The program being fully funded, it became a straight forward presentation to convince his superiors that they could offer the dealer or the retail buyer special terms.

Noel said he understood that too, but there were other complexities facing many of their Australian dealers, who had for many varied reasons, locked themselves into the country's marginal areas. Often these businesses were small, caught between the vagaries of the season and trying to manage cashflow often hit by the volatility of governments State and Federal monetary policy.

Gordon agreed, but recommended Noel investigate the different exchange rates as there was often opportunity hiding there.

Understanding that the following morning they would go through all the evaluation processes and costs incurred to bringing the 8401 this far, Noel wanted to divert the subject. He asked Gordon to elaborate his part in the introduction of the 6X series of tractors. Although Noel had heard most of it before, it would give Gordon time to talk about himself and might even toss up an idea or two that he could utilise later. Asking people to talk about themselves had always worked at putting them at ease and he knew it would have the desired effect on Gordon too.

Arriving at the farm, Ian Griffith met them in the house yard and after introductions, Gordon and Noel squeezed into the front of farm utility and Ian drove them to his field on the Goulburn River where the tractor was hauling a 36-foot, wideline cultivator. Watching Gordon climb into the tractor and take the controls, he felt even more confident of his colleague's support. However, he couldn't be too, sure there would still be a presentation they had to go through in the morning.

Having laid out all the financial planning for the 8401, Noel had watched as from under a mop of grey hair, a pair of eagle eyes searched for any cost that might have been omitted, any price adjustment that had been discounted too far. Gordon's fingers drew invisible lines down and across the pages stopping occasionally to allow his internal calculator to total and take stock.

Noel craned forward when Gordon flicked back to the beginning of the presentation and began scouring the papers again. It started becoming too much for him and he asked what it was the older man had been searching for.

Gordon had a bemused look on his face and wiping a white handkerchief across his brow, told Noel he could not find any mention of the Tractor Bounty Scheme, or a figure applied to it.

For Noel, this came as a surprise as while he knew his superiors were aware a bounty scheme had been in place since the early fifties, he didn't think they would have asked Gordon to look for it. Putting that thought to the back of his mind, he explained that as the company representative to both the Tractor and Machinery Association and the Advisory to Treasury, he had always argued for the abolishment of the scheme, it would be hypocritical to apply for the same assistance. Besides, the profit in the 8401, along with the increased margin in the 6700 and 7700 platform tractors when fitted with the new Gason built cabins, more than satisfied the company guidelines for managing a margin and setting a retail price.

With the explanation accepted, Gordon said his goodbyes to the individuals that made up the nucleus of people in Ford Australia whom he had had dealings with during his time with Ford Tractors. Eventually he made his way out to the car where Noel was waiting to drive him to the airport. This would be Gordon's last official visit to Australia as he was now preparing to retire from Ford Tractor Operations, a company where he had served for most of his working career.

Noel found a certain sadness in saying goodbye to Gordon and before he passed into the Customs Lounge, they shook hands and wished each other well. Gordon would wait for his flight and close the pages on his career, Noel would go back to building his place in the world of Ford and its myriad of subsidiaries. Gordon had often been an ally in many boardroom tussles when passions had reached a white-heat intensity, able to cut through jargon and politicking to bring the proponents back to whatever the issue had been. It was a skill Noel admired, something to emulate.

CHAPTER FORTY-EIGHT

QANTAS Boeing 747 Flight: QF 4: Honolulu Time: -- 12 05 pm

The stewardess smiled at Noel as he handed her his jacket, she asked if he wanted her to take his briefcase. However, he shook his head thinking how hard it had been to save it yesterday. Those images were still playing out each vivid nano-second in his memory and taking his place in the front of the aircraft, everything seemed as it had been the night before.

Sitting in almost the same seat as he had on United Airlines Flight 811, he clipped his belt into its place and set his case in between his feet where it would stay until they were airborne. The day had found him reliving those events and they would come in sharp bursting terror, interrupting his thoughts when he least expected it. Although when it counted, he had embraced those fears and they had given him enough strength to help the stewardess. It had become a warmth he embraced whenever those unwelcome flashes ripped the implosion back to the front of his consciousness. Even though moments of horror would tear great jagged ravines through his memory and as quickly as they came, they were gone. He looked again at his hands and stretched his fingers; the hands were still steady.

He felt the mechanics of a door lock closing somewhere in the hold and pictured the officers on the flight deck going through their preparations to take control of the aircraft. It helped that he had chosen to fly QANTAS tonight rather than accept the United Airlines offer.

The Australian company's safety record stood out as a leader in air safety, other airlines seemed to have a more relaxed attitude to procedure. He supposed the crew knew about the events of Flight 811 and wondered if they were more alert tonight because of it. Dismissing the idea, he imagined the chatter between the crew, running down the myriad of checks and cross checks before proceeding to the next stage of their routine.

In his mind he ran through the multitude of pre-flight checks he had completed in the Ansett Airlines Boeing 727 simulator. Smiling at another memory, it took him back to the times when he had been invited into the cockpit on several 727 flights while serving on assignment to Ford of South Africa. Remembering those journeys, he didn't notice how his head nod was now complementing his smile.

Sure, he had things in common with Boeing pilots other than a love of flying. However, that love had been pulling the corners of his mouth for a few minutes now and it felt good to register that he was smiling. Another memory stirred, and it banished any concerns he harboured about tonight's trip home.

Ford Australia's relationship with the founder, Sir Reg Ansett and it had facilitated Noel an opportunity to fly Boeing's

727 Tri Jet simulator in Melbourne. At the time, Sir Reg's many business interests included several Ford tractor dealerships across the central and western districts of Victoria along with vehicle dealerships in other important markets throughout Australia. The benefits of the relationship had always been reciprocal.

QANTAS -Flight: QF 4 -- Honolulu Time: -- 12 35 pm

Lost in his thoughts he felt a hand tap his shoulder and turned to see the stewardess smiling and addressing him by name, held out a blanket and a pillow. She began assuring him QANTAS took safety seriously, adding they would get him home happy and in one piece. He wanted to make a quick quip, but could only say thank you and take the offerings.

Closing his eyes, he remembered seeing joy in dealers' faces when Gor Cowl and Glen Neuenschwander, a finance specialist, had announced during a whistle-stop tour of dealer meetings, that the tractor company was in the process of buying a harvesting machinery manufacturer. During that week, many hints had been dropped and dealers may have made their guesses, but New Holland was never confirmed as the target by either man.

Noel had been told by his contact in World Head Quarters, that the reason for building a solid share of the tractor and machinery market was to position Ford's tractor division for sale. Although hearing the news a few years before, in his mind he believed it would reward those dealers who had remained loyal to Ford. He had often taken calls from dealers stating their loyalty when it made good business sense to change allegiances. After Cowl and Neuenschwander's tour, dealers began believing they would be able to meet International Harvester, Massey Ferguson and John Deere with a complete product range.

He felt the thump of the front cargo door close and wondered if they had checked the hold for explosives, then dismissing the thought, chastised himself for allowing such a negative assumption to surface.

His mind returned to Gor Cowl; the man was a traitor to every dealer that had trusted him during that round of dealer meetings. Now, very few of those dealers had any time for him. Some had lost more than just their businesses. They had lost their position in their community and as paint peeled and cracked on the Ford Tractor signs, buildings closed and as their staff began to leave, small country communities hollowed out.

Skilled people followed the work and their little towns that were once glued together by the Ford tractor dealership, slowly decayed. For those who remained, their options were thin and knowing that, had cut through his desire to see his dealer network flourish. Noel would never forget Gor Cowl, and as for forgiveness, if it were to happen at all, it would take place in another realm.

He felt the aircraft push back, then they stopped for less than a minute, then the 747 started moving forward and he imagined *hands on the throttles. Eyes on the gauges and the chatter in the cabin as the crew continued checking and crosschecking each other. He found it fascinating that the best aircrews had no ego, everyone competent and self-assured enough to undergo the rigors of working together in a small space. There had been times when he experienced friction and jealousies among the people at Ford Tractor Operations, but for the main part everyone got on with their work without the need to interfere in other areas. With the New Holland purchase, for his colleagues in Australia that had changed. Their tractor world had gone to Hell in a handcart and unpleasantness hung in the air like a fog.*

The huge aircraft started gaining speed as it taxied toward the runway and without realizing, Noel began clamping his case between his legs, his hands searching for something solid to hold onto. Only becoming aware when he felt his toes curling, grabbing at the innersoles of his shoes, did Noel notice how the effect of the events of yesterday had assaulted his subconscious. Stretching his fingers, he unbuckled the seat belt, stood and shook the creases out of his trousers. Easing back into the seat again and clipping the belt into place he gave the loose end a security tug. He reminded himself of the safety record of Boeing aircraft and how accidents like the one he had experienced twenty-four hours ago, were extremely rare.

Sinking into his seat as the wings took the weight of the plane and hearing the landing gear folding back into their holds, he looked at his watch. He would wait for thirty minutes or so and only when they had achieved cruising altitude, would he would try to get some sleep on the way home. Within minutes the captain had extinguished the seatbelt light and the stewardess appeared with a blanket. Thanking her he took it.

She asked if he needed a drink, running through a menu of the airline's offerings. Telling her he was okay; Noel again thanked her, declining anything until later.

Laying the seat back, it took several attempts until he had positioned the pillow. Pulling the blanket around his shoulders, he closed his eyes and began waiting until they levelled out. However, the next movement surprised him, he felt his weight shift forward and woke to the engines winding down. They were on approach, shaking his arm out from the blanket he shook his wrist to move his watch from under the cuff of his shirtsleeve. It took a moment to focus and it surprised him to think he had slept for almost nine hours. He stood, folded the blanket and put it and the pillow on the seat alongside him, all they had to do now was land.

CHAPTER FORTY-NINE

Noel believed planning and rehearsing every detail of the marketing team's 80s Action release would be essential. He wanted to build unprecedented excitement level among dealers. Everyone on the team willing the desired effect to have them submitting orders before the new products came on line. Addressing the marketing team and asking to begin raising anticipation levels among zone managers, by giving them only a few tantalizing facts that they could tease their dealer salespeople with. Maybe a photo or specification to hint that the company had something special coming.

Rumours of new products had to come first, nothing official, just of few remarks would get the word out without disclosing the whole product line and, in particular, the 8401.

While a big dealer reveal might be important. It could only happen after every little part of their plan had been approved. Until then it would only be marketing personnel who could feed hints to National and Rural Press.

Noel understood the value of pre-release gossip and it certainly outweighed that of any paid advertising. So, in February 1978, Noel's handpicked marketing team began planning a release that would set their theme for the new decade.

Making it known that he desired this dealer release be something special. Introducing the new Australian designed and built 8401 to the sales group, he handed around a few glamour photos of the new tractor. When the murmurs died down, he followed them with a few similar photos of a 7700 fitted with the same Australian built cabin. Again, he waited for the chatter to die down and outlined how he saw the marketing team's first responsibilities unfold. They agreed that this release required an inordinate amount of secrecy and were definite in deciding that only the people in the room would be aware of the project until more help would be required. Until that time arrived, some of them would need to cover one or more roles.

Taking on the role of Acting Merchandising Manager, John Henchy would return from his appointment in Western Australia. Setting a release date and working back from it had not always been the favoured method, however, the importance of this particular dealer meeting demanded it. The month chosen had been debated around harvest and Autumn farm activity. Selecting March meant the dealers would be between cane, cereal and vine harvests. Horticulture and fruit growing regions would fall into the same six-month activity cycle.

Choosing the last weekend in summer would be the best opportunity to introduce sales teams across Australia to the new TW Series, the Gason cabins on the 700 series and the 8401. Other releases would follow, but this one would set the tone for Ford in this decade.

Agreeing that on Friday the 28th of March, dealers and their sales people would begin arriving at Melbourne Airport called for a co-ordinated approach. An action plan began taking shape, tasks identified and responsibilities allocated. With the date in place their next priorities were product allocation, dealer accommodation, venues and promotional activities.

Noel believed for the reveal on the Friday night, he could get Brian Ingles to agree to it being staged at the Design Centre, adjacent to Ford Australia's Headquarters. However, knowing it had always been an extremely secret area, its last design success, had been Ford Australia's XD Falcon. Although the centre had always been out of bounds for all but a select few, Noel believed the importance of the tractor and the trust he had built within the management group would carry the day. Telling the team to work on the program as if permission had been given, he also asked them to find an alternative in case the Design Centre would be declared out of bounds.

Even as they began planning in 1978, on paper it appeared that for Ford Australia, it looked as if this would turn out to be a great decade. New Australian designed products were set against and exceeded best in class bench marks and everyone from the janitor to the managing director had an optimistic air about them. In Australia, for Ford Tractor Operations, the eighties were poised be the best decade yet.

Noel, having set goals for the sales and marketing team also required them to build a teaser campaign. A campaign that would ensure that not only the dealer principals attended but their key sales staff too. Having tossed around several ideas to call the release, they finally agreed on Noel and John Henchy's suggestion, 80's Action, believing it set the right theme for the coming decade.

Following the lead of his business heroes, Noel believed the most successful ideas were the simple ones and if this release proved their plans, then future dealer meetings could follow the same template. While the marketing people pressing ahead with launch preparations, Noel now had to work with engineer Keith Pincott and the factory managers in Perth, Brisbane and Campbellfield. A target for the first number of units to build had to be set and, he needed to be sure the managers were ready with trained assembly staff who could produce the new tractor.

From their own sales data and best guess predictors, a figure of seven hundred, 8401 tractors had been planned for the first build run. Gason had lifted its production capacity to build the first cabins and Les Gason had expressed an enthusiasm for the project, telling Noel they would do whatever it took to meet the production schedule.

The plan had always been to take advantage of the vast size of Australia and to ship containers of skid units into three ports, Freemantle, Melbourne and Brisbane. By doing this they could alleviate some of the freight costs associated with shipping completed tractors out of a single assembly point. It would require transporting completed cabins to Queensland

and Perth, but as a truck could carry more cabs than tractors, it became one of their most cost-efficient decisions.

At the same time as it fell to John Henchy to come up with a design for the program and associated propaganda material, Noel's contact in World Headquarters had been keeping him abreast of their moves to sell off the tractor business and return to an automotive focused manufacturer. However, as unsettling as the news had been, there was still a company to run and making his part of it profitable, would help Ford in its efforts to attract a suitor.

While the 80's Action team began assembling and assessing a roster of suitably skilled staff to present each different product, they needed to decide which was the best way to present this expanded product line. Did they choose to keep everyone in a large group, or divide them and use *round-robin* stations, or was there another way? The choices were endless, as was the pressure to set this up as the best release to date.

Gordon Guthrie's return to Michigan and his report on the 8401, brought attention from other quarters as more senior managers sought information on this parts bin special from down-under. Bob Nassau General Operations Manager, of Affiliates and Direct Markets, Ford Tractor Operations had flown in on a whistle stop tour and requested that Noel take him to see the tractors working in the field. Only one of the mules remained on test, the other being brought back into the factory and torn down for inspection.

Chartering a suitable aircraft, the pilot flew Noel and Bob from Essendon Airport to Bendigo. There they were to be met by Arthur Cowling who was working as the Manager of Tractor Business for Provincial Motors, an Ansett owned company. Arthur's family had been the Ford dealers in the Bendigo area since 1927 and had enjoyed unparalleled sales success during the days of the Fordson E27N. Continuing it with later models, until closing the dealership in 1969.

Arthur met them at the airfield and drove them to where the tractor was working. Bob quizzed the operator about the tractor's suitability, before getting behind its controls and driving the machinery himself. Within an hour they were back in Arthur's car and heading to their waiting aircraft. Bob enthused about the tractor while returning to Essendon, but reminded Noel there were reports to review.

At the end of his tour, Bob wished Noel well with the tractor and telling him that because of what he had witnessed during this visit, he held high-hopes for Ford Australia's tractor business. He continued saying, that with the 8401 and his knowledge of other models due for release in the coming decade, Ford Tractors were poised to enjoy unparalleled success in all markets, particularly Australia and New Zealand

As encouraging as his words were for Noel to hear, he knew they only had a short time to ready themselves for what he and his team were planning to become the benchmark for every subsequent Australian tractor release. Scheduled for March 1980, several tasks had already begun to meet the deadlines set, one of these had been the name and now 80s Action would be the clarion call for Ford Tractors during the next ten years.

More than anything, Noel wanted the success of the Australian designed 8401 to prove to his Canberra colleagues, that there was no need for the Australian Government to pour taxpayers' money into propping up multinational tractor companies. Although making Ford Australia profitable would always be his first purpose, Noel knew that by demonstrating Ford Australia could build a tractor without relying on government money, would be the beginning of the end for International and Chamberlain John Deere bounty models.

Now it would be up to his sales and marketing teams to prove to the dealer body that the 8401's sales demand would extend right across cereal growing areas, into the cane and on to other markets. Everything was falling into place and he couldn't wait to take his part in the dealer release where he could imagine the TW tractors meeting with elated applause. However, it would be the 8401 and Gason cabin 6700 and 7700 tractors, which he expected would raise the excitement even further. These were the machines that the dealer body had been calling for, and during the last weekend of March in 1980 they would not only get to see them, but be able to operate them in field conditions too.

Leaving his team to get on with it, once again Noel had found himself in a 747 and crossing the Pacific to attend a round of meetings which would culminate in the Las Vegas release of the new Ford TW range. Until now he had only seen photos of the tractors testing in European or American situations. This would be the moment he could run his eyes over every facet of Ford's latest heavy tractor offering. The six-cylinder 700 series had been building a strong following and he hoped the new tractors which now included the 190 horsepower TW30 would incentivise dealers to compete in the market segment.

The more he saw the opportunities the more he could see how everything came back to finance. Under current conditions he understood the reasons why smaller dealers often showed reluctance to selling bigger tractors. Taking a late model high priced trade in that may need to be financed on overdraft or floor plan would suck their cashflow into negative if it could not be sold on in hurry. The more he thought about it, the more Gordon's words echoed circles in his thinking. The opportunity to resolve the issue was in there somewhere and he knew by applying some time to research all avenues of pricing he would find it.

Ford Credit Australia had been slowly building its influence among Ford Tractor dealers. More than at any time before, a bigger percentage of dealers had been switching their allegiance from companies like AGC and Esanda to the Finance arm of Ford. Before anymore could be done with retail finance he needed to understand how much the new machines would cost the Australian company and then ask the marketing team to set about calculating a selling figure that included a set aside for a Free of Interest Floor Plan scheme. They had made interest free periods available in the past but these were always for a short term, or dedicated to stock that had been proving slow to sell.

Having ruminated over the issue for a few weeks before heading to the Las Vegas for a round of meetings and the new model release, Noel had been searching to an answer for a new dealer floor plan initiative which would put Ford Australia dealers on if not an equal stocking scheme at least something similar that they could fight with.

Even though the bounty on International Harvester and Chamberlain tractors had reduced, it still gave them an advantage. His continual reference to the disadvantage by other manufacturers at the Treasury Advisory meetings was beginning to tire his colleagues, but it hadn't dampened his desire to see it gone and so every report continued to carry a reference to it. Interest rates and inflation were beginning a slow spiral upwards and he figured that as the company set its exchange rate four times a year there might be a bit of wriggle room in which he might find an advantage.

Having time to kill while he waited for his flight to Los Angeles to be called, he wandered into a newsagent and scanned through the racks of magazines on display. A copy of Power Farming, it was the September issue and having Howard written in bold on the cover it caught his eye. Sure, it might have nothing to do with him but being the middle of hay season, it wouldn't hurt to know what might suit the family farm at Kilmore and for the princely sum of a dollar, it would entertain him for a few minutes during the flight.

CHAPTER FIFTY

SYDNEY: The day after United Airlines Flight 811 was due to make Auckland and

> *happy to be back on Australian soil, all Noel wanted now was to catch the flight back to Melbourne and go home. There he could access his future based on fact, rather than allow emotions to rule his decision. To hold Joan would be all he needed to let the events of the last forty-eight hours wash their way out of his mind.*

Waiting for his connecting flight, Noel had time to look back over his life until now, remembering the risks around the development of the 8401 took him back to the Las Vegas release of the then new TW Series. It brought a raft of welcome images and he saw himself smiling reflected in the windows of the QANTAS Lounge. Life had always had its difficulties, having survived a mid-air disaster and if he took the Case IH offer, Gor Cowl would soon be out of his life.

His thoughts went back to the meeting schedule prior to heading for the TW release. It had become an exhausting round of what-ifs, Noel slapped the dog-eared schedule down onto the coffee table of his suite in Troy, Michigan hotel. Every question he had asked about heavy tractors had been fobbed off. His superiors telling him to wait. everything would be revealed during the TW Launch. This trip across the Pacific, the ongoing travel and Cowl's cronies mocking his small run of Australian built tractors, had combined to make him uncharacteristically edgy. He could feel himself spoiling for a *knock-em-down and drag-em-out* fight. However, this was Troy and Tractor Head Office and not the place, but there would be plenty of other opportunities this week, when he could mount a counter attack by presenting his sales and profitability report. The pre-release orders for the 8401 alone, should be enough to put all these little taunts and niggles to bed.

What he had wanted to know from this trip was whether Troy had negotiated for Australia to receive supply of articulated, high horsepower tractors. To him it was unbelievable that no-one knew, or was able to tell him. The grapevine was rife with rumours of a range of tractors supplied by Steiger and being available for sale around the time of the TW release. They may have been North American and British rumours, but they had Australian dealers ready, itching to take farmers' orders. Australian dealers in broadacre areas pleas for this type of machine, had been growing louder and, they did not miss a chance to tell him that Ford Tractor Operations Australia were missing sales and profits because of it.

Sure, the 8401 would go a long way to filling one market segment, but Allis Chalmers, International Harvester and John Deere were blowing new markets wide open and it seemed to him as if the people at the top of this company he loved, just didn't consult.

He too, had heard the whispers of an arrangement with Steiger and that by 1979, they would be supplying Ford with a range of high horsepower products. The rumours coming from inside the company held the line that the tractors would be distributed into some markets in North America and Europe. Knowing that Ford Tractor Operations' goal would be to fend off challenges from White, Allis Chalmers and John Deere in their home market, Noel found it hard to understand why they considered the growing Australian tractor market would be any different.

Steiger had not been resistant to supplying others, International Harvester had a supply agreement with them. If Ford of Europe had supply of these giants from Steiger, all in blue and carrying the Ford name, then for Noel it was logical Australia should be privy to a similar arrangement.

The struggle for market supremacy was fast becoming a gunfight and for the eighties and, a tractor company needed every bullet it could find. For Noel, this had become personal, he had spent almost fifteen years building the business in Australia into a dynamic and profitable arm of the multinational that employed him. Yes, he enjoyed the rewards that being General Manager offered. Yet there were times too when he could have been at home with Joan and the family, but had been called to work somewhere in the world and investing much of himself into the welfare of Ford Motor Company.

He could imagine his competition polishing a full armoury in readiness for the battle of the eighties, while Ford Australia were more than a few rounds short in the magazine. Knowing that cut deep, so he thinking there had to be another way, put his energy into finding it.

Just as every concern Gordon Guthrie raised had evaporated when he first saw the 8401, it proved good results often come when the risk taken is mitigated through good research and strict controls. Noel smiled at the recollection and it rushed another memory forward, it came from a remark one of zone managers had used to cover an unauthorised expenditure. His colleague had reasoned, that if his plan had proved successful, then it was far better to ask Noel for forgiveness than permission he would not have received otherwise.

Still smiling, another thought seeded, the first church of the Ford Motor Company was cashflow and even prior to the twenties, it had always worshipped profit before every other thing. If a supply of FW models from Steiger could not be realised, then it might be time to change his tactics and start investigating another means for a supply of a heavy tractor range.

For now, he would continue to seek an answer to the question he had been asking of Troy Headquarters and, if he did become a pest, that would be okay too. Being told no had never bothered Noel, all it meant was he hadn't phrased his question right, or his timing had been off. Nothing would ever be off the table, there were many ways to find a solution and flipping through the pages of the magazine, he began constructing a list of other small Australian articulated tractor manufacturers. Most of them were based in West Australia and from his memory, some had been enjoying minor success. He had worried after hearing a few reports about them being financially weak, or having supply agreements with a competitor and that crossed the West Australians of his list.

Until now, staying on top of the 8401 program had consumed the team at Campbellfield. Although there were still a few things to do, he understood that keeping the excitement and energy high, new projects had to be found. Some of it would come through the Blueline range and the new contract with Alfarm Industries of Albury.

Albury, had started life as a small border settlement on the road between Sydney and Melbourne and where it crossed the Murray River. Albury in New South Wales with its sister city Wodonga on the river's southern bank, had grown from busy country towns serving the river trade along with primary and government industries to a major industrial hub and manufacturing centre. The highly skilled labour required to serve the military based there, proved a major benefit to firms like Alfarm Industries who could draw highly skilled and progressive people to their company.

Noel knew Alfarm Industries owner Bob Meredith well, having met him as a business consultant at an earlier meeting in Perth. At that time Alfarm were Ford tractor dealers and due to the high number of tractors they sold an important one. For Noel, he valued the trust they had built between them and a relationship that developed into a lasting friendship.

Starting his business Alfarm, Brian O'Donoghue had been the area's tractor dealer since the early Fordson Days, but it was the partnership with Arthur Holloway who had begun importing Claas combine harvesters from Germany, that saw the business grow. It had been this growth that saw the firm becoming one of the most important players in the Australian farm machinery industry during the late sixties right through to the company eventually going into receivership in 1986.

Noel had known Brian O'Donoghue from the time he had first started working as an apprentice with Charlie Milthorpe and it was this continuing association that made Alfarm Industries an easy fit for Blueline products. However, as Brian was readying to retire, Bob Meredith purchased the company, later installing Leo Mulkearns as managing director.

With negotiations for a supply arrangement aimed at a modern range of tillage equipment for Ford dealers to sell under the Blueline brand underway, Noel foresaw the mutual interests of Ford Tractor Operations Australia and Alfarm continuing to grow.

During one of his many visits to Alfarm, Leo showed Noel sketches for a new Airseeder. It would be a complete departure from the machines they were selling under their own name, which were identical to those they had been supplying for New Holland.

Patents had been submitted and while the machines would follow the same Albert Fuss patent for the distribution system, these machines were completely modern. Alfarm had an association with Australian Training Aids, a secretive company with arrangements for developing bespoke automated target systems for military and police forces across the world. It had been their hi-tech computing capability that first attracted Alarm's engineers. If ATA designed a unique system to make these seeing machines stand apart from very other machine on the market. By employing a computerised metering system Alfarm would seize the initiative and dominate the market.

The fan that provided an airflow to carry seed and fertilizer to the secondary distribution heads, would use a variable speed hydraulic motor to maintain a high volume of air at a low pressure. This low-pressure airflow would be less likely to damage the large variety of grains which could be planted. It differed too in the way the seed and fertilizer distribution would be managed. Separate high-torque hydraulic motors would drive the metering system shafts over an infinite speed range. A sensor on one of the seed-cart wheels sent an electronic signal to a console mounted computing system in the cabin of the tractor. This would allow for the operator to increase or decrease seeding and fertilizer rates on the move if necessary. The engineering would be years ahead of its time, taking until the end of the century and the emergence of modern computing, for it to be repeated.

Knowing it would be a challenge to release these machines at the same time as the 8401 the team worked on the assumption of it being possible. As he left Melbourne the artwork for the brochures had been signed off. Now it was in the hands of marketing to ensure enough pamphlets and pre-release material would be on hand.

Before leaving for the Las Vegas tractor launch, his Marketing Manager Roy Pinney and National Sales Manager David Hosking had agreed they should ask the sales team to bring potential buyers to Melbourne. This would put these people in a position to see touch and feel the new tractor, creating a groundswell of positivity and from which, orders should come.

Ken Fyffe, Sales Manager for Southern Region, John Henchy in Western Australia and Ian Blow in Northern Region had primed their zone managers to select dealers who they thought could identify candidates as potential buyers. This group of zone managers recommend the best dealer salespeople who would then invite only their most positive candidates as buyers. Their brief was to headhunt the people, who if invited to the Campbellfield factory for an exclusive viewing, would order a tractor on the day, or within the following couple of weeks.

One of these dealers taking up the offer was Curtis's of Tumby Bay. Richard had by now taken over as dealer principal after buying the business from his father Laurie Curtis in the seventies. Noel had worked with both Laurie and Richard during his time as State Sales Manager for South Australia and at the height of Ford 5000 Select-o-Speed sales.

Richard, having served with Ford Tractors as a tractor demonstrator for the launch of the 6Y range in 1968 and again for the launch of the 7000, proved the ideal candidate. Richard could sell, and he convinced eight Eyre Peninsula grain growers to join him on an exclusive flight to Melbourne where they would see the new tractor weeks before it was to be released to the farming public.

The twin engine aircraft landed at Essendon airport after flying direct from Port Lincoln. From there, Richard and his prospective buyers were ferried to the Sydney Road office to be met by Ken Fyffe and a few members of the sales team. After a round of introductions and presentation about the tractor's inception, they were escorted to a cordoned off area behind the factory where their hosts had arranged a row of 8401s ready for inspection.

By the time Richard and his group had returned to Port Lincoln, he had shaken hands securing an order for an 8401 with six individual farmers. The early order program had been proving so successful that the sales and marketing teams were reporting the initial production runs may be sold out by the time the dealers would gather for the March 28th 80s Action launch.

Confident 1980 would set the decade on an upward sales curve, Noel made a mental list of why he should be so bullish when reporting how the tractor business would reward Ford in the coming year. They had cemented their relationship with Robot Industries and dealers were selling the South Australian company's product in record numbers. Napier Bros still produced Blueline products which continued to sell steadily. The arrangement with A F Gason for the supply of Blueline cabins kept growing and with the 8401, he believed this close affiliation would continue.

On the service front, Noel had agreed Service Manager, Ian O'Rourke could produce a ninety-page service schedule to be known as the RPM Manual. Taking his lead from similar servicing plans that accompanied the vehicle company's products, Ian believed this book would live on well beyond a tractor's warranty period.

It had been designed to help an owner eliminate avoidable minor breakdowns, which during the process would grow Ford an enviable reputation for building reliable and easy to maintain tractors. If each item on the scheduled page was followed as printed, then the regular servicing would find frayed fanbelts, slow coolant leaks and ensure oils and filters were replaced after the correct period, adding to the sale of consumable spare parts.

Embedded on the Installation Service Report (page 19) were a series of data fields that once completed and returned to Campbellfield, identified the tractor to its first owner. When this information had been uploaded into the Head Office computing system, it had the ability to identify warranty fraud and vehicle theft.

The introduction of the RPM manual ushered in a totally computerised Warranty Claim processing system. It replaced the arduous and manual task of approving and accounting service claims. With only an occasional referral from the computer back to Tractor Service for correction, or return to the Dealer for further information. All claims were processed on the day received, thereby reducing payment delays to the Dealer.

Another advantage the service and planning departments came to experience was that the computer produced a monthly printout identifying trends, or failures needing investigating by Service personnel. Designed as a complete system in a book format that should stay with the tractor during its lifetime, the Regular Planned Maintenance Manual would become envied by other Ford Tractor Operations across the globe.

While Noel hoped the products he would see in Las Vegas, would be everything the early release information had promised, his mind continued sifting through the hundreds of images his memory held. Searching for anything related to the articulated White and John Deere tractors he had seen during visits to field days over the past seven years.

CHAPTER FIFTY-ONE

Ready for the flight back to Australia and after having seen the new six-cylinder tractor range, Noel started reminiscing. Remembering just how far he had come from that early start he had enjoyed as an apprentice motor mechanic in Albury. As far back as he could remember, there had been *"grandfathers,"* guiding him throughout his career. Apart from blood relatives like Walter Quonoey, these were people like George Reid, the man who had taught him to fly and then by encouraging to dive down and fly the Tiger Moth under the Bethanga Bridge, proved to him that rewards would come from well-judged risk.

Then there was Mister R.J. (Bob) Brown, a man patient and fatherly enough to speak to him as an equal and always with time to fill his youthful curiosity. Teaching Noel practical trade skills that would follow him through every aspect of life, Bob Brown was the first of many such grandfathers.

However, it had been Ron Wilde, Charlie Milthorpe's brother-in-law who, while Noel had still been working as a junior ambulance officer, explained the benefits of cultivating mentors, or as he suggested, grandfathers. These he explained, were people within the senior management of an organization or company, who just as a grandfather would, take a young man or woman under their wing and guide their career. Some of that almost parental nurturing during his first years at Ford, had played a big part in helping Noel to reach General Manager status.

Now as he considered where Ford Tractor Operations would be in the next five years, he could hear, Harry Watson, Clarence Ragan, and Gordon Guthrie encouraging him, calling on him to have faith, urging him to take the company forward.

He could almost hear Paul Gillis' slow and considered Texan drawl saying, "So, what does it matter if you ruffle a few feathers? A strong profit will always preen many a doubting rooster."

Listening to the subtext that came from the different speakers during the many rounds of meetings he had attended during this trip, Noel had learned that this series of TW tractors would enjoy a four-year production run. After that, the same engines and transmissions would receive minor updates, benefitting each tractor with a power increase. A TW5 would replace the TW10, the TW20 would change to a TW25 and the TW30 would become a TW35. While the power increases were minimal, the company was planning to offer front wheel assisted tractors with radial tires across the range, aimed at boosting each tractor's productivity.

Knowing his team in Australia had learned many valuable lessons while developing the 8401, if he couldn't get supply of a high horsepower tractor range from North America, there was always an alternative way. All week Noel had put every spare moment directing his mind into exploring ways Tractor Operations in North America or Europe could fill the holes in the Australian model range. Hoping that whatever it was, would be ready long before the five series tractors arrived. Missed opportunity was causing the dealer body to look elsewhere for product and if the people based in Troy remained indifferent, he knew how Tractor Operations Australia could remedy the situation.

This trip had been exciting, the new tractors were an update of the existing range and now by offering the TW30 the company had supplied many of the features Australian farmers had been asking for. Although, as with their predecessors, these tractors came with three-point-linkage and would need a drawbar redesign to make them suitable for Australian conditions. Ford now had a serious competitor in the up to 200-horsepower market. The cabin had remained mostly unchanged, the immediate benefit being existing parts inventories would be adequate; the downside would still prove to be the air-conditioning filter. The item had not changed, something the sales and service teams had been calling for. The Ford Parts and Accessories' after-market, external filter kits did little to halt the complaint level.

Fielding questions about the 8401 from his colleagues and superiors alike had been tiring and yet, all but a known few had been supportive. He had even detected an element of envy in their voices when they asked about it and explaining how keen they were to get their hands on something before the still in development 7810 would be available. Noel's only response was to smile and say he had the tractor they were looking for, they only had to raise an order and he could supply them.

Flying from one continent to another offered long periods of waiting in airport lounges, some executives spent their time at the bar and then there were others like Noel, who finding a quiet corner would choose to work. Waiting for his flight to be called for the journey home, he opened his diary and finding the page where he had written down the names of his contacts within the Australian heavy tractor manufacturers studied the names and addresses, there were only three who he might consider to partner in a project.

Taking a lead from an article he had found in a farming magazine, on a new sheet of paper, he drew a grid and named each of the three manufacturers across the top and in the first column listed his desired specifications and features. For the time being he would keep the idea to himself, only carrying out preliminary research before making official contact with anyone. As part of his role with the Tractor and Machinery Association, he had met the men behind these companies and held all in high regard.

However, it had been James Nagorcka whose approach to business had impressed him most. James had been open and possessed a farsighted vision of how he wanted his company, Waltanna to develop. Noel thought him articulate, possessing great engineering skills and a man who had built his business on a sound financial footing. All were strengths required of a company doing business with Ford Tractors Australia.

Working across the gridlines of his table, there were many items he left blank along with others he had written in pencil, things he would need to verify back in Melbourne. Even if the exercise didn't come to much in the airport lounge, it would settle his mind during the flight home and give him a strong reference to begin with when back at work.

Of the three recognised heavy articulated tractor builders in Australia in 1979, two of them were in wheat belt of Western Australia. Not that visiting them would be a problem, however it had been their distance from Perth and Melbourne that posed a problem. Confirming Noel's thinking that Waltanna should be the first company he approached when the time came, was later justified when at the Melbourne Air Show, James Nagorcka took out the most coveted First Prize for a homebuilt aircraft. The quality of build and attention to detail of his scale replica of a WWII P51 Mustang setting it apart from other entrants. However, it had been the scale and depth of James' engineering ability along with his combination of skill and passion that endorsed him to Noel.

Reinforcing his mindset, had been a chance conversation with the general manager of Ansett Motors tractor and machinery operations. The Western District of Victoria had become a good hunting ground for both Ansett Motors and the Ford Motor Company for over half a Century and as graziers were turning to farming, the larger land owners saw high horsepower tractors and wide machinery as an answer. It concerned him that if Ford were unable to supply tractors over the 250-horsepower market then they would choose to look elsewhere, one of those options being the home grown Waltanna.

For now, all he could do would be to wait for answers from his superiors regarding supply of the Steiger built FW series, only when he received confirmation it would be declined, could he begin to move forward. Although, he had a confidant inside Ford Headquarters and would be able to discuss the issue with to get a definitive answer, respect of the proper channels would always be his first approach.

In 1979 Australia's cereal farming industry had moved from being a follower of overseas trends, to one that embraced technology and as chemical weed controls improved, so the machinery adapted. John Deere and others were leading the with articulated tractors of over 200 horsepower and in the light sandy soils of West Australia, machinery working widths had increased to monstrous proportions. Small engineering and welding shops in these areas took up the challenge. The tractor builders among them had found by adapting a modular approach and treating the machine as an assembly of readily available components they could engineer a practical, powerful and workable product to manage their own, or customers' requirements.

Laurie Phillips of Merredin in Western Australia, had been meeting this need for around five years before Noel had even started thinking about how he could source a supply of this type of machine. Laurie's company Acremaster, had steadily been building a loyal customer base. His tractors were as powerful as anything coming from the United States of America and while basic, the ergonomics of the cabins could be tailored to the client's specification during manufacture.

Tillage and seeding machinery had grown in width and many broadacre farmers were utilizing trailed hitch units to multiply the number of seed drills hauled. Drills like John Shearer's 60 Row Wide-Seeder which on their own could sow a 36-foot strip and when coupled to a tandem hitch, one tractor could double that. However, the seed drill with its close tine arrangement easily blocked with trash and weeds making it inefficient as farming practices changed. The trend of using a tandem hitch had proved that large areas could be worked and by applying higher horsepower, hauling wider machinery, costs were being driven down.

The introduction of spray applied herbicides now began driving the speed of change and demanding a different approach to agriculture. Air-seeder design was beginning its transformation, no longer would the seed and fertilizer delivery system only be mounted to a secondary cultivation tool.

The Australian wheat farmers' thinking had changed, with products like Sprayseed from ICI, chemical companies were beginning to demonstrate how moisture could be stored in the ground by leaving the dry stubble and dead weeds to shade and protect the topsoil. One demonstration would be to lay out an old grain sack in the paddock and remove it after a month or two. Lifting it, a farmer would see that there would be moisture and life under the bag, the demonstration worked. It proved that if he could spray out the weeds and flatten the residue by introducing sheep or cattle to level the organic material, moisture could be conserved and waiting for the next crop.

Machinery however, was not adapting as fast as the moisture conservation message and manufacturers had to catch up. Priority now focused on getting the trash through the tines, for generations the seed had been sown across the centreline of the drill's wheels. Now to get trash to flow through the seeding tines, the layout had grown from three, to four, and up to five rows on heavier tined scarifier style machinery.

Wheat growers across Australia were taking notice of the methods used by their West Australian counterparts and the success they were finding by employing a minimum till mindset. Where once ground may have been left fallow and therefore subjected to wind and rain erosion compounded by continual passing over the field with cultivation machinery, weed sprays like Sprayseed by ICI, demonstrated that they could minimise the number of times they cultivated. The cost savings were plenty and, improved crop returns gathered many converts to the cause.

Noel had learned from his colleagues that the company's research teams had received notice of a glyphosate-based chemical named Round-up would soon be released by Monsanto to combat ICI's Sprayseed. Trials had shown that this product all but eliminated mechanical weed control and at its worst, reduced repetitive cultivation.

Knowing this advancement was gaining ground, proved the predictions of Gordon Guthrie and his colleagues. It also made sense of the reason why anyone would wish to start a tractor manufacturing business at Merredin in Western Australia. However, that was where demand was.

The majority of equal wheel 4WD tractors being sold in Australia were International Harvester, White and John Deere models in the 150-200 horsepower market segment. as His private research told him these could be broken down to being 2WD tractors with an extra rear axle housing grafted to an articulated steering system. Although it looked a simple engineering exercise and gave him a lot to think about, the lessons learned with the 8401 proved an Australian FW tractor might just be possible. While it gave him a lot to consider, the idea lit a flame inside him and he knew until he explored every possibility, it would not let him rest.

For Noel it was almost the end of the year and time to turn his attention to holidays and time with the family. Before the new year began, new tractors would begin arriving on the docks and the marketing team should have their dealer meeting plans almost squared away. It may be some time before he could assure the dealers of an articulated high horsepower range, but a promise he vowed to keep.

CHAPTER FIFTY-TWO

At home Noel relaxed into the warmth of the family, Summer is always hot in Australia, even in the southern state of Victoria. However, sitting alongside Joan on the veranda with a cold beer in one hand and having her hand in his, life felt good. A breeze rustled through the grapevine while they sat, silence enfolding them. Life had found them this way for some years now, quiet almost knowing what the other was thinking and without need of speech. He wondered how long their peace would last. With four kids in the house, all full of spirit and bursting with energy, he suspected not long enough.

Letting her free hand caress her belly she squeezed his hand, in June their fifth child would come into the world and as she had with the others, Joan could feel herself glowing with excitement. As a girl she had wanted a family and had vowed to give her children equal status in it, determined never to allow herself or her children to be subjected to a misogynist regime like the one she had been born into. She had chosen well, Noel came from a family where every child had been offered the same opportunity. From the earliest days of their relationship, she realised he shared her views and they soon become an easy match. Now with the farm running on routine and the children settled in their schools, she hoped to relax into the last six months of her time.

To her she had never seen Noel so happy and although it would be rare that he brought his troubles home, she could always sense when things were not right at work. This last trip to Troy and then on to Las Vegas brought him home in a great mood, she could hear him whistling or singing when he thought no one was listening, and that had always lifted her spirits. Happy man, happy wife.

Joan had plenty to deal with on the farm and taking care of a home full of growing children took up almost all her time. However, sometimes between chores, she had begun thinking about their future, or more to the point: her future. At first, she dismissed the idea. However, the more she tried to push it into the background, the more appealing it become. Eventually deciding when their finances were on a more solid footing, she would like to further her education. Maybe taking it as far as gaining a diploma or degree. Knowing there would only be encouragement from Noel when the time came, for now she would keep it as her little secret. A dream for later.

Noel drained the last sip of his beer, stood up and rolling his shoulders stretched their muscles. Somewhere a kookaburra called over the chatter of the wattlebirds, while magpies and honeyeaters joined the chorus. The cattle, sheep and geese, were hardly moving and looked content just to be lazing on the pasture. On the mirror-flat surface of the house-paddock dam, cotton-ball clouds reflected in puffs of white against the changing colours of a flaming sunset.

Highfield Park had always been a sanctuary, a place to shield him from the pace of an industrial world he would inhabit for at least forty-eight weeks of the year. Just standing there, Noel underscored the importance of having a place to retire to at the end of his annual schedule. He could feel his batteries being boosted, recharging, building strength for what he imagined would be a busy time to come.

Their solitude didn't last, the children were becoming restless, sticking their heads through doors or windows and asking what was for dinner. Joan had made sure all of them could cook and clean well enough to look after themselves, but even so, it fell to her to provide. Saying he would light the barbeque; Noel offered his hand to help her out of the lazy chair.

A life of organised chaos would be Joan's description when asked about the family. However, it was a life of their design and she and Noel had always been more than happy with the way it had been panning out.

CHAPTER FIFTY-THREE

Having seen the tractor presentation in Las Vegas and thinking about how his team would be planning to achieve a similar impact with the Australian Ford TW launch. What he had seen coming from the many planning meetings the team had conducted prior to him leaving for the USA, had his approval.

While on holidays, and in a quiet moment, Noel had often found himself thinking about the ease with which the company's different teams went about their work. These thoughts may have only come at random moments, however they always left him smiling. Remembering how he only had to set an objective and give competent people the space to complete it without him. Nobody needed him hovering in the background micromanaging their every move. Sure, there were management meetings designed to follow everyone's progress, but these were always driven by reports and questions from the floor. Protocol demanded a review of current actions before a new set of tasks would be added to the rolling action plan, leaving everyone to their act on their own initiative until the next meeting.

The best way to lead a team of talented people had never been drilled into him at a management school. Although many different leaders of business and their different styles had been discussed during his Ford Marketing Institute and Harvard training, no one way had ever been pointed out as being superior. The more the thought kept rolling around in his mind, the more he recognised his methods had come from observing and following the example of others. He had been drawn to people like Harry Watson and Brian Inglis, these were passionate men with the company's interest at heart and yet they never led with anger or fear. Each one of these men had offered him an open-door policy, had an ability to adjudicate in a decisive manner, never allowed anger to influence their judgement and held a cool head during a meeting.

These men had always been encouraging, filling their visions with passion and enthusiasm, never directing open anger at an individual, always reserving those matters to be resolved in private.

Until now, Noel hadn't thought too much about his own leadership style. However, he had always been guided by his experience of the past and taking a lesson from the arrogance shown to him by the elitist management at Massy Ferguson when he joined them, vowed never to treat people the same way.

The marketing team had completed many of their goals for the 80s Action dealer release before the company had broken-up for the Christmas Holiday period. Tractors had been ordered and were due to arrive well before the event. John Henchy in his role as Acting Merchandising Manager had ordered the agreed marketing material and his enthusiasm for the role showed in the correspondence being forwarded to dealers.

A few weeks before the end of year break-up, regional and zone managers had been briefed on the importance of the program. Their efforts in calling dealers to follow up their RSVPs, had started building excitement among the retail sales force and as the replies were beginning to flood in, it showed.

Formula One champion, Jack Brabham had agreed to become an ambassador for the TW series and being a farmer himself, ordered two TW30s for his farm near Mount Galore in New South Wales. Without ever knowing it, from this early meeting, Noel and Jack would begin a lasting friendship surviving that survived outside the boundaries of business. Noel had felt a mutual trust from the earliest of discussions and although Jack's ability to trade on his fame had rewarded him well, when the business had been concluded the friendship resumed.

Noel's idea of using Jack to spruik the benefits of the new tractor range was almost a bridge too far in the mind of the team, how would they go about approaching a Formula One champion. Noel thought their concerns unnecessary, as he knew through his many management meetings at Ford Australia, Jack had once been the major shareholder in an automotive dealership, Jack Brabham Ford at Bankstown close to central Sydney. Failing that, there could always be an approach made through the company's advertising agency.

Offering his support, Brian Inglis said he knew Jack and being a spokesman for the TW series should be a good fit with his farming ambitions. Giving Noel a letter of introduction, he went on to say that he thought using the Formula One wizard would be an advertising masterstroke. However, he reminded Noel it would be up to him to fit any expenses surrounding such an ask into his advertising budget. Ford Australia were helpful, not generous.

It took only a couple of phone calls to find Jack at home on the farm and once an agreement had been reached, their business concluded. However, the call changed and they explored each other's interests in a *"getting to know you"* manner. From that first moment, it was always business first and then onto the pursuits they each enjoyed. Over the next few years business became secondary to their friendship where both men explored ideas and entertained each other with tales of flying, farming and the mechanical world.

The Brabham family were generous with the time they were giving to the marketing team. David Hosking reporting back, that Betty Brabham had spoiled his young daughters with chocolate cake while the crew were off filming television commercials. Jack joking to the girls, that chocolate cake was very much a rarity in the Brabham house, when it was only him and Betty at home.

As plans for the 80s Action release were now well underway, Noel received confirmation that Ford's eastern region manager for tractor operations, Bert Pinard would accept his invitation to attend. Pressure had been building on the team to make sure the press release material would be ready to reach their targeted media outlets in time. A special invitation only press presentation would be run a few days before the dealer presentation. Every invited rural reporter showed their eagerness to attend, by the early return of their RSVP and their agreement to the terms of an embargo.

On Friday the 28th of March 1980, aircraft carrying dealer principals, their wives and their sales staff started landing at Tullamarine. These dealer personnel soon found themselves being ushered onto busses to take them to their city hotels. Before evening, the busses would again line up in front of the hotels and load with their guests before taking them to the Ford Research Centre at Broadmeadows. The mood in the busses felt light as everyone had now changed into evening wear and being filled with eager anticipation, were waiting for the night's activities to commence.

The highpoint for the dealers and the company staff had been the reveal of the new models featuring the 8401. Every speech, every feature, every speaker receiving healthy applause had made the night remarkable. However, for Noel one of his proudest moments had nothing to do with tractors, but it did come during the speeches. Official guest speaker, Ford Tractor Operations Regional Manager Bert Pinard spoke about the worldwide sales, Ford's marketing position and their forward predictions and ambition. When he had concluded the official part of his duties, Bert said he wanted to talk about Noel and his part in a ruckus during a flight from San Paulo to Rio in Brazil. To oohs, ahs and encouragement from the assembly, Bert entertained the dealer group with his story.

NOEL   An Authorised Memoir

Noel had been flying to Rio from Peru and on to Sau Paulo for negotiating a secure supply of an eighty-horsepower straddle mount tractor. He had found what would meet the specification in a Brazilian version of the much-loved Ford 5000, the 6600. This tractor would be a perfect fit between the platform mount 6700 and straddle mount 4600, making it a great platform to fit a front-end-loader for farmers who were looking for a second tractor. Other than a second tractor, it would have many applications in horticulture, also for cane growers and dairying.

Noel and his fellow passengers were about an hour into the flight when the stewards on the plane began calling for a doctor. A woman had started into labour and the urgency coming from the crew's calls began having an air of panic about them. Since his days in the ambulance, Noel hadn't come close to delivering anything more than a parcel and he knew his skills were rusty, but it seemed that he was the only person onboard with any experience in medicine.

He volunteered and, after giving the crew a list of needs he would require to help the mother and her baby. Soon he began to calm the mother, settling her on a mountain of cushions in the aisle of the aircraft. Before they landed in Rio, another little passenger was ready to disembark. The mother promising everyone onboard she would call her baby Nola, in respect of an Australian man who had come to her aid in the skies between Peru and Brazil.

While it was a memory for Noel to treasure, Bert's telling made sure the press would be enthralled by the event. Tonight, was the second time Noel had heard the speech and again none of Bert's enthusiasm for its telling had been lost.

By the end of the evening, even the security guards had an air of excitement about them and that bode well for the following two days. On the Saturday, dealers would travel on the same coaches to Ian Griffith's property at Nagambie where they would split into groups and begin a round-robin of presentations by different company personnel.

To cater for this many people to have lunch at the same time, the catering arm of Ansett Airlines had provided a, *Lunch in a Box*, like those served as an inflight meal. For dealers and their sales staff, having a mid-day meal would not be as important as getting back to the presentations that would continue from the morning's touch and feel, six position sales tours. Every dealer and salesman desperate to operate each tractor in a work situation.

While Noel had no role in the running of the event on the day, he did have Bert and other dignitaries to attend to, speeches to make and questions to answer. He had learned from the people who had offered him an open-door policy, that this was one of those events where he should offer the same to others.

He maintained the belief that if you had worked your way into the upper echelons of business you should hold the door ajar, making the way available for others to follow. Only on days like these it would be dealers, their staff, or the press he had to make time for and mostly people were supportive and pleasant.

At 4.30pm the busses would load again and take the dealers back to their hotels where they would have a free night to let loose in Melbourne. Noel and Bert would drive home to Highfield Park where Joan would welcome their guest with a home cooked meal. When tired by the men's tractor stories she would retire, leaving them to their discussions.

While dealers enjoyed a day of tractors dust and flies, their partners and wives having taken advantage of a company tour to the Dandenongs and lunch at Dorsett Gardens, returning to their hotels just before the men arrived. Company practice had always been to follow this schedule and a personally addressed information pack would be waiting in each room before the busses returned.

Inside would be pricing, brochures, suggested parts and accessory lists. Included would be a list of the capitalised tractors and machinery from the launch itself. These machines of varying specification and offered at discounted pricing, made them attractive for dealers to order for stock.

The following morning, dealers would rush order forms to their respective zone managers ensuring discounts and bonus offers be secured. Field training would continue until 11.30am, busses were boarded. Everyone meeting again at Mitchelton Winery where their wives would be joining them for a barbeque luncheon.

Speeches had, Noel imploring the assembly to take advantage of the bumper harvest their wheat farmers had just brought in. He gave a summary of the financial forecast that he had been a party to, in his work in Advisory to Treasury. One of the biggest drivers of his optimism had been the sale of 3.2 million metric tonnes of wheat to Russia. It meant the wheat stockpile had reduced, and Australian farmers had never been in a better position to replace their machinery.

For dealers in cane growing areas, knowing the Sugar Industry had recently completed its conversion to 100% mechanization, he believed this also opened an opportunity for dealers to sell the benefits of upgrading their tractors to their canegrowers. 1980 was in every way, looking like a wonderful year for Ford.

Two weeks after the tractor release, Noel was at his desk in his Campbellfield office and following up on reports when Wendy Orchard, his secretary, knocked on his door. Handing Noel the morning mail, she drew his attention to a letter resting on the top of the pile. It was from Bob Nassau the General Operations Manager, of Affiliates and Direct Markets and it read:

R. H. Nassau
General Operations Manager
Affiliates and Direct Markets
Ford Tractor Operations
Ford Motor Company

2500 East Maple Road
Troy, Michigan 48084 U.S.A.
Cable Address: Ford TO TRMI
Telephone: 313/643-2451
Telex Number: 23-6772

April 17, 1980

Mr. R. N. Howard
Ford Motor Company of Australia, Ltd.
Tractor & Equipment Division
Private Bag 14
Campbellfield, Victoria 3061
Australia

Dear Noel,

Just a short note to congratulate you on the outstanding job you did with the National Dealer Meeting and the launch of your new products. Bert has given me a "blow-by-blow" description of the National Dealer Meeting and has told everyone here that it was the finest one he has ever seen, including the Las Vegas meetings that you attended.

You and all of our Tractor people in Australia should be very proud of all the actions we have accomplished and brought together in a dramatically stronger sales and marketing effort to capture a big share of the Australian market.

The first quarter results are great! I can't wait to see what the results are going to be like when you have all the product you have just introduced in full availability. Keep up the momentum and hope to see you in the next few months.

Sincerely,

Bob Nassau

Terry L Probert

TRANSCRIPT:

Dear Noel,

Just a short note to congratulate you on the outstanding job you did with the National Dealer Meeting and the launch of your new products. Bert has given me a blow-by-blow description of your National Dealer Meeting and has told everyone here that it was the finest he has ever seen, including the Las Vegas meetings that you attended.

You and all of our tractor people in Australia should be very proud of all the actions we have accomplished and brought together in a dramatically stronger sales and marketing effort to capture a big share of the Australian market.

The first quarter results are great! I can't wait to see what the results are going to be like when you have all the product you have just introduced in full availability. Keep up the momentum and I hope to see you in a few months.

Sincerely,

Bob Nassau

Noel's gamble on building a tractor in Australia had paid off and it would continue to bring results until the model run had completed. For Noel, there had only been one moment of disappointment and that was fleeting. On his return to West Australia, John Deere had approached John Henchy with an offer to manage their sales operation and John found the opportunity to good to pass up. For Noel, he took comfort in knowing that losing John to a fierce competitor, again proved his ability to spot and train talent. Although, he would be a force for Noel to contend with, he was sure John would now have an opportunity to also nurture and grow his own people.

CHAPTER FIFTY-FOUR

The results for 1980 would see the impact of the Shibaura built 1000 Series range of compact tractors begin to grow Ford's overall market share. Not only did Australian Tractor Operations have models to combat the bounty tractors from International Harvester and Chamberlain, they had now firepower in other markets too. It still left the company without a harvest machinery line. However, Noel understood the Ford Motor Company board in Detroit still harboured an intention to buy one of several international harvest equipment manufacturers that could fill that gap, and it had been his understanding that this purchase would come within the next five years.

Enthusiasm among the people at Ford Tractor Operations on Sydney Road continued growing, and with each morning's incoming mail, SIDO (Single Item Dealer Order) numbers kept increasing and as the monthly reports from Alan Murray's Agriview came in, Ford's market share showed growth.

The tractor business in Australia was changing, no longer were statistics dominated by cereal growers. Dairy farming had also started to look for efficiencies of scale and through the adaption of advanced mechanization, milking sheds of the fifties and sixties had all but disappeared. The average size of farms had increased and as farmers children began leaving in search of a better education, their parents clung on until reaching pension age. Having to sell their small dairy properties to other farmers to fund their retirement may have been heart breaking, but for most, it was their only way to financial freedom. The dairying landscape followed an emerging trend of "*Get Bigger, or Get Out*" and it had been growing every year.

Part of it could be attributed to the Whitlam Government opening Universities to all. Tertiary education no longer became the entitlement of the wealthy, everyone who applied themselves could attend. Unlike those before who had worked while attending university as Noel had done, now a college education was possible for all. Just as his parents had seen their children able to secure a better future than their own by sending them to the city. Australia's dairy farmers knew that selling their properties at retirement age, had become a way for them to relax into their last years.

For many men and women of the land, these were hard places to leave and often a sense of failure drove their demons to find another purpose. Hard working hands need action and the contribution these people made to their communities can easily be seen in the many projects that retired farming families have built in thousands of Australia's country towns.

Drip-irrigation started replacing flooding the trees in wine and fruit growing regions, everywhere primary producers began employing similar trends. Australia's separate primary industries were developing new technologies and adopting them at an increasing pace. For struggling small acre farms that couldn't afford to adapt, they began surrendering to progress. A farm that had raised a large family after the war could no longer provide the same level of cash flow and listing it for sale became the only avenue for a family's survival.

Change started working its way into every crevice of primary production and, at the beginning of the eighth decade, within this change came an opportunity for tractor sales. Because of a consolidation of small operations, meant a change of farming methods for those who had increased their acreage. Merging of properties meant the new owners began aggressively exploring options to streamline their operation. Just as the wheat farmers began increasing their horsepower requirements, so did the dairies. Operator comfort and safety had become a priority for their business model too.

Horticultural operations were early adopters of automated harvest and planting machinery that was being imported from Europe. Some even began mimicking those machines, or in saving costs, started building designs of their own. For others they had copied and adapted machinery on show in Europe or North America and then had similar equipment built by Australian manufacturers. Precision planting equipment became wider, tool bars grew and these trends meant a tractor's linkage now had to be able to cater for these changes. Rotary hoes too, had grown and the increased weight had horticulturists demanding improved hydraulic linkage and flow capacities.

Nothing remained as it had, everywhere across Australia change had been gaining momentum through the desire to remove drudgery from farming.

Against this background of change, Kubota had become the most successful compact tractor importer in Australia. Now with the Shibaura range, Ford began clawing some of that market for themselves. Dealers had to adapt and not all Ford tractor dealers would make great compact tractor marketers. Therefore, the sales and marketing teams had to adapt too. With Noel's blessing, they decided to aggressively recruit new dealers, people with experience in, or having a passion for, making a healthy living from the compact tractor and machinery market.

While the 8401 had been bringing customers from other brands across to the Ford fold, so too did the usability of the TW series. The 401CID six-cylinder engine had built its reputation for reliability, genuine lugging horsepower and high torque. Those characteristics were bolstered by the easy availability of spare parts and the mechanical back-up of well-trained service technicians. If that hadn't proved enough, then the rising cost of diesel ensured the tractor's reputation for fuel economy had the TW series gaining much of the ground that had previously been enjoyed by their North American competitors.

In June, Noel had an even bigger reason to celebrate. At Melbourne's Women's and Children's Hospital, Joan gave birth to their fifth child. Kathryn, a strong and healthy baby, she came into the world just over ten years after her sister. However, not to go unnoticed, it seemed Kathryn had convened a touch of drama surrounding her birth.

Joan, heavily pregnant and being the beginning of winter, was sliding around in the sheep yards. Catching and cradling a mob of five hundred lambs for Noel to mark and tail -- felt her waters break. Shouting to make herself heard above bleating of lambs and return calls from the ewes, Joan shouted at him trying to get his attention.

Noel looked up and continued turning this young ram into a wether. Not quite hearing her, he cocked his head to one side and yelled above the chaos for her to repeat what she had just tried to tell him.

Getting more desperate and pointing first to her belly and then to her feet. Joan waved her hands in an exaggerated bow, giving what she hoped he would interpret as a whooshing sign.

The message received, Noel pointed to the house, finished the lamb he had been working on and hurried her past the shearing shed and toward the garage.

Joan insisted she shower and change, this child was not coming into the world with its mother covered in blood and sheep yard grime. Demanding Noel change out of his overalls, clean up and grab the case she had packed a few weeks before. Telling him to wait in the car, she would not be long.

Deputizing Geoffrey and Peter to turn out the finished lambs and their mothers. He called for Tania to watch over

Natalie telling her he would be back as soon as their own mother was settled in the hospital.

Joan rolled her eyes at the bath towel Noel had used to cover the leather passenger seat of his Ford LTD company car. The towel was one of her best. Waving her finger and admonishing him for fussing she lifted a hand at him as he tried to explain. Finding it difficult to click her seat belt into its buckle, Joan sighed with resignation as Noel took it from her hands and managed it for her. Right now, she just wanted to get to the hospital, have this baby and bring their new son, or daughter home to the farm.

Noel tried holding her hand to discreetly feel for a pulse, however she had lived with him long enough to understand and sighing again, surrendered her wrist. Her pulse normal and with contractions about to start. He turned the car out onto the highway and pushed up to the speed limit. Watching his wife about to deliver, he started running different scenarios and options through his mind. How to call for an ambulance, where the public phone boxes were and would he have enough change. He knew if it came to helping her to give birth on the side of the road he could. If he delivered a baby in an aircraft between Peru Brazil, then on the side of the road in Victoria should be a doddle. He pictured the items he kept in the car's first aid kit and ran through each one in his mind. They would be fine; everything was there and he had a plan to manage the birth if they didn't make it on time.

Arriving at The Women's and Children's Hospital, Noel and Joan found the staff already waiting for her. Geoffrey as asked, had phoned ahead to let the hospital know Joan's condition and given them the approximate time they would arrive. A porter held a wheel chair for Joan and the moment she was seated he began whisking her toward the elevators while Noel completed the last of the hospital's paperwork. Catching the next elevator car, he found Joan's private room and seeing she was settled, relaxed for a moment.

Reminding him that he still had lambs waiting to be tailed and reaching up to draw his lips to hers said goodbye, adding she or someone would call with news of the baby.

No drive had been longer, no lambs more difficult to tail than those on that Sunday afternoon. However, as the last lamb searched for its mother Natalie and Tania raced down to the sheep yards to break the news, they had a sister. Noel instructed the boys to shower and clean up for dinner, the girls were to heat the stew Joan had prepared for the occasion and after dinner they would all drive to the hospital and meet their baby sister.

By the time the family arrived back at Highfield Farm, everyone had had a long and tiring day.

Home life would be more settled for Kathryn than it had been for Natalie, as she had been born a month before the family boarded a ship bound for South Africa. This time, there was nothing as dramatic as a wild and stormy sea passage, this was a short fifty-minute trip home to Highfield Park in the back of Noel's LTD. She did however, become the first child to claim the new nursery of the Howard's Kilmore Home.

Their family complete, Noel and Joan found the farm a haven, an easy place to welcome their baby. Kathryn would spend her first years safe from the noise and bustle of suburbia, surrounded by animals and space to grow.

For the Ford Motor Company in Australia, even though there had been slight increases in petrol pricing during the second quarter of 1979, the effect of the revolution in Iran had not truly been felt in vehicle sales results. However, from his Advisory to Treasury meetings, Noel understood the ramifications had been reducing World oil supply and this would have a negative effect on on-farm fuel prices. It made him cautious when trying to predict market share, or profitability.

In Michigan fears were growing too, the company felt the tensions in Iran would escalate into a far bigger problem than those between the new regime and its relationship with the United States of America. By 1980 those same fears had escalated, and although oil production had only reduced by four percent, the price of crude oil doubled. This spike in pricing rippled through industry, only to be further impacted by the Iran-Iraq War as oil production in the Middle East fell dramatically.

Newton's third law states, that for every action, there is an equal and opposite reaction. Having been outgunned by a lack of models in the past, now it was time to explore the strengths and weaknesses of Ford Tractor's competitors. The marketing team began searching for strategies to take advantage of the changes facing Australian farmers and build sales initiatives to exploit their reputation for fuel efficiency. Tractors designed primarily to suit the European markets and because of the region's higher on-farm fuel costs, had made fuel efficiency a key priority. For Ford, fuel efficiency had been a tenet of their diesel engine design since the very first Fordsons. Now that the designers had coupled constant-mesh transmissions and modern epicyclic reduction rear axles in a unitary construction, their machines now more than ever, offered a big sales advantage of their North American counterparts.

Economy of operation would still be fought out with European manufacturers. However, the big Fords would still have an advantage through their cabins and transmission reliability. Noel and his team worked hard at getting the message out to the salesforce, highlighting the differences and making sure everyone had the information they needed to close a deal.

To encourage dealers in broadacre areas to stock a range of TW tractors, Noel and his team designed a new way for them to make an interest-free floor-plan scheme available. These terms would extend for almost a year, where Ford Tractor

Operations would pay the floor-plan interest during that term. It became another masterstroke.

If the tractors sold early, the company incurred a lower interest impost. If the tractor stayed on the floor plan for a longer term, the dealer would have to pay those accruing fees monthly. At three monthly intervals, curtailments were paid by the dealer reducing the overall amount borrowed and ensuring they had *"some skin in the game."* This arrangement was new and a means of giving dealers plenty of confidence to show the world that their dealership was in business to sell tractors. And sell tractors they did.

From those first dealer reveals, the 8401 started taking sales from the Australian built Chamberlains and now, even more tractor buyers were gravitating toward the Ford dealer. John Deere owners who had bought the early quiet cabin tractors and who were now seeking to upgrade had a real option. The TW was comfortable, had a one-piece windscreen and superior fuel economy. They knew about the Ford's fuel economy, because they talked to neighbours who had owned 8600 and 9600 tractors. Often it didn't even take a demonstration to convince them, Ford dealers had been building on their service reputation and most of the John Deere owners had a Ford 5000, or Fordson Power Major in their past that had offered them a baseline for a decision. For many it was like returning home to something they already knew and trusted.

Sales of TW series tractors in cane growing areas mimicked those of the dealers selling to cereal farmers. With three-point-linkage and high horsepower, not only could the TWs haul cane wagons, their ability to lift and power larger rotary hoes, made them an easy machine for the dealer's staff to sell. Soon, dealer sales teams across Australia and New Zealand were selling to customers loyal to Massey Ferguson, International Harvester, Case, Fiat and Deutz. It wasn't only John Deere who were losing to Ford, every distributor was losing ground.

While Ford's six-cylinder tractors were making their presence felt, the three and four-cylinder models were about to undergo a facelift. A new 10 Series with added features and fresh looks, would offer dealers a new opportunity to ensure Ford would dominate the market share in their horsepower segments too. Ford's range of industrial tractors had always enjoyed a good portion of the markets they sold into too. Although the company had divested of its Richier sourced machines and supply of an articulated heavy loader or an excavator line were no longer available, the 555 Loader/Backhoe continued to hold its share of that market segment, adding to a solid bottom line.

Overall, by the end of 1981 the fuel crisis and talk of an all-out escalation of the Iraq-Iran war had little impact on primary production, sales of new tractors had increased. Although the price of fuel had seen the car buying public turn away from the large V8 engine cars they had bought in years past, Ford Australia's automotive arm had gained market share due to customer preference for the new 6-cylinder Falcon and Fairlane models. Ford Australia were riding high on confidence across all departments, making everyone bullish about the future.

It had been almost two years since Ford's more than successful release of the TW Series, their compact range and the 8401. Therefore, around the middle of 1981 Noel called his teams together, instructing them to begin planning for a reveal of the new 10 Series range. For this extravaganza Australian dealers would be joined by their counterparts from New Zealand and distributors from across the Asia Pacific would fly in too. With excellent links to airports in Brisbane and Coolangatta, Surfers Paradise appeared to be the best place to hold the release.

During his management meetings at Ford House, Noel had started advising caution, because the word back from Treasury was predicting a need to implement higher interest rates to stem rising inflation. However, while caution was necessary there were still factories to run and people to employ, so Ford's need to sell tractors was foremost in his mind.

By early December 1981, David Hosking had received approval to go ahead with the artwork for invitations to, *"Tractors in Paradise,"* the theme chosen for the coming dealer meeting. Scheduled for the end of March, there would be much to do after the company's traditional Christmas break. With only a few days before the annual factory shutdown, glossy four-page brochures were in the mail and on the way to dealers. It was time to relax and holiday, before work on the dealer meeting began.

CHAPTER FIFTY-FIVE

In 1982 and buoyed by the promise of early opening rains across Australia's wheat and barley growing areas, farmers pushed into Ford dealerships inquiring about their latest machinery innovations. A great vintage appeared to be rewarding wine growers too and with the cane showing promise, Ford tractor dealers were flooding into Queensland for the 1982 Tractors in Paradise event.

For the first time since the mid-1960s, Ford Tractor Operations in Australia had finally broken through to achieve market leadership and, against the efforts of the bounty tractor builders, had held it for several months during 1981. Ford's dealers and the company staff had a confident cockiness about the prospects for the coming year. Everyone becoming caught up in the excitement a promise of forward sales and were looking forward to seeing the tractors revealed at an extravaganza in blue at one of Australia's premier convention centres, The Chevron Paradise Hotel.

Feeling pleased with the tractors appearance and the way everyone had mastered their roles at practice for their Saturday release tasks, Merchandising Manager, David Hosking figured everything was on track for a great weekend. He had double checked the roster of rooms; every dealer and their companions had reservations. The Ford Tractor staff who were scheduled to fly in that day had accommodation, along with outside crew contracted for the event. It was only when General Manager Noel Howard arrived at check-in did David have a problem.

After a few frantic phone calls and shuffling of reservations, a suite for the General Manager of Tractor Operations was found. While Noel thought it amusing and, although he wanted to reassure David there was nothing to worry about, he left him to ponder the situation. Noel had not wanted to be cruel, but he knew it would be a good lesson for David to learn. Once checked in and settled in his room, Noel hoped that one little hiccup would be the only upset of the weekend.

The evening began with dinner in the ballroom where the weekend's festivities would be explained, notables introduced and tractor presentations made. Introduced by the master of ceremonies, Noel welcomed the guests and dignitaries, before outlining a new program of incentives designed to drive the business forward.

FORDGING AHEAD in 82, introduced an initiative Noel had wanted to bring on-line, years before. However, resistance from a conservative finance department had prevented it, for *Tractors in Paradise* he had found another way. Late in 1981 and with sales numbers riding high, Noel had convinced the company finance department and other partners that the time was right to bring a retail "Interest Free Purchase Plan" on-line.

Now in the glitzy Chevron ballroom, he could capitalize on the dealers' enthusiasm to reinforce their acceptance for the plan. After introducing it, he waited for the applause to die to a hush. Then outlined a schedule of bonusses the company would pay to each salesperson for every tractor sold during the life of the program. The applause from the dealers in the ballroom becoming louder with every point he made. Continuing above the noise, he shared more details of the program. Going on to say, that as easy as it was to understand, he knew there may be doubters among the dealer body and invited those individuals who may still be unconvinced, to work with their zone manager who would help clarify any doubts they might have

During dinner the level of anticipation had been building, now as dessert started being served the formal part of the night was over, it was time to relax and be entertained. The room darkened and the Master of Ceremonies introduced a little-known entertainer.

Marketing Manager Roy Pinney agreed with the Chevron's management to engage Jackie Love as the support act for the release. Love, an up-and-coming variety star whose strong voice and opening act only whet the dealers and their wives' appetite for more. Captivated, many of the those gathered that night began following her career, first as she took her place among Australia's greats. Then as she toured the world, Jackie repaid her Australian followers by singing up a storm with performers like Sir Cliff Richard, John Denver and Peter Allen.

Excited to see, touch and feel the new 10 Series tractors, dealers milled around waiting to board busses to take them to the demo site. They knew from experience that Ford Australia personnel would have their sales pitches polished to a premium and couldn't wait to see what they had to show them.

What they couldn't know, had been the frantic re-establishing of the demonstration site during the previous day. When the Ford Tractors demo team arrived on the Friday morning a river separated the carpark where the dealers would assemble and the tractors all ready to work on the hill opposite. From the moment Noel and the team saw their predicament, everyone started swinging into action, moving the whole display to a dry area near the carpark.

Spilling from their busses, dealers found innovation everywhere, some of the weaknesses evident in earlier models had been addressed. Between the back window and the three-point-linkage on a 7710, the company's technicians had mounted a digital screen much like an airport arrivals and departures board. Demonstrating the Load-Monitor feature, it measured and displayed the horsepower being used by the rotary hoe in real time. Mounted to the three-point-linkage and driven by the power-take-off, the Load-Monitor would adjust the linkage depth to maintain an even load on the engine and save fuel. The screen may have been a gimmick, but more than one dealer principal or salesman had been intrigued by the digital readout and a demonstration of power being used.

Following Ford's tradition of a free Saturday night, dealers were let loose to sample the delights of Surfers Paradise. However, it came with the warning of an early start the following day. By Monday afternoon the last of the dealer staff had departed, heading home and all eager to put what had been shared into practice. With great products and a top year behind them everyone had been expecting big things for 1982

With late summer rain in early 1982, the promise of an early start across Australia's cereal growing areas saw tractor sales continue and as expected, machinery orders flowed at a similar rate. Approaching May, for the farming community the season had a good feel, almost as if another bumper crop hovered on the horizon. A promise so real, it felt as though they could reach out and touch it.

However, everything the farmers saw, everything they felt had all been an illusion. A heart breaking and soul-destroying mirage disappearing into fine red dust. Winter rains didn't come and as spring approached, hot north winds began their endless sweeping of parched red soil paddocks. The sun bleaching every remaining strand of vegetation. Australian farming was now wilting under a clear blue sky.

Spring became summer and cracks opened deeper in fertile farming ground. Station owners de-stocking their land as fast as they could find markets for starving stock. From across the Darling Downs and right through New South Wales, to the wheat belt of Victoria, the drought continued tightening its grip. West into South Australia, the big dry was taking hold. In all but irrigated areas and as river flows decreased, even those began to suffer. Through the sand belt of West Australia, the drought took its price and, in excellent cropping areas only next year's seed was reaped. Elsewhere, farmers would not even try to harvest. All they could do was ask their banks to allow them enough credit so they could hang on in readiness for sowing again in the coming year.

Right across Australia, the land cracked deep. It was as if the sun was reaching down into Earth's core and taking its fee. Where green had always painted the landscape going into July, yellow strands of perennial grasses fought to hold against the daily grit being blasted at them. This fine sand was being driven by hot north westerly winds that swept endlessly across the plain, nothing was spared. Even those irrigators who had water available, had a hard time keeping crops alive. Deep furrows were ploughed across bare ground as landowners began desperately searching for better ways to stop the constant drift sand blasting the bark from everything in its path. Resisting the consequence of drought is relentless and eventually restrictions to the irrigator's original water allocation cut in, further reducing their farm's capacity.

Machinery open to the weather, faded in dealers' yards. The dowdy paint and peeling decals giving them a saddened and sombre look. By the end Australia's financial year, June 30th 1982, dealers across the country were reporting a crisis on the land. Sales they could have relied on had not come that year. Farmers had even stopped talking about buying. Others were cancelling machinery they had placed on order and rural workshops increasingly found it hard to keep busy.

The market was falling, but Ford Tractor Operations Australia had been building sales and gaining market share. However, for Noel it would be a time of caution. A time to promise Michigan little, and deliver as much as possible. While Tractor Operations continued to fight inflation on one front and rising interest rates on another, Noel found his political view at odds with those of the incoming government.

Bob Hawke had led the Labor Party to victory in the March 5th election and at Government House in Canberra, on the 11th of March 1983, Hawke's Labor Government had been sworn in. For Noel and those Australians who didn't rate Labor, they were left wondering what was in store for those making a living from the land.

More than once Noel had heard a farmer say. "*If Labor win government, we'll put the cheque book up on the mantlepiece and leave it there until they're out of office again.*"

Anticipating a fiery first meeting with new Treasurer Paul Keating and that it might bring his role within the group to an end, Noel had steeled himself to the idea of leaving the advisory board. Believing he had to approach Keating before he wasted too much time sitting in a chair he'd soon be asked to vacate; Noel took the first opportunity to introduce himself. The Treasurer surprised him, Keating knew who he was, along with the company he worked for, and the other industries he represented.

Almost as if caught off guard, Noel had been expecting the man's persona to be frosty, and yet he found himself warming to his personality. Paul told him while they might not always agree, he had been briefed about the board and saw Noel as being integral to it. Further stating that he had always expected Noel to retain his position.

Noel had thought about how he would use the time available if his position to the Advisory Board was revoked. Being an unpaid position, and although he had enjoyed his time on the board, Noel had thought if he lost it, it would offer him more time to concentrate on Ford's tractor business. Any extra time he could have at home would be a bonus. However, being able to ensure that the interests of manufacturing and primary producers were always represented when it came to influencing policy, made his choice to stay an easy one.

Within weeks of winning government, new Prime Minister Bob Hawke announced a "National Economic Summit." This he claimed, was to be more than a talkfest. Both Hawke and Keating when talking to the press were adamant that the outcome of this summit should underpin policy reform in Australia during their first term in government. While the population watched on and the sceptics among them talked the idea down, the summit did go ahead. As Hawke predicted, representatives from industry, union and public service rubbed shoulders to have their say in an unprecedented forum.

While Hawke and Keating were designing a Prices and Income Accord to put before parliament, the divide between rural and city interests widened. Australian farmers in grain growing areas having seen their 1982 winter crops wither from the lack of follow up rain, predicted another tight year and a sense of pessimism began to take hold. Repairs to tractors and machinery ceased and it was a rare event whenever a rural cheque book opened. Not because of the Labor government, but because people living from the land were conserving cash, waiting for the new season to unfold.

Although there had been good rains in part of the country, across wide areas of station country and dry farmland, sheep were still being slaughtered and dumped into pits because it would have been in humane to have let them starve. Sheep, which would have made ten to twenty dollars per head mere months before, would now be shot as they were unable to be sold for the cost of the freight. In some cases, because money was so tight and they could save the cost of bullets, farmers opted for the gruesome task of individually cutting the throats of their stock instead. They were to become days when blood mixed with unseen tears as tough-as-nails men succumbed to stress. Then there were worse days too. Days when a farmer's wife and family stared at the cooling meal in front of his empty chair.

For a woman on the land, every day agony deepened the lines across her brow, always waiting for her stress beaten husband and relief descending when he did. In these times, her morning would begin by kissing adieu to a man who she could never be sure would come home again. In these driest of years, bush suicide is always stalking the farmyard.

Cattle prices suffered a similar fate and thousands of beasts in poor condition had to be shot, often leaving them where they stood. Native animal numbers suffered too, apart from carrion enjoying a fast bounty from the animals caught in the drought's cruel harvest.

In South Australia, the cause of the wildfires in February 1983, had been put down to the effect of the drought creating abnormally dry conditions. In another fickle weather twist, Mother Nature rubbed a little salt into the wounds of the Ash Wednesday tragedy by dumping widespread rains over most of the drought affected country from March to May.

The sowing season may have been on again, but cash wouldn't begin to flow into small business until the first wheat payments reached farmers bank accounts.

The city and country could not have been more different. Real wages had started to climb under the new government and the optimism it created drove consumer sales. This euphoria proved to be unsustainable and as Australia's Unions thought they were working with friendly government, pressed their case for better terms and conditions for their members. Sometimes they won, sometimes they elected to strike to get their way.

At the beginning of the new decade, inflation that had bubbled at an even rate during the early seventies began escalating. For Ford Australia and other manufacturers, crippling union disputes began having a serious effect on productivity. Sometimes the disruptions started with workers seeking better conditions, or because of demarcation disputes between the various unions working within the manufacturer and other organizations.

Urgency expressed by those reporting to the monthly management meetings at Ford Australia had intensified. Everyone now focusing on strategies that could provide enough product to feed the demand, while creating new formulas to monitor a less predictable cost of operation.

By the end of 1983, Hawke and Keating's, Prices and Income Accord appeared to be working. Designed to bring stability to manufacturing and try to put an end to wildcat and rolling strikes, the new government convinced Australia's union membership to forego dramatic wage increases in return for a compulsory superannuation scheme. It had been a tough ask and yet Hawke had put a strong argument to both sides, believing if Australia was to present itself as a powerhouse economy, industrial relations had to change. At the same time, they argued that Australia's closed and closely regulated banking system had to be abandoned. Another part of their financial reform would be floating the dollar on the world economy, only then would it find its true value.

In his position on the advisory board, Noel represented more than just the interests of Ford Australia, he had the Tractor and Machinery Association to consider and made sure his voice was heard. When he had agreed to taking on the role of chairman at the Tractor and Machinery Association, he took it seriously and in doing so, felt the expectation of every member not just those of the Ford Motor Company. Concerned that the dollar being bought and sold in world currency markets would see sharp swings in its value, Noel sought advice from Treasury officials to prepare Ford and the TMA members for the volatility of the first trading months.

Looking over the tractor company's end of year income and expenses report for 1982, although the year had been profitable, it was easy to see it had been tough on everyone too. If the current drought persisted revenues would fall. If that happened then he would have to lose staff and that was a decision he did not want to make. Finding another strategy to increase tractor sales had become a matter of urgency.

CHAPTER FIFTY-SIX

QANTAS Sydney to Melbourne: 25th February 1989

The overnight flight from Honolulu had relaxed him and although not ready to forgive Gor Cowl for his present situation, it had allowed Noel time to reflect on his career at Ford Tractor Operations. Time to review everything against the disappointment and anger he felt about the creation of Ford New Holland Australia and to put his not so insignificant achievements into perspective.

Overall, there had been many more successes than failures. He had rubbed shoulders with the good and great of the world during his career, something he would never have thought possible as an apprentice working with Charlie Milthorpe. Life at Ford had offered opportunity and he had made the most of it, always accepting the challenge, always willing to find new and better ways to put the company on a stronger footing. The release of the 8401 and seeing Chamberlain disappear, being the most dramatic of those highpoints, but setting the wheels in motion to create an Australian Ford 4WD tractor range had even eclipsed that.

Chance meetings with people from all walks of life and pictures of those random moments caused him to smile. The face of Prince Phillip filled the memory and he remembered the kerfuffle happening in the Singapore QANTAS lounge as seat allocations were being re-organised. Noel had no idea why until when taking his seat, when he noticed he had been chosen to share the trip with a member of British Royal Family. He couldn't remember how the conversation flowed as it had been mostly chit-chat about farming, living in Australia and life as an emissary for a power greater than themselves. During the eight-and-a-half-hour flight Noel found an easiness in his seating companion, that and the beers they shared ensuring their time in the air seemed short. He had thought about exchanging contact details and extending an invitation to Highfield Park, but not quite sure of the protocol, thought better of the idea. Besides it was unlikely the offer would be reciprocated. He remembered repressing the desire to ask the prince why his big toes were poking through a pair of well-worn socks, to Noel it seemed laughable a man of his position would be wearing clothing that would have been better off tossed into the trash.

As the aircraft took to the sky, Noel's thoughts returned to his search for product to fill the gaps in Ford Tractor Operations Australia's tractor line.

The decision by the Labor government to float the Australian Dollar had started a ripple charging through the value of the currency. Once set adrift, its value bobbed and fell to the speculation of the world's money traders, the unpredictability making it difficult for importers and exporters alike. Within three to four years, that ripple had become a tsunami throwing many Australian manufacturer's long-term economic plans into turmoil.

For Noel, when the factory started back at work in January 1983, the Australian dollar being tied to the Pound Sterling could be relied on to maintain its value. After the Economic Summit, the new government argued this policy was an impediment to world trade and thought the only way to establish the actual value of Australian currency would be to allow it to find its price in an international forum. Until December, it had been easy to establish a rate of exchange when negotiating a contract, because A$1.01 could be traded for US $1.00. By the end of November 1986, it would take another fifty-one cents Australian, to buy an American Greenback, making the cost of tractors and imported machinery much more expensive. The upside came in the form of better prices when selling wool, wheat, barley and livestock exports to international markets.

There were casualties of a deregulated banking system, another of the Hawke/Keating government's initiatives. The idea had been to bring rivalry into the banking system and make Australian banks more competitive. Until then banks had operated under the guarantee of the Federal and State Governments, a guarantee that would not be required if banks once privatised. The privatisation of the Commonwealth Bank of Australia being a priority and one which would bring much needed cash to fund the new government's social initiatives.

During the early months of the decade, a relaxing of some rules around foreign capital had allowed Australia's banks to finance large purchases using overseas capital. Capital that had been bought at a similar rate of exchange, but at a much more favourable interest rate. For primary producers wishing to expand, the product came as a Godsend and large tracts of Queensland scrub country had been purchased. The belief being that clearing it and planting more than one crop a year should more than cover the cost of purchase. It had been a great plan while the Australian currency held its value, but by November 1986 with a rise in Worldwide interest rates and the Aussie dollar only buying about sixty-four cents American, many farming businesses failed. Even with the resources of the Ford Motor Company, managing to maintain cashflow and keep a business in profit during the eighth decade had become extremely difficult. At a dealer and small manufacturer level, it was becoming near impossible to remain afloat.

However, against this background of change, business was finding a way. Noel believing it was now time to explore the difficulties involved in building two new tractors based on the

TW25 to take care of the lower end of the power range and another utilizing the TW 35. He reasoned that the same practices Frank Pitney had employed to create the 8401, could be employed in sourcing genuine Ford parts to construct the mainframe of the tractors. Parts bin specials they might be, but as his Harvard studies had proved on more than one occasion, there had never been a good reason to re-engineer the wheel.

Taking a lead from the articulated John Deere and White models, the engine power would be driven through a similar transmission arrangement to that of their two-wheel drive counterparts. Another transaxle, hollow of its gearset completed the mainframe. Noel understood that while the idea he had sketched onto a sheet of paper looked easy, putting it into practice would be a much more difficult proposition.

The idea had been swirling around in his mind for well over two years, and the more he thought about it the more it made sense. If they could find enough parts within the myriad of options of SKID units, then it had to be possible and just as he had done with the 8401, it came time to call on others to duplicate Frank Pitney's skills. The first challenge they would encounter would be setting the right specification, one thing Noel demanded they keep front of mind was affordability, the tractor had to be profitable for both the company and the dealers.

Frank had worked hard to create the 8401 and now others in his team had to take charge by combing through the thousands of different build specifications. They saw Noel's idea as another challenge to prove Aussies could build quality products capable of matching any competitor any day and a challenge to be tackled with relish. Taking dimensions from the homologation drawings of the different SKID units the company produced, they began searching parts lists for other sub-assemblies to complete the chassis and powertrain. For the parts they couldn't find, the team felt confident they could outsource their manufacture to the same companies who had been involved in the 8401.

Noel had set specific design parameters for these below two-hundred horsepower tractors had to look like a continuation of the TW line. The desired axle weights would follow a sixty/forty ratio giving the tractor a nose forward stance, adopting a similar tradition to their competitors. The same cabin design as used on the 8401 would again come from Gason and if necessary be modified to suit.

Unlike the 8401 where SKID units were prepared in England prior to shipping, these machines would be built by a third party in Australia, taking away the need for a separate assembly line in Campbellfield for managing the complexity. While the idea bubbled away Noel made it his mission to understand who in Australia had a capacity for taking thousands of parts and assembling them into an all-new Australian tractor.

There was a large amount of risk involved, but he knew from contacts inside the inner circle at the company's board level, that they were intent on building a complete agricultural machinery line and, preparing to sell that whole side of Ford Motor's business as a stand-alone establishment. The Aussie Dollar having fallen against the US Dollar and the Pound Stirling, had made the imported component's more expensive, meant there could be an export opportunity once the tractors had proven themselves. For Ford Tractor Operations in Australia, Noel believed there would be more rewards waiting for the brave, than risk for the timid.

His opportunity to meet with the creator of these Waltanna tractors came when both were asked to speak at a seminar arranged during the 1982 Elmore Field Days. Noel had been billed as the main speaker. As Chairman of the Tractor and Machinery Association as well as representing the Ford Motor Company, he based his talk on the proliferation of makes and models in the Australian tractor market, calling for regulation of the tractor and machinery industry.

Former Liberal Politician and columnist for The Bulletin magazine, Bert Kelly had taken exception to the premise of Noel's speech and questioned Noel's reasoning.

Peter, Noel's son had travelled with him to take in and enjoy the evening. Noel took little offence to Bert's scrutiny, batting away his questions with candour, answers and reasons. Peter agitated by the attacking way Bert Kelly framed his questions, began shifting in his seat ready to protect his father. All the time wondering if this *has-been* politician cared, or even had enough nous, to understand the thrust of what his father was saying.

Noel seeing Peter's uneasiness, gave a subtle hand movement, it was just the lift of his finger, but it was enough to assure him that he had the situation in hand and Bert's objections although intrusive were welcome. Besides they would have more to talk about on the way home.

Bert followed his appearance at the seminar with an article titled, "Watchdog Barking Up the Wrong Tree," in his column for The Bulletin newspaper on November 2nd 1982. Noel however, unaware of Bert or his column, had met with James Nagorcka.

Noel found James a man of intellect, his company personable, and interesting. As their discussions deepened, a common inquisitive mindset began an acquaintance that neither man realised would develop into a lasting business relationship.

James had first come to the attention of Victorian and the wider Australian farming community, when as a Hamilton farmer, he displayed a 175-horsepower articulated tractor at the Wimmera Field days. Branded a Waltanna, James had named it after the family farm and was the second tractor to carry name. Waltanna tractors came into being because James had been searching for a tractor of similar specification and finding imported machines were outside of his budget.

Working from concepts he had thought about for some time, he set about engineering his own machine, buying a new 120-horsepower engine from Caterpillar to power his idea. A second machine followed the same lines and incorporated lessons learned from building the original. For James and his wife June, to have a customer buy the tractor on the second day of the field days for the tidy sum of $33,000 was not unlike mana from Heaven. That one sale changed the way the Nagorckas believed their farming business would operate. More orders followed and within a few years, the farming tinkerer come self-taught engineer, had become a bona-fide tractor manufacturer.

For Noel, thinking about approaching James to source a supply of articulated tractors above 200-horsepower was more than just a meeting of similar minds. From that initial meeting in Elmore, he knew both were ambitious for the Australian tractor industry. Desperately wanting it to thrive. If Waltanna were able to build a new range of Ford FW tractors in Hamilton, there would be an opportunity to deliver tractors that could not be sourced from Dearborn or Basildon. Key to making this work would be convincing James and June how they could benefit from such a venture.

For Ford Tractor Operations Australia they would profit by having the new models to fill out their range and as an OEM supplier, Waltanna would reach wider sales distribution through a larger and well financed dealer network.

CHAPTER FIFTY-SEVEN

Howard Park, Tallangatta, Victoria

Returning home had always been a time of sanctuary. No matter which property they owned, Noel and Joan created a place where he could put the week-to-week work worries behind him, finding peace in the menial work that came from toiling with his hands. Being at their new farm felt like putting a full stop at the end of his trip to Honolulu and any tension he had felt over the past week, dissipated. He had much to discuss with Joan and the rest of the family. However, all he wanted now was walk the farm.

To be alone, dodging cow pats while walking through the pastures and thinking about different ways to manage the farm. Listening to their animals and hearing the breeze whispering through the trees, allowing nature to reveal itself while he prepared for his next career move. Noel had always believed that to properly understand the opportunities of tomorrow, he needed to appreciate the lessons of his past. As if in a continuous loop, days drifted past. His mind continuing to find the highs and lows of his time with Ford Tractor Operations.

He needed to settle this and the only way forward would be to open-up to Joan, but telling her about the cargo door failure on the United Airlines Jumbo might heighten her fears of flying. He knew he had to tell her and keeping it to himself was like having a secret and the longer he left it, the heavier it would be to carry.

In 1981 when sources inside Ford's Global Headquarters had confided to Noel that the company had been looking for a harvesting company to purchase a company that would provide a full product line, he thought New Holland would be a perfect fit for Australia. Disturbed in the knowledge that the acquisition was driven by Ford's intention to create a viable entity capable of trading without the Ford name, he felt assured that whatever happened, the people loyal to Ford Tractor Operations would be assured of their positions. After cogitating on the knowledge for some time, he realised his team at Campbellfield had the expertise to make a silk purse from a New Holland Australia's pig's ear.

If the purchase was to be New Holland, it offered truckloads of opportunity for those wanting to be part of the new Australian operation. Rumours about New Holland Australia haemorrhaging cash filtered through the gossip channels and, so too stories of their losses continuing to build every year. If Ford Tractors did manage to buy New Holland, the same sources inside Dearborn had led him to believe that as General

Manager of Ford Tractor Operations in Australia, installing him as Managing Director would be their most logical decision.

Having heard the many rumours about New Holland Australia's business processes being deficient, it tripped an excited sense of urgency in his mind. Although he couldn't wait to begin sorting the mess out, he understood other opportunities existed too, but nothing felt as good of a fit as New Holland. For now, the whole matter was little more than mere speculation and a confidence for his keeping.

After the drought of 1982, more speculation about New Holland's viability intensified. Financial market watchers had been keeping a close check on the Australian operation for the best part of a decade and feared for its future. Rural media and independent observers speculated that the Cranbourne company's stock control of whole-goods had become dismal. Stories of grass growing through tillage machinery, new combines and hay machinery surrounded the industry. Other sources pointed to the state of the stock in the company's Cranbourne holding yards and although Noel had verified such things, it confirmed beyond doubt the deficiency of the current management's competency.

As delicious as the prospect of a having a harvest line seemed and, even if the two companies were to become one in the future, he had more than the purchase of New Holland to busy his mind with.

By the end of 1982, Noel understood the company had to innovate and search for new and different marketing initiatives to increase tractor and machinery sales. He believed that because of the Federal Government's changing of banking laws, there would be plenty of opportunity within the turmoil, they only had to discover them. At the tractor company's next marketing meeting, he asked the team to hunt for new methods of exploiting the Nation's rising interest rates.

By the end of 1983, Treasurer Paul Keating's desire was driving fiscal policy to halt Australia's rampant inflation. Searching for an advantage among these changes to the country's economy, Noel approved extending the Interest Free Floor Plan terms.

However, the marketing team believed there was still an element missing and after distilling research from their Zone Managers, they had become convinced they had discovered a world beater. It was audacious in its simplicity and a break from Ford Australia's promotional traditions.

They recommended trialling a pilot plan for only TW tractor sales only. The company would grab sales they would otherwise miss to their competitors and, if it worked, they could extend the plan to other regions.

Dealers had long been reporting sales were being lost. Nearly every other manufacturer had been offering a range of terms and conditions outside those available through Ford Tractor Operations' normal financing policies. The 1982 drought was still with them and by mid-1993, Zone Managers were reporting it would remain that way until after harvest payments began to flow in late December to January 1984.

For most of 1983, they had been reporting that dealers were losing sales and there were some who were not even quoting because they were concerned that high value trade-ins would strangle their financial viability. To combat this and inject enthusiasm into dealership sales teams, the team proposed that at the next marketing meeting they select a few dealers to trial a new "*Interest Free, Used Tractor Floorplan Scheme*".

To encourage a dealer to fully engage in the sale of TW tractors, the balance of the current interest free term of a new tractor ordered for stock, would apply to the used machine being traded in. The team had faith that their proposal would work. The company's exposure to interest charges would be less than if the new tractor remained unsold and with quarterly curtailments further reducing the balance outstanding. Negotiating an agreement with Ford Credit would be crucial to seeing the tractor company contain its costs within the original tractor's free of interest floor plan budget.

Deciding the plan would only be available to a few South Australian dealers in broadacre areas, a trial date was set for late 1984/early 1985. It would mean more work for the dealer's finance company as they would be completing stock checks and verifying the trade-in value. Ford Credit in Adelaide agreed to handle the finance and of the dealers approached, most saw merit in the scheme and were happy to agree to trial it.

Several, *Buy Now-Pay Later*, schemes competitors had been offering had been around for many years, however, it was the *No Deposit,* purchase arrangements that Ford dealers found the most difficult to beat. In the early eighties, Vic Mallin from Ford Credit in Adelaide had found a way for Ford Credit's dealers to not only compete, but to win retail finance business on many fronts.

Vic proposed a plan that if its method could be kept confidential, would offer the dealer a head start on the competition until it became common knowledge though the farming grapevine. He understood it would take only a few months before other finance institutions caught on and found a way to bring a similar product on line. Time was of the essence.

Leasing had benefits for all parties, the customer, Ford Tractors, the dealer and Ford Credit. Rather than purchasing the tractor or machine outright, there was taxation relief for a farming community who were now seeing benefits of the reduced value of Australia's currency. By agreeing to take out a lease agreement, the investment allowance available to such purchases could be assigned to the finance company and, an adjustment made to the fee structure. After signing the terms and conditions of the usual annual payment structured lease agreement, the client would apply for a *Variation of Lease*. This variation would better reflect lease payments that could take advantage of current Australian taxation laws. With the variation accepted and if their trade-in covered the first annual payment, the client would be free to drive the new machine away having made no initial cash outlay for the transaction. It was legal and saved the client plenty.

For Ford Credit Australia dealers, the ability to vary the lease agreement opened a new way to combat early order plans by Massey Ferguson, John Deere and Case IH, without exposing Ford Tractor Operations, or Ford Credit to unnecessary risk. Only the vagaries of a Labor Government trying to halt inflation with an ever-increasing interest rate could see the scheme falter.

Leasing had never been a concept that broadacre farming communities had wanted to employ until the nineteen-eighties, due in part because of the rigid monthly repayment terms demanded by the financier. With the new changes to Australia's banking system, annual payment schemes were now fitting within the harvest cycle, making a cereal farmer's cash flow easier to manage. Resistance to Ford Credit's scheme came from traditional bankers who advised their clients, that they would never be the owners of leased equipment. However, it was a short sighted and short-lived strategy. The old argument used by bankers who had enjoyed a long-held monopoly on farm finance was changing. This was the eighties and machinery sales people hungry for market share cleaned up.

Often to make a large capital purchase like a truck or tractor, the farmer's bank would extend a current mortgage, or demand a new title be added to an existing account. For many farmers, the opportunity to lease a tractor or other machinery that could improve their efficiency without adding extra security to the mortgage, they were discovering new ways to financial freedom. That was now becoming a freedom that offered options. Should a neighbour be selling a parcel of land that they were interested in and because there was no security attached to a loan that secured the purchase of the new tractor, it became an easier proposition to secure a new bank loan when needed. No longer held back by limited mortgage security, traditional lending options for land purchases became more easily available to them.

Every dealer's arrangement for floor plan was different, but annual repayment leasing had now begun changing the way they sold machinery to farmers. Floor plan interest rates had continued to rise and had almost doubled in the ten years to 1985. Against this background, and although applying a credit for any retail finance paper a dealer prepared had always been a way of reducing the Floor Plan costs, tractor and machinery leasing added another product Ford tractor dealers to sell. By the end of 1982, finance and insurance had become an essential part of a dealership's operation and by the middle of 1986, those who had embraced it saw their business continuing to grow.

CHAPTER FIFTY-EIGHT

Standing to welcome James Nagorcka into his Campbellfield office, Noel could have never predicted how similar were the ambitions of James' for his tractor business, to those he held for Ford's Tractor Operations in Australia.

Recognizing that a tractor owner's experience with a certain brand often coloured their next purchase decision, James had been searching for a means of providing a below 200 horsepower tractor range to introduce smaller farmers to his Waltanna brand.

His views came from his own farming purchase decisions, if he'd had a good experience with a particular brand and had built a relationship with the dealer. It would be them he turned to next. He'd also witnessed the success of companies like Yamaha, a business that offered a range of small capacity motorcycles aimed at children, with gradual increases in engine size they could cater for every buyer's experience. John Deere employed a similar strategy with tough, diecast tractor models capable of surviving the children's sandpit, or looking great when displayed with pride in the brand loyal farmer's office.

Toy or model tractors may have been a thought for the future, but his knowledge felt sound and moving into a new market made sense. Farming practices were changing and being able to engineer an articulated tractor with a steering front axle would be a leap forward for large scale row-crop farmers. Controlled traffic farming had become a new buzz phrase used by agronomists at the time and having attended seminars on the subject, James could see a benefit for farmers in areas where soil compaction had become a major problem.

Seeking a meeting with Noel had been on James' mind for some time and if a deal could be reached with Ford, then building a Waltanna using either of the two larger TW series would begin another phase of his business. James, believing he could engineer a product that would better his competitors in this emerging market, took care in explaining his idea to Noel. When their meeting concluded, Noel not only understood the way James saw the products developing, but had agreed to providing Waltanna with a TW25 tractor and the axle from an industrial loader as the basis for development. A handshake sealing their deal, they left it up to each of their legal departments to formulate the contracts and until it came time for signing, they could begin thinking about the next developmental phase.

In Hamilton the Waltanna engineers split the TW25 at the clutch housing and wheeled the six-cylinder engine assembly away. Deciding to house the clutch in the rear transaxle assembly, James and his engineering team set about designing the main tractor framework. This chassis needed to provide bracketry to attach the front axle assembly, provide mountings for the drop box and a myriad of other components required to facilitate an articulated steering system.

Within a few months the Waltanna engineering team had managed to build a working prototype and testing showed it outperformed the front-wheel-assist TW35. An easier machine to drive than the original two-wheel drive tractor, it built confidence among the workforce and soon had everyone at Waltanna believing they were onto a winner. It may have seemed a simple exercise when first discussed, however, taking the lessons they learned when building the prototype and turning them into engineering drawings took time. They had a tractor that would work, now it had to be scaled for production and that may take even more time.

Confident with the planning involved in managing the conversion, Noel and James needed to set their specification for SKID units which would be shipped from Basildon to Melbourne. Again, as surrounded the development of the 8401, this project had to be cloaked in secrecy. The last thing Noel needed, was another Tom Rae incident.

Making arrangements for James and his wife June to visit the Basildon assembly plant, Noel gave him a letter of introduction to Jim Thomas the senior executive in charge of OEM sales. Consuming a mountain of time and working alongside Jim, James eventually made sense of the options available. Then it came to selecting the right component from the millions of variations available and finalizing their desired specification. However, James had been willing to study the build sheets, double checking everything, ensuring that there would be no or little need for change when the units arrived in Australia.

James and June, found Jim a wonderful host as they had with other British executives. Jim differed from some by having a love of fine dining and ensuring his clients were well catered for when in his care. Belgian Mussels had been a delicacy Jim enjoyed and he was determined that while in his company, James and June should experience them at least once on their trip to The Continent. The dining experience is one that has changed James' palate, and whenever he sees Belgian Mussels on a menu, he orders them. For him it has become a delicacy to be savoured and they also made a nice memory from a very successful trip.

The Waltanna team had used a clean sheet to draw many tractors before; however, these FWs would be the first time they had taken an existing tractor and re-imagined it into an equal-wheel, four-wheel-drive model. The task ahead would mean reverse engineering a tractor primarily designed as a front wheel assist machine. Some of the original features would be removed while others would be redesigned. Having a SKID assembled to suit their specification may have made the build easier, but they still prepared for plenty of hurdles they would encounter as the project progressed. It would only be during the first build run that they would find any challenges requiring an engineering solution to push the project on. This would be a task to grab with both hands, as the benefits of building a product unlike anything else from Australia, had a better than good potential for export sales.

Wanting to keep the image in line with that of the 8401, a Gason cabin would be modified to fit. Noel and James both agreed, that using it in lieu of the original imported unit, would separate these machines from their TW cousins, providing these tractors with a more modern look. Redesigning the Gason built cabin meant removing the wheel well and this decision gave the design team extra space in which to design the operator platform.

During the eighties, operator comfort had become a point of difference with buyers and understanding their challenge, Waltanna engineers were extremely diligent when designing the operating ergonomics. A lesson in tractor evolution, ergonomics had been an important consideration for James when designing the very first Waltanna and these new machines would be no different. He declared to his team that every control had to be in a position determined by a repetitive use priority, if the Ford system was the best keep it, if not, find an inexpensive way to modify or replace existing components. His team had been challenged to take every opportunity to make the operator more comfortable.

Les Gason had agreed that his company would re-engineer the design of the 8401 cabin and supply it exactly to the Waltanna engineer's drawings. He also agreed to manufacturing the front chassis and articulation system for the new tractors. For all three companies, this was an exciting time and one destined to create a big mark on Australia's tractor landscape.

Progress continued at pace and to ensure a look and style followed that of the 8401, Noel arranged for the Ford Design Centre stylist, Andy Jacobsen to travel to Hamilton. At Waltanna, Andy used large colour films and tapes of different shades before determining an appropriate design to show James and June. After three days, Andy's flair had added another dynamic to the tractor range, leaving only the Nagorckas and Noel to sign off on this new image for the complete FW Series.

Finding a way to circumvent Ford's strict rules around using the Ford oval or any other trademark, Noel and James knew finding a name to tie the tractors to Ford Tractor Operations Australia would have its difficulties. While setting them apart from the Waltanna brand would not be high priority, having a distinctive logo to market them through Ford tractor dealers had its complications.

From those early days of the tractors' development, the conundrum of a suitable name had plagued their thoughts. Reluctant to waste precious time seeking approval to put a Ford name on a Waltanna tractor, James and Noel thought it easier to find an alternative. At the end of many discussions, they agreed to name them FW Series. The "F" being for Ford and the "W" for Waltanna.

Andy's graphics made sense and by developing a designation like that used on the TWs to separate the power ranges, meant it kept the model message parallel. The engine cowling and general style of the tractor, along with Gason's "E" cabin, tied the look of the tractors to both the TW Series, and the 8401. However, it was when the tractors had made it through final assembly and driven into the sunshine, that the brilliance of Andy Jacobsen's colour scheme offered them something more. They were more than just a new tractor flashing with life, these smaller FWs stood out as a beacon of rural Australian ingenuity and farmers loved them.

Unmistakably a Ford sourced product, the FW25 was nothing like anything Ford had made in either Europe or the USA. Sure, these models had their roots in the same design and development as the first Ford 8000. However, this tractor had been imagined in Australia, built in Hamilton, Victoria and would in years to come, be desired by every tractor collector who came across them.

A bright future beckoned.

While in Hamilton, Noel and James had challenged Andy to continue the same colour scheme through to the high horsepower Waltannas. His efforts had managed to create a look that Noel had been aiming for.

Keeping the script for the FW Series insignia a similar yellow to the colour James had used for his Waltanna logos and numeration, turned out to be a masterstroke. It took only a glance for loyal Waltanna owners to know who built these tractors and it soon became a magnet, drawing buyers to the FW Series brand.

For Noel and James, knowing some farmers would see the Waltanna built tractors as being a completely different product, they also believed that by employing the FW designation, it may help to attract buyers making a logical link to the tractor's North American cousins. Confident that Ford dealers understood their customers had a good understanding of North American farm machinery, it offered a possibility to trade on the success Ford FWs had enjoyed in European and North American Markets.

Now Noel had a full line of articulated tractors to release to the rural press and the marketing team concluded that the ideal venue would be in front of the Exhibition Building in the Carlton Gardens of Greater Melbourne. Being the scene of Australia's first Federal Parliament proved the perfect backdrop and most appropriate for a venture of such promise. The FW Series now involved six models with ratings from 160hp to 400hp and being developed in Australia added to their international mystique.

Dowerin Field Days had been earmarked for the official public release and after an excellent reception from Rural and National Press in Melbourne's Carlton Gardens, Noel was keen to see the public's reaction for himself. Dick Kiefer the Regional Manager for Asia Pacific would arrive in time to attend and participate in the FW Series release program.

Dick's reaction on seeing the tractors boosted Noel's confidence and listening to the enthusiasm in his voice as he talked about their sales potential, had both Noel and James brimming with pride.

Dick had always been a tractor man loyal to the Ford Motor Company and as he had been visiting Australia for some years now, Noel again looked forward to his visit.

Believing, should journalists take an aggressive approach when interrogating any of the sales team at the media release, it would be best if James took Dick through each part of the development cycle. James began by describing the need of both his business and that of Ford Tractor Operations Australia. How finding a product to compete in this market segment had been a challenge and how Noel and he both wanted to find an efficient way to produce something other than a *"Me-Too"* product. From the outset, they declared that whatever they designed, it had to be uniquely Australian.

The rural press sounded enthusiastic and the conference presentation followed a six position sell format. The team giving reasons why these tractors were important to Victorian Manufacturing, Australian farmers and rural employment too. Using Ford's well developed presentation technique to explain the features and benefits of each tractor model, the marketing team offered an overview of the tractor's original design parameters.

James' followed by describing the process of bringing their idea to life. To offer an engineering insight into the difficulties they encountered, he explained the challenges and the myriad of solutions his engineering team found. Commissioning AF

Gason to build the front chassis and articulation system, meant providing complete sets of drawings for every part, including those the Gason parts attached to. Every part required its own production number and a parts catalogue produced.

Early in the design phase, Waltanna engineers decided on moving the engine forward of the TW tractor's transaxle. This would be a departure from a design used by some of their competitors, but it did offer a simple way to take advantage of the engineering already available, providing for three-point-linkage and power-take-off. An added bonus would be eliminating some of the noise, vibration and harshness that came from mounting the cabin directly over the transmission as in the donor tractor's design.

The cabin, first designed as part of the 8401 had now been modified by Waltanna's engineers to accept the instrument console and steering equipment from Ford's industrial tractor parts bin. Although the door now hinged at the front, it gave excellent access to a spacious operating platform, yet kept a design that had been familiar with Ford tractor owners since 1979.

As the company and press personnel drifted away, Noel could sense pride and enthusiasm emanating from his sales team, if these tractors fulfilled their promise, a bright future lay before Ford Tractor Operations and Waltanna Australia.

CHAPTER FIFTY-NINE

After a successful media release in the Carlton Gardens, Noel, accompanied by Dick Kiefer and Ford Australia's Marketing Manager, Max Grandsen, flew to Perth. Max had joined the tractor men on this trip, to support Ford's rural vehicle dealers in West Australia and participate in any official duties on behalf of Broadmeadows Head Office. The following morning all three boarded the twin engine Beechcraft Baron Noel had chartered to fly them to the Dowerin Field Days.

Being late August, the day had the promise of an early start to Spring about it and the morning weather forecast predicted clear skies all the way to the field day site. Always employing an experienced pilot, Noel insisted on the safety of the aircraft and for the passengers in it. If the opportunity arose, he would wear the second radio headset whenever flying. About thirty-five minutes out from the Dowerin Airfield, he became concerned when incessant chatter from an unusual source began filling his headset.

Whenever chartering a plane, Noel insisted it be fitted with full instrumentation and the full array of radio equipment. In this machine however, it took time for him to recognise which radio was picking up the signal. Watching the pilot flicking through each, one at a time he eventually isolated it to the Citizens Band Unit. A group of farmers flying a single engine

Cessna had become lost on their way to Dowerin and the chatter that had been filling his ears was being driven by their desperation for directions from anyone with reception.

Using proper protocols, the pilot was able to reckon the lost farmer's location and asked if they were heading for the same airfield. Giving them his position and suggesting they look for him he pulled the throttles back slowing his speed and asking that they call back when they had him in sight.

For several nervous minutes the Beechcraft's passengers kept watching above and to each side, searching the skies for these lost agrarians. After what had felt like ten-minutes the pilot called them again, only to find they were closing in right behind him. A determined left hand pushed both throttles forward and the Beechcraft stretched into the distance. After a few stern words from the pilot, he slowed again encouraging them to come alongside and remain where he could see them. On approach to the airfield, he suggested that they land first while he went around, and after completing another full circuit, followed them in. After landing, Max and Dick, both keen to know why their pilot's words to the other flyers had been so heated were brushed away with Noel's easy candour.

Ford Tractor Operations, West Australian Regional Manager, and his team supported by dealership volunteers had taken every opportunity to present the new FW Series Ford at the very front and centre of their exhibit. The strategy worked. When Noel arrived, he could see what seemed to be a mix of farmers, tyre kickers, pinstripe-whistlers and interested school children milling around each of the FWs. All morning, people had been peppering Ford's field day staff with questions about the machines. Having Dick present to witness the excitement would demonstrate that the Australian arm of the company had both vision and capability and Noel knew Dick would relay this in his report. Having him at Dowerin representing World Headquarters added an element of prestige to Ford's West Australian effort.

It had only been three years since Dick last visited Dowerin for the 1983 event where when interviewed by a member of the Field Day Committee, he claimed then that the West Australian display, "was as fine as anything I have seen anywhere in the world."

The organisers being advised that he would be attending again, made him a welcome guest on this occasion too.

Now the buzz created around this Ford Australia FW Series tractor release three years later, found him even more upbeat. Dick, always keen to interact with farmers simply beamed with positivity. Moving easily around the crowd he spent a good part of his day being introduced to farmers by dealers, who brimming with pride, watched on as the man from Ford shook the hand of many a farmer from remote Western Australia. He answered their questions in easy fashion, while pointing out features and offering benefits. He proved his product knowledge on all the new tractors, including these new FWs.

Dick had done his homework; his enthusiasm was contagious and seeing it rub off onto their sales team, for Noel and Max this could only be a good thing. Sales leads were being created and appointments to value trade-ins made. The 1986 Dowerin Field Days offered much to be excited by.

By 1987, rural Australia had been enjoying the benefits of a falling exchange rate for over three years. However, while the lower price of the Aussie dollar against the US greenback had been boosting farming income for wheat, wool and other primary exports, for producers who relied on overseas sales, some were doing better than others.

Over the past ten to fifteen years, soil sciences had proved that better yields could be achieved by dramatically shortening the sowing times. Availability of more powerful tractors and the wider machinery had been driving improvements made to seeding and spraying machinery. The challenge to take advantage of emerging dryland farming practices had been driven by the availability of heavier high-horsepower tractors and the pace of development could be best described as feverish.

Financially secure farmers were more readily able to buy such machines and by employing these new techniques their crop yields had increased, as a result these cereal growers were enjoying unprecedented profits.

For their less-financial neighbours, or those who had borrowed foreign currency prior to floating the Aussie dollar, their prospects were dire. For the latter, their balance outstanding had for some, more than doubled and at best, had probably risen by at least half. The rapid increase in interest rates applied even more pressure, often causing the account to be further over drawn. These situations would automatically flag them as being outside the lending criteria and a phone call or visit from the bank manager just added to their ever-rising stress level.

A relentless arrival of overdue notices continued grinding away at even the strongest of men and women, piling on worry and taking its toll. Only a year or two before, their bankers had been arriving with an offer of cheap overseas money, suggesting they could use it to buyout a struggling neighbour, or invest into a cheap rural development scheme somewhere interstate. The sales pitch was good and many took the chance. Now they were finding themselves in a worse position than the farmer they bought out.

It was the bank that had turned from lapdog to lion that was tearing them apart now. The earlier offer had its origins seeded in greed and in typical fashion, *"Greed is Good,"* seemed written in stone, and the catch-cry of the eighties.

In the years that Noel had worked for Ford, Australian interest rates had risen from around 5% in February 1964 to 15.50% in September 1986. For those who had extended their mortgages to cover the changing value of the Australian dollar, their situation was becoming worse than grim. Banks always applied an extra percentage for borrowers who had drawn outside their overdraft limit, in some cases they even applied an extra fee for late payment. These were perfect conditions for creating intense levels of stress and many farmers couldn't cope.

Small business found itself having trouble too, the business model that had worked well for the last fifty years had broken and rural Australia found itself at the forefront of a collapsing economy.

Treasurer Paul Keating had been trying to slow the economy, but nothing seemed to be working and, across Australia rural areas were suffering. For a small number farmers often without a crop to harvest, the weight of unpaid creditor's demands and an aggressive bank chewing at their overdue accounts, couldn't face their friends. And because of pride, stupidity, or some sense of failure, it seemed for them, suicide had become their only answer.

Similar problems began seeping into every farming community and across Australia once thriving small towns now had For Sale signs filling boarded up shopwindows. This wasn't anything like the *dust-bowl* described in John Steinbeck's *"The Grapes of Wrath,"* for the most part, Australia's farming processes were sound. This recession had its roots in a government willing to sacrifice rural people, rural manufacturing and almost every struggling farm, business or tradesman. Many lost everything and most of it because of the government's desperation to manage the country's inflation.

It didn't feel right and in Advisory to Treasury meetings, Noel stood his ground to say his piece. As a seasoned advisor who represented Australian manufacturing, rural business, and that of the Murray Darling Basin Board, he saw it as his duty to highlight how the current monetary policy was devastating country Australia. While his voice had been joined by others, the Hawke and Keating government had let a tiger loose on the Australian economy and they couldn't even catch its tail, let alone put it back in its cage.

As the treasurer, Paul Keating had become so hated by one tractor dealer in the Barossa Valley, that he had bought a small statue of the man so that his customers had somewhere to aim in his urinal. For some, Keating was every bit a genius, for others a man to be reviled.

Across Australia, rural people who still hadn't recovered from the 1982 drought, began leaving their homes for the city. All of them desperate to provide and protect their family's future. And because of the collapsing economy, their finances depleted further.

Their hurt cut deep and by the end of 1986, desperation started creeping closer to the houses of power.

Overstocked machinery dealers who had traded somewhat aggressively after a couple of good years, were now finding the stock in the used yard was no longer wanted. Farming had changed and their second-hand stock aged, rusting where it stood.

Against this very fluid background, Noel had to reset his goals and prioritise models to navigate Ford Tractor Operations Australia through this coming recession. The purchase of New Holland became just another complexity adding to the mix.

The ambitions Noel held for the Waltanna built FW Series hadn't diminished. Teneco, the company that bought Case and International Harvester a few years earlier, acquired Steiger. Knowing he had to move quickly, he understood there would be a wonderful opportunity for Ford Tractor Operations Australia to export the Hamilton built tractors to Ford distributors in every corner of the globe. Australia's economy may have been failing, but opportunity is hidden among change and this was one of those times. He only needed to find a way to exploit it. Encouraged by his contact inside World Headquarters, Noel pressed on with the sales and marketing in Australia while building an export proposal he could present the next time he visited Troy in Michigan.

At the time, the only complication he envisaged would be if the deal to buy New Holland fell in favour of buying Case IH from Teneco. If that happened, Ford Headquarters in Michigan would have a complete machinery line including high-horsepower tractors. From the many casual discussions he'd had, he understood the board were uncertain, which company offered them the best option for profit now and its saleability later.

His own research told him that the purchase of Case IH would be subject to more Anti-Trust Laws in the USA than buying New Holland. Making the harvester company the easy option for the Ford Motor Company's board. If the deal with New Holland was to be the direction the board took, then for Ford Tractor Operations in Australia and James Nagorcka's Waltanna, a bright future lay before them

CHAPTER SIXTY

During the last months of 1985, Noel's assumption that Ford's board would buy New Holland had been confirmed to the dealer body and Australia's Ford dealers, began rejoicing. After years, decades of having been held back without a full product line, they now had something to celebrate, something to shout about and the mood sweeping over the whole group becoming one of elation.

Sure, everyone knew there would be problems to iron out and a lot of New Holland dealers would be unhappy, but spoils always go to the victor, don't they?

Now that the deal had been agreed and made public, Noel understood a formal announcement would be coming in a few months. Leading up to the statement, he had started thinking of ways and developing new strategies for transforming New Holland Australia. Since the late seventies, its lacklustre profitability resulting from their combine harvester business, had taken much of the shine off a company once considered an industry best. A reputation deservedly earned through hay equipment was now waning. Preparing for transition, he began writing and rewriting drafts of those initiatives he thought would inject much needed cash and opportunity into a business he believed was clearly failing under Nels Craig's watch.

Buoyed and encouraged by the rumours and other information dripping out of World Headquarters, Noel felt satisfied that if the General Manager's role didn't fall to him, he expected it would go to an overseas appointment. At the very least he would assume a major position within the new hierarchy. An overseas appointment, he was sure would come from the tractor company. It would likely be a person he knew; someone he could work with and between them they would lay the foundations for becoming formidable teammates.

After news of the takeover become public, a countless number of unpredicted tasks began, all having to be prioritised and all requiring delegating or scheduling as an action point for a planning meeting. During these months, his time was not his own. The task of bringing the two organizations together was huge and even though he would have welcomed the opportunity to work on the planning from a Worldwide point of view, he relished the task of making the Australian region tick.

For the Howards, their family life underwent a huge change with the New Holland buyout. Soon after the announcement, Noel received a directive to close out Ford's tractor operations in Campbellfield and shift the company to the more modern New Holland facility at Cranborne. Organising and negotiating to place the right staff in control of managing both parts distribution and tractor assembly requirements was crucial for a smooth transition. Outside suppliers required new contracts or be terminated if a clash of such agreements demanded it.

Working from Cranborne meant at least an extra hour of travelling for Noel if the traffic proved favourable, two if the lights were against him. Until deciding if they were ready to put Highfield Park at Kilmore on the market and move closer to his new office in Cranborne, Noel would stay at the Ambassador Hotel in Frankston. This allowed him an easier journey to Ford New Holland Australia. Their two boys now apprenticed to Caterpillar distributor William Adams at their Clayton Head Office and living away from home, their commute had simplified their commute by living close to their work. Tania was now living in close to her study at Australian Catholic University in Melbourne. It made the school run duties Joan had held for the past twenty years a little easier.

Around the same time Noel's brother, an Albury / Wodonga Solicitor advised him that one of the best farms in the Tallangatta district would soon be coming on the market. It was a substantial operation, but the cherry in the pie was another dairy farm in the nearby Mitta Valley. A move there would mean some parental sacrifice, but between them Joan and Noel agreed to buy both farms. Joan would move north with their youngest child to manage the farms and Noel would join her to help around the property on weekends. For Kathryn, they would transfer her education to a primary school at close by Tallangatta.

Moving hadn't been an easy decision. Their home at Highfield Park had become a place of sanctuary, somewhere

Noel could walk away the continuing demands of a seventy-five-hour working week. Just to tighten the fence wires, slash a firebreak around a paddock, or fill holes in the driveway allowed him to concentrate on other things. These may have been menial tasks, but it refreshed him, making him ready for another high-charged week of being the General Manager of Ford Tractor Operations in Australia.

Being told to wait until a clearer picture had been thrashed out in Troy, started fraying nerves of the people under his command. To keep from getting caught up in the rumours, Noel set tasks, asking various teams to define new goals and refine Australia's plans for a successful Ford New Holland Australia. Much of their time had been spent determining schedules, applying analysis, building charts and redrafting recommendations.

Noel took it upon himself to research the employee history of both companies and although he knew his own staff as well as any manager could, he needed to know the ability and ambition of people who made up New Holland's sales and management personnel. Find New Holland's natural leaders, identify those people whose skills were lacking and make subsequent training recommendations on their record. Assisting with his own research, his secretary Wendy Orchard would compare it to any that he had asked for, making notes to confirm or deny any comments on the file. At his desk he would make any changes and send it back to her for further collation.

In late October, his attention had been drawn to an article published October 11th 1985 by James Risen, a staff writer at the TIMES. James was commenting on Ford's purchase of New Holland.

Risen wrote that Phillip E Benton Jnr, a senior Ford executive in Detroit had claimed that to survive in agricultural farm machinery manufacturing, products had to be integrated to ensure profitability. Although currently profitable, both Ford and New Holland as separate entities faced a future of being squeezed out by companies like John Deere. The purchase of New Holland would mean the creation of a new company that was efficient, integrated, employing over 18,000 employees and capable of enjoying an annual turnover in excess of $2 billion US. It seemed the Ford New Holland star, was set to continue rising.

A few months before any formal news of the buyout had leaked to the international press, Noel had been asked to oversee a valuation of the Cranbourne manufacturing and distribution facility. What he found distressed him. This state of the art, less than ten-year-old premises appeared tired and unloved. However, it was when given access to New Holland Australia's accounts that began concerning him most, convincing him that this particular set of accounts appeared to be a lesson for driving a business onto the rocks. Account Receivable write-offs for each of the last few years had seemed unconscionable, particularly as the offending dealers would repeatedly be extended further credit with favourable terms. He could confirm in the figures, just what his sales teams had been telling him.

New Holland Australia appeared to be driven by sales first and its collection of overdue accounts became secondary. This had never been the way the Ford Motor Company operated and an ability to pay had always driven decisions when extending credit. A policy that to him, had been lost somewhere within the tangled and winding open office halls of New Holland Australia.

Stock Control, flagged another area of concern, often insolvent dealers were being shipped product without pre-payment or arrangement through a finance company. Used floorplan valuations were often those of fairy tales. Everywhere Noel looked, he found opportunity for improvement. Outgoing expenses could be trimmed, duplication of services eliminated, stock control easily improved. Even a shake-up of manufacturing and spare parts could use attention, nothing would be spared or overlooked. It was like the work he had done in South Africa, only this time, he wouldn't need to leave the country to do it.

Something had shifted over the past weeks, the communication between the old tractor company and its new regime. Unsettling him. Information that had once flowed freely between departments before, now had an element of secrecy about it. The feeling and memories of the last few months of his time at Massey Ferguson flowed back. Could it be that Ford New Holland were shifting the roles of people, interviewing candidates, or something different entirely? Seeking answers from World Headquarters in Dearborn would seem needy and not something someone in his position should explore.

And it wasn't just him, everyday dealers phoned asking for an update to their status and when could they expect to begin quoting New Holland products? When would brochures and price lists be available? He could have asked the zone managers to field these questions, but he knew if the smallest dealer from outback Australia was phoning, he was obliged to listen. It didn't matter the size of the dealership, each principal who phoned was both excited and concerned.

New Holland Australia's dealers were being given a different story, it seemed that Cranborne had assumed they would be leading every aspect of the operation. It was hard to find anything in writing, but their sales team were confident and assuring when it came to telling their dealers of the new company's approach to their dealer status.

Surprising him, was an internal communique advising that Ford New Holland had made an offer for Canadian heavy tractor manufacturer, Versatile. It was late 1986 and he had been blindsided, any hopes he and James Nagorcka may have held for a collaboration, now seemed doomed. They would build out the last tractors, by which time the Versatile units would begin arriving on Australia's docks.

James proved his mettle as a man when Noel delivered him the news, saying that from the moment he had heard rumours that a deal with Versatile was being considered, he had expected as much. It fell to Noel to formally cut the ties, this was something he had to do on his own, the friendship they had built determined it. However, Noel always regretted not pursuing Waltanna as an OEM Distributor for Ford New Holland, which would have ensured the future of Waltanna FNH.

The power in Michigan decreed that the new company would operate out of Cranbourne and Noel should offer Campbellfield staff an opportunity to relocate, or apply for new roles within the automotive business. It would be inevitable some would retire and others be retrenched. A difficult task, but he was the General Manager and it fell to him to close out the Tractor office on Sydney Road Preparing for a move eighty kilometres south east would not be an easy ask for anyone.

However, before closing out the Sydney Road office, Noel in extending a welcome to his new colleagues, invited all New Holland staff to a meet and greet at Ford's Broadmeadows Headquarters. About thirty New Holland personnel from Cranbourne took in an extensive tour of the factory. These were to be fellow Ford people who Noel believed should be included in everything. Beginning with the assembly plant, the group took in the Engineering and Product Development departments followed by a tour of Ford's modern Parts Distribution Centre.

During his initial welcome he had stressed that while he understood there could be difficulties as the companies melded together, to achieve the best outcome there was also a need for staff from both companies to work together. Something important for both employee groups, it wasn't Ford Tractors anymore and it wasn't New Holland either, the new Company was Ford New Holland and its success or failure depended on everyone.

The tour, followed by drinks and finger food in the Management Dining Room had been well received by all. Noel had mingled and chatted with his guests over the course of the day and that evening he felt that those in the room had reached a positive feeling toward the future of the new company.

At the end of the function Noel Presented Nels Craig with a handmade cricket bat. The bat representing a new beginning and an emphasis of, "Fair Play for All."

What should have been an easy transition, quickly became a clash of cultures. Where at Campbellfield Noel had openly discouraged smoking in the office until it had become a non-smoking workplace, anyone at Cranbourne could light up, anytime and anywhere. He assumed if Nels Craig, a high-ranking manager did so in his own office, then the smoking culture at New Holland was determined by the man at the top. By 1987 tobacco was taboo, quickly becoming shunned in Australia's offices and some public spaces. If it was allowed at New Holland, then to him, it indicated that a few sloppy habits may be trickling from the top down and through the ranks.

Moving to Cranbourne meant a change of secretary for Noel as Wendy had elected to stay and transfer to Ford Australia. Capable and efficient, Anne Purdie now stepped into the role that Wendy Orchard had held since his appointment as

General Manager. The mix of staff mainly due to the change of location, remained heavily biased toward New Holland, somewhat stifling Noel's vision of a cohesive and balanced workforce.

Irritations were many in those first months and the Cranbourne staff's indifference to their Ford Tractor colleagues, seemed to hamper an easy Ford New Holland future. This difference in personalities grating within the organization began permeating into the dealer body too, hard decisions were coming and Noel understood he would have to lead the process.

News of the buyout had filled front pages of rural papers across the country and he was determined to take advantage of the frequency of these reports be they positive or otherwise. Noel knew the importance of good publicity and by bringing the teams together the company could halt speculation and grab a lead on their competitors.

Establishing new operating systems at Cranbourne became a priority, but having everyone working toward a common goal became his first order of business. He had always believed it was people who created business success, no matter the size of the operation. Believing this mixed group could with the right leadership, find a way to make their company number one. They only needed to focus on what was important.

He understood Ford systems, having studied them during the Ford Marketing Institute courses when he first joined, then having those lessons reinforced during his time at Harvard, but it had been living with them for twenty-five years that reinforced the lessons. He had invested a lot of his time working for the company. As Joan often said, he'd been with Ford long enough to make him bleed blue. He had expected New Holland executives would have been exposed to similar systems, but now in the Cranbourne office, it appeared as if the checks and balances of those systems sat low on the current management's list of priorities.

Seeing shabby stock in the holding yard and fading out of date Sperry signs, Noel became even more determined to force change. The business was crying for another approach and with change in mind, he invited four of New Holland Australia's key personnel to his office. These were the men who he believed had the power to repair the position and bring the harvest and tillage business into profit. Together they would have their challenges, but he wanted to believe that if he demonstrated faith in their ability, outlined a plan and invited commentary on his ideas, then as a team they would find a way through this present malaise.

As Noel put forward ideas, he invited input and with every uneasy silence he found himself sensing animosity emanating from his colleagues. An "us and them" mentality, paramount in the room. To him, Sales Manager Peter Morris, Northern Branch Manager Tony Tait, Sales and Marketing and Spare Parts Manager Neil Stirling, and Engineering Manager Ian Vale, were not taking the meeting seriously and their investment for finding a different strategy... nil.

These were simple issues, signage had remained Sperry New Holland, as had the stationary and promotional material. Nowhere in Cranbourne could he find a FNH logo, or anyone with any interest in the world-wide Dealer and Company Appearance Program. Nowhere was there an item that made reference to the new company and it shocked him.

All of Cranbourne should have been glistening with Ford New Holland branding, letterheads for dealer bulletins and all other communication printed and distributed, but the people responsible for those actions appeared to be showing no inclination to do so.

Coming from a brand conscious company like Ford, the lack of discipline irritated and although he knew it would take time to get this team focused, he understood. Preparing for structural changes to the management team as would be required, but as frustrating as the current situation was, he knew it was best to keep those thoughts to himself.

Standing at the meeting's end and ushering the team from his office, as they walked away, he overheard them becoming more animated. He watched them, not fully understanding why they had resisted his suggestions in the way that they did. However, before disappearing into the beige maze of this modern open planned workplace, Noel overheard Peter Morris telling Vale and Stirling that, "this is the biggest load of bullshit that Noel is calling for action on."

The purpose for his requests had been to address selling the ever-increasing number of completed Combine Harvesters in stock. At the time, New Holland had four hundred unsold new machines on their books and at a time when Australia's interest rates had hit their highest in history. During the twenty-four years he had worked with the Tractor and Machinery Association gave him an insight into all aspects of the combine harvester market. The total harvester market for that year had only recorded retail sales of a similar number and, New Holland's oversupply concerned him.

That the people in charge had not planned better and appeared blasé required immediate action, which had been the whole reason for meeting with these men. Ford New Holland Australia were drowning in stock and it looked as if no one employed at Cranbourne seemed to be taking it seriously.

An excess of the New Holland tillage range built by Alfarm was even worse, these machines having been purchased from an outside supplier, were deep in weeds and withering under at least one of Australia's harsh summers. if not two. To his thinking nobody cared, but Ford money had bought this company and he owed it to his friends at Ford's World Headquarters in Dearborn to make it work. It would take time, but Noel had become determined to find a way to change attitudes and lift the company into consistent profitability.

Following the meeting with Morris, Tate, Vale and Stirling, Nels Craig summoned Noel to his office. When he arrived, he found Craig and Vale waiting. Then for what seemed an age, Noel found himself being berated by Craig in front of the less senior executive, Vale.

To him, the meeting had been most unprofessional for a man in Craig's position and it took an even more unpleasant twist when he invited Vale to comment. Vale stated Noel had no experience and he believed it would be better for the Ford New Holland if Noel just stayed away

The conceit of Vale's comment floored him, had they not done their homework? Did they not read the resume of the man who could one day soon become their boss? Was it another show of arrogance driven by fear or ignorance? Whatever it was, they were wrong and Noel didn't care, nothing either man said was worth a memory?

Inside information from his contact at World Headquarters had indicated that Ray Cartade would be installed as general manager displacing Nels Craig as CEO. Cartade had excellent credentials and Noel believed, that between them, they could forge Ford New Holland Australia into a strong and profitable company. However, Craig and Vale's attack on his ability required reporting. Over the next few days, he wrote to senior management in Troy Michigan USA seeking clarification of his own position.

Noel took time to report his own findings into the state of New Holland Australia's position to trusted colleagues in Michigan. Addressed to Bernard Sarfras, Gary Tessitore, and Dick Kiefer, he reported that all the Ford staff who had transferred to Cranbourne were under siege from an outdated culture driven by their New Holland colleagues. Concluding his report, Noel advised he would be unable to work with Craig without some meeting of the minds.

Where did he fit within Cowl's scheme?

The answer to that question soon became clear when he heard an office rumour that a meeting had been called at Cranborne. The scuttlebutt had travelled, being first picked up from Western Australia where the Regional Manager for New Holland, Alan Montgomery had told Noel's confidant that New Holland's senior executives and their Ford Tractors counterparts would gather at Cranbourne to terminate his tenure. Noel's sacking would make way for Nels Craig to assume the role of General Manager. His information claimed Peter Morris as being the conduit to Montgomery's statements. The toxic rift among Ford and New Holland staff continued its ever-widening mental chasm.

The rivalry within Ford New Holland had to cease and it seemed that under Craig's stewardship that was unlikely. Noel knew he had the skills to end it and to set the company on a course to a harmonious future, he would say as much. The opportunity never came, he hadn't been invited to defend these accusations in front of Cowl and Vice President Moglia. After his last meeting with Craig and his cronies, he had believed an apology from Nels was in order, only then could they work together.

He and Gor Cowl had faced each other before, often arguing different points of view, but behind their passion had always been a desire to drive the company forward and make it profitable for all concerned. Moglia too, should see the current human resources situation as being untenable and very much in need of a new visionary. Someone who could unite the team and drive a FNH restructure into profitability.

This Australian gathering consisted of Vice President Moglia who had by now transferred from Troy Michigan to FNH World Headquarters in Pennsylvania, Gor Cowl also now working out of FNH Pennsylvania and Nels Craig were to meet in the latter's Cranbourne office in the morning. Noel expected, that as a director of Ford New Holland Australia, he would be given the courtesy of an invitation. No such notice came, and the silence stung. He, a director of the new company and who's signature appeared on the purchase contract, never received an invitation.

The only way he could justify Gor Cowl approving or instigating any of this treatment, was to see it as a revenge attack, and all because of a difference of opinion that led to the success of the Ford 8401 in Australia.

Were they just two high power executives who banged heads, or was Cowl that weak and insecure in this advancement, that he only wanted yes men around him? Millions of questions searched for answers in Noel's psychic, nothing made sense and even though the time felt ripe for change, there was still more to do with Ford New Holland.

Earlier he had scheduled a round of meetings to face dealers from both New Holland and Ford, who until now had been starved of any official communication. It had been a factual communique that invited dealers from both Ford Tractors and New Holland to meetings in respective capital city locations and Noel had signed it as a director of the new company, Ford New Holland Australia.

The dealers believing that at last there would be some chance for clarity, seemed uneasy at first. It felt strange to see old adversaries seated together in dining halls. While an obvious tension seemed to creep into every corner of the room, everyone remained civil.

For Noel, wanting to convey a positive message had been priority one, and he approached each meeting with confidence. Gladhanding each crowd afterwards, it was easy to see that he'd been welcomed by all but a few.

More than one dealer described these meetings as a lesson in dignity.

Representing FNH Noel stood alone before the crowd in each room, taking control and responsibility as he always had. Outlining the steps that had brought the companies together, was a history lesson in mergers and acquisitions. Not everyone could understand why it was necessary to go over old ground, but Noel had taken it upon himself to describe the strengths of the new organization and how it, and its dealers would profit from Ford's buy out. It may have been stating the obvious, but he confirmed there would-be dealers who would end up without an agreement and others who would benefit from the sales park of those who had lost out. Nothing in Ford New Holland's next few years would be pretty, staffing would change, methods of communication would evolve and everyone would be asked for even more tolerance.

No-one left these meetings euphoric; it was a dirty business handled in a blunt but supportive way.

The tight knit South Australian dealers were particularly shocked. Ian Denton representing a Ford Dealership of over sixty years and with sales volumes consistently above State Market Share, stood to congratulate Noel on the frank way he faced the crowd. Saying on behalf of the SA Ford Tractor dealers how much they appreciated the courage and confidence he showed that night.

By April 1988 only a handful of Ford dealers had been offered new dealership agreements, Denton's of Farrell Flat, was not one of them.

The tension in the Cranbourne office the morning of Cowl and Moglia's visit crushed the chatter and only an occasional whisper drifted around the open-space office complex. Noel stayed at his desk, the hours groaning by, he waited steadfast, desperate to face his challengers and call out to his accusers.

For a man who solved problems, being excluded from the discussion had been made worse when his Personal Assistant, Anne Purdie told him that all New Holland personnel and six Ford people who had transferred from Campbellfield had been summoned to a meeting. It amazed him, he had served twenty-five years with Ford and the lack of respect shown, if not to him but to the position he held, was at its best insulting and at worst an act of treachery by Craig and Cowl.

He remained in his office; hours of nothingness dragged by until Gor Cowl appeared at his door.

"Come for a walk," Cowl said.

They walked the yard, through the mess of tillage machinery sinking into weeds and thistle that clambered through the fading frames. As grass seeds grabbed at their socks, Cowl spoke. There was no apology or reason given for the non-communication, nothing. All he said was that Noel had an office waiting at Ford Tractors Asia Pacific for him. He was to report to St Kilda Road and take up a role within Merv Manning's group. He would be responsible for closing out Ford Tractors New Zealand and find a suitable candidate to act as a Direct Dealer for New Zealand, just as Massey Ferguson had done a few years before. Noel knew Merv Manning and understood his team in Australia was drafting a plan for Ford to enter China and South Korea, but as sweet as the promise of a new challenge appeared the bitterness of betrayal lingered, blackening his mood.

In an instant he had been cut from Ford New Holland Australia and the sting from that unhealing wound would run deep, festering for years.

It now fell to Gor Cowl, former manager Ford's operations in Europe to oversee the distribution of roles within the new company. Being promoted to manage a large region including

Australia and New Zealand, he appeared to take more than a little glee when declaring most of the Ford Tractor Operations Australia positions would be terminated. Only those Ford positions he and the New Holland management considered vital to operations would be absorbed into the new organization. As everyone was now working out of Cranborne, without Noel any tractor specific roles that were left open would now be covered by New Holland staff.

Cowl's sword may have been swift, it may have been sharp, but even three decades later, the blood of betrayal continues to flow deep. People who had served most of their careers with Ford's tractor operations, or as dealer staff and even loyal customers soon found themselves confounded by emotion.

The whole premise of a buyout, would in a sane man's world, prune out the ineffective, the loss makers. Cowl provided Ford New Holland Australia with a new challenge, Noel Howard would go, but the bitterness of his departure would linger to haunt them.

NOEL   An Authorised Memoir

CHAPTER SIXTY-ONE

Still aggrieved with the way Gor Cowl had manipulated him out of Ford New Holland Australia, Noel straightened his shoulders and taking the Norwood folder from Merv Manning, smiled. His expression hiding feelings for the man who had purposely brushed him aside. Believing Cowl's action had proved himself a dishonest associate, Noel couldn't comprehend that someone trained by Ford Motor Company had favoured appointing New Holland Australia's people at the expense of competent and long-serving, loyal Ford Tractor staff. Every effort to reach out to his superiors had been rebuffed. Unable to reverse Cowl's act left him feeling helpless and it hurt.

Locking horns with Gor Cowl had begun over a decade before coming to a head when Cowl had threatened Noel with a poor Performance Review in 1989. He had been quite vocal in his opposition to the development of the 8401 for the Australian broadacre market Now with complete control over the Australian arm of Ford New Holland, Cowl had exacted his revenge.

Noel knew he had bigger fish to fry and, for now, finding a New Zealand distribution chain beckoned. The assignment came with a sense of excited anticipation that he didn't expect. Thoughts rippled through him, his mind shuffling through all the permutations of negotiating a supply deal with a competitor's distributor.

Reaching back to the introduction of the "Little Grey Fergy," Norwood had forged a long history of distributing Massey Ferguson products for New Zealand. Enticing them to tear up their current distribution agreement with Massey would not be an easy task.

Asking Norwood to replace their long-standing agreement with Massey Ferguson and replacing them with Ford New Holland would be quite a challenge. Experience told Noel that a deal of this size would come down to something more than money.

He knew that finding a way to successfully conclude a new arrangement would mean presenting them with a point, or points of difference to their current supplier. An offer so dissimilar to those presented by other prospective suitors, that it would be irresistible and it was important to him that the whole agreement be presented in full. There would be no hidden agenda and everything offered, was to be backed by solid accounting.

Working on the strong links Norwood had forged with New Holland since 1985, and demonstrating how they could timely extricate themselves from Massey Ferguson would be key. However, their business had changed hands less than three years ago and presenting his case to a new board of directors offered promise.

The key point for Noel was the recent ownership change. He knew Walter Norwood's descendants had in 1978, sold their interest to Dalgety whose primary interest was being a stock and station company. Now it was Dalgety who had in less than ten years, divested themselves of the tractor and machinery business to the Zeullig Family. Although there would be a lot of history associated with the company to consider, this change presented an opportunity around which, Noel could craft his Ford New Holland presentation.

Establishing the exact year that the Norwood company was formed, depended on which iteration of the automotive and agricultural conglomerate was founded. The date Noel decided to begin his presentation with, was 1938, the year C.B. Norward had been founded. His research discovered that from day one, Walter Norwood's drive to find new products for the company was insatiable. Distributing a wide range of tractors and machinery throughout New Zealand easily confirmed theirs as a business that embraced change, and decades later other tractor and machinery businesses were still following the Norwood supply model.

Fifty-one years later and the company were now distributing several market leading agricultural brands including New Holland. Noel prided himself as being on top of the factors that would affect the market at any time, and his knowledge of the New Zealand market was no different. His time as General Manager of Ford Tractor Operations Australia, had highlighted Norwood as being a business intent on owning any market they entered, and this would be another of the key points to structuring his presentation.

As Massey Ferguson and New Holland had several comparable hay and harvest machines competing for the same business, Noel knew there were monies to be saved through rationalizing inventories. To add emphasis to the report, he set about developing a comparable spreadsheet, which would highlight both the advantages and disadvantages in each manufacturer's product. Where he saw opportunities, he predicted conservative FNH market share targets. He also had his team include other manufacturer's lines where Norwood's Ford New Holland distribution could eliminate duplications too.

Knowing Massey Ferguson had an emotional connection with their customers going back to the grey Ferguson models, he thought it prudent to demonstrate a similar percentage of Ford and Fordson owners that had product loyalty too.

However, for Norwood a bigger threat to their existing Massey Ferguson arrangement was that the tractor and machinery manufacturer had seen their share valuation plummet over the years. Noel could relate to these difficulties due to his own Massey Ferguson experience working with them for eighteen months during 1962 and 1963. The continuing losses and a lack of faith in the company's management had seen over a decade of financial problems and a reluctant Canadian Government becoming a major shareholder.

For Massey, this financial turbulence had sometimes led to design deficiencies across different models. Whereas with

Ford Worldwide, their company finances were more than secure, and the money the company put aside for product development, grew every year.

Staying with a desire to be one of the major tractor and machinery manufacturers, Ford had recently introduced replacements for the TW range with more modern and ground breaking Genesis tractors. As Ford New Holland the product offering now included the updated TR range of combine harvesters and although not a big part of New Zealand operations, they were poised to grow an even larger market share.

For Norwood, New Holland were a known product and retaining their line of hay and harvesting machinery would be important to them. Therefore, Noel believed that agreeing to distribute the whole FNH range would made sense to the Norwood board.

At the beginning of the nineties, distributing the complete FNH range proved an attractive package. Along with the recently released Genesis models, the tractor range would further be supported by the soon to be released 10 Series tractors out of Europe. A facelifted compact tractor series from Japan would complete the line-up. For Norwood, a change to Ford New Holland was an offering of much promise.

After the highs of several successful meetings with Norwood management and its board representatives, he had another task to complete. Closing out Ford Tractor's New Zealand office had a great number of low points; it had always been task that nobody enjoyed. Telling people that they no longer have a job is difficult and for Noel it would be no different.

Equipped with relevant personnel files, Noel worked with each employee on staff to find the best option open to them. Whether it be transferring to automotive, working for a dealer or a competitor, each meeting became a dignified affair.

Flying home after each of these interviews he would relive the discussions, studying his memory for each person's facial expression and what had been said. These same faces continued to haunt him for long after the flight had landed back in Australia.

He had however, added each of their names and addresses, including other relevant information into his contact book, a journal which over the ensuing years would become as valuable as his diary.

Having secured a handshake agreement with Norwood's management for the distribution of Ford New Holland products in New Zealand, Noel handed the framework to the company's legal team to refine and structure the necessary documents. Part of this agreement included Noel's draft of a plan that would allow Norwood to gradually withdraw from Massey Ferguson. It would take more than three years before they parted ways with Massey.

Nearing the end of Noel's brief stay in Cranbourne, he began feeling ripe for a change and as the New Zealand assignment had proved, ready too. This new role with Ford Asia Pacific was brimming with interesting challenges and his next assignment had him drawing parallels to his time in South Africa with Harry Watson. Only this time Harry had retired, he would be able to do things his way and although there would be times that would test him, the challenge was more than exciting.

From his contact inside World Headquarters in Troy, Michigan, Noel understood Ford Tractors had been harbouring a desire to take advantage of the predicted emergence of China's agricultural industry. China's presence was growing and with it, its farming communities need to mechanise would create huge demand.

The Western World were beginning to understand, that as "The Sleeping Giant" began moving away from subsistence farming and into industrialization. China would need vehicles of all description to drive its economic growth. To achieve this, they would need foreign investment to improve their technical base and the date for allowing western companies to invest in manufacturing was moving ever closer. For the Ford Motor Company, it fell to Merv Manning and his group at Ford Asia Pacific in Australia to draft plans for launching Ford New Holland into this emerging superpower.

Noel would still need to oversee the Norwood business in New Zealand until it concluded, but now his whole focus would be on China. In his mind, establishing an office for Ford New Holland should be straight forward.

After engaging a tight-knit team who were seconded from various Ford Asia Pacific offices, Noel's next task would be selecting a suitable site in which to base the research team and begin work on the Chinese operation.

These people were chosen for their integrity, their various skill sets and their ability to communicate in several languages. The whole project was top secret and although he had to report to several heads of departments in Detroit, his team had carte-blanch when it came to building the data on which to base their recommendations.

Noel had always maintained that it made no sense to give someone a task and then micro manage them, it just wasn't his way. He had found that by showing trust, people always went that little bit further than expected. This trust would be rewarded by the quality and effort each team member showed to research and present their findings.

Now with a team in place and the preliminary research completed Noel found himself, flying to Michigan for a meeting with members of the board again. He held in his briefcase, a presentation nominating Seoul in South Korea as the ideal place to base their operations. The few senior colleagues who understood what he was proposing, implied he was mad, suggesting setting up an office outside China might just be enough to get him fired. The prospect of being fired didn't bother him, his presentation came from considerable research by the best people in the region and the data he had to support the recommendation within the proposal, gave him confidence.

Before opening his address to the proposal, he looked at the assembly of different managers and board members. In that moment all he could see in his mind was a wall of senior colleagues openly chanting, "No way!"

He looked at his notes and when he raised his again, the room was as it should be. The solid oak boardroom table surrounded by the good and the great of Ford Tractor Operations.

At the end of the presentation Noel handed each member a hardcopy of his presentation. It came complete with notes supporting each image, every page building a compelling argument for basing their expansion into China, via a business office in South Korea.

When the questions came, Noel explained that while he liked the Chinese people, the same could not be said for the regime who governed the country. Siting recent copyright and intellectual property infringements regularly being covered by western news reports. It seemed that if an aggrieved party challenged an infringement in China's court system, they did little to protect the rights of the western manufacturer. He argued that any, or every aspect of Ford New Holland's intellectual property could be easily copied from inside China without recourse through the Chinese legal system. To him, that was an impediment and too big of an issue to ignore.

During the meeting it was evident to Noel every man and his dog had a differing opinion, but when the assembly finally concluded, his proposal had been accepted in its entirety. Now the work of setting up an office in Seoul could begin,

Assigning new roles to staff he had assembled from his original taskforce and employing other specialists. He had, after several flights to and from South Korea, the right people in place to spearhead Ford New Holland into China. Confident his work with Ford had been completed with dignity, he was ready to say goodbye to his friends in the St Kilda offices and begin the next chapter of his life within the agricultural machinery industry.

CHAPTER SIXTY-TWO

Having sold his farm at Kilmore, Noel had time to look back over his almost two years at Cranbourne. The battle to help FNH emerge from ineffective management had drained his energy. His every suggestion being met with a challenge; every question fielded with resistance; at every turn he felt a door shutting before him.

He knew it wasn't just him, the Ford Tractor Operations people who had transferred from Campbellfield to Cranbourne felt it too. Some stayed, but at the end of three years most had left, feeling they had been forced out and it pained him to see loyal Ford Motor Company staff treated with such disregard. And yet despite all his efforts, he was powerless to confine the toxicity shown them. If the Ford New Holland had no use for him, he would work happily with Ford Asia Pacific until ready to announce his new appointment.

He remembered walking into the Ford Asia Pacific offices on St Kilda Road to a sense of welcoming and being among friends, something he had missed at Cranbourne. Energy started bubbling from each pore and he found himself becoming excited to be at work again.

Appointing Norwood and the successful closing out of Ford's New Zealand tractor operations, showed his superiors that he had lost none of his effectiveness. The China project too, had given him an opportunity to show the management in Dearborn that his skill for evaluating the pitfalls and benefits of an idea were still sharp. It was now time to fly to Honolulu, hand over the completed China project to Gor Cowl and whoever World Headquarters had assigned. Assuming Cowl would have a plan in place to move him as far from the Australian operation as possible, Noel decided to hear them out before committing to anything.

In Honolulu, Noel listened as Gor Cowl and the manager responsible for Ford New Holland's American West Coast, presented an offer that would see Noel moving to the United States and taking a management role based in Dallas. As words poured from Cowl's mouth, all Noel could think about was where he would be going next. This proposition was only one of two and moving his family to the other side of the word didn't appear to carry enough reward with it. At the end of the meeting, he said his goodbyes, knowing that this would probably be the last time he would talk to them as a Ford employee.

Back in his Honolulu hotel suite, he began packing for the flight to Auckland where he would visit Norwood. Then it would be a quick hop across the Tasman Sea and home to Joan and their family. Everything suitably packed, he made a few notes in his diary and checked the airline ticket for the time of departure. Given a choice, he would be on a QANTAS flight, however tonight, he would be flying with United Airlines on their regular Honolulu to Auckland run, Flight 811, he checked his watch and picking up the phone, booked a taxi to arrive with time enough to ferry him to the airport.

CHAPTER SIXTY-THREE

Howard Park, Tallangatta, Victoria, March 10th 1989

Admiring the organised way Joan went about preparing tea and biscuits for them both, he still felt the same as he had all those years ago when he met her at the Rosebud Town Hall. In his eyes she 'd remained the same as that evening on those front steps. Beautiful with just a pinch of sass. His mind wandered back to their courting days and the fun they'd had. The years hadn't diminished the feelings he felt for her and though their love had grown stronger, for them, affection had always been a private thing. He wanted to reach out and take her in his arms and hold her until he felt her melt into him, but there would be time for that later, for now the pressure of his career would intervene once again.

From the moment his flight touched down in Sydney, he had planned and rehearsed what he would say to her over a hundred times. Because of her fears around flying, he would need to be careful about the way he explained the flight out of Honolulu and why he cancelled his visit to New Zealand. Then there was the rationale behind accepting a position with Case IH instead of accepting Cowl's offer of a position with Ford New Holland in Dallas Texas. Even a role with Ford Australia Limited had been offered, but it would require a lot of study and many *hours of grind to reach something similar to that being offered by Case IH.*

For all his cogitating and playing different scenarios in his mind, as usual Joan threw her trust in him, they worked through the costs and benefits as they had done with every new stage of their lives and a move to Case would just be another leap of faith.

Noel found himself leaning back against the sink his interlocked fingers in the middle of her back while she stared into his eyes. Gradually he felt Joan lean back until she was pushing her forearms and fists against his chest.

She pushed away, letting him take most of her weight saying nothing for what seemed to be an age. Then her sobbing started, tears building until rivers ran down her cheeks.

For a brief minute she succumbed to quiet, then unable to contain her anger, a lifetime of being alone to raise their family erupted. Raging at him, she described just how hard it was for her to stay behind, holding their family together while he flew to God knows where in the world. And for what, the Ford Motor Company? The very people who had now cast him aside like an empty cigarette packet. With each sentence a fist hammered into his chest and marking the full stop. He understood her sacrifice, he understood her fear, he knew how much she loved him and every tear, every fist-fall proved it.

Long before Noel had moved to the Ford Asia Pacific office on Melbourne's St Kilda Road, he had been keeping another opportunity secret. Months before his walk in the weeds with Gor Cowl, he been approached by a head hunter from Egon Zehnder, an executive recruitment company. Their officer outlined a desire to engage him for the position of General Manager of Case Australia and New Zealand. Noel, flattered by the interest, knew something this important had more than a small element of risk to his present career. As the meeting proceeded his mind flashed back to Massey Ferguson and how they terminated his employment just for talking to Ford. Knowing that people like these regularly canvassed high ranking executives to fill prime positions, it would not seem strange. However, he also knew how quickly rumours spread and decided it better to keep their interest to himself. At the time he had been fully invested in the New Holland buyout and the work needed to merge both organizations. Until he understood exactly what his role in FNH would be, he would leave the Egon Zehnder offer open.

Within a few days Egon Zehnder made contact again, this time offering him a contract with the terms of employment including a substantial renumeration package. A package he found very tempting. However, even while his loyalty to Ford remained the offer on behalf of Case IH, far exceeded anything he could expect from the new Ford New Holland business.

His source within Ford World Headquarters in Michigan had been keeping him appraised of the company's plan to acquire Case IH from Tenneco, believing it should fall their way soon. Noel understood the USA's anti-trust laws and, that the creation of such a large conglomerate would require selling off many parts of both companies to satisfy the regulators. However, if the deal did happen, his being on the Case IH payroll might offer him an opportunity to renew old Ford Motor Company associations.

Before he said yes to the proposal he needed to know more, could he find ways to improve Case IH's current position, not only in the market place, but their financial stability as well. Not yet an employee, it became important to know what he would be walking in to if they did agree to terms. With the information he had gathered during his time with Ford Tractors and using any data that came through the Tractor and Machinery Association, Noel knew he had enough to build a marketing strategy for Case IH. A winning strategy that could meet New Holland and John Deere head on.

One of the first tasks would be building a database that focused on the dealer body, independently rating them by volume, viability, liquidity, attitude and willingness to grow. While the Case IH dealers were important, he wanted to understand the products better. Which lines were profit makers and which were loss takers? The sales and service history of each and to find those that offered the best opportunity for increasing market share.

Then there were the Case IH company stores to consider. His experience with both Massey Ferguson and to a lesser extent, Ford Tractors South Africa had shown him that company stores became lame ducks in tough times. Places to where excess production can be hidden. Moving it from the parent company's wholesale inventory and burdening their branches with excessive and unwarranted expense. Company stores too often retained burgeoning used inventories. Stock values were written down and sold off in auctions as fire-sale lots. Until he had access to the financial records of company stores, he could only consider them as break even operations at best.

Should he take the job?

Noel believed it would take at least twelve months before he could begin turning Case IH Australia's fortunes around. Although he understood how the products sat within the Australian market, it was time to look at the strengths and weaknesses in both industrial and agricultural operations. It would not be enough to compare the machinery to their competitors, he needed to establish what machines were best in class.

Where a machine had been so identified, he would begin focusing the advertising and sales teams' efforts into developing a strong and vibrant argument for buyers to choose Case IH before all others. Create a desire had always been a pillar of any sales team, but it needed to be founded on fact. In a new role at Case, he would be reverse engineering the information his sales and marketing teams at Ford Tractors had created.

Starting with the new Magnum tractor range, Noel found that since its introduction, Case had conquered many of the transmission reliability issues that had dogged the previous higher horsepower International Harvester tractor offering. Marketing had decided replacing the older white over orange colour scheme of the Case tractor range with a deep red livery on the bodywork, a black chassis with silver for the wheel rims and silver script on black highlights. The new livery set this range apart from everything offered before it. The red, almost the same as that used by International Harvester melded the two manufacturers together. A strategy aimed to bring loyal buyers from both marques to the new Case IH brand. The Magnum range was something Noel believed he could work with.

Most tractor companies in Australia understood that brand loyalty had always been strong among the farming community and they spent huge amounts making sure they retained their customers. For the young Case IH conglomerate, Noel saw this as virgin ground to mine.

All the projections and plans he had commissioned in preparation for merging Ford New Holland may have stayed in his office, but the raw data from which it was drawn could not be forgotten. Finding a way to now use a similar strategy against the very company whose senior executive had said his ideas were all bullshit, now became a priority and Noel wasted no time drafting a new plan. One he knew would make Peter

Morris rue the moment he had uttered those disparaging words.

The more he investigated Case IH, the more he could see that most of the hard work to ratify product-lines had been done. Outdated broadacre tractors having been replaced an all-new Magnum series had just been one such move. Just as Ford had also considered the purchase of Steiger Tractors a priority, Case IH's parent company Tenneco had too and beating Ford to the deal late in 1984.

With a corporate red now claiming Steiger's high horsepower range as part of Case IH, the company had almost completed their tractor line-up, all that they required now would be a new lower horsepower tractor range.

Applying the lessons their engineers had learnt designing the Magnum, Case IH's Maxxum range again set a bench mark for the company. This time they had built in a new factory located in the city of Neuss in Germany for the project. From the first day of its introduction, the Maxxum series had been tested in the competitive European market and each year had the company been taking huge chunks out of Ford New Holland, John Deere and Massey Fergusons' market share in many regions across the continent.

Noel could see a similar trend happening in Australia and New Zealand. However, it would be defeating Cowl in the Australian Combine Harvester and hay machinery markets that would bring him most pleasure.

International Harvester's Axial Flow combine had started its life as an on-paper idea in the late 1950s when a rotary separator concept, as described by Elof Karlsson and drawn by Mel Van Buskirk found its way onto paper as <u>Rotary Separator</u> <u>Fig. No. 1</u> on the 23rd of November 1956. At the time, most of the harvester manufactures had been based in the Mid-West of USA. John Deere had its factory next door to the International Harvester facility and as an industry, engineers and line workers from competitive factories would often mingle at church services and other social events. It proved that keeping projects a secret would always be a challenge.

Understanding how espionage and rumour could undermine their development, and because their concept would be so advanced, the engineering team needed a place to develop their ideas away from curious on lookers. A super-secret garage where they could take the project forward without any fear of their research being sold to, or copied by a competitor.

That is how the early development of the axial flow concept began life, inside a locked garage within the East Moline Plant. The building had frosted windows and for the next seventeen years, access would be strictly limited to a few key engineers. Elof Karlsson and Mel Van Buskirk working with a small team of engineers, developed a wide variety of concepts and for years no evidence of their work appeared on the company's books.

By 1969 all that had changed the project was now showing in the accounts and its development was proving potential to Harvester management. Development now continued at pace with test versions being run as far away as Australia until the release of the 1440 and 1460 models in 1977.

For New Holland however, it would take another two years before they were able to release their version of a rotary combine, the twin rotor TR70.

By the time Noel Joined Case IH, rotary combines had become harvesters of choice for farmers in dryland areas. Soil conservation groups driving a message that supported an idea of leaving a mat of fine straw behind the harvester would conserve moisture. With rotary combines the operator could run the header knife as close to the ground as possible and the thresher would pulverise the straw into small lengths. While conventional combines could manage the same effect, when it came to sowing next year's crop minimum and tillage being practiced, a shorter chopped straw residue would be required. The longer straw leaving the walkers of a conventional combine would build up around tines, clogging narrow spaced seeding and harrowing implements. It was the advent of prickle chains replacing a conventional or finger harrow systems, that saw seeding times reduced leading to better germination and higher crop yields

Satisfied the Case IH product offering would equal or better the competition and, the having scope of his authority noted in his employment contract, Noel embraced the challenge. The time was ripe to take the fight to the hierarchy at Ford New Holland Australia.

CHAPTER SIXTY-FOUR

Noel pleased that by mid-1990 he could see solid results coming from the solid sales and marketing initiatives that his teams had put in place. The Case IH Magnum range had been steadily increasing its market share and was threatening market leadership. A knock-on effect, being the ever-increasing number of targeted marketing campaigns coming from their competitors. Axial-Flow combine harvesters had always proven to be a solid competitor and because these machines were of high value and profit leaders for the company, Noel had made himself readily available to dealers who were requiring assistance in getting a deal over the line and the strategy worked.

Noel's easy nature and a great capacity for work was bringing with it a camaraderie not experienced by his colleagues in his recent past. Dealers too, had a new found confidence in the company's willingness to listen and, for him it felt more like his days at Ford when enthusiasm drove the workplace. Enjoying regular contact with dealer principals and their staff, Noel looked forward to assisting a salesperson to clinch a deal and his regular visits to dealerships continued, building solid relationships.

However, his help was never at the expense of profit. Initiatives to cut costs were having a positive effect on the business and programs to reduce aging stock levels all counted toward his goal of market leadership.

Halfway into his contract with Case IH Australia, as was routine, he had been going through the morning's mail when his secretary, Sue Hardman knocked and passed him a fax from Robert Carlson. It was not unusual to receive communications from Case IH's Headquarters in Racine Wisconsin USA, however Bob had scrawled a message in his own hand across the top of a letter he had received from Gor Cowl.

Receiving communiques from competitors had always been commonplace in the tractor industry and normally Noel would not have given it priority. This however, was not a sales bulletin it came from Robert Carlson the CEO of Case IH and attached was a Ford New Holland Letterhead, its message to the head of the company from Cowl and for Noel, its message most personal.

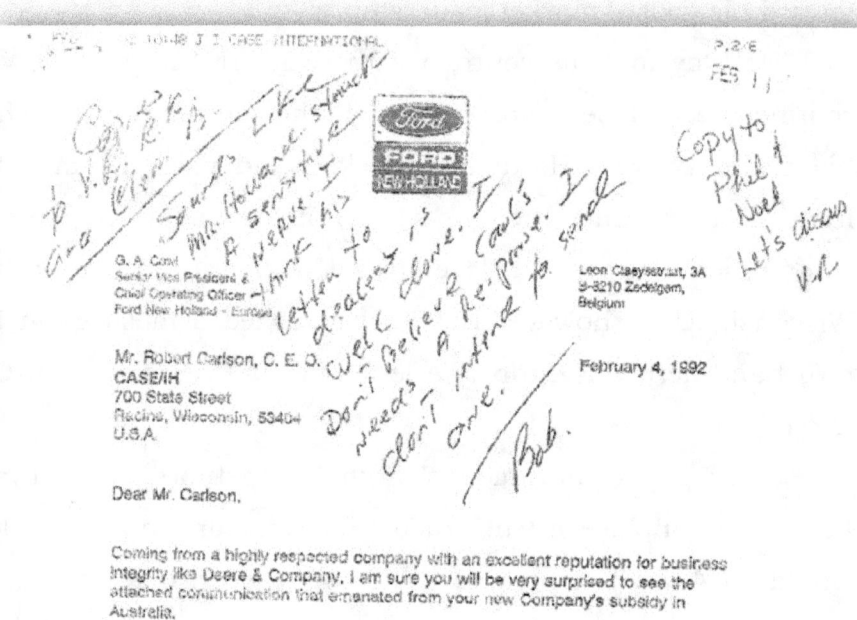

Ford New Holland

G. A. Cowl
Senior Vice President &
Chief Operating Officer
Ford New Holland - Europe

Leon Claeysstraat, 3A
B-8210 Zedelgem,
Belgium

Mr. Robert Carlson, C.E.O.
CASE/IH
700 State Street
Racine, Wisconsin, 53404
U.S.A.

February 4, 1992

Dear Mr. Carlson,

Coming from a highly respected company with an excellent reputation for business integrity like Deere & Company, I am sure you will be very surprised to see the attached communication that emanated from your new Company's subsidy in Australia.

For your information, the author, Mr. R. N. Howard is a very bitter ex-employee of Ford (New Holland).

We take very strong objection to Mr. Howard's inference that "the Ford name is being phased out". At a minimum, I ask you to consider requesting a retraction of this statement by either Mr. Howard or his superior in Case-Australia.

I find this type of "mud-slinging" even in "internal" communications very unprofessional and I would hope not to see a continuation of this type of conduct. I believe it would be desirable if in the future, our Companies can compete on a basis of product value and after sales supports to the Australian agricultural community.

You may wish to obtain other viewpoints on the business conduct and ethics of Case Australia from some of your former colleagues in Chamberlin-Deere.

Sincerely,

G. A. Cowl

mdv

Handwritten annotations:
- "Copy to R.B. / Circulation"
- "Sounds like Mr. Howard struck a sensitive nerve. I think his letter to dealers is well done. I don't believe Cowl needs a response. I don't intend to send one. Bob."
- "Copy to Phil & Noel. Let's discuss. V.R."

G.A. Cowl
Senior Vice President & Chief
Operating Officer
Ford New Holland – Europe

Leon Claesstraat, 3A
B-8210   Zedelgem
Belgium

Mr. Robert Carlson C.E.O.
CASE/IH
700 State Street,
Racine, Wisconsin, 53404
U.S.A.

February 4, 1992

Dear Mr. Carlson,

Coming from a highly respected company with an excellent reputation for business integrity like Deere & Company, I am sure you will be very surprised to see the attached communication that emanated from your new Company's subsidy in Australia.

For your information, the author, Mr. R.N. Howard is a very bitter ex-employee of Ford (New Holland).

We take very strong objection to Mr. Howard's inference that "the Ford name is being phased out"! At a minimum, I ask you to consider requesting a retraction of this statement by either Mr. Howard or his superior in Case-Australia

I find this type of "mud-slinging" even in internal communications very unprofessional and I would hope not to see a continuation of this type of conduct. I believe it would be desirable if in the future, our Companies can compete on a basis of product value and after sales supports to the Australian agricultural community.

You may wish to obtain other viewpoints on the conduct and ethics of Case Australia from some of your former colleagues in Chamberlin-Deere.

Sincerely,
G.A. Cowl

Gor Cowl's letter, accompanied by a copy of Noel's letter to Case IH dealers, a copy of Ford New Holland's media release referring to the recent sale of New Holland by Ford Motor Company to Fiatagri introducing a new name for the company was attached. It had been supported by a single page, two column, twenty-one dot-point listing of each company's history, and to emphasise the name change employed the tagline: **"More Than Ever"**.

The thrust of Noel's letter to Case IH staff, the company's dealers and sales and service teams had been an opportunity to take advantage of the confusion created by NH Geotech's press release and its tagline:

THIS IS NH GEOTECH

A MAJOR COMMITMENT TO FARMERS.

A NEW IMPETUS FOR PRODUCT INNOVATION.

IMPRESSIVE FINANCIAL BACKING.

A GLOBAL GROUP SUPPORTING TWO SEPARATE NETWORKS DEDICATED TO THE SALE AND SERVICE OF OUTSTANDING PRODUCTS.

THIS IS N.H.GEOTECH.

N.H.GEOTECH

It was in Turin in 1919 that Senator Giovanni Agnelli launched the production of the 702 tractor, the first "machine for the land". This marked the birth of Fiat's Farm Equipment Sector, which came to be known as Fiat Trattori.

In 1881, Pietro Laverda, a mechanic, opened a workshop in Breganze to make machines for the combined cultivation of grapes and grains; this was the origin of Laverda.

In 1947, Lyle Yost, a farmer from Hesston, invented new technical solutions for hay-handling and forage-harvesting equipment and created the base for the future development of Hesston.

In 1988, the integration of these organisations – to which Braud and Agrifull were later added – resulted in the creation of Fiatagri, with its full line of machinery and equipment to serve the needs of farmers.

To focus on every aspect of soil engineering, FiatGeotech was formed in 1988 by uniting Fiatagri, a specialist in farm machinery, with Fiatallis, a specialist in earth-moving equipment. FiatGeotech has 9 factories, where it manufactures the products it sells in 130 countries, through more than 2,600 dealers.

But today's competitive challenges make it necessary to constantly anticipate the needs of farmers internationally, offering leading-edge innovation.

With this goal in mind, Fiatagri and Ford New Holland have decided to work together within one organisation – N.H. Geotech – while preserving their separate identities.

The optimisation of the two systems will release added resources, which will be dedicated to product development and innovation for the benefit of farmers around the world.

It is also certain that Fiatagri will retain its own brands, its own distribution and service networks and its own marketing policies.

From 1991 ...... – Fiatagri, more than ever!

It was in 1907, even before he turned his attention to automobiles, that Henry Ford realised the importance of agricultural machinery and paved the way for the production of tractors on an industrial scale. In 1917 the first of many Ford tractors came off the line.

In 1895, Abe Zimmerman, a mechanic, decided to establish a workshop in Pennsylvania to design machinery more responsive to the needs of farmers. That's how New Holland was born.

In 1906, Leon Claeys started his workshop dedicated to agriculture in Zedelgem, Belgium. His son, Louis Claeys, developed the first self-propelled combine 1952. This company merged with New Holland in 1964.

In 1945, in a Toronto basement, Peter Pakosh invented a portable grain auger and laid the foundation for the company that was to grow into Versatile.

In 1987, these organisations were integrated to Ford New Holland, a group that is one of the world leaders in the manufacture of farm machinery.

In plants located on three continents, Ford New Holland manufactures products that it sells in 150 countries, through more than 3,900 dealers.

But today's competitive challenges make it necessary to constantly anticipate the needs of farmers internationally, offering leading-edge innovation.

With this goal in mind, Fiatagri and Ford New Holland have decided to work together within one organisation – N.H. Geotech – while preserving their separate identities.

The optimisation of the two systems will release added resources, which will be dedicated to product development and innovation for the benefit of farmers around the world.

This also means that Ford New Holland will retain its own brands, its own distribution and service networks and its own marketing policies.

From 1991 ...... – Ford New Holland, more than ever!

When Noel had first read the N.H. Geotech press release, it made him smile. For a marketing man, this was manna from heaven for a competitor and he thought others in the tractor and machinery industry would think the same. If he found it confusing in its message, then he had no doubt others would find it difficult to make sense of too, and it was a weakness he believed Case IH Australia could exploit.

Making it a priority, on August 28th 1991 he had drafted his dealer letter to the dealers, which Cowl had also attached in his communication to Robert Carlson, it read:

Case (Australia) Pty. Ltd.

AUGUST 28, 1991

TO: ALL DEALERS

FIATAGRI/FORD ADVERTISING

Attached is a copy of a recent advertisement and in an endeavour to understand what the advertisement is intended to communicate we came up with 4 main observations:-

1. Whilst it gives both "Ford New Holland and Fiatagri" most impressive pedigrees it does not explain who or what N.H. Geotech is.

   - Is it a new company formed by the two?

   Is it a management company, an R & D company, a distribution company or a finance company or all of the above? Does N.H. mean New Holland and if so what about Fiat?

   The ad does not answer these questions and is therefore confusing. Confusion always creates anxiety in a customers mind.

2. It repeats the proposition for both brands that ..... the optimisation of the two systems will release added resources ..... dedicated to product development and innovation for the benefit of farmers ..... Ford/Fiatagri will retain its own brands, distribution and service networks and its own marketing policies .....", and in so doing compounds the confusion.

   How can this new organisation N.H. Geotech help the two companies since they will continue to carry all the expensive operating items separately and individually?

   Apart from a rhetorical promise of product development and innovation it gives no explanation of the reason for the two companies to co-operate. Confusing! Worrying!

3. The key points of finance - "impressive financial backing" - and support - "a global network supporting two separate groups dedicated to the sale and service of outstanding products" - that would provide a great deal of reassurance are simply statements with no explanation.

   We would have thought that these are critical and worth expanding to make sense of the whole idea.

P.T.O.

Accredited to AS3901 Standard for Quality Control

Page Two

Maybe Ford/Fiat haven't worked out the details yet? Thats what it sounds like. However, as Phillip Benton said "The new company will sell Ford branded equipment for at least a period of time", does the ad mean that the Ford name will disappear soon?

4. For those who like to read between the lines - and pretty well every farmer is smart enough to do that - the line "But today's competitive challenges make it necessary to constantly anticipate the needs of farmers internationally, offering leading edge innovation", suggests that the two companies couldn't afford to invest in the future on their own so they got together.

    Still it begs the question about who or what N.H. Geotech is.

    Lets assume its a jointly funded R & D group. Since the ad does not say that we consider the customers will assume that the Ford name is being phased out.

Our summary in short is:
- A golden opportunity missed to set all the rumours to rest with an aggressive explanation.
- too much rhetoric and not enough fact.
- confusing and therefore worrying.
- they would have been better off doing nothing.

Well, its only this type of advertisement which will, in our opinion, continues the unrest which exists in the Ford New Holland and Fiat dealerships.

Yours faithfully,
J I CASE (AUSTRALIA) PTY LTD.

N. NOONAN
GENERAL MANAGER, AG DIVISION

Terry L Probert

TRANSCRIPT:

August 28, 1991

TO: ALL DEALERS
<u>FIATAGRI/FORD ADVERTISING</u>

Attached is a copy of a recent advertisement and in an endeavour to understand what the advertisement is intended to communicate we came up with 4 main observations: -

1. Whilst it gives both Ford New Holland and Fiatagri most impressive pedigrees it does not explain who or what N.H. Geotech is.

    Is it a new company formed by the two?

    Is it a management company, an R&D company, a distribution company or a finance company or all of the above? Does N.H. mean New Holland and if so what about Fiat?

    The ad does not answer these questions and is therefore confusing. Confusion always creates anxiety in a customer's mind.

2. It repeats the proposition for both brands that .... the optimization of the two systems will release added resources ..... dedicated to product development and innovation for the benefit of farmers ..... Ford/Fiatagri will retain its own brands, distribution and service networks and its own marketing policies .....", and in so compounds the confusion.

    How can this new organization N.H. Geotech two companies since they will continue to carry all the expensive operating items separately and individually?

Apart from a rhetorical promise of product development and innovation it gives no explanation of the reason for the two companies to co-operate. Confusing!   Worrying!

3. The key points of finance - "impressive financial Backing" - and support - "a global network supporting two separate groups dedicated to the sale and service of outstanding products" - that could provide a great deal of reassurance are simply statements with no explanation.

We would have thought that these are critical and worth expanding to make sense of the whole idea.

Maybe Ford/Fiat haven't worked out the details yet! That's what it sounds like. However, as Phillip Benton said, "The new company will sell only Ford branded equipment for at least a period of time" does the ad mean the Ford name will disappear soon?

4. For those of us who like to read between the lines - and pretty well every farmer is smart enough to do that - the line "But today's competitive challenges make it necessary to constantly anticipate the needs of farmers internationally, offering leading edge innovation", suggests that the two companies couldn't afford to invest in the future on their own so they got together.

Still it begs the question about who or what N.H.Geotech is, the ad does not say that we consider will assume that the Ford name is being phased out.

Our summary in short is: - a golden opportunity missed to set all the rumours to rest with an aggressive explanation.

    Too much rhetoric and not enough fact

    Confusing and therefore worrying.

    They would have been better off doing nothing.

Well, it's only this type of advertising which will, in our opinion, continues the unrest which exists in the Ford New Holland and Fiat dealerships.

Yours faithfully,

JI CASE (Australia) PTY LTD
R.N. HOWARD
GENERAL MANAGER,   AG DIVISION

Receiving a copy of Gor Cowl's letter had been a surprise to Noel, he had done no more than alert his dealers to a situation he thought they could exploit and gain sales from. The fact that it had unearthed some deep-seated bitterness from its author was evidenced in the venom contained in the second paragraph. That it had caused his old foe to respond, was a thing to savour.

However, it was Bob Carlson's hand scrawled note across the top of Cowl's letter that gave him most joy:

> *Copies to V.R, R.K. and Clem B.*
> *Sounds like Mr. Howard struck a sensitive nerve. I think his letter to dealers is well done. I don't believe Cowl's needs a response.*
> *<u>I don't intend to send one.</u>*
>
> *Bob*

In the far-right top corner, a date stamp marking the fax as FEB 11, below that a copy list to Phill & Noel with the message: Let's discuss and it was initialled by VR.

Taking a few days to think through how he would respond. Gor Cowl's statement about him being a bitter ex-employee had grated with Noel and believed he should explain where the reference had come from. His personal integrity had been questioned and needed explaining. Deciding a hand written letter to be the most honest way of answering Cowl's assertions, he faxed the following to Robert Carlson in Wisconsin:

Bob

Gen Cowel was the head of the overseas operations and it was Gen who along with Moglia presided over the "Sham Meeting" in Cranbourne when Alan Montgomery & Peter Morris attacked my integrity as part of a campaign to shore up Nels Craig the then Managing Director.

I was excluded from the meeting which was conducted with appox 50 staff members and my sub was never discussed. Cowl pandered to the incompetence of Craig and did not support his old Ford team. I had advised my then Director Bernard Sarges that there were many problems at Cranbourne and that I was not prepared to work with Craig.

I had already been approached to join Case therefore the die was cast as far as my future was concerned, and Cowl arranged a meeting in Maui for me to move to the USA for which I was pleased, but I was aware the business was on the market. Cowl did not give me credit or did he accept that I had a conduit to the top management at Dearborn telling me that FNH will be sold, and that advise received in 1988.

I stand by my letter to Case dealers and state that Cowl is misleading you in his attempt to "play the man".

Bob I am prepared to wager my <u>Family</u> and <u>Farm</u> that the FNH or Ford name will "disappear" and the Automotive News Extract of April 30th 1990 attached will vindicate the future of Ford in this debate.

As Case Australia is now <u>leading the Australian market</u> and <u>Gens Geotech is floundering</u>, you be the judge on Cowls credibility. My period at Ford was a privilage but FNH a disaster.

Regards

Mark Harris

Feb 18th 1992

Bob

Gor Cowl was the Head of the overseas operation and it was Gor who along with Moglia presided over the "Sham Meeting" in Cranbourne when Alan Montgomery & Peter Morris attacked my integrity as part of a campaign to shore up Nels Craig the then Managing Director.

I was excluded from the meeting which included approx 50 staff members and my side was never discussed. Cowl pandered to the incompetence of Craig and did not support his old Ford Team. I had advised my then Director Bernard Sarfras that there were many problems at Cranbourne and I was not prepared to work with Craig.

I had already been approached to join Case therefore the die was cast as far as my future was concerned, and Cowl had arranged a meeting in Maui for me to move to the USA for which I was pleased, but I was already aware the business was on the market. Cowl did not give me credit or did he accept that I had a conduit to the top management in Dearborn telling me that FNH will be sold and that advise I received in 1988.

I stand by my letter to Case dealers and state that Cowl is misleading you in his attempt to "Play the Man."

Bob I am prepared to wager my <u>family</u> and <u>farm</u> "that the FNH or Ford name will disappear" and the Automotive News extract of April 30th 1990 attached will indicate the future of Ford in this debate."

As Case is now <u>leading the Australian market</u> and Gor's Geotech <u>is floundering,</u> you be the judge on Cowl's credibility. My period at Ford was a privilege, but FNH a disaster.

Regards,
Noel Howard
Feb 18th 1992]

Pressing the send button on the facsimile machine and watching the paper roll through its internals felt like a cleansing. Gor Cowl had proved it was he, who was charged with bitterness. Moving to Case IH and the success they were experiencing filled Noel with pride. Cowl had underestimated Noel's reach and influence, the power of the contacts he had built relationships with while working in various roles within Ford and, the esteem those people held him in.

Because of his ability to cut through to the source of problems and allowing his teams to find solutions, results improved, Case IH enjoyed a steady rise in the markets where Noel worked. His diligence and his easy-going nature had seen him still being welcomed as a confidant to North America's automotive royalty, the Fords.

CHAPTER SIXTY-FIVE

Being with Case for well over two thirds of his three-year contract and having led the agricultural products sales team to market leadership. Noel now felt it was time to shine a light on an internal problem that had troubled him from the very moment he started in Sydney.

Scuttlebutt and rumours of corruption continued to persist even after the company transferred its offices from Northmead to St Marys. Noel believed it important for it to stop and only having inuendo to work from, he began planning a strategy to gather solid evidence, to prove or refute the persistence of these stories.

Case IH products continued to sell well against the increased efforts of their competition. He credited a reinvigorated sales and marketing team who had under his leadership won their dealer's confidence. Trust in the company had been restored and that trust was flowing into their customers. Compared to the final days of International Harvester, a Case IH Australia dealership was now an agreement to be coveted.

Joan and their son-in-law Mark had the family farming operation working like a well-oiled machine and every flight home from Sydney reinvigorated him. His eagerness to drive around the cattle, or stroll along a fence line clearing sticks from fallen trees, took him back to his childhood roots on his grandfather's farm in the Kiewa Valley. While he often made management suggestions, deep down he knew the two of them had everything in hand. Joan and Mark had attended Dookie Agricultural College and managed each of their roles expertly.

He now had time to socialise with old friends and family in and around the twin cities of Albury and Wodonga. Downtime had always been a precious commodity in the past, always spent working in and around their farming projects. Doing everything they could themselves, had made both he and Joan quite resourceful. By only relying on each other over the years, they had made sure they were always working efficiently.

A consummate networker, Noel began thinking about a life after tractors and questioned if he wanted to continue his life flying around the world. Never being home and missing time with his children takes its toll, but had it taken enough? It didn't take long for him to him realise that if he didn't extend his contract with Case IH, it would not be the end of his world. A new challenge away from tractors and machinery may be just what he needed. The more he thought about it, the more enticing the idea became, but what could he do?

Thinking about a new challenge brightened his mood and yet there were still some loose ends at Case that needed attending to. He would take his holidays as usual and give the matter some thought before he returned to work.

Reading a through "The 1990 Case Report" over Christmas, Noel stopped at page thirteen, and stared into the faces of his colleagues. Tapping his index finger on a director's face, he asked himself if the rumours had substance, or was it all a matter of sour grapes?

Company travel arrangements had been one area that fuelled rumours. Staff from different areas of the business pointed out that the business engaged to handle all of Case IH's flight and car hire bookings was an agency owned by a manager's sister. Noel, believing this could easily be construed as a conflict of interest and an opportunity for fraud to exist, knew the business transacted could be legitimate. However, the stories persisted, but without any proof everything was gossip and hearsay.

Another story that came to his attention was the purchase of a luxury power yacht that he understood had been ordered for a wealthy client. The vessel had been built in Western Australia, but by the time the yacht had been completed, the deal had fallen through. Leaving the company with a multimillion-dollar asset they couldn't sell.

The rumours being relayed to Noel inferred that the "client" never existed. Those telling him adamant that the story had been concocted. A ruse to cover the company's upper management fraudulent desire to own a piece of marine luxury at an attractive and grossly written down price.

On more than one occasion the same statement had been reported to him and, by people in positions of access right across Case IH Australia. To Noel, if there was something wrong at the top of the company he now represented and if it were true, just thinking about it made him feel dirty through association. Over the years he had been involved in several small issues with employees, but if these accusations proved true, it would have been thievery on a grand scale.

It may have been paranoia, or his senses may have been in tune to the scuttlebutt, but he thought there may be an answer to his questions found as his contract neared its completion date. Then, when a high-ranking company official approached him to become part of the cartel interested in buying a share in this luxury yacht, did he fear this approach would have him party to these persisting accusations.

The executive who made the approach had said this yacht having no chance of a sale in Western Australia would be freighted overland from Perth to Sydney at the company's expense. Noel understood the cost of freight from one side of the country to another and quickly calculated a rough figure. It alarmed him. Having written down its value, it was now showing on the company's balance sheet for a small portion of its original cost. The deal would have been a wonderful investment if it didn't reek of corruption. His colleague persisted saying, if Noel could stump his portion of the cash required, he too could join their consortium. Laughingly, he agreed with the proposer that it was a deal of a lifetime and as it sounded too good to be true, he would decline their invitation.

Having passed on the opportunity to join their boat buying cartel, Noel started to feel like a marked man. He knew his decision had consequences and with his senses keen, soon started feeling a target for his superiors. Those fears becoming even more acute when he received a terse phone call from a Case Construction Equipment dealer in Melbourne. Angry and looking for answers, he claimed that after a visit from the company's finance director, who told him that Noel had ordered the termination of his dealer agreement. Something that was clearly not the case.

With nothing more said about the boat buying proposal, Noel continued to focus on sales and strengthening the dealer network. However, after coming back from an appointment in Sydney he found an unknown man sitting in the lounge leading to his office. Wondering who he was and asking why he'd be there had Noel's senses primed. When the gentleman said he was from an employment team and was there to conduct a routine psychology test with Noel, he felt his hair stand on end. To him if this were routine, he'd have been made aware of it happening. Could this be another attempt to undermine him?

Copies of his recent tests from Ergon Zehnder and those by Chandler McLeod, (both of whom had recommended his engagement) he'd supplied Case IH prior to his appointment. Angry that he'd had no prior notice of the intruder's mission, Noel explained there would be no need for such a test, opened the door and waving his hand toward it, ordered the man out.

Believing this proposed test being part of an orchestration to be rid of him, by those who had tried to sell him a share in a suspect boat purchase, it was time to leave Case IH Australia. Before Noel resigned, one of his last acts at Case Australia was to forward a report to a superior in Wisconsin. Included was a record of the approach of the boat buying cartel and the evidence that suggested the deal was corrupt.

It would now be up to those at the head of Case IH in the USA to make a decision on how they would proceed and, if an internal investigation was warranted. Taking a last look around after packing his personal items including messages and cards from well-wishers, without regret, he left the building for the last time. A new life in Wodonga beckoned.

CHAPTER SIXTY-SIX

Noel knew Joan and Mark had the farming operation under control, and it might take some time to slip back into life after tractors. He wondered what career move might come next. For the first time since he left the ambulance service, could he live a life without tractors? While becoming excited for a change, he quietly canvassed his close contacts for another opportunity, whatever it might be.

Sitting at a table in the dappled shade of the vine covered veranda, Noel mentally listed his skills and how they might transfer to another industry. Without thinking, for the first time in years he found himself creating a resume and immediately forced the thought from his mind. He knew his value and believed someone from his many contacts would soon come calling with a new project to tackle, or knowledge of a management position that would suit him.

The first few weeks at home energised him and feeling his batteries now charged, he began pursuing that new challenge. He found himself missing the cut and thrust of managing a large and diverse organisation. As the weeks passed, Noel became restless, burning with need of an office routine to quell his desire to be busy. Through the wide range of people, he could be call on, Noel knew he would soon find something of a National or International nature.

However, now in his early fifties, he thought something closer to home might be a better option. He had many influential friends in the twin cities of Albury and Wodonga and it being a major centre for government business, primary production with many major manufacturing opportunities, there may be an opportunity close to home. Canvassing for employment would seem desperate, however if he let people know he was available, interest would come.

If not actively looking for a position within the tractor industry, he knew it would bring a chance to show how skills and experience he had accumulated could transfer into other industries. Quietly, he let his local colleagues know he was available should any of them be looking for a general manager.

An approach did come and from an unexpected source. James Nagorcka and his Waltanna engineering company had been developing a new product set to revolutionise the way tractors would transmit engine power to the ground.

James, his wife June and their company Warane Pty Ltd had been building a successful range of Waltana Tractors at their Hamilton Factory since 1977. For Noel, the approach had similar synergies to his pre-Ford New Holland days when James and his company had built the Australian FW Series tractors for Ford dealers to market.

The Western Victoria based company had developed a rubber track Waltanna and had released to the Australian market in 1989, coinciding with the southern Australian introduction of Caterpillar's Challenger range. Noel understood the irony of these niche models hitting the market in the same year and he knew Caterpillar dealers had been distributors of Waltanna tractors in the early years.

Proving Waltanna was a company of innovation, during 1988 they launched their Timber Forwarder to compete for sales to logging contractors in Australia's extremely competitive forestry equipment market. However, while the sales team were working on ways to attract buyers to existing products, the Waltanna research and development team had an extremely secret design in development. One they knew would revolutionise and influence farm tractor design during the next decade.

Understanding how soil compaction caused by tractors impacted the growing cycle of grain crops, James and his engineers, workshopped ideas. Their goal would be reducing the downforce applied to the ground as the tyres passed over it. From those early conversations, the "Waltanna Drive" system was born. Before long patents were being applied for and partners sought to adapt the system to competitive products.

Waltanna Drive is a positive drive track system with a rolling drive bar that incorporates track alignment and air ride suspension. The working design proved a significant technological breakthrough.

By 1993 Waltanna track systems incorporating their Waltanna Drive had been sold in Japan, Europe, the United

States of America and the United Kingdom. However, James and June knew taking it to the world would require new thinking if they were to exploit the opportunities that an international market might bring.

Ongoing talks with Deere and Company, had by the end of February 1993, seen both parties entering into a Joint Development Agreement to further develop the Waltanna track system for both agricultural and industrial equipment. This agreement was the culmination of discussions that had begun in 1990.

Development continued and by the time Noel had declared his interest, James and June's, Warane Pty Ltd, entered partnership with Doug Shears' company ICM Australia Pty Ltd, his interest making him a principal shareholder of their new company, The Waltanna Corporation.

Investment from Doug Shears meant the new company had everything required to increase production. Part of the plan they had workshopped, was to partner with either Deere and Company, or one or more original equipment manufacturers. In the vernacular of the day, they had planned for a "Moon-Shot" and, were prepared to jettison any ideas that may have proved fanciful. If their presentation would be accepted in full, it could have reinvigorated tractor manufacturing in Australia.

Warane held, International Patent Applications for a "Positively Driven Elastomeric Tracked Work Vehicle," and with applications filed in Australia, USA, Canada, Japan, the United Kingdom and Europe, they believed their proposal strong.

Patent rights had already been assigned to Deere and Company for the Australian Patent Application. This alone signalled to Noel their approach to him had been serious and reinforcing his confidence that their mission might succeed.

By Monday the 13th of December 1993, Noel, James, June along with Doug Shears and Tim Healy representing The Waltanna Corporation met with senior executives of Deere and Company. The Australians had thoroughly researched and rehearsed their presentation. So polished were they, that as a group, they had tried to anticipate every curly question that might arise and have an answer ready.

Halfway into the first day's presentation by their colleagues from ICM Australia, Noel and James felt they were losing the interest of Deere and Company's people. Understanding the need to stay on message, James proposed they adjourn proceedings and reconvene the following day. John Deere's executive agreed and the Australians retired to their hotel.

For the next meeting, James took control of the Waltanna presentation. Explaining how he saw the synergies between a large organisation like Deere & Company and a much smaller operation like Waltanna could benefit both. Noel believed James had from their early development agreements earned the respect of Deere's executive. In his opinion, their colleagues, by leading the first day's discussions had come across as brash Aussies pitching a complex raft of scenarios and by doing so, had lost their audience.

James and June were tractor people, they understood the ebb and flow of a business that lives within the rhythm and seasonal nature of agriculture. Noel, when asked for his opinion, leaned on his years of managing a large and structured organisation to deliver informed and considered answers.

Boarding the plane for the return journey to Australia, Noel believed that when it looked as if their sojourn to the United States may have been a wasted trip, James and June's presence of mind had rescued the day.

Since that early meeting, the popularity of Waltanna's positively driven elastomeric drive systems has grown. Farmers across the globe are now enjoying the benefits of lower ground compaction because of it. Noel has always spoken with pride about his friendship with the Nagorckas and delighted in seeing the system employed on an increasing number of John Deere products each year.

At home, Noel continued working with Joan and Mark, making improvements to the infrastructure of their farming operations until 1996. Then he joined Mann, a large hardware firm, as their managing director. This appointment provided a new challenge laced with responsibility.

Likening it to his grandfather's much smaller Kiewa operation, he relished the opportunity to guide this long-established family business into a new Century and a new chapter of Rhoden Noel Howard's life began.

CHAPTER SIXTY-SEVEN – The Eulogies

During our many discussions, Noel often said that his desire to tell his story was his way for letting his children know what he did while he was absent in their early lives. He emphasised the importance of recording the events that drove the improvements in mechanising Australia's farmers. He asked the text be simple to read and entertaining.

Only two days before he died in August 2021 one of three proof copies of this book was delivered into his hands. As his body surrendered to the tumours, his children took turns reading parts of it to him.

His funeral as deemed by the laws during Victoria's Covid 19 lockdowns could only be attended by a few of his family. However, with the advances in communication, the celebration of his life had been livestreamed. The family's initiative allowing a huge number of mourners to share the service in real time and others availing themselves of the video later. Because of the impositions Covid 19 forced on Australia at the time, transcripts of the eulogies given by Ian O'Rourke and his daughter Tania can be found below.

Eulogy written by Ian O'Rourke and read by Tania Foss, Noel's daughter.

Terry L Probert

Good morning, friends,

First, I would like to thank the Howard family for the opportunity to relay my thoughts during this sad time.
For over 59 years I feel very privileged to have been a close friend and work colleague of Rhoden Noel Howard.
I can still clearly remember a sunny day in September 1962, when I drove into a Massey Ferguson dealer's yard in Woomelang, in north Victoria. As I alighted from my vehicle, a sprightly young man with a crop of black hair walked toward me with his right hand outstretched.
"You must be Ian; my name is Noel Howard and I have joined Massey Ferguson and I have been told to help you in looking after the dealers in the top half of your territory."
Little did I realise at the time how long this personal friendship would grow, also meeting Joan and seeing their family as they grew up. I was so proud, when noel asked if I would be Geoffrey's godfather. For over two and a half years noel and I worked alongside each other, with Massey Ferguson.
However, in January 1964 I decided to sever ties with Massey and joined ford Australia Tractor Operations, and two weeks later Noel came over to Ford. In time, we were both elated to have made the move.

We worked in different sections of Ford Tractors, we crossed paths on many occasions, but our personal friendship continued.

Noel had an amazing brain and it did not take long for Detroit management of ford to recognise his talent. About the late 60's, he, together with Joan and his young family were posted to South Africa having been handed the task of revitalising the dealer network there.

NOEL   An Authorised Memoir

As expected, Noel completed the assigned task with great success, returning to Australia in 1973 to take over the position of General Manager of the local Ford tractor division.

Three months after he returned home, he 'phoned me one weekend [*asking me*]to visit him at home; whilst over a couple of beers he said he had advised Detroit I was to be the new tractor service manager, working directly with him.

Our personal friendship continued; whilst at work we came together on a constant basis dealing with the day-to-day situations.

Noel clearly showed his capability in running a business operation covering Australia. He always had new ideas on every facet of the company's operation that definitely impressed Detroit.

Noel was a very active member of the Australian Tractor and Machinery Association; assisting in gaining Federal Government approval to legislate the fitting ROP [*roll over protection*] frames to all makes of tractors, thus saving driver's lives.

He also recommended to Detroit Engineering to review the possibility of designing plastic fuel tanks to the tractors to eliminate the delays in refilling the small fuel tanks on our tractors. His suggestion was immediately taken on board. One of his greatest achievements was the development of the Ford 8401 model tractor. In Australia Ford did not have a model in the 100 hp range and competition was having a right-royal field-day selling their 100hp models. To say Noel was not happy was an understatement.

One day in the late 70's Noel took me out into the assembly area to see where he had arranged to split one of our 115hp tractors and join it onto the rear end of one of our 90hp models.

He said, "we will cut the power down from 115 to 100 hp, and that will be our answer to competition."

I was taken back and did comment, "there will be troubles here, as Detroit will create a hornet's nest against you.'

But Noel said quite strongly, "we have to do it in the interests of the company."

There were a few stumbling blocks, but Noel covered them all; particularly by approaching and gaining the strong backing of Brian Ingles, the CEO of Ford Australia. This was a classic [*Noel*] move.

When Detroit became aware of Noel's project, they were not pleased with Noel's ingenious plan being made without their approval and sent a senior manager out to Australia, obviously to kill Noel's brilliant idea. Noel's incredible planning and determination, plus the backing of Brian Ingles and his automotive team resulted in total approval of the project. Manufacturing the, '*new model*,' in the UK plant, proved to be a winner in Australia, with 739 units sold prior to the Ford takeover of New Holland.

Noel never let boundaries bother him if he could see opportunities in exploring new avenues. His continual thoughts centred on the improvements that could be made on any project he became involved in. He did not tolerate fools, or slackers and, any such people in that category found great pressure being exerted on them. So, they usually moved away to avoid the pain.

One particular day, Noel handed me a letter and asked, "do you know this guy?"

And my reply was, "do you want to review the file, as the dealer and I have been working together for weeks trying to resolve all his complaints that are of no real substance."

Noel knew the customer well, having met him some years before and said, "we had better go see him tomorrow."

At 11.30am next day we arrived at the huge property and as we walked in there were a number of old competitor units in the yard, not looking in good condition, possibly due to a poor maintenance plan.

After pleasantries were exchanged, we went out to a shed to inspect the 6600 tractor. The owner began a real outburst stating that some bolt slots were not in line, the holes in the grille were too big and let too much dust in, the battery was too hard to pull out, the seat cannot be adjusted easily and when my staff change shifts, it is so difficult to reset.

Despite trying to calm the owner down, he just kept on talking without raising any major service troubles with the unit.

Finally, Noel said, "the dealer and our service department have been more than generous to your complaints and from now on, the normal warranty will continue, but no more attention will be given to your comments."

As we got into our car Noel said, "there is nothing worse than an academic having too much spare time on his hands."

As I described earlier, Noel always had a solution in mind, which I found out when he told me, "I've been talking to Detroit a few minutes ago and I have given approval for you to go to Japan on a project, leaving next Saturday. You had better get up to the bank in the Glass House and get airline tickets and travelers cheques."

Believe it or not, this was on a Wednesday, in September at 2.45pm, and the bank closed at 3.00pm.

I told Noel, "I do not have a visa for Japan."

But he said, "go down to the embassy tomorrow morning and get one."

I said, "Sorry Noel, but tomorrow is Melbourne Showday and everything is shut."

His reply [*was to*] just go down there and sit there until they give you one.

I fronted the embassy at 9.00am the next day and the receptionist told me to return on Monday as the ambassador, due to protocol, will not sign it under 2 days' notice. Around

10.00 am the ambassador arrived with his entourage; within 5 minutes a bewildered receptionist handed me my signed passport with a visa.

On return to our office, Noel said, "did you get it?"

So, I showed him the visa and asked him, "How did you arrange it."

Simple he said, "I phoned the Minister for Trade in Canberra, and told him where you were and if he does not get his visa, today, then 250 [*Japanese*] Ford tractors sitting on the wharf in Japan, will not be leaving for Thailand, so the minister said, 'no worries, I will fix it.'"

The old saying, "it is not what you know, but who you know, that counts." Noel definitely thought of that.

Noel had been opening every door in Canberra for years whenever he needed advice or assistance. Sometimes he even made suggestions to government ministers. He never ever gave up on anything worthwhile

These stories are only a couple of the incredible accomplishments from a brilliant man whom I was privileged to know for 59 years, and I cannot recall us ever having bad words on anything we discussed, even when we both had strong views on certain matters.

He will be remembered and dearly missed by many, many people in Australia and around the world.

Knowing Noel for so long, leads me to believe he will already be rearranging the big tractor park up in the sky and my wish, one day, is to join him again for another great journey.
I miss you mate, but I will never forget your strengths and your style.

NOEL   An Authorised Memoir

FAMILY EULOGY

<u>Eulogy of Rhoden Noel Howard – Conway Funeral Home Friday 6th August 2021 at 3:00 pm</u>

We gather here to celebrate the life of Rhoden Noel Howard aged 82 years. Also known as Dad or Pop.
Unfortunately, due to the current COVID restrictions there are many people who could not be here today, however, I know that you are here in spirit, watching locally and from afar. Thank you all.
Dad was born in Wodonga, Victoria on 5th January 1939 during the bushfires which ravaged Wodonga. He was named after Sir Rhoden Cutler a WWII hero. He was the son of Elizabeth Lilian and Finlay Brien and they lived in Kiewa where they owned the bakery. He is survived by his siblings and their partners – Graeme and Doreen, Yvonne and Peter and Barrie and Dorothy.
Dad grew up in the Kiewa Valley, Victoria. He spent time at the Kiewa Dairy where he was taken under the wing of Mr R G Brown who explained the intricacies of how to repair mechanical things. He also met George Reid who taught him to fly a Tiger Moth air plane he owned. With his siblings he would work on his mother's father's farm and often helping out in the stock and station business attached to the farm. He often travelled to the Mornington Peninsula to spend holidays with the Grants who were the Howard's neighbours and had moved there from Kiewa where they had been orchardists. Dad often joined May, one of Frank Grant's sisters on her rounds as a district nurse and it was here that he gained an insight into various medical procedures and his early dreams of becoming a doctor. He attended Wangaratta Tech and Albury High School and boarded at the Anglican Hostel.

In 1953 he commenced an apprenticeship with Charlie Millthorpe in Albury. In 1954 with the help of Charlie he received his certificate for Junior Qualification with the NSW Ambulance. For Dad this meant he could complete his education and was able to gain his matriculation, an achievement he was very proud of. Working with Charlie and volunteering with the Ambulance, Dad would use Ron Wild's (the ambulance Managers) office to study when the station was not busy. He ended the year of 1956 with an A Grade mark in his motor mechanic apprenticeship.

In 1957 Ron Wild pulled a few strings and Dad was able to start with the Victorian Civil Ambulance. It was then he commenced mechanical engineering at Melbourne University. He started by attending 3 lectures per week whilst working 35 hours for the ambulance, he spent as much free time as he could manage studying.

In December 1959 Dad met Mum. He was on a holiday to stay with the Grants on the Mornington Peninsula. After repairing Frank's big Nash, he drove it to the dance at the Town Hall in Rosebud where he parked it right in front of the large building's steps. Seeing a young lady there who took his eye, he tried to talk to Miss Joan Brown, who told him she thought him a poser! By night's end they became inseparable.

In 1960 he completed his Mechanical Engineering course via correspondence after moving to Wangaratta to take up a position as Machinery Department Manager with Gippsland and Northern Co-operative. He continued working as a volunteer ambulance officer part time. Dad and Mum were married at the Holy Trinity Cathedral in Wangaratta on 10th September in 1960 aged 21. They were married for over 60 years before Mum passed in March this year – a mere 5 months ago.

In 1962 he was he was approached by Cliff Freer of Massey Ferguson to take up the role as a sales and service representative where he spent two years in Sunshine which is where he met Jack Thake. Uncle Jack, (as) he was known to us as, was a very special person in all our lives. This same year in April 1962, Geoffrey Noel, the first of Mum and Dad's children were born.

In 1964 Dad was head-hunted from Massey Ferguson to Ford Australia Tractor Operations with others including Ian O'Rourke. He was admitted into the Ford Marketing Institute where he studied Accounting and Dealer Management.

In 1965 he was appointed Zone Manager to Riverina Territory so, Dad, Mum and Geoffrey moved to Vermont. In January 1966, Peter John their second child was born. In 1967, the family moved to Adelaide as Dad transferred as State Manager South Australia - during this time in June 1967 I, Tania Joan, was born. In 1968 he was promoted to Market Representation which meant moving back to Victoria. They sold the home in Vermont and bought land to build on in Templestowe.

Dad worked alongside Harry Watson to prepare a plan for restructuring the South African Tractor Operation. Harry become a close friend and mentor to Dad until Harry passed away on the Isle of Wight. In 1970 their fourth child Natalie Anne was born. So, with four young children, the youngest, Natalie being 6 weeks of age our family moved to South Africa where Dad took the position to implement the plan he and Harry devised.

In 1973 we returned to Melbourne holidaying in a number of countries along the way. Mum and Dad then designed and built a home on the land in Templestowe. During this time, Clarence E Ragan organised and paid through the internal systems for Dad to study International Business at Harvard, USA.

In 1973 he was appointed as the Sales and Marketing Manager Ford Tractor Australia Operations and then to General Manager Ford Tractor Operations in Australia and he held that position for 15 years.

In 1976 they sold the Templestowe home and purchased their first large scale farm being Highfield Park in Broadford, a sheep and cattle property.

In 1978 he was responsible for the incorporation of Blueline as a Ford Australia Tractor Operations subsidiary which opened the opportunity to do things outside the Ford Corporate bubble. In 1980 at the 80's release, his name is synonymous for the introduction of the Ford 8401 Tractor which was a huge success and the orders for the compact tractors and new Blueline equipment models poured in. A turning point in the tractor and machinery industry.

In June 1981 their fifth child Kathryn Faye was born. In 1982 Australia faced one of the worst droughts in history and Dad had to navigate the tractor industry through it.

In 1987 he was appointed Director of Ford New Holland Australia and moved to Ford Asia Pacific concentrating on New Zealand, Singapore, Korea, China and Japan. In 1987 they relocated to the North East of Victoria to a beef cattle property at Nooroongong in the Mitta Valley and then also acquired an additional irrigated dairy property in Tallandoon – where Natalie and Mark worked with them as share farmers.

Dad then transferred to J.I. Case in Sydney – he would travel from Albury Airport on Sunday night and return from Sydney on Friday night. Whilst in Sydney during the week he would stay with Vicki (his niece), Greg and their family for which he developed a lasting close relationship with.

He was Director and General Manager of Case International Headquarters Australia, New Zealand for four and half years. Achieving market leadership with tractors, combine harvesters and back hoes.

In 1996 Dad retired but not for long as he then joined the Mann Mitre 10 business where he was catalyst to the eventual sale of the business to Bunnings and the redevelopment of the company's central Wodonga site into the Mann Central Shopping centre.

At various stages during these years Dad was an adviser on rural matters to the Federal Treasury, member of the Dairy Board, Water Board chair, National rural independents, councillor with the Tractor and Machinery Association for Australia along with being a board member for 6 years with Aware Industries, Wodonga, an employer of 108 persons with disabilities.

After this he began some joint ventures with his sons Geoffrey and Peter, building, leasing and selling factories in Wodonga.

By 2003 both farms were sold and they moved into Wodonga to a property they designed and built. Dad then began to write his biography which he did not get to finish. We will however with the assistance of Terry Probert the author, endeavour to complete this for him.

After Mum's diagnosis of Ovarian Cancer and her decline in health they moved to Westmont Aged care where they had lived for the past two years. Dad cared for Mum during this time and in 2019 was unfortunately diagnosed with metastatic lung cancer. His will to live would see him take every treatment possible to keep on living. All of which would pose him some life-threatening side effects which required hospitalisation and shocking reactions. Dad fought as hard as he could but lost his battle on Saturday 24th July.

We are sure as the sun has set upon the long and fulfilled life of Dad, we all pause for moment to thank those who provided love, comfort and medical care to Dad during his final journey to his resting place.
The Family would like to thank:
The medical staff involved in his care.
All our family members for your endless, loving care during such a difficult time.
Uncle Barrie and Auntie Dorothy for being such a wonderful support.
Terry Probert – Author of Dad's book – for your assistance, friendship and providing hope to Dad.
Thank you to Conway Funeral Home – Heath
Thank you to Barbara for such a caring job.
Ben and Claire of The Wodonga Golf Clubhouse for all present to partake in drinks and food afterwards.
So, Dad, in the words that you have spoken at the many Eulogies that you have been honoured to deliver,
"It is one of the most beautiful compensations of this life, that no man or woman, can seriously help another without helping him or herself.
If it is to be ……it is up to me!"
Dad was a dear husband, father, grandfather, colleague and friend and by his own making an accomplished, determined and generous person who left a lasting memory on everyone he met. We love you, Dad.

As he said to us in the last days of his life, "I've had a good life!"
Dad you will be greatly missed by all who were blessed to have known you.

MAY YOU REST IN PEACE!

Terry L Probert

ACKNOWLEDGEMENTS

A project like this memoir has many parents, people who help the author. It is through their confirming of events, offering opinion and support, that they were always helping to drive the story on. My wife Ruth, in the words of writer Steven King my is, "ideal reader." However, by being my soundboard, proof reader and honest critic she helps to keep the story flowing. Suggesting where I rid the text of any verbose phrasing, or my habitual use of complicated wording, always makes the reading experience better.

Ian O'Rourke, who during his time working with Noel, kept service concerns to a minimum and had always been a right-hand man offered me an insight into Noel few of his colleagues would know. And, in no particular order: Richard Curtis, David Hosking, Graeme Ainsworth, Bill McKie, Bernard Sarfras, Steve Officer, Peter Barrie, Brenton Byerlee, Anne Purdie, Brian Matchett and Alan Kirsten are but a few who gave their time freely. To others I have missed, your help has been just as important and I apologise for not mentioning you here. Some contributing by sharing their memories of working with Noel, others by reminding me about Australia's rich farm machinery heritage and, the development in dryland farming practices that drove the Australian horsepower race during the eighties.

Thank you all.

NOEL   An Authorised Memoir

AUTHOR'S NOTE

From the earliest of my Fordson/Ford tractor memories, Noel Howard's name had been in my conciseness. My parents started as Ford dealers in the tiny farming community of Orroroo in 1955. We were a proud Ford family and during school holidays I sometimes accompanied Dad to Ford meetings in Adelaide. I have always been drawn to associating names to places and their assigned titles, wondering if someday I might meet these people. Noel being the South Australian state manager, has always been among the Probert's household heroes.

During February 1979 I had been attending management training in Adelaide when my father died of a massive heart attack. My world changed in an instant. My guide, my mentor, my hero was gone and it was time too, to take the reins of our family business.

Change appeared all around us and with change comes opportunity. With the new decade opening, our Goyder's Line farmers began demanding more powerful tractors and wider machinery. Their rules of operation began changing too and as the number of less powerful tractors declined, higher horsepower tractors and wider machinery was being delivered and our used yards began bulging with trade-ins.

Terry L Probert

In 1982, the season opened well and then it didn't rain enough to see a crop survive. By 1986 rising interest rates and inflation had decimated the country's agricultural machinery dealers. Some walked away, some suicided and others clung on until they could no more. Marriages failed and families drifted apart, my business suffered too. Having a high inventory of new tractors and Blueline equipment, I reached out to Noel and asked if there were something he could do to save me from declaring bankruptcy. Within a few minutes he agreed to take back the unsold Ford and Blueline stock. Because of that one act by Noel bringing relief to a family fraught with worry we survived and continued fighting bankruptcy until we sold the business in 1996.

In 1990 I had been working with the team at Melbourne Tractors to build an agricultural arm of their successful industrial machinery business. Noel working for Case often called to see Vin Smith his colleague from the Ford days and always called past my desk. I felt humbled as he always enquired about me and my family Later our paths would cross at a couple of Ford Tractor reunions and at Christmas, he would call to offer season's greetings, a gesture that always left me feeling valued.

In 2019 he called, inviting me to write his story. It caught me by surprise, as I believed his life too important to be trusted to a self-published author of novels. Sure, I had encouraged people from the tractor and machinery industry to participate in my AgList website by telling their story.

NOEL   An Authorised Memoir

With AgList I would help craft the text, but it was always their life we were recording. Pleased that a group of my Ford colleagues had used the platform to record their yarns, I thought no more about who those invitations reached. However, when we met with Noel and Joan in Wodonga, I was astounded that he produced the original AgList question-and-answer template for the website.

From that first meeting we began a close relationship, me asking questions, Noel answering them honestly, opening-up about the often-secret twists and turns of his business life. During the next two years I listened and laughed with him and, every now and then, noticed his memory beginning to slide. His body slowly succumbing to the tumour that had invaded his brain, and taking its toll.

I expected the manuscript to be completed within eighteen months, but Noel lived a big life. Not something to be completed in a two-hundred-page yarn. His is a life deserving of the time to record the influences on him and, the influences he had on the life of others too. Unfortunately for all who knew him, Noel died far too early. He did however, see a draft of this book a few days before he passed away and I'm extremely grateful to a lot of people who helped make that happen.

Rest in Peace my friend, I miss you.

Terry L Probert

NOEL   An Authorised Memoir

About the Author

Born in Orroroo South Australia, after leaving his role as a demonstrator with Ford Tractor Operations Australia, Terry ran a successful Ford Tractor and vehicle business until joining AGCO Massey Ferguson in 1996. With AGCO, his travels as an Area Service Manager and other management roles, gave him an innate understanding of the tractor and machinery industry, both within Australia and overseas.

Promoted into sales and marketing roles at first with AGCO and then with other tractor and machinery businesses within the Australian industry, exposed him to a wide variety of products.

Diagnosed with muscular dystrophy in 2012, a change was needed and he decided to focus his energy into writing and putting the skills he had gained, to another use.

Today, while he now lives in Bendigo, he and Ruth can often be found traveling Australia seeking locations, characters, and plots for his next novel. Everything and anyone, becoming a possible subject for another story.

www.ingramcontent.com/pod-product-compliance
Lightning Source LLC
Chambersburg PA
CBHW081351290426
44110CB00018B/2346